MW01125666

Neuro-Dynamic Programming

Dimitri P. Bertsekas and John N. Tsitsiklis

Massachusetts Institute of Technology

WWW site for book information and orders

http://www.athenasc.com

Athena Scientific, Belmont, Massachusetts

Athena Scientific
Belmont, Mass.
U.S.A.

Email: info@athenasc.com
WWW: http://www.athenasc.com

Cover Design: *Ann Gallager*

Publisher's Cataloging-in-Publication Data

Bertsekas, Dimitri P.
Neuro-Dynamic Programming
Includes bibliographical references and index
1. Neural Networks. 2. Mathematical Optimization.
3. Dynamic Programming. I. Title.
QA76.87.B47 1996 519.703 96-085338

ISBN-10: 1-886529-10-8, ISBN-13: 978-1-886529-10-6

3rd Printing (2016)

To the memory of Dimitri's parents

To John's parents

Contents

Ὦ Χαιρεφῶν, πολλαὶ τέχναι ἐν ἀνθρώποις εἰσὶν ἐκ τῶν ἐμπειριῶν ἐμπείρως ηὑρημέναι· ἐμπειρία μὲν γὰρ ποιεῖ τὸν αἰῶνα ἡμῶν πορεύεσθαι κατὰ τέχνην, ἀπειρία δὲ κατὰ τύχην.

O Chaerephon, many arts among men have been
discovered through practice, empirically;
for experience makes our life proceed deliberately,
but inexperience unpredictably.

(Plato, Gorgias 448c)

Preface

A few years ago our curiosity was aroused by reports on new methods in reinforcement learning, a field that was developed primarily within the artificial intelligence community, starting a few decades ago. These methods were aiming to provide effective suboptimal solutions to complex problems of planning and sequential decision making under uncertainty, that for a long time were thought to be intractable. Our first impression was that the new methods were ambitious, overly optimistic, and lacked firm foundation. Yet there were claims of impressive successes and indications of a solid core to the modern developments in reinforcement learning, suggesting that the correct approach to their understanding was through dynamic programming.

Three years later, after a lot of study, analysis, and experimentation, we believe that our initial impressions were largely correct. This is indeed an ambitious, often ad hoc, methodology, but for reasons that we now understand much better, it does have the potential of success with important and challenging problems. With a good deal of justification, it claims to deal effectively with the dual curses of dynamic programming and stochastic optimal control: Bellman's *curse of dimensionality* (the exponential computational explosion with the problem dimension is averted through the use of parametric approximate representations of the cost-to-go function), and the *curse of modeling* (an explicit system model is not needed, and a simulator can be used instead). Furthermore, the methodology has a logical structure and a mathematical foundation, which we systematically develop in this book. It draws on the theory of function approximation, the theory of iterative optimization and neural network training, and the theory of dynamic programming. In view of the close connection with both neural networks and dynamic programming, we settled on the name "neuro-dynamic programming" (NDP), which describes better in our opinion the nature of the subject than the older and more broadly applicable name "reinforcement learning."

Our objective in this book is to explain with mathematical analysis, examples, speculative insight, and case studies, a number of computational ideas and phenomena that collectively can provide the foundation for understanding and applying the NDP methodology. We have organized the book in three major parts.

(a) The first part consists of Chapters 2-4 and provides background. It includes a detailed introduction to dynamic programming (Chapter 2), a discussion of neural network architectures and methods for training them (Chapter 3), and the development of general convergence theorems for stochastic approximation methods (Chapter 4), which will provide the foundation for the analysis of various NDP algorithms later.

(b) The second part consists of the next three chapters and provides the core NDP methodology, including many mathematical results and methodological insights that were developed as this book was written and which are not available elsewhere. Chapter 5 covers methods involving a lookup table representation. Chapter 6 discusses the more practical methods that make use of function approximation. Chapter 7 develops various extensions of the theory in the preceding two chapters.

(c) The third part consists of Chapter 8 and discusses the practical aspects of NDP through case studies.

Inevitably, some choices had to be made regarding the material to be covered. Given that the reinforcement learning literature often involves a mixture of heuristic arguments and incomplete analysis, we decided to pay special attention to the distinction between factually correct and incorrect statements, and to rely on rigorous mathematical proofs. Because some of these proofs are long and tedious, we have made an effort to organize the material so that most proofs can be omitted without loss of continuity on the part of the reader. For example, during a first reading, a reader could omit all of the proofs in Chapters 2-5, and proceed to subsequent chapters.

However, we wish to emphasize our strong belief in the beneficial interplay between mathematical analysis and practical algorithmic insight. Indeed, it is primarily through an effort to develop a mathematical structure for the NDP methodology that we will ever be able to identify promising or solid algorithms from the bewildering array of speculative proposals and claims that can be found in the literature.

The fields of neural networks, reinforcement learning, and approximate dynamic programming have been very active in the last few years and the corresponding literature has greatly expanded. A comprehensive survey of this literature is thus beyond our scope, and we wish to apologize in advance to researchers in the field for not citing their works. We have confined ourselves to citing the sources that we have used and that contain results related to those presented in this book. We have also cited a few sources for their historical significance, but our references are far from complete in this regard.

Finally, we would like to express our thanks to a number of individuals. Andy Barto and Michael Jordan first gave us pointers to the research and the state of the art in reinforcement learning. Our understanding of the reinforcement learning literature and viewpoint gained significantly from interactions with Andy Barto, Satinder Singh, and Rich Sutton. The first author collaborated with Vivek Borkar on the average cost Q-learning research discussed in Chapter 7, and with Satinder Singh on the dynamic channel allocation research discussed in Chapter 8. The first author also benefited a lot through participation in an extensive NDP project at Alphatech, Inc., where he interacted with David Logan and Nils Sandell, Jr. Our students contributed substantially to our understanding through discussion, computational experimentation, and individual research. In particular, they assisted with some of the case studies in Chapter 8, on parking (Keith Rogers), football (Steve Patek), tetris (Sergey Ioffe and Dimitris Papaioannou), and maintenance and combinatorial optimization (Cynara Wu). The joint researches of the first author with Jinane Abounadi and with Steve Patek are summarized in Sections 7.1 and 7.2, respectively. Steve Patek also offered tireless and invaluable assistance with the experimental implementation, validation, and interpretation of a large variety of untested algorithmic ideas. The second author has enjoyed a fruitful collaboration with Ben Van Roy that led to many results, including those in Sections 6.3, 6.7, 6.8, and 6.9. We were fortunate to work at the Laboratory for Information and Decision Systems at M.I.T., which provided us with a stimulating research environment. Funding for our research that is reported in this book was provided by the National Science Foundation, the Army Research Office through the Center for Intelligent Control Systems, the Electric Power Research Institute, and Siemens. We are thankful to Prof. Charles Segal of Harvard's Department of Classics for suggesting the original quotation that appears at the beginning of this preface. Finally, we are grateful to our families for their love, encouragement, and support while this book was being written.

Dimitri P. Bertsekas
John N. Tsitsiklis
Cambridge, August 1996

Learning without thought is labour lost;
thought without learning is perilous.
(Confucian Analects)

1

Introduction

Contents

This book considers systems where decisions are made in stages. The outcome of each decision is not fully predictable but can be anticipated to some extent before the next decision is made. Each decision results in some immediate cost but also affects the context in which future decisions are to be made and therefore affects the cost incurred in future stages. We are interested in decision making policies that minimize the total cost over a number of stages. Such problems are challenging primarily because of the tradeoff between immediate and future costs. Dynamic programming (DP for short) provides a mathematical formalization of this tradeoff.

Generally, in DP formulations we have a discrete-time dynamic system whose state evolves according to given transition probabilities that depend on a decision/control u. In particular, if we are in state i and we choose control u, we move to state j with given probability $p_{ij}(u)$. The control u depends on the state i and the rule by which we select the controls is called a *policy* or *feedback control policy* (see Fig. 1.1). Simultaneously with a transition from i to j under control u, we incur a cost $g(i,u,j)$. In comparing, however, the available controls u, it is not enough to look at the magnitude of the cost $g(i,u,j)$; we must also take into account the desirability of the next state j. We thus need a way to rank or rate states j. This is done by using the optimal cost (over all remaining stages) starting from state j, which is denoted by $J^*(j)$ and is referred to as the optimal *cost-to-go* of state j. These costs-to-go can be shown to satisfy some form of *Bellman's equation*

$$J^*(i) = \min_u E\big[g(i,u,j) + J^*(j) \mid i,u\big], \qquad \text{for all } i,$$

where j is the state subsequent to i, and $E[\cdot \mid i,u]$ denotes expected value with respect to j, given i and u. Generally, at each state i, it is optimal to use a control u that attains the minimum above. Thus, controls are ranked based on the sum of the expected cost of the present period and the optimal expected cost of all subsequent periods.

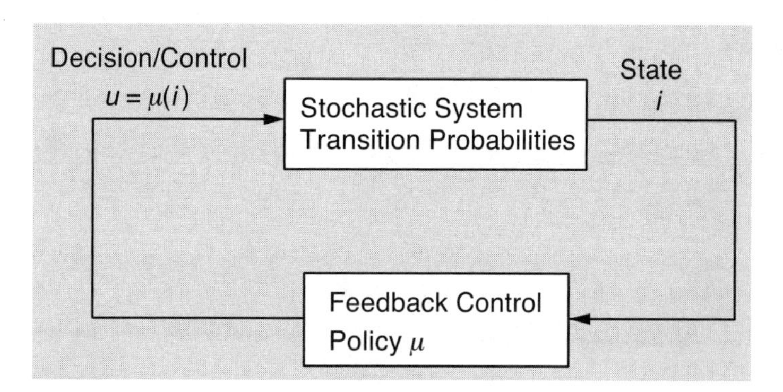

Figure 1.1: Structure of a discrete-time dynamic system under feedback control.

The objective of DP is to calculate numerically the optimal cost-to-go function J^*. This computation can be done off-line, i.e., before the real system starts operating. An optimal policy, that is, an optimal choice of u for each i, is computed either simultaneously with J^*, or in real time by minimizing in the right-hand side of Bellman's equation. It is well known, however, that for many important problems the computational requirements of DP are overwhelming, because the number of states and controls is very large (Bellman's "curse of dimensionality"). In such situations a suboptimal solution is required.

1.1 COST-TO-GO APPROXIMATIONS IN DYNAMIC PROGRAMMING

In this book, we primarily focus on suboptimal methods that center around the evaluation and approximation of the optimal cost-to-go function J^*, possibly through the use of neural networks and/or simulation. In particular, we replace the optimal cost-to-go $J^*(j)$ with a suitable approximation $\tilde{J}(j,r)$, where r is a vector of parameters, and we use at state i the (suboptimal) control $\tilde{\mu}(i)$ that attains the minimum in the (approximate) right-hand side of Bellman's equation, that is,

$$\tilde{\mu}(i) = \arg\min_u E\big[g(i,u,j) + \tilde{J}(j,r) \mid i,u\big].$$

The function $\tilde{J}$ will be called the *scoring function* or *approximate cost-to-go function*, and the value $\tilde{J}(j,r)$ will be called the *score* or *approximate cost-to-go* of state j (see Fig. 1.2). The general form of $\tilde{J}$ is known and is such that once the parameter vector r is fixed, the evaluation of $\tilde{J}(j,r)$ for any state j is fairly simple.

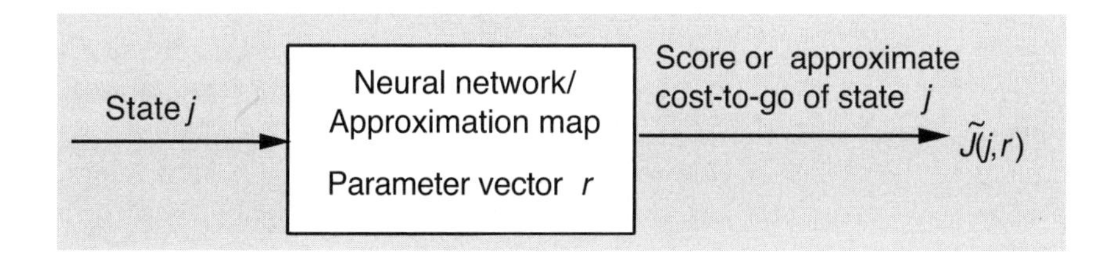

Figure 1.2: Structure of cost-to-go approximation.

We are interested in problems with a large number of states and in scoring functions $\tilde{J}$ that can be described with relatively few numbers (a vector r of small dimension). Scoring functions involving few parameters will be called *compact representations*, while the tabular description of J^* will be called the *lookup table representation*. In a lookup table representation, the values $J^*(j)$ for all states j are stored in a table. In a typical compact representation, only the vector r and the general structure of the scoring function $\tilde{J}(\cdot, r)$ are stored; the scores $\tilde{J}(j, r)$ are generated only when needed. For example, if $\tilde{J}(j, r)$ is the output of some neural network in response to the input j, then r is the associated vector of weights or parameters of the neural network; or if $\tilde{J}(j, r)$ involves a lower dimensional description of the state j in terms of its "significant features," then r could be a vector of relative weights of the features. Naturally, we would like to choose r algorithmically so that $\tilde{J}(\cdot, r)$ approximates well $J^*(\cdot)$. Thus, determining the scoring function $\tilde{J}(j, r)$ involves two complementary issues: (1) deciding on the general structure of the function $\tilde{J}(j, r)$, and (2) calculating the parameter vector r so as to minimize in some sense the error between the functions $J^*(\cdot)$ and $\tilde{J}(\cdot, r)$.

We note that in some problems the evaluation of the expression

$$E\big[g(i, u, j) + \tilde{J}(j, r) \mid i, u\big],$$

for each u, may be too complicated or too time-consuming for making decisions in real-time, even if the scores $\tilde{J}(j, r)$ are simply calculated. There are a number of ways to deal with this difficulty (see Section 6.1). An important possibility is to approximate the expression minimized in Bellman's equation,

$$Q^*(i, u) = E\big[g(i, u, j) + J^*(j) \mid i, u\big],$$

which is known as the *Q-factor corresponding to* (i, u). In particular, we can replace $Q^*(i, u)$ with a suitable approximation $\tilde{Q}(i, u, r)$, where r is a vector of parameters. We can then use at state i the (suboptimal) control that minimizes the approximate Q-factor corresponding to i:

$$\tilde{\mu}(i) = \arg\min_u \tilde{Q}(i, u, r).$$

Much of what will be said about the approximation of the optimal costs-to-go also applies to the approximation of Q-factors. In fact, we will see later that the Q-factors can be viewed as optimal costs-to-go of a related problem. We thus focus primarily on approximation of the optimal costs-to-go.

Approximations of the optimal costs-to-go have been used in the past in a variety of DP contexts. Chess playing programs represent an interesting example. A key idea in these programs is to use a *position evaluator* to rank different chess positions and to select at each turn a move that results in the position with the best rank. The position evaluator assigns a numerical value to each position according to a heuristic formula that includes weights for the various features of the position (material balance, piece mobility, king safety, and other factors); see Fig. 1.3. Thus, the position evaluator corresponds to the scoring function $\tilde{J}(j, r)$ above, while the weights of the features correspond to the parameter vector r. Usually, some general structure is selected for the position evaluator (this is largely an art that has evolved over many years, based on experimentation and human knowledge about chess), and the numerical weights are chosen by trial and error or (as in the case of the champion program Deep Thought) by "training" using a large number of sample grandmaster games. It should be mentioned that in addition to the use of sophisticated position evaluators, much of the success of chess programs can be attributed to the use of multimove lookahead, which has become deeper and more effective with the use of increasingly fast hardware.

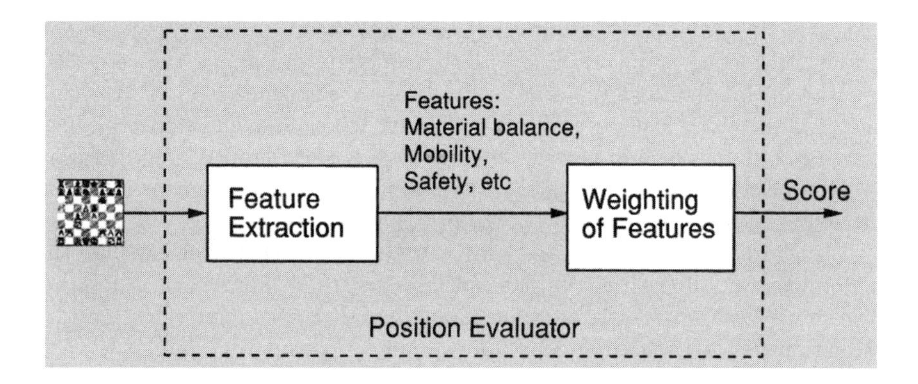

Figure 1.3: Structure of the position evaluator of a chess program.

As the chess program paradigm suggests, intuition about the problem, heuristics, and trial and error are all important ingredients for constructing cost-to-go approximations in DP. However, it is important to supplement heuristics and intuition with more systematic techniques that are broadly applicable and retain as much as possible of the nonheuristic characteristics of DP. This book will focus on several recent efforts to develop a methodological foundation for a rational approach to complex stochastic decision problems, which combines dynamic programming, function approximation, and simulation.

1.2 APPROXIMATION ARCHITECTURES

An important issue in function approximation is the *selection of an architecture*, that is, the choice of a parametric class of functions $\tilde{J}(\cdot, r)$ or

$\tilde{Q}(\cdot,\cdot,r)$ that suits the problem at hand. One possibility is to use a neural network architecture of some type. We should emphasize here that in this book we use the term "neural network" in a very broad sense, essentially as a synonym to "approximating architecture." In particular, we do not restrict ourselves to the classical multilayer perceptron structure with sigmoidal nonlinearities. Any type of universal approximator of nonlinear mappings could be used in our context. The nature of the approximating structure is left open in our discussion, and it could involve, for example, radial basis functions, wavelets, polynomials, splines, aggregation, etc.

Cost-to-go approximation can often be significantly enhanced through the use of *feature extraction*, a process that maps the state i into some vector $f(i)$, called the *feature vector* associated with i. Feature vectors summarize, in a heuristic sense, what are considered to be important characteristics of the state, and they are very useful in incorporating the designer's prior knowledge or intuition about the problem and about the structure of the optimal controller. For example, in a queueing system involving several queues, a feature vector may involve for each queue a three-valued indicator that specifies whether the queue is "nearly empty," "moderately busy," or "nearly full." In many cases, analysis can complement intuition to suggest the right features for the problem at hand.

Feature vectors are particularly useful when they can capture the "dominant nonlinearities" in the optimal cost-to-go function J^*. By this we mean that $J^*(i)$ can be approximated well by a "relatively smooth" function $\tilde{J}\big(f(i)\big)$; this happens for example, if through a change of variables from states to features, J^* becomes a (nearly) linear or low-order polynomial function of the features. When a feature vector can be chosen to have this property, it is appropriate to use approximation architectures where features and (relatively simple) neural networks are used together. In particular, the state is mapped to a feature vector, which is then used as input to a neural network that produces the score of the state (see Fig. 1.4). More generally, it is possible that both the state and the feature vector are provided as inputs to the neural network (see the second diagram in Fig. 1.4).

1.3 SIMULATION AND TRAINING

Some of the most successful applications of neural networks are in the areas of pattern recognition, nonlinear regression, and nonlinear system identification. In these applications the neural network is used as a universal approximator: the input-output mapping of the neural network is matched to an unknown nonlinear mapping F of interest using a least-squares optimization, known as *training the network*. To perform training, one must

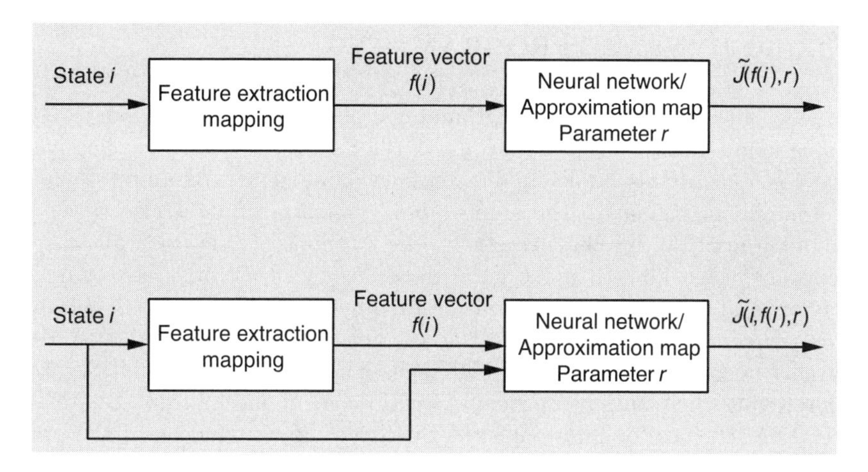

Figure 1.4: Approximation architectures involving feature extraction and neural networks.

have some training data, that is, a set of pairs $\big(i, F(i)\big)$, which is representative of the mapping F that is approximated.

It is important to note that in contrast with these neural network applications, in the DP context there is no readily available training set of input-output pairs $\big(i, J^*(i)\big)$ that could be used to approximate J^* with a least squares fit. The only possibility is to evaluate (exactly or approximately) by simulation the cost-to-go functions of given (suboptimal) policies, and to try to iteratively improve these policies based on the simulation outcomes. This creates analytical and computational difficulties that do not arise in classical neural network training contexts. Indeed the use of simulation to evaluate approximately the optimal cost-to-go function is a key new idea that distinguishes the methodology of this book from earlier approximation methods in DP.

Simulation offers another major advantage: it allows the methods of this book to be used for systems that are hard to model but easy to simulate, i.e., problems where a convenient explicit model is not available, and the system can only be observed, either through a software simulator or as it operates in real time. For such problems, the traditional DP techniques are inapplicable, and estimation of the transition probabilities to construct a detailed mathematical model is often cumbersome or impossible.

There is a third potential advantage of simulation: it can implicitly identify the "most important" or "most representative" states of the system. It appears plausible that these states are the ones most often visited during the simulation, and for this reason the scoring function will tend to approximate better the optimal cost-to-go for these states, and the suboptimal policy obtained will on the average perform better.

1.4 NEURO-DYNAMIC PROGRAMMING

In view of the reliance on both DP and neural network concepts, we use the name *neuro-dynamic programming* (NDP for short) to describe collectively the methods of this book. In the artificial intelligence community, where the methods originated, the name *reinforcement learning* is also used. In common artificial intelligence terms, the methods of this book allow systems to "learn how to make good decisions by observing their own behavior, and use built-in mechanisms for improving their actions through a reinforcement mechanism." In the less anthropomorphic DP terms used in this book, "observing their own behavior" relates to simulation, and "improving their actions through a reinforcement mechanism" relates to iterative schemes for improving the quality of approximation of the optimal costs-to-go, or the Q-factors, or the optimal policy. There has been a gradual realization that reinforcement learning techniques can be fruitfully motivated and interpreted in terms of classical DP concepts such as value and policy iteration.

In this book, we attempt to clarify some aspects of the current NDP methodology, we suggest some new algorithmic approaches, and we identify some open questions. Despite the great interest in NDP, the theory of the subject is only now beginning to take shape, and the corresponding literature is often confusing. Yet, there have been many reports of successes with problems too large and complex to be treated in any other way. A particularly impressive success that greatly motivated subsequent research, was the development of a backgammon playing program by Tesauro [Tes92] (see Section 8.6). Here a neural network was trained to approximate the optimal cost-to-go function of the game of backgammon by using simulation, that is, by letting the program play against itself. After training for several months, the program nearly defeated the human world champion. Unlike chess programs, this program did not use lookahead of many stages, so its success can be attributed primarily to the use of a properly trained approximation of the optimal cost-to-go function.

Our own experience has been that NDP methods can be impressively effective in problems where traditional DP methods would be hardly applicable and other heuristic methods would have limited potential. In this book, we outline some engineering applications, and we use a few computational studies for illustrating the methodology and some of the art that is often essential for success.

We note, however, that the practical application of NDP is computationally very intensive, and often requires a considerable amount of trial and error. Furthermore, success is often obtained using methods whose properties are not fully understood. Fortunately, all of the computation and experimentation with different approaches can be done off-line. Once the approximation is obtained off-line, it can be used to generate decisions fast enough for use in real time. In this context, we mention that in the artificial intelligence literature, reinforcement learning is often viewed as an "on-line" method, whereby the cost-to-go approximation is improved as the system operates in real time. This is reminiscent of the methods of traditional adaptive control. We will not discuss this viewpoint in this book, as we prefer to focus on applications involving a large and complex system. A lot of training data are required for such systems. These data often cannot be obtained in sufficient volume as the system is operating; even if they can, the corresponding processing requirements are often too large for effective use in real time.

We finally mention an alternative approach to NDP, known as *approximation in policy space*, which, however, we will not consider in this book. In this approach, in place of an overall optimal policy, we look for an optimal policy within some restricted class that is parametrized by some vector s of relatively low dimension. In particular, we consider policies of a given form $\tilde{\mu}(i,s)$. We then minimize over s the expected cost $E_i\big[J^{\tilde{\mu}(\cdot,s)}(i)\big]$, where the expectation is with respect to some suitable probability distribution of the initial state i. This approach applies to complex problems where there is no explicit model for the system and the cost, as long as the cost corresponding to a given policy can be calculated by simulation. Furthermore, insight and analysis can sometimes be used to select simple and effective parametrizations of the policies. On the other hand, there are many problems where such parametrizations are not easily obtained. Furthermore, the minimization of $E_i\big[J^{\tilde{\mu}(\cdot,s)}(i)\big]$ can be very difficult because the gradient of the cost with respect to s may not be easily calculated; while methods that require cost values (and not gradients) may be used, they tend to require many cost function evaluations and to be slow in practice.

The general organizational plan of the book is to first develop some essential background material on DP, and on deterministic and stochastic iterative optimization algorithms (Chs. 2-4), and then to develop the main algorithmic methods of NDP in Chs. 5 and 6. Various extensions of the methodology are discussed in Ch. 7. Finally, we present case studies in Ch. 8. Many of the ideas of the book extend naturally to continuous-state systems, although the NDP theory is far from complete for such systems. To keep the exposition simple, we have restricted ourselves to the case where the number of states is finite and the number of available controls at each state is also finite. This is consistent with the computational orientation of the book.

1.5 NOTES AND SOURCES

1.1. The origins of our subject can be traced to the early works on DP by Bellman, who used the term "approximation in value space," and to the works by Shannon [Sha50] on computer chess and by Samuel [Sam59], [Sam67] on computer checkers.

1.4. The works by Barto, Sutton, and Anderson [BSA83] on adaptive critic systems, by Sutton [Sut88] on temporal difference methods, and by Watkins [Wat89] on Q-learning initiated the modern developments which brought together the ideas of function approximation, simulation, and DP. The work of Tesauro [Tes92], [Tes94], [Tes95] on backgammon was the first to demonstrate impressive success on a very complex and challenging problem. Much research followed these seminal works. The extensive survey by Barto, Bradtke, and Singh [BBS95], and the overviews by Werbös [Wer92a], [Wer92b], and other papers in the edited volume by White and Sofge [WhS92] point out the connections between the artificial intelligence/reinforcement learning viewpoint and the control theory/DP viewpoint, and give many references. The DP textbook by Bertsekas [Ber95a] describes a broad variety of suboptimal control methods, including some of the NDP approaches that are treated in much greater depth in the present book.

Philosophy is written in this grand book, the universe,
which stands continually open to our gaze.
But the book cannot be understood
unless one first learns to comprehend the language
and read the letters in which it is composed.
It is written in the language of mathematics.

(Galileo)

2

Dynamic Programming

Contents

In this chapter we provide an introduction to some of the dynamic programming models that will be the subject of the methodology to be developed in the subsequent chapters. We focus primarily on infinite horizon models and on algorithmic issues relating to the basic methods of value and policy iteration. We also discuss some examples, both for the purpose of orientation, and also as an introduction to the case studies that will be discussed in more detail in Ch. 8.

2.1 INTRODUCTION

The principal elements of a problem in dynamic programming (DP for short) are:

(a) *A discrete-time dynamic system whose state transition depends on a control.* Throughout this chapter, we assume that there are n states, denoted by $1, 2, \ldots, n$, plus possibly an additional termination state, denoted by 0. When at state i, the control must be chosen from a given finite set $U(i)$. At state i, the choice of a control u specifies the transition probability $p_{ij}(u)$ to the next state j.

(b) *A cost that accumulates additively over time* and depends on the states visited and the controls chosen. At the kth transition, we incur a cost $\alpha^k g(i, u, j)$, where g is a given function, and α is a scalar with $0 < \alpha \leq 1$, called the *discount factor*. The meaning of $\alpha < 1$ is that future costs matter to us less than the same costs incurred at the present time. For example, consider the case where the cost per stage g is expressed in dollars, and think of kth period dollars depreciated to initial period dollars by a factor of $(1+r)^{-k}$, where r is a rate of interest; then we naturally have a discounted cost with $\alpha = (1+r)^{-1}$.

We are interested in *policies*, that is, sequences $\pi = \{\mu_0, \mu_1, \ldots\}$ where each μ_k is a function mapping states into controls with $\mu_k(i) \in U(i)$ for all states i. Let us denote by i_k the state at time k. Once a policy π is fixed, the sequence of states i_k becomes a Markov chain with transition probabilities

$$P(i_{k+1} = j \mid i_k = i) = p_{ij}\big(\mu_k(i)\big).$$

We can distinguish between *finite horizon problems*, where the cost accumulates over a finite number of stages, say N, and *infinite horizon problems*, where the cost accumulates indefinitely. In N-stage problems the expected cost of a policy π, starting from an initial state i, is

$$J_N^\pi(i) = E\left[\alpha^N G(i_N) + \sum_{k=0}^{N-1} \alpha^k g\big(i_k, \mu_k(i_k), i_{k+1}\big) \;\Big|\; i_0 = i\right],$$

where $\alpha^N G(i_N)$ is a terminal cost for ending up with final state i_N, and the expected value is taken with respect to the probability distribution of the Markov chain $\{i_0, i_1, \ldots, i_N\}$. This distribution depends on the initial state i_0 and the policy π, as discussed earlier. The optimal N-stage cost-to-go starting from state i, is denoted by $J_N^*(i)$; that is,

$$J_N^*(i) = \min_{\pi} J_N^{\pi}(i).$$

The costs $J_N^*(i)$, $i = 1, \ldots, n$, can be viewed as the components of a vector J_N^*, which is referred to as the *N-stage optimal cost-to-go vector.*

In infinite horizon problems, the total expected cost starting from an initial state i and using a policy $\pi = \{\mu_0, \mu_1, \ldots\}$ is

$$J^{\pi}(i) = \lim_{N \to \infty} E\left[\sum_{k=0}^{N-1} \alpha^k g\big(i_k, \mu_k(i_k), i_{k+1}\big) \;\Big|\; i_0 = i \right].$$

(If there is doubt regarding the existence of the limit, one can use "lim inf" in the above expression.) The optimal cost-to-go starting from state i is denoted by $J^*(i)$; that is,

$$J^*(i) = \min_{\pi} J^{\pi}(i).$$

We view the costs $J^*(i)$, $i = 1, \ldots, n$, as the components of a vector J^*, referred to as the *optimal cost-to-go vector.*

Of particular interest in infinite horizon problems are *stationary policies*, which are policies of the form $\pi = \{\mu, \mu, \ldots\}$. The corresponding cost-to-go is denoted by $J^{\mu}(i)$. For brevity, we refer to $\{\mu, \mu, \ldots\}$ as the stationary policy μ. We say that μ is optimal if $J^{\mu}(i) = J^*(i)$ for all states i. The vector J^{μ} that has components $J^{\mu}(i)$, $i = 1, \ldots, n$, is referred to as the *cost-to-go vector of the stationary policy* μ.

2.1.1 Finite Horizon Problems

In this subsection, we develop the DP algorithm for the case of finite horizon problems and when the special termination state 0 is absent. Consider first the case where there is only one stage, that is, $N = 1$. Then the optimal cost-to-go is by definition

$$J_1^*(i) = \min_{\mu_0} \sum_{j=1}^{n} p_{ij}\big(\mu_0(i)\big)\Big(g\big(i, \mu_0(i), j\big) + \alpha G(j)\Big).$$

For any fixed state i, the minimization over μ_0 is equivalent to a minimization over $u \in U(i)$, so we can write the above formula as

$$J_1^*(i) = \min_{u \in U(i)} \sum_{j=1}^{n} p_{ij}(u)\big(g(i, u, j) + \alpha G(j)\big). \tag{2.1}$$

The interpretation of this formula is that the optimal control choice with one period to go must minimize the sum of the expected present stage cost and the expected future cost $G(j)$ appropriately discounted by α. The DP algorithm expresses a generalization of this idea. It states that the optimal control choice with k stages to go must minimize the sum of the expected present stage cost and the expected optimal cost $J_{k-1}^*(j)$ with $k-1$ stages to go, appropriately discounted by α; that is,

$$J_k^*(i) = \min_{u \in U(i)} \sum_{j=1}^n p_{ij}(u)\big(g(i,u,j) + \alpha J_{k-1}^*(j)\big), \qquad i = 1, \ldots, n. \tag{2.2}$$

Thus, the optimal N-stage cost-to-go vector $J_N^*(i)$ can be calculated recursively with the above formula, starting with

$$J_0^*(i) = G(i), \qquad i = 1, \ldots, n. \tag{2.3}$$

To prove this, we argue as follows. Any policy π_k for the problem with k stages to go and initial state i is of the form $\{u, \pi_{k-1}\}$, where $u \in U(i)$ is the control at the first stage and π_{k-1} is the policy for the remaining $k-1$ stages. Thus,

$$\begin{aligned} J_k^*(i) &= \min_{u \in U(i),\, \pi_{k-1}} \sum_{j=1}^n p_{ij}(u)\big(g(i,u,j) + \alpha J_{k-1}^{\pi_{k-1}}(j)\big) \\ &= \min_{u \in U(i)} \sum_{j=1}^n p_{ij}(u)\big(g(i,u,j) + \alpha \min_{\pi_{k-1}} J_{k-1}^{\pi_{k-1}}(j)\big) \\ &= \min_{u \in U(i)} \sum_{j=1}^n p_{ij}(u)\big(g(i,u,j) + \alpha J_{k-1}^*(j)\big). \end{aligned}$$

This argument can also be used to prove that if $\mu_{N-k}^*(i)$ attains the minimum for each k and i in the DP equation (2.2), then the policy $\pi^* = \{\mu_0^*, \mu_1^*, \ldots, \mu_{N-1}^*\}$ is optimal for the N-stage problem; that is, $J_N^*(i) = J_N^{\pi^*}(i)$ for all i.

Ideally, we would like to use the DP algorithm (2.2) to obtain closed-form expressions for the optimal cost-to-go vectors J_k^* or an optimal policy. Some interesting models admit analytical solution by DP, but in many practical cases an analytical solution is not possible, and one has to resort to numerical execution of the DP algorithm. This may be very time-consuming, because the computational requirements are at least proportional to the number of possible state-control pairs as well as the number of stages N.

2.1.2 Infinite Horizon Problems

Infinite horizon problems represent a reasonable approximation of problems involving a finite but very large number of stages. These problems are also interesting because their analysis is elegant and insightful, and the implementation of optimal policies is often simple. For example, optimal policies are typically stationary; that is, the optimal rule for choosing controls does not change from one stage to the next. We will consider three principal classes of infinite horizon problems. In the first two classes, we try to minimize $J^\pi(i)$, the total cost over an infinite number of stages given by

$$J^\pi(i) = \lim_{N\to\infty} E\left[\sum_{k=0}^{N-1} \alpha^k g\big(i_k, \mu_k(i_k), i_{k+1}\big) \;\Big|\; i_0 = i\right].$$

(a) *Stochastic shortest path problems.* Here, $\alpha = 1$ but we assume that there is an additional state 0, which is a cost-free termination state; once the system reaches that state it remains there at no further cost. The structure of the problem is assumed to be such that termination is inevitable, at least under an optimal policy. Thus, the objective is to reach the termination state with minimal expected cost. The problem is in effect a finite horizon problem, but the length of the horizon may be random and may be affected by the policy being used. We will consider these problems in the next section.

(b) *Discounted problems.* Here, we assume that $\alpha < 1$. Note that the absolute one-stage cost $|g(i, u, j)|$ is bounded from above by some constant M, because of our assumption that i, u, and j belong to finite sets; this makes the cost-to-go $J^\pi(i)$ well defined because it is the infinite sum of a sequence of numbers that are bounded in absolute value by the decreasing geometric progression $\alpha^k M$. We will consider these problems in Section 2.3.

(c) *Average cost per stage problems.* Minimization of the total cost $J^\pi(i)$ makes sense only if $J^\pi(i)$ is finite for at least some policies π and some initial states i. Frequently, however, we have $J^\pi(i) = \infty$ for every policy π and initial state i (think of the case where $\alpha = 1$ and the cost for every state and control is positive). It turns out that in many such problems the *average cost per stage*, given by

$$\lim_{N\to\infty} \frac{1}{N} J_N^\pi(i),$$

where $J_N^\pi(i)$ is the N-stage cost-to-go of policy π starting at state i, is well defined as a limit and is finite. We will consider these problems in Section 7.1.

A Preview of Infinite Horizon Results

There are several analytical and computational issues regarding infinite horizon problems. Many of these revolve around the relation between the optimal cost-to-go vector J^* of the infinite horizon problem and the optimal cost-to-go vectors of the corresponding N-stage problems. In particular, let $J_N^*(i)$ denote the optimal cost-to-go of the problem involving N stages, initial state i, one-stage cost $g(i,u,j)$, and zero terminal cost. For all states $i = 1,\ldots,n$, the optimal N-stage cost-to-go is generated after N iterations of the DP algorithm

$$J_{k+1}^*(i) = \min_{u\in U(i)} \sum_{j=1}^{n} p_{ij}(u)\big(g(i,u,j) + \alpha J_k^*(j)\big), \qquad k = 0,1,\ldots \tag{2.4}$$

starting from the initial condition $J_0^*(i) = 0$ for all i. (We are assuming here that the termination state 0 is absent.) Since the infinite horizon cost of a given policy is, by definition, the limit of the corresponding N-stage costs as $N \to \infty$, it is natural to speculate that:

(1) The optimal infinite horizon cost-to-go is the limit of the corresponding N-stage optimal costs-to-go as $N \to \infty$, that is,

$$J^*(i) = \lim_{N\to\infty} J_N^*(i), \tag{2.5}$$

for all states i. This relation is very valuable computationally and analytically, and fortunately, it typically holds.

(2) The following limiting form of the DP algorithm should hold for all states i,

$$J^*(i) = \min_{u\in U(i)} \sum_{j=1}^{n} p_{ij}(u)\big(g(i,u,j) + \alpha J^*(j)\big), \qquad i = 1,\ldots,n, \tag{2.6}$$

as suggested by Eqs. (2.4) and (2.5). This is not really an algorithm, but rather a system of equations (one equation per state), which has as a solution the optimal costs-to-go of all the states. It will be referred to as *Bellman's equation*, and it will be at the center of our analysis and algorithms. An appropriate form of this equation holds for every type of infinite horizon problem of interest to us in this book.

(3) If $\mu(i)$ attains the minimum in the right-hand side of Bellman's equation for each i, then the stationary policy μ should be optimal. This is true for most infinite horizon problems of interest and in particular, for the models discussed in this book.

Most of the analysis of infinite horizon problems revolves around the above three issues and also around the efficient computation of J^* and an optimal stationary policy. In the next three sections we will provide a discussion of these issues. We focus primarily on results that are useful for the developments in the subsequent chapters.

2.2 STOCHASTIC SHORTEST PATH PROBLEMS

Here, we assume that there is no discounting ($\alpha = 1$) and, to make the cost-to-go meaningful, we assume that there exists a *cost-free termination state*, denoted by 0. Once the system reaches that state, it remains there at no further cost, that is,

$$p_{00}(u) = 1, \qquad g(0, u, 0) = 0, \qquad \forall\, u \in U(0).$$

We are interested in problems where reaching the termination state is inevitable, at least under an optimal policy. Thus, the essence of the problem is how to reach the termination state with minimum expected cost.

Two special cases of the stochastic shortest path problem are worth mentioning at the outset:

(a) *The (deterministic) shortest path problem*, obtained when the system is deterministic, that is, when each control at a given state leads with certainty to a unique successor state. This problem is usually defined in terms of a directed graph consisting of n nodes plus a destination node 0, and a set of arcs (i, j), each having a given cost c_{ij}. The cost of a path consisting of a sequence of arcs is the sum of the costs of the arcs of the path. The objective is to find among all paths that start at a given node and end at the destination, a path that has minimum cost; this is also called a *shortest path*. If we identify nodes with states, the outgoing arcs from a given node with the controls associated with the state corresponding to the node, and the costs of the arcs with the costs of the corresponding transitions, we obtain a special case of the problem of this section.

(b) *The finite horizon problem* discussed earlier, where transitions and cost accumulation stop after N stages. We can convert this problem to a stochastic shortest path problem by viewing time as an extra component of the state. Transitions occur from state-time pairs (i, k) to state-time pairs $(j, k + 1)$ according to the transition probabilities $p_{ij}(u)$ of the finite horizon problem. The termination state corresponds to the end of the horizon; it is reached in a single transition from any state-time pair of the form (j, N) at a terminal cost $G(j)$. Note that this reformulation is valid even if the finite horizon problem is nonstationary; that is, when the transition probabilities depend on the time k.

The above two special cases will often arise in various analytical contexts and examples, and while they possess special structure that simplifies their solution, they also incorporate some of the key ingredients of the generic stochastic shortest path problem structure.

2.2.1 General Theory

In order to guarantee the inevitability of termination under an optimal policy, we introduce certain conditions that involve the notion of a *proper policy*; that is, a stationary policy that leads to the termination state with probability one, regardless of the initial state.

Definition 2.1: A stationary policy μ is said to be *proper* if, using this policy, there is positive probability that the termination state will be reached after at most n stages, regardless of the initial state, that is, if

$$\rho_\mu = \max_{i=1,\ldots,n} \mathrm{P}\big(i_n \neq 0 \mid i_0 = i, \mu\big) < 1. \tag{2.7}$$

A stationary policy that is not proper is said to be *improper*.

With a little thought, it can be seen that μ is proper if and only if in the Markov chain corresponding to μ, each state i is connected to the termination state with a path of positive probability transitions. Note from the condition (2.7) that

$$\begin{aligned} \mathrm{P}(i_{2n} \neq 0 \mid i_0 = i, \mu) &= \mathrm{P}(i_{2n} \neq 0 \mid i_n \neq 0, i_0 = i, \mu) \\ &\qquad \times \mathrm{P}(i_n \neq 0 \mid i_0 = i, \mu) \\ &\leq \rho_\mu^2. \end{aligned}$$

More generally, under a proper policy μ, we have

$$\mathrm{P}(i_k \neq 0 \mid i_0 = i, \mu) \leq \rho_\mu^{\lfloor k/n \rfloor}, \qquad i = 1, \ldots, n; \tag{2.8}$$

that is, the probability of not reaching the termination state after k stages diminishes to zero as $\rho_\mu^{\lfloor k/n \rfloor}$, regardless of the initial state. This implies that the termination state will eventually be reached with probability 1 under a proper policy. Furthermore, the limit defining the associated total cost-to-go vector J^μ exists and is finite, since the expected cost incurred in the kth period is bounded in absolute value by

$$\rho_\mu^{\lfloor k/n \rfloor} \max_{i,j} \big| g\big(i, \mu(i), j\big) \big|. \tag{2.9}$$

Throughout this book and whenever we are dealing with stochastic shortest path problems, we assume the following.

Assumption 2.1: There exists at least one proper policy.

Assumption 2.2: For every improper policy μ, the corresponding cost-to-go $J^\mu(i)$ is infinite for at least one state i.

When specialized to the classical (deterministic) shortest path problems, Assumption 2.1 is equivalent to assuming the existence of at least one path from each node to the destination in the underlying graph, while Assumption 2.2 is equivalent to assuming that all cycles in the underlying graph have positive cost. These are the typical conditions under which deterministic shortest path problems are analyzed.

Note that a simple condition that implies Assumption 2.2 is that the expected one-stage cost $\sum_{j=0}^n p_{ij}(u)g(i,u,j)$ is strictly positive for all $i \neq 0$ and $u \in U(i)$. Another important case where Assumptions 2.1 and 2.2 are satisfied is when *all* policies are proper, that is, when termination is inevitable under all stationary policies.

Some Shorthand Notation

The form of the DP algorithm motivates the introduction of two mappings that play an important theoretical role and provide a convenient shorthand notation in expressions that would be too complicated to write otherwise.

For any vector $J = \big(J(1), \ldots, J(n)\big)$, we consider the vector TJ obtained by applying one iteration of the DP algorithm to J; the components of TJ are

$$(TJ)(i) = \min_{u \in U(i)} \sum_{j=0}^n p_{ij}(u)\big(g(i,u,j) + J(j)\big), \qquad i = 1, \ldots, n,$$

where *we will always use the convention that $J(0) = 0$ instead of viewing $J(0)$ as a free variable.* Note that T can be viewed as a mapping that transforms a vector J into the vector TJ. Furthermore, *TJ is the optimal cost-to-go vector for the one-stage problem that has one-stage cost g and terminal cost J.*

Similarly, for any vector J and any stationary policy μ, we consider the vector $T_\mu J$ with components

$$(T_\mu J)(i) = \sum_{j=0}^n p_{ij}\big(\mu(i)\big)\Big(g\big(i,\mu(i),j\big) + J(j)\Big), \qquad i = 1, \ldots, n.$$

Again, $T_\mu J$ may be viewed as the cost-to-go vector associated with μ for the one-stage problem that has one-stage cost g and terminal cost J.

Given a stationary policy μ, we define the $n \times n$ matrix P_μ whose ijth entry is $p_{ij}\big(\mu(i)\big)$. Then, we can write $T_\mu J$ in vector form as

$$T_\mu J = g_\mu + P_\mu J,$$

where g_μ is the n-dimensional vector whose ith component is

$$g_\mu(i) = \sum_{j=0}^{n} p_{ij}\big(\mu(i)\big) g\big(i, \mu(i), j\big).$$

Note that the entries of the ith row of P_μ sum to less than 1 if the probability $p_{i0}\big(\mu(i)\big)$ of a transition to the termination state is positive.

We will denote by T^k the composition of the mapping T with itself k times; that is, for all k we write

$$(T^k J)(i) = \big(T(T^{k-1}J)\big)(i), \qquad i = 1, \ldots, n.$$

Thus, $T^k J$ is the vector obtained by applying the mapping T to the vector $T^{k-1}J$. For convenience, we also write

$$(T^0 J)(i) = J(i), \qquad i = 1, \ldots, n.$$

Similarly, the components of $T_\mu^k J$ are defined by

$$(T_\mu^k J)(i) = \big(T_\mu(T_\mu^{k-1}J)\big)(i), \qquad i = 1, \ldots, n,$$

and

$$(T_\mu^0 J)(i) = J(i), \qquad i = 1, \ldots, n.$$

It can be seen from Eq. (2.2) that $(T^k J)(i)$ *is the optimal cost-to-go for the* k*-stage stochastic shortest path problem with initial state* i*, one-stage cost* g*, and terminal cost* J. Similarly, $(T_\mu^k J)(i)$ *is the cost-to-go of a stationary policy* μ *for the same problem.*

Finally, consider a k-stage policy $\pi = \{\mu_0, \mu_1, \ldots, \mu_{k-1}\}$. Then, the expression $(T_{\mu_0} T_{\mu_1} \cdots T_{\mu_{k-1}} J)(i)$ is defined recursively by

$$(T_{\mu_m} T_{\mu_{m+1}} \cdots T_{\mu_{k-1}} J)(i) = \big(T_{\mu_m}(T_{\mu_{m+1}} \cdots T_{\mu_{k-1}} J)\big)(i), \ \ m = 0, \ldots, k-2,$$

and represents *the cost-to-go of the policy* π *for the* k*-stage problem with initial state* i*, one-stage cost* g*, and terminal cost* J.

Some Properties of the Operators T and T_μ

We now derive a few elementary properties of the operators T and T_μ that will play a fundamental role in the developments of this chapter, as well as in later chapters.

Lemma 2.1: (Monotonicity Lemma) For any n-dimensional vectors J and $\overline{J}$, such that

$$J(i) \leq \overline{J}(i), \qquad i = 1, \ldots, n,$$

and for any stationary policy μ, we have

$$(T^k J)(i) \leq (T^k \overline{J})(i), \qquad i = 1, \ldots, n, \quad k = 1, 2, \ldots,$$

$$(T_\mu^k J)(i) \leq (T_\mu^k \overline{J})(i), \qquad i = 1, \ldots, n, \quad k = 1, 2, \ldots$$

Proof: This follows from the interpretations of $(T^k J)(i)$ and $(T_\mu^k J)(i)$ as k-stage costs-to-go: an increase of the terminal cost can only result in an increase of the k-stage cost-to-go. A formal proof is easily obtained by induction on k. **Q.E.D.**

For any two vectors J and $\overline{J}$, we write

$$J \leq \overline{J} \quad \text{if} \quad J(i) \leq \overline{J}(i), \qquad i = 1, \ldots, n.$$

With this notation, Lemma 2.1 is stated as

$$J \leq \overline{J} \quad \Rightarrow \quad T^k J \leq T^k \overline{J}, \qquad k = 1, 2, \ldots,$$

$$J \leq \overline{J} \quad \Rightarrow \quad T_\mu^k J \leq T_\mu^k \overline{J}, \qquad k = 1, 2, \ldots$$

Let us also denote by e the n-vector with all components equal to 1:

$$e(i) = 1, \qquad i = 1, \ldots, n.$$

The following lemma can be verified by induction using the definition of T and T_μ, and the monotonicity property.

Lemma 2.2: Consider the stochastic shortest path problem. Then, for every k, vector J, stationary policy μ, and positive scalar r, we have

$$\big(T^k(J + re)\big)(i) \leq (T^k J)(i) + r, \qquad i = 1, \ldots, n,$$

$$\big(T_\mu^k(J + re)\big)(i) \leq (T_\mu^k J)(i) + r, \qquad i = 1, \ldots, n.$$

If the scalar r is negative, the inequalities are reversed.

Main Results

The following proposition gives the main results regarding stochastic shortest path problems. The proof is fairly sophisticated and can be found in [BeT89], [BeT91a], or [Ber95a].†

Proposition 2.1: Consider the stochastic shortest path problem under Assumptions 2.1 and 2.2.

(a) The optimal cost-to-go vector J^* has finite components and satisfies

$$J^* = TJ^*.$$

Furthermore, J^* is the only solution of the equation $J = TJ$.

---

† We provide a quick proof of the Bellman equation $J^* = TJ^*$ and also that $T_{\mu^*}J^* = TJ^*$ implies optimality of μ^*, using slightly different assumptions than those of Prop. 2.1. In particular, we use Assumption 2.1 and we also use the assumption $\sum_{j=0}^{n} p_{ij}(u)g(i,u,j) \geq 0$ for all (i,u) (in place of Assumption 2.2, which will not be used). Let J_0 be the zero vector, and consider the sequence T^kJ_0. Our nonnegativity assumption on the expected one-stage cost implies that $J_0 \leq TJ_0$, so by using the monotonicity of T (Lemma 2.1), we have $T^kJ_0 \leq T^{k+1}J_0$ for all k. Hence,

$$J_0 \leq T^kJ_0 \leq T^{k+1}J_0 \leq \lim_{m\to\infty} T^mJ_0,$$

where each component of the limit vector above is either a real number or $+\infty$. By the definition and the monotonicity of T, we have for any policy $\pi = \{\mu_0, \mu_1, \ldots\}$

$$\lim_{m\to\infty} T^mJ_0 \leq \lim_{m\to\infty} T_{\mu_0}T_{\mu_1}\cdots T_{\mu_m}J_0 = J^\pi.$$

Since the cost-to-go vector of a proper policy has finite components and there exists at least one proper policy, it follows that the vector $J_\infty = \lim_{m\to\infty} T^mJ_0$ has finite components. By taking the limit as $k \to \infty$ in the relation $T^{k+1}J_0 = T(T^kJ_0)$ and by using the continuity of T, we obtain $J_\infty = TJ_\infty$. Let μ^* be such that $T_{\mu^*}J_\infty = TJ_\infty = J_\infty$. Then, since $J_0 \leq J_\infty$, we have

$$J^{\mu^*} = \lim_{m\to\infty} T_{\mu^*}^mJ_0 \leq \lim_{m\to\infty} T_{\mu^*}^mJ_\infty = J_\infty.$$

Since we showed earlier that $J_\infty \leq J^\pi$ for all policies π, we obtain $J^{\mu^*} = J_\infty = J^*$. The preceding argument also proves that J^* is the unique solution of the equation $J = TJ$ within the class of vectors J with $J \geq 0$.

(b) We have

$$\lim_{k \to \infty} T^k J = J^*,$$

for every vector J.

(c) A stationary policy μ is optimal if and only if

$$T_\mu J^* = T J^*.$$

(d) For every proper policy μ and every vector J, the associated cost-to-go vector J^μ satisfies

$$\lim_{k \to \infty} T_\mu^k J = J^\mu. \tag{2.10}$$

Furthermore,

$$J^\mu = T_\mu J^\mu,$$

and J^μ is the only solution of this equation.

The DP operator T often has some useful structure beyond the monotonicity property of Lemma 2.1. In particular, T is often a *contraction mapping with respect to a weighted maximum norm*. By this we mean that there exists a vector $\xi = \big(\xi(1), \ldots, \xi(n)\big)$ with positive components and a scalar $\beta < 1$ such that

$$\|TJ - T\bar{J}\|_\xi \leq \beta \|J - \bar{J}\|_\xi,$$

for all vectors J and $\bar{J}$, where the weighted maximum norm $\|\cdot\|_\xi$ is defined by

$$\|J\|_\xi = \max_{i=1,\ldots,n} \frac{|J(i)|}{\xi(i)}.$$

We have the following result:

Proposition 2.2: Suppose that all stationary policies are proper. Then, there exists a vector ξ with positive components such that the mapping T and the mappings T_μ, for all stationary policies μ, are contraction mappings with respect to the weighted maximum norm $\|\cdot\|_\xi$. In particular, there exists some $\beta < 1$ such that

$$\sum_{j=1}^n p_{ij}(u)\xi(j) \leq \beta \xi(i), \qquad \forall\ i,\ u \in U(i).$$

Proof: We first define the vector ξ as the solution of a certain DP problem, and then show that it has the required property. Consider a new stochastic shortest path problem where the transition probabilities are the same as in the original, but the transition costs are all equal to -1 (except at the termination state 0, where the self-transition cost is 0). Let $\hat{J}(i)$ be the optimal cost-to-go from state i in this new problem. By Prop. 2.1(a), we have for all $i = 1, \ldots, n$, and stationary policies μ,

$$\begin{aligned} \hat{J}(i) &= -1 + \min_{u \in U(i)} \sum_{j=1}^{n} p_{ij}(u) \hat{J}(j) \\ &\leq -1 + \sum_{j=1}^{n} p_{ij}\big(\mu(i)\big) \hat{J}(j). \end{aligned} \tag{2.11}$$

Define

$$\xi(i) = -\hat{J}(i), \qquad i = 1, \ldots, n.$$

Then for all i, we have $\xi(i) \geq 1$, and for all stationary policies μ, we have from Eq. (2.11),

$$\sum_{j=1}^{n} p_{ij}\big(\mu(i)\big)\xi(j) \leq \xi(i) - 1 \leq \beta\xi(i), \qquad i = 1, \ldots, n, \tag{2.12}$$

where β is defined by

$$\beta = \max_{i=1,\ldots,n} \frac{\xi(i) - 1}{\xi(i)} < 1.$$

For any stationary policy μ, any state i, and any vectors J and $\overline{J}$, we have using Eq. (2.12),

$$\begin{aligned} \big|(T_\mu J)(i) - (T_\mu \overline{J})(i)\big| &= \left| \sum_{j=1}^{n} p_{ij}\big(\mu(i)\big)\big(J(j) - \overline{J}(j)\big) \right| \\ &\leq \sum_{j=1}^{n} p_{ij}\big(\mu(i)\big)\big|J(j) - \overline{J}(j)\big| \\ &\leq \left(\sum_{j=1}^{n} p_{ij}\big(\mu(i)\big)\xi(j) \right) \left(\max_{j=1,\ldots,n} \frac{\big|J(j) - \overline{J}(j)\big|}{\xi(j)} \right) \\ &\leq \beta\xi(i) \max_{j=1,\ldots,n} \frac{\big|J(j) - \overline{J}(j)\big|}{\xi(j)}. \end{aligned}$$

Dividing both sides by $\xi(i)$ and taking the maximum over i of the left-hand side, we obtain

$$\max_{i=1,\ldots,n} \frac{\big|(T_\mu J)(i) - (T_\mu \overline{J})(i)\big|}{\xi(i)} \leq \beta \max_{j=1,\ldots,n} \frac{\big|J(j) - \overline{J}(j)\big|}{\xi(j)},$$

so that T_μ is a contraction with respect to the weighted maximum norm $\|\cdot\|_\xi$.

The preceding calculation also yields

$$(T_\mu J)(i) \le (T_\mu \overline{J})(i) + \beta\xi(i) \max_{j=1,\ldots,n} \frac{\big|J(j) - \overline{J}(j)\big|}{\xi(j)},$$

and by taking the minimum of both sides over μ, we obtain

$$(TJ)(i) \le (T\overline{J})(i) + \beta\xi(i) \max_{j=1,\ldots,n} \frac{\big|J(j) - \overline{J}(j)\big|}{\xi(j)}.$$

By interchanging the roles of J and $\overline{J}$, we also have

$$(T\overline{J})(i) \le (TJ)(i) + \beta\xi(i) \max_{j=1,\ldots,n} \frac{\big|J(j) - \overline{J}(j)\big|}{\xi(j)},$$

and by combining the last two relations, we obtain

$$\big|(TJ)(i) - (T\overline{J})(i)\big| \le \beta\xi(i) \max_{j=1,\ldots,n} \frac{\big|J(j) - \overline{J}(j)\big|}{\xi(j)}.$$

Dividing both sides by $\xi(i)$ and taking the maximum over i of the left-hand side, we obtain that T is a contraction with respect to $\|\cdot\|_\xi$. **Q.E.D.**

We now discuss a number of different methods for computing the optimal cost-to-go vector J^*.

2.2.2 Value Iteration

The DP iteration that generates the sequence $T^k J$ starting from some J is called *value iteration* and is a principal method for calculating the optimal cost-to-go vector J^*. Generally, the method requires an infinite number of iterations. However, under special circumstances, the method can terminate finitely. A prominent example is the case of a deterministic shortest path problem, but there are other more general circumstances where termination occurs. In particular, let us *assume that the transition probability graph corresponding to some optimal stationary policy μ^* is acyclic.* By this we mean that there are no cycles in the graph that has as nodes the states $0, 1, \ldots, n$, and has an arc (i, j) for each pair of states i and j such that $i \neq 0$ and $p_{ij}\big(\mu^*(i)\big) > 0$. Then it can be shown (see [Ber95a], Vol. II, Section 2.2.1) that *the value iteration method will yield J^* after at most n iterations when started from the vector J given by*

$$J(i) = \infty, \qquad i = 1, \ldots, n.$$

The preceding acyclicity assumption requires in particular that there are no positive self-transition probabilities $p_{ii}\big(\mu^*(i)\big)$ for $i \neq 0$, but it turns out that under our assumption, a stochastic shortest path problem with such self-transitions can be converted into another stochastic shortest path problem where $p_{ii}(u) = 0$ for all $i \neq 0$ and $u \in U(i)$. In particular, suppose that $p_{ii}(u) \neq 1$ for all $i \neq 0$ and $u \in U(i)$. Then, it can be shown that the modified stochastic shortest path problem that has one-stage costs

$$\tilde{g}(i,u,j) = g(i,u,j) + \frac{p_{ii}(u)g(i,u,i)}{1 - p_{ii}(u)}, \qquad i = 1,\ldots,n,$$

in place of $g(i,u,j)$, and transition probabilities

$$\tilde{p}_{ij}(u) = \begin{cases} 0, & \text{if } j = i, \\ \frac{p_{ij}(u)}{1-p_{ii}(u)}, & \text{if } j \neq i, \end{cases} \qquad i = 1,\ldots,n,$$

instead of $p_{ij}(u)$, is equivalent to the original in the sense that it has the same optimal costs-to-go and policies. This becomes intuitively clear once we note that $p_{ii}(u)/\big(1-p_{ii}(u)\big)$ is the expected number of self-transitions at state i, so the cost $\tilde{g}(i,u,j)$ is the total expected cost incurred in reaching a state j other than i, counting the cost of intermediate self-transitions. Furthermore, $\tilde{p}_{ij}(u)$ is the probability of transition from state i to a state $j \neq i$, conditioned on the fact that the next transition is not a self-transition.

Gauss-Seidel Methods and Asynchronous Value Iteration

In the value iteration method, the estimate of the cost-to-go vector is updated for all states simultaneously. An alternative is to update the cost-to-go at one state at a time, while incorporating into the computation the interim results. This corresponds to what is known as the *Gauss-Seidel method* for solving the nonlinear system of equations $J = TJ$.

For n-dimensional vectors J, define the mapping F by

$$(FJ)(1) = \min_{u \in U(1)} \sum_{j=0}^{n} p_{1j}(u)\big(g(1,u,j) + J(j)\big), \tag{2.13}$$

and, for $i = 2,\ldots,n$,

$$\begin{aligned}(FJ)(i) = \min_{u \in U(i)} \Bigg[\sum_{j=0}^{n} p_{ij}(u)g(i,u,j) &+ \sum_{j=1}^{i-1} p_{ij}(u)(FJ)(j) \\ &+ \sum_{j=i}^{n} p_{ij}(u)J(j)\Bigg].\end{aligned} \tag{2.14}$$

In words, $(FJ)(i)$ is computed by the same equation as $(TJ)(i)$ except that the previously calculated values $(FJ)(1),\ldots,(FJ)(i-1)$ are used in place of $J(1), \ldots, J(i-1)$. Note that the computation of FJ is as easy as the computation of TJ (unless a parallel computer is used, in which case TJ may potentially be computed much faster than FJ; see [BeT91b], [Tsi89]).

The Gauss-Seidel version of the value iteration method consists of computing J, FJ, $F^2J, \ldots$. This method is valid, in the sense that F^kJ converges to J^* under the same conditions that T^kJ converges to J^*. In fact the same result may be shown for a much more general version of the Gauss-Seidel method. In this version, it is not necessary to maintain a fixed order for iterating on the cost-to-go estimates $J(i)$ of the different states; an arbitrary order can be used, as long as $J(i)$ is iterated infinitely often for each state i. We call this type of method the *asynchronous value iteration method*, and we prove its validity in the next proposition. The method of proof is typical of convergence proofs of asynchronous algorithms that satisfy monotonicity conditions (see [BeT89], Section 6.4).

Proposition 2.3: (*Asynchronous Value Iteration Convergence*) Consider the algorithm that starts with an arbitrary initial vector J_0 and at the kth iteration, chooses an index $i_k \in \{1, \ldots, n\}$, replaces the i_kth component of the current vector J_k with $(TJ_k)(i_k)$, and leaves all other components of J_k unchanged; that is,

$$J_{k+1}(i) = \begin{cases} (TJ_k)(i), & \text{if } i = i_k, \\ J_k(i), & \text{otherwise.} \end{cases}$$

Assume that all components are chosen for iteration infinitely often. Then, the generated sequence of vectors J_k converges to J^*.

Proof: Assume first that the initial vector J_0 satisfies $TJ_0 \leq J_0$. Let $k_0, k_1, \ldots$ be an increasing sequence of iteration indices such that $k_0 = 0$ and each component $J(1), \ldots, J(n)$ is updated at least once between iterations k_m and $k_{m+1} - 1$, for all $m = 0, 1, \ldots$. Then it can be proved by induction on k that

$$J^* \leq J_k \leq T^m J_0, \qquad \text{if } k_m \leq k.$$

Since by Prop. 2.2, we have $T^m J_0 \to J^*$, it follows that $J_k \to J^*$. Similarly, we can prove that if $J_0 \leq TJ_0$, then $J_k \to J^*$.

Let now J^+ and J^- be vectors of the form $J^+ = J^* + \delta e$ and $J^- = J^* - \delta e$, where δ is a positive scalar such that $J^- \leq J_0 \leq J^+$, and $e = (1, \ldots, 1)$. Let J_k^- and J_k^+ be the vectors obtained after k iterations of the algorithm starting from the vectors J^- or J^+, respectively, instead of starting from J_0 (but with the same sequence of component choices). Then it can be seen that we have $TJ^+ \leq J^+$ and $J^- \leq TJ^-$, so the preceding argument implies that $J_k^+ \to J^*$ and $J_k^- \to J^*$. On the other hand, the relation $J^- \leq J_0 \leq J^+$ implies that $J_k^- \leq J_k \leq J_k^+$ for all k. It follows that $J_k \to J^*$. **Q.E.D.**

Sequential Space Decomposition and Gauss-Seidel Methods

In some stochastic shortest path problems there is favorable structure, which allows decomposition into a sequence of smaller problems. In particular, suppose that there is a sequence of subsets $S_1, S_2, \ldots, S_M$ of $\{1, 2, \ldots, n\}$ such that each of the states $i = 1, \ldots, n$ belongs to one and only one of these subsets, and the following property holds:

> For all $m = 1, \ldots, M$, and states $i \in S_m$, and under any policy, the successor state j is either the termination state or else belongs to one of the subsets $S_m, S_{m-1}, \ldots, S_1$.

Then it can be seen that the solution of this problem decomposes into the sequential solution of M stochastic shortest path subproblems. Each subproblem is solved using the optimal solution of the preceding subproblems. The state space of the mth subproblem is the subset S_m plus a termination state which corresponds to a transition to one of the states in the subset $\{0\} \cup S_1 \cup \cdots \cup S_{m-1}$. The cost of transition from a state $i \in S_m$ of the mth subproblem to the termination state via a state $j \in \{0\} \cup S_1 \cup \cdots \cup S_{m-1}$ is $g(i, u, j) + J^*(j)$.

A more favorable form of decomposition can be obtained when the following property holds, which requires in addition that transitions from states in a subset S_m cannot lead to states in the same subset:

> For all $m = 1, \ldots, M$, and states $i \in S_m$, and under any policy, the successor state j is either the termination state or else belongs to one of the subsets $S_{m-1}, \ldots, S_1$.

For this case, consider the Gauss-Seidel method that iterates on the states in the order of their membership in the sets S_m; that is, it first iterates on the states in the set S_1, then iterates on the states in the set S_2, etc. It can be seen by using induction on the index m, that after a single iteration on the states in the set S_m, the Gauss-Seidel iterates will be equal to the corresponding optimal costs-to-go because they will satisfy Bellman's equation, which takes the form

$$J^*(i) = \min_{u \in U(i)} \sum_{j \in \{0\} \cup S_1 \cup \cdots \cup S_{m-1}} p_{ij}(u)\big(g(i, u, j) + J^*(j)\big), \qquad \forall\, i \in S_m.$$

Thus, only one Gauss-Seidel iteration per state is needed.

An important example of such a decomposition is the finite horizon DP algorithm, which first finds the optimal costs-to-go of the states corresponding to the last stage, then uses these optimal costs-to-go to find the optimal costs-to-go of the states corresponding to the next-to-last stage, etc. For another interesting example, suppose that the states are K-tuples of the form $i = (i_1, i_2, \ldots, i_K)$, where each i_k is a nonnegative integer. The state $(0, 0, \ldots, 0)$ is the termination state. Assume that the transition probabilities have the property

$$p_{ij}(u) > 0 \quad \text{only if } \sum_{k=1}^{K} j_k < \sum_{k=1}^{K} i_k.$$

For all nonnegative integers m, consider the sets of states

$$S_m = \left\{ i = (i_1, \ldots, i_K) \;\middle|\; \sum_{k=1}^{K} i_k = m \right\}.$$

Then it can be seen that all transitions from states in a set S_m lead to the termination state or to states in the set $S_1 \cup \cdots \cup S_{m-1}$, and the preceding decomposition method applies.

2.2.3 Policy Iteration

In general, value iteration requires an infinite number of iterations to obtain the optimal cost-to-go vector. There is an alternative to value iteration, called *policy iteration*, which always terminates finitely. In this algorithm, we start with a proper policy μ_0, and we generate a sequence of new proper policies $\mu_1, \mu_2, \ldots$. Given the policy μ_k, we perform a *policy evaluation step*, that computes $J^{\mu_k}(i)$, $i = 1, \ldots, n$, as the solution of the (linear) system of equations

$$J(i) = \sum_{j=0}^{n} p_{ij}\big(\mu_k(i)\big)\Big(g\big(i, \mu_k(i), j\big) + J(j)\Big), \qquad i = 1, \ldots, n, \tag{2.15}$$

in the n unknowns $J(1), \ldots, J(n)$ [cf. Prop. 2.1(d)]. [Recall that $J(0)$ is set to zero and is therefore not an unknown.] We then perform a *policy improvement step*, which computes a new policy μ_{k+1} as

$$\mu_{k+1}(i) = \arg \min_{u \in U(i)} \sum_{j=0}^{n} p_{ij}(u)\big(g(i, u, j) + J^{\mu_k}(j)\big), \qquad i = 1, \ldots, n, \tag{2.16}$$

or equivalently, as the policy satisfying

$$T_{\mu_{k+1}} J^{\mu_k} = T J^{\mu_k}.$$

The process is repeated with μ_{k+1} used in place of μ_k, unless we have $J^{\mu_{k+1}}(i) = J^{\mu_k}(i)$ for all i, in which case the algorithm terminates with the policy μ_k. The following proposition establishes the validity of policy iteration.

Proposition 2.4: The policy iteration algorithm generates an improving sequence of proper policies [that is, $J^{\mu_{k+1}}(i) \le J^{\mu_k}(i)$ for all i and k] and terminates with an optimal policy.

Proof: In the typical iteration, given a proper policy μ and the corresponding cost-to-go vector J^μ, one obtains a new policy $\overline{\mu}$ satisfying $T_{\overline{\mu}} J^\mu = T J^\mu$. Then we have $J^\mu = T_\mu J^\mu \ge T_{\overline{\mu}} J^\mu$, and using the monotonicity of $T_{\overline{\mu}}$, we obtain

$$J^\mu \ge T_{\overline{\mu}}^k J^\mu = P_{\overline{\mu}}^k J^\mu + \sum_{m=0}^{k-1} P_{\overline{\mu}}^m g_{\overline{\mu}}, \qquad k = 1, 2, \ldots$$

If $\overline{\mu}$ is not proper, Assumption 2.2 implies that some component of the sum in the right-hand side of the above relation diverges to ∞ as $k \to \infty$, which is a contradiction. Therefore $\overline{\mu}$ is proper, and the preceding relation and Prop. 2.1(d) imply that

$$J^\mu \ge \lim_{k\to\infty} T_{\overline{\mu}}^k J^\mu = J^{\overline{\mu}}. \tag{2.17}$$

Note that strict inequality $J^{\overline{\mu}}(i) < J^\mu(i)$ holds for at least one state i, if μ is nonoptimal; otherwise we would have $J^\mu = T J^\mu$, and by Prop. 2.1(a) it would follow that $J^\mu = J^*$, so that μ would be optimal. Therefore, the new policy is strictly better if the current policy is nonoptimal. Since the number of proper policies is finite, the policy iteration algorithm terminates after a finite number of iterations with an optimal proper policy. **Q.E.D.**

In practice, the policy iteration algorithm typically requires very few iterations, although there are no theoretical guarantees for this.

Multistage Lookahead Policy Iteration

There is a variation of the policy iteration algorithm that uses a lookahead of several stages in the policy improvement step. The method described above uses single stage lookahead for generating an improved policy starting from a given policy μ: at a given state i, it finds the optimal decision for a one-stage problem with one-stage cost $g(i, u, j)$ and terminal cost (after the first stage) $J^\mu(j)$. An m-stage lookahead version finds the optimal policy for an m-stage DP problem, whereby we start at the current state i, make the m subsequent decisions, incur the corresponding costs of the m stages, and pay a terminal cost $J^\mu(j)$, where j is the state after m stages. This may be a tractable problem, depending on the horizon m and the number of possible successor states from each state. If $\overline{u}$ is the first decision of the m-stage optimal policy starting at state i, the improved policy in the multistage policy iteration method is defined by $\overline{\mu}(i) = \overline{u}$. This policy is then evaluated in the usual way, i.e., by solving the linear system $J^{\overline{\mu}} = T_{\overline{\mu}} J^{\overline{\mu}}$.

It can be shown that the multistage lookahead policy iteration terminates with an optimal policy under the same conditions as the ordinary version. To see this, let $\{\overline{\mu}_0, \overline{\mu}_1, \ldots, \overline{\mu}_{m-1}\}$ be an optimal policy obtained from the m-stage DP algorithm with terminal cost J^μ; that is,

$$T_{\overline{\mu}_k} T^{m-k-1} J^\mu = T^{m-k} J^\mu, \qquad k = 0, 1, \ldots, m-1. \tag{2.18}$$

Since $TJ^\mu \leq T_\mu J^\mu = J^\mu$, we have using the monotonicity of T,

$$T^{k+1} J^\mu \leq T^k J^\mu \leq J^\mu, \qquad k = 0, 1, \ldots \tag{2.19}$$

By combining this equation with Eq. (2.18), we have

$$T_{\overline{\mu}_k} T^{m-k-1} J^\mu = T^{m-k} J^\mu \leq T^{m-k-1} J^\mu, \qquad k = 0, 1, \ldots, m-1,$$

from which we obtain for all $l \geq 1$, using the monotonicity of $T_{\overline{\mu}_k}$,

$$T^l_{\overline{\mu}_k} T^{m-k-1} J^\mu \leq T_{\overline{\mu}_k} T^{m-k-1} J^\mu = T^{m-k} J^\mu, \qquad k = 0, 1, \ldots, m-1.$$

By taking the limit as $l \to \infty$ and by using Eq. (2.19), it follows that

$$J^{\overline{\mu}_k} \leq T^{m-k} J^\mu \leq J^\mu, \qquad k = 0, 1, \ldots, m-1.$$

In particular for the successor policy $\overline{\mu}$ generated by the m-stage policy iteration, which is $\overline{\mu} = \overline{\mu}_0$, we have

$$J^{\overline{\mu}} \leq T^m J^\mu \leq J^\mu.$$

It follows that $\overline{\mu}$ is an improved policy relative to μ. Furthermore, if $J^{\overline{\mu}} = J^\mu$, in view of Eq. (2.19), we have that $TJ^\mu = J^\mu$ and μ is optimal. Similar to the proof of Prop. 2.4, it follows that the algorithm terminates with an optimal policy.

Note that the expression $T^m J^\mu$, which is the result of m successive value iterations applied to J^μ, provides an upper bound on the cost-to-go function $J^{\overline{\mu}}$. This upper bound improves as m increases [cf. Eq. (2.19)], thereby suggesting that for many problems, the multistage policy iteration method may perform better than its single stage counterpart. This is particularly so when cost-to-go function approximations are introduced (see the discussion of Section 6.1.2). On the other hand the computation required to perform the policy improvement step increases rapidly with the size of the lookahead.

Modified Policy Iteration

When the number of states is large, solving the linear system in the policy evaluation step by direct methods such as Gaussian elimination is time-consuming. One way to get around this difficulty is to solve the linear system iteratively by using value iteration, as suggested by Prop. 2.1(d). In fact, we may consider solving the system approximately by executing a limited number of value iterations. The resulting method is called the *modified policy iteration algorithm.*

To formalize this method, let J_0 be an n-dimensional vector such that $TJ_0 \leq J_0$. Let $m_0, m_1, \ldots$ be positive integers, and let the vectors $J_1, J_2, \ldots$ and the stationary policies $\mu_0, \mu_1, \ldots$ be defined by

$$T_{\mu_k} J_k = TJ_k, \qquad J_{k+1} = T_{\mu_k}^{m_k} J_k, \qquad k = 0, 1, \ldots$$

Thus, a stationary policy μ_k is defined from J_k according to $T_{\mu_k} J_k = TJ_k$, and the cost-to-go J^{μ_k} is approximately evaluated by $m_k - 1$ additional value iterations, yielding the vector J_{k+1}, which is used in turn to define μ_{k+1}.

Note that if $m_k = 1$ for all k in the modified policy iteration algorithm, we obtain the value iteration method, while if $m_k = \infty$ we obtain the policy iteration method, where the policy evaluation step is performed iteratively by means of value iteration. Analysis and computational experience suggest that it is usually best to take m_k larger than 1 according to some heuristic scheme. A key idea here is that a value iteration involving a single policy (evaluating $T_\mu J$ for some μ and J) is much less expensive than an iteration involving all policies (evaluating TJ for some J), when the number of controls available at each state is large. The convergence properties of the modified policy iteration method will be discussed in the context of a more general algorithm, which we now introduce.

Asynchronous Policy Iteration

We may consider a more general policy iteration algorithm, whereby value iterations and policy updates are executed selectively, for only some of the states. This type of algorithm generates a sequence J_k of cost-to-go estimates and a corresponding sequence μ_k of stationary policies. Given (J_k, μ_k), we select a subset S_k of the states, and we generate the new pair (J_{k+1}, μ_{k+1}) in one of two possible ways: either we update J_k according to

$$J_{k+1}(i) = \begin{cases} (T_{\mu_k} J_k)(i), & \text{if } i \in S_k, \\ J_k(i), & \text{otherwise}, \end{cases} \tag{2.20}$$

while we leave the policy unchanged by setting $\mu_k = \mu_{k+1}$, or else we update μ_k according to

$$\mu_{k+1}(i) = \begin{cases} \arg\min_{u \in U(i)} \sum_{j=0}^{n} p_{ij}(u)\big(g(i,u,j) + J_k(j)\big), & \text{if } i \in S_k, \\ \mu_k(i), & \text{otherwise}, \end{cases} \tag{2.21}$$

while we leave the cost-to-go estimate unchanged by setting $J_k = J_{k+1}$.

Note that by combining value and policy updates in various ways, we may synthesize more structured methods. In particular, if a policy update is followed by a value update for the same set of states S_k, the latter update is equivalent to the value iteration update

$$J_{k+1}(i) = \begin{cases} (TJ_k)(i), & \text{if } i \in S_k, \\ J_k(i), & \text{otherwise.} \end{cases} \tag{2.22}$$

Furthermore, if $S_k = \{1, \ldots, n\}$ and an "infinite" number of value updates (2.20) are done before doing a single policy update (2.21), we essentially obtain the policy iteration method. If $S_k = \{1, \ldots, n\}$ and m_k value updates (2.20) are done followed by a policy update (2.21), we obtain the modified policy iteration method (or the value iteration method if $m_k = 1$). On the other hand, if S_k always consists of a single state and each policy update (2.21) is followed by a value update (2.20) [or equivalently by an update (2.22)], the algorithm reduces to the asynchronous value iteration method.

The asynchronous policy iteration algorithm also contains as a special case a version of the policy iteration method that uses a partition of the state space into subsets, and performs "partial" policy iterations that are restricted to one of the subsets at a time. In particular, at the kth iteration, this method selects one of the subsets, call it S_k, obtains a new policy μ_{k+1} via the policy update (2.21), and "evaluates" this policy only for the states in S_k. The evaluation can be performed by using an infinite number of value updates for all states in S_k, with the cost-to-go values of the states $j \notin S_k$ held fixed at $J_k(j)$. Alternatively, the iteration can be performed by solving a restricted system of linear equations whose variables are the cost-to-go values of the states $i \in S_k$ only; this can be viewed as an evaluation step of the policy μ_{k+1} for a stochastic shortest path problem that involves the states $i \in S_k$ plus an artificial termination state which is reached at cost $J_k(j)$ whenever a transition to a state $j \notin S_k$ occurs.

Finally, let us consider the convergence of the asynchronous policy iteration method. For this, we will need some assumptions, including that the initial conditions J_0 and μ_0 must be such that $T_{\mu_0} J_0 \leq J_0$ (one possibility to obtain such initial conditions is to select μ_0 in some way and then take $J_0 = J^{\mu_0}$). We have the following proposition.

Proposition 2.5: (*Asynchronous Policy Iteration Convergence*) Assume that the value update (2.20) and the policy update (2.21) are executed infinitely often for all states, and that the initial conditions J_0 and μ_0 are such that $T_{\mu_0} J_0 \leq J_0$. Let (J_k, μ_k) be the sequence generated by the asynchronous policy iteration algorithm. Then J_k converges to J^*.

Proof: The idea of the proof is to first use the assumption $T_{\mu_0} J_0 \leq J_0$ to show that $J^* \leq J_{k+1} \leq J_k$ for all k, so that the sequence J_k converges to a limit $\overline{J}$. We will then use the assumption that the value and policy updates are executed infinitely often for all states to show that $\overline{J} = T\overline{J}$, so that $\overline{J} = J^*$.

We first show that for all k, we have

$$T_{\mu_k} J_k \leq J_k \qquad \Rightarrow \qquad T_{\mu_{k+1}} J_{k+1} \leq J_{k+1} \leq J_k. \tag{2.23}$$

Indeed assume that at iteration k we have $T_{\mu_k} J_k \leq J_k$, and consider two cases:

(1) *The value update (2.20) is executed next.* Then we have

$$J_{k+1}(i) = (T_{\mu_k} J_k)(i) \leq J_k(i), \qquad \text{if } i \in S_k, \tag{2.24}$$

and

$$J_{k+1}(i) = J_k(i), \qquad \text{if } i \notin S_k, \tag{2.25}$$

so that $J_{k+1} \leq J_k$. Using this inequality, the monotonicity of T_{μ_k}, and the fact $\mu_{k+1} = \mu_k$, we obtain

$$T_{\mu_{k+1}} J_{k+1} = T_{\mu_k} J_{k+1} \leq T_{\mu_k} J_k. \tag{2.26}$$

On the other hand, from Eq. (2.24) we have

$$(T_{\mu_k} J_k)(i) = J_{k+1}(i), \qquad \text{if } i \in S_k,$$

while from Eq. (2.25) and the hypothesis $T_{\mu_k} J_k \leq J_k$, we have

$$(T_{\mu_k} J_k)(i) \leq J_k(i) = J_{k+1}(i), \qquad \text{if } i \notin S_k.$$

The above two relations imply that $T_{\mu_k} J_k \leq J_{k+1}$, which when combined with Eq. (2.26), shows that $T_{\mu_{k+1}} J_{k+1} \leq J_{k+1}$ and completes the proof of Eq. (2.23).

(2) *The policy update (2.21) is executed next.* Then we have $J_{k+1} = J_k$, and by using the relation $T_{\mu_k} J_k \leq J_k$, we obtain

$$\begin{aligned}(T_{\mu_{k+1}} J_{k+1})(i) &= (T_{\mu_{k+1}} J_k)(i) = (T J_k)(i) \leq (T_{\mu_k} J_k)(i) \\ &\leq J_k(i) = J_{k+1}(i), \qquad \text{if } i \in S_k,\end{aligned} \tag{2.27}$$

and

$$\begin{aligned}(T_{\mu_{k+1}} J_{k+1})(i) &= (T_{\mu_{k+1}} J_k)(i) = (T_{\mu_k} J_k)(i) \\ &\leq J_k(i) = J_{k+1}(i), \qquad \text{if } i \notin S_k,\end{aligned} \tag{2.28}$$

so that $T_{\mu_{k+1}} J_{k+1} \leq J_{k+1}$, and Eq. (2.23) is shown.

Equation (2.23) and the hypothesis $T_{\mu_0} J_0 \leq J_0$ imply that

$$J_{k+1} \leq J_k, \qquad T J_k \leq T_{\mu_k} J_k \leq J_k, \qquad \forall\, k. \tag{2.29}$$

By using the monotonicity of T, we also have $T^m J_k \leq J_k$ for all m, and by taking the limit as $m \to \infty$, we obtain

$$J^* \leq J_k, \qquad \forall\, k.$$

From this equation and Eq. (2.29), we see that J_k converges to some limit $\overline{J}$ satisfying

$$T\overline{J} \leq \overline{J} \leq J_k, \qquad \forall\, k. \tag{2.30}$$

Furthermore, from Eqs. (2.26)-(2.28), we have

$$T_{\mu_{k+1}} J_{k+1} \leq T_{\mu_k} J_k, \qquad \forall\, k. \tag{2.31}$$

Suppose, to arrive at a contradiction, that there exists a state i such that $(T\overline{J})(i) < \overline{J}(i)$, and let $\overline{k}$ be such that $(TJ_k)(i) < \overline{J}(i)$ for all $k \geq \overline{k}$ (such an integer $\overline{k}$ exists by the continuity of T). Let k be an iteration index that satisfies $k \geq \overline{k}$ and is such that the policy update (2.21) is executed for state i, and let k' be the first iteration index with $k' > k$ such that the value update (2.20) is executed for state i. Then we have

$$\begin{aligned} J_{k'+1}(i) &= (T_{\mu_{k'}} J_{k'})(i) \\ &\leq (T_{\mu_{k+1}} J_{k+1})(i) \\ &\leq (T_{\mu_{k+1}} J_k)(i) \\ &= (T J_k)(i) \\ &< \overline{J}(i), \end{aligned} \tag{2.32}$$

where the first equality follows from the value update (2.20), the first inequality follows from Eq. (2.31), the second inequality follows from the relation $J_{k+1} \leq J_k$ and the monotonicity of $T_{\mu_{k+1}}$, and the second equality follows from the policy update (2.21). The relation (2.32) contradicts the relation (2.30). Thus we must have $(T\overline{J})(i) = \overline{J}(i)$ for all i, which implies that $\overline{J} = J^*$, since J^* is the unique solution of Bellman's equation by Prop. 2.1(a). **Q.E.D.**

We note that without the assumption $T_{\mu_0} J_0 \leq J_0$, one may construct counterexamples to the convergence of the asynchronous policy iteration method (see Williams and Baird [WiB93]).

Policy Iteration as an Actor-Critic System

There is an interesting interpretation of policy iteration methods that will prove useful when we will consider approximate versions of policy iteration in subsequent chapters. In this interpretation, the policy evaluation step is viewed as the work of a *critic*, who evaluates the performance of the current policy, i.e., calculates an estimate of J^{μ_k}. The policy improvement step is viewed as the work of an *actor*, who takes into account the latest evaluation of the critic, i.e., the estimate of J^{μ_k}, and acts out the improved policy μ_{k+1} (see Fig. 2.1). Note that the exact, the modified, and the asynchronous policy iteration algorithms can all be viewed as actor-critic systems. In particular, modified policy iteration amounts to an incomplete evaluation of the current policy by the critic, using just a few value iteration steps. Asynchronous policy iteration can be similarly viewed as a process where the critic's policy evaluation and the critic's feedback to the actor are irregular and take place at different times for different states.

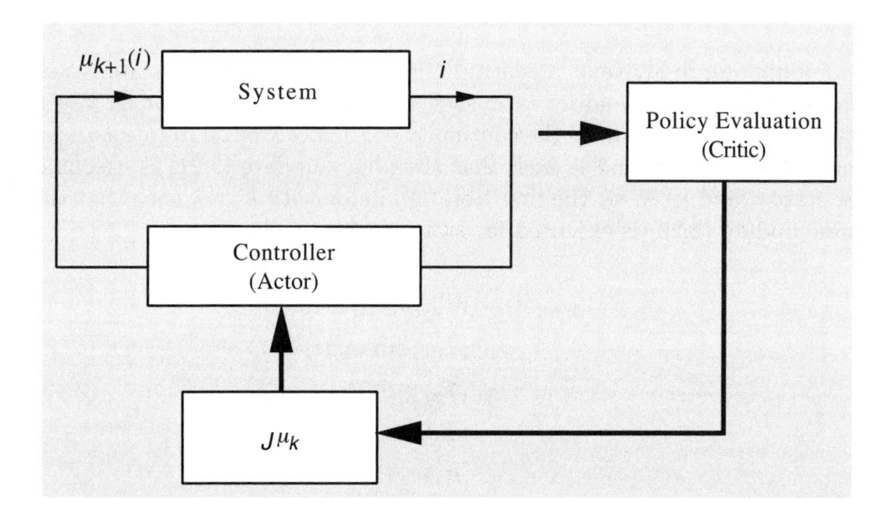

Figure 2.1: Interpretation of policy iteration as an actor-critic system.

2.2.4 Linear Programming

Another method to solve the stochastic shortest path problem is based on linear programming. Since $\lim_{k\to\infty} T^k J = J^*$ for all J [cf. Prop. 2.1(b)], we have

$$J \leq TJ \qquad \Rightarrow \qquad J \leq J^* = TJ^*.$$

Thus J^* is the "largest" J that satisfies the constraint $J \leq TJ$. This constraint can be written as a finite system of linear inequalities

$$J(i) \leq \sum_{j=0}^{n} p_{ij}(u)\big(g(i,u,j) + J(j)\big), \qquad i = 1, \ldots, n, \quad u \in U(i),$$

and delineates a polyhedron in $\Re^n$. In particular, $J^*(1), \ldots, J^*(n)$ solve the following problem (in the variables $\lambda_1, \ldots, \lambda_n$):

$$\text{maximize} \quad \sum_{i=1}^{n} \lambda_i$$

$$\text{subject to} \quad \lambda_i \leq \sum_{j=0}^{n} p_{ij}(u)\big(g(i,u,j) + \lambda_j\big), \quad i = 1, \ldots, n, \ u \in U(i),$$

where λ_0 is fixed at zero. This is a linear program with n variables and as many as $n \times q$ constraints, where q is the maximum number of elements of the sets $U(i)$. As n increases, its solution becomes more complex. For very large n and q, the linear programming approach can be practical only with the use of special large-scale linear programming methods.

2.3 DISCOUNTED PROBLEMS

We now consider a discounted problem, where there is a discount factor $\alpha < 1$. We will show that this problem can be viewed as a special case of the stochastic shortest path problem of the preceding section. We will then obtain the associated analytical and algorithmic results by appropriately specializing the analysis of the preceding section.

We first modify the definition of the mappings T and T_μ to account for the absence of the termination state. For any vector $J = \big(J(1), \ldots, J(n)\big)$, we consider the vector TJ obtained by applying one iteration of the DP algorithm to J; the components of TJ are

$$(TJ)(i) = \min_{u \in U(i)} \sum_{j=1}^{n} p_{ij}(u)\big(g(i,u,j) + \alpha J(j)\big), \qquad i = 1, \ldots, n.$$

Similarly, for any vector J and any stationary policy μ, we consider the vector $T_\mu J$ with components

$$(T_\mu J)(i) = \sum_{j=1}^{n} p_{ij}\big(\mu(i)\big)\Big(g\big(i,\mu(i),j\big) + \alpha J(j)\Big), \qquad i = 1, \ldots, n.$$

Again, $T_\mu J$ may be viewed as the cost-to-go vector associated with μ for the one-stage problem that has one-stage cost g and terminal cost αJ.

If P_μ is the $n \times n$ matrix whose ijth entry is $p_{ij}\big(\mu(i)\big)$, then $T_\mu J$ can be written in vector form as

$$T_\mu J = g_\mu + \alpha P_\mu J,$$

where g_μ is the n-dimensional vector whose ith component is

$$g_\mu(i) = \sum_{j=1}^{n} p_{ij}\big(\mu(i)\big) g\big(i, \mu(i), j\big).$$

The following two lemmas parallel Lemmas 2.1 and 2.2 of the preceding section.

Lemma 2.3: (Monotonicity Lemma) For any n-dimensional vectors J and $\overline{J}$, such that

$$J(i) \le \overline{J}(i), \qquad i = 1, \ldots, n,$$

and for any stationary policy μ, we have

$$(T^k J)(i) \le (T^k \overline{J})(i), \qquad i = 1, \ldots, n, \quad k = 1, 2, \ldots,$$

$$(T_\mu^k J)(i) \le (T_\mu^k \overline{J})(i), \qquad i = 1, \ldots, n, \quad k = 1, 2, \ldots$$

Again we denote by e the n-vector with all components equal to 1. Then, for any vector J and scalar r

$$\big(T(J + re)\big)(i) = (TJ)(i) + \alpha r, \qquad i = 1, \ldots, n,$$

$$\big(T_\mu(J + re)\big)(i) = (T_\mu J)(i) + \alpha r, \qquad i = 1, \ldots, n.$$

More generally, the following lemma can be verified by induction.

Lemma 2.4: For every k, vector J, stationary policy μ, and scalar r, we have

$$\big(T^k(J + re)\big)(i) = (T^k J)(i) + \alpha^k r, \qquad i = 1, \ldots, n,$$

$$\big(T_\mu^k(J + re)\big)(i) = (T_\mu^k J)(i) + \alpha^k r, \qquad i = 1, \ldots, n.$$

We define the *maximum* norm $\|\cdot\|_\infty$ on $\Re^n$ by

$$\|J\|_\infty = \max_i |J(i)|.$$

We then have the following contraction property of the DP operators T and T_μ.

Lemma 2.5: For any vectors J and $\overline{J}$ and any policy μ, we have

$$\|TJ - T\overline{J}\|_\infty \leq \alpha \|J - \overline{J}\|_\infty, \tag{2.33}$$

$$\|T_\mu J - T_\mu \overline{J}\|_\infty \leq \alpha \|J - \overline{J}\|_\infty. \tag{2.34}$$

Proof: Denote

$$c = \max_{i=1,\ldots,n} \big|J(i) - \overline{J}(i)\big|.$$

Then we have

$$J(i) - c \leq \overline{J}(i) \leq J(i) + c, \qquad i = 1, \ldots, n.$$

Applying T in this relation and using the Monotonicity Lemma 2.3 as well as Lemma 2.4, we obtain

$$(TJ)(i) - \alpha c \leq (T\overline{J})(i) \leq (TJ)(i) + \alpha c, \qquad i = 1, \ldots, n.$$

It follows that

$$\big|(TJ)(i) - (T\overline{J})(i)\big| \leq \alpha c, \qquad i = 1, \ldots, n,$$

which proves Eq. (2.33). Equation (2.34) follows by applying Eq. (2.33) to the modified problem where the only control available at state i is $\mu(i)$. **Q.E.D.**

We will now show that the discounted cost problem can be converted to a stochastic shortest path problem for which the analysis of the preceding section holds. To see this, let $i = 1, \ldots, n$ be the states of the discounted problem, and consider an associated stochastic shortest path problem involving the states $1, \ldots, n$ plus an extra termination state 0, with state transitions and costs obtained as follows. From a state $i \neq 0$, when control u is applied, a cost $g(i, u, j)$ is incurred with probability $p_{ij}(u)$; the next state is j with probability $\alpha p_{ij}(u)$, and 0 with probability $1 - \alpha$; see Fig.

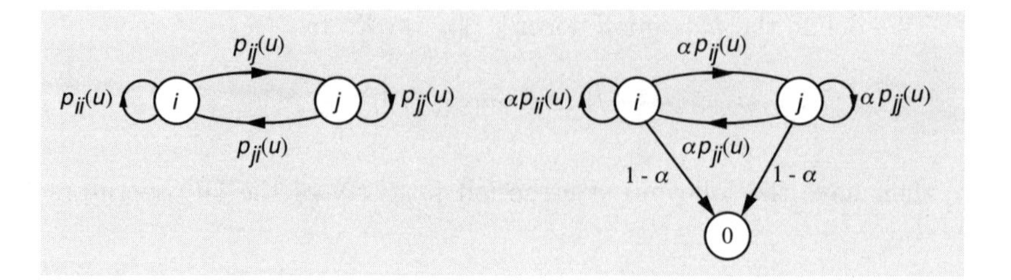

Figure 2.2: Transition probabilities for an α-discounted problem and its associated stochastic shortest path problem. In the latter problem, the probability that the state is not 0 after k stages is α^k.

2.2. Note that all policies are proper for the associated stochastic shortest path problem, so the assumptions of the preceding section are satisfied.

Suppose now that we use the same policy in the discounted problem and in the associated stochastic shortest path problem. Then, as long as termination has not occurred, the state evolution in the two problems is governed by the same transition probabilities. Furthermore, the expected one-stage cost at each state i is

$$\sum_{j=1}^{n} p_{ij}(u)g(i,u,j)$$

for both problems, but it must be multiplied by α^k because of discounting (in the discounted case), or because it is incurred with probability α^k, when termination has not yet been reached (in the stochastic shortest path case). We conclude that the cost-to-go of any policy starting from a given state, is the same for the original discounted problem and for the associated stochastic shortest path problem. Since in the associated stochastic shortest path problem all policies are proper, we can apply the results of the preceding section to the latter problem and obtain the following result.

Proposition 2.6: The following hold for the discounted cost problem:

(a) The optimal cost-to-go vector J^* satisfies

$$J^* = TJ^*.$$

Furthermore, J^* is the only solution of this equation.

(b) We have

$$\lim_{k \to \infty} T^k J = J^*,$$

for every vector J.

(c) A stationary policy μ is optimal if and only if

$$T_\mu J^* = TJ^*.$$

(d) For a stationary policy μ, the associated cost-to-go vector J^μ satisfies

$$\lim_{k \to \infty} T_\mu^k J = J^\mu,$$

for every vector J. Furthermore,

$$J^\mu = T_\mu J^\mu,$$

and J^μ is the only solution of this equation.

(e) The policy iteration algorithm generates an improving sequence of policies [that is, $J^{\mu_{k+1}}(i) \leq J^{\mu_k}(i)$ for all i and k] and terminates with an optimal policy.

There are also discounted versions of the multistage lookahead, the modified, and the asynchronous policy iteration algorithms that we discussed earlier in connection with stochastic shortest path problems. In particular, the convergence result of Prop. 2.5 admits a straightforward extension to the discounted case.

2.3.1 Temporal Difference-Based Policy Iteration

We mentioned earlier that when the number of states is large, it is usually preferable to carry out the policy evaluation step of the policy iteration method by using value iteration. A drawback of this approach, however, is that the value iteration algorithm may converge very slowly. This is true for many stochastic shortest path problems with a large number of states, as well as for discounted problems with a discount factor that is close to 1. We are thus motivated to use an approach whereby the discount factor is effectively reduced in order to accelerate the policy evaluation step.

The idea underlying the approach is that the discount factor can be reduced without altering the cost-to-go of a given policy only if the expected value of the one-stage cost under that policy is equal to 0. We thus introduce a transformation that asymptotically induces this one-stage cost structure. The transformation is based on the notion of *temporal differences* (TD for short), which will be of major importance in the context of the simulation-based methods of Chapters 5-7.

We now describe a policy-iteration like algorithm that maintains a policy-cost vector pair (J_k, μ_k). We can view J_k as an approximation to the cost-to-go vector J^{μ_k}. In the typical iteration, given (J_k, μ_k), we calculate μ_{k+1} by the policy improvement step

$$T_{\mu_{k+1}} J_k = T J_k. \tag{2.35}$$

To calculate J_{k+1}, we define the TD associated with each transition (i, j) under μ_{k+1}:

$$d_k(i, j) = g\big(i, \mu_{k+1}(i), j\big) + \alpha J_k(j) - J_k(i). \tag{2.36}$$

We also consider a parameter λ from the range $[0, 1]$. We view $d_k(i, j)$ as the one-stage cost of policy μ_{k+1} for an $\alpha\lambda$-discounted DP problem with the transition probabilities $p_{ij}\big(\mu_{k+1}(i)\big)$ of the original problem, and we calculate the corresponding cost-to-go vector. This vector, denoted by Δ_k, has components given by

$$\Delta_k(i) = \sum_{m=0}^{\infty} E\big[(\alpha\lambda)^m d_k(i_m, i_{m+1}) \mid i_0 = i\big], \qquad \forall\ i. \tag{2.37}$$

The vector J_{k+1} is then obtained by

$$J_{k+1} = J_k + \Delta_k. \tag{2.38}$$

We refer to the method as the *λ-policy iteration method.*

Note that when $\lambda = 1$, by using the TD definition (2.36) in the expression (2.37), we obtain

$$\begin{aligned}
\Delta_k(i) &= \sum_{m=0}^{\infty} E\big[\alpha^m d_k(i_m, i_{m+1}) \mid i_0 = i\big] \\
&= \sum_{m=0}^{\infty} E\Big[\alpha^m g\big(i_m, \mu_{k+1}(i_m), i_{m+1}\big) \\
&\qquad\qquad + \alpha^{m+1} J_k(i_{m+1}) - \alpha^m J_k(i_m) \mid i_0 = i\Big] \\
&= \sum_{m=0}^{\infty} E\Big[\alpha^m g\big(i_m, \mu_{k+1}(i_m), i_{m+1}\big) \mid i_0 = i\Big] - J_k(i) \\
&= J^{\mu_{k+1}}(i) - J_k(i).
\end{aligned}$$

Thus, by using Eq. (2.38), we see that $J_{k+1} = J^{\mu_{k+1}}$, so *the 1-policy iteration method coincides with the standard policy iteration method*. However, for $\lambda < 1$, the two methods are different, and in fact it can be seen from Eqs. (2.36)-(2.38) that *the 0-policy iteration method coincides with the value iteration method* for the original problem.

The motivation for the λ-policy iteration method is that the $\alpha\lambda$-discounted policy evaluation step [cf. Eqs. (2.37)-(2.38)] can be much easier when $\lambda < 1$ than when $\lambda = 1$. In particular, when value iteration is used for policy evaluation, convergence is faster when $\lambda < 1$ than when $\lambda = 1$. Similarly, when policy evaluation is performed using a simulation approach, as in the NDP methods to be discussed in Chapters 5-7, the variance of the cost samples that are averaged by simulation is typically smaller when $\lambda < 1$ than when $\lambda = 1$, as can be seen from the definition of Δ_k [cf. Eq. (2.37)]. As a result, the number of cost samples that are required to evaluate by simulation the cost-to-go of a given policy within a given accuracy may be much smaller when $\lambda < 1$ than when $\lambda = 1$.

On the negative side, we will show shortly that the asymptotic convergence rate of the sequence J_k produced by the λ-policy iteration method deteriorates as λ becomes smaller. However, this disadvantage may not be very significant within the NDP approximation context to be discussed later (cf. Chapter 6). In particular, when the policy evaluation step cannot be performed with high accuracy, a reasonable practical objective is to obtain a policy whose cost is within the best "achievable" tolerance from the optimum, rather than to obtain the optimal cost-to-go function and an optimal policy. Under these circumstances, the asymptotic rate of convergence of J_k may not be crucial, and using $\lambda < 1$ often requires a comparable number of policy iterations to attain the same performance level as using $\lambda = 1$.

The following proposition introduces a mapping that underlies the λ-policy iteration method, and provides some basic results. The proposition applies to both cases where $\alpha < 1$ and $\alpha = 1$.

Proposition 2.7: Given $\lambda \in [0, 1)$, J_k, and μ_{k+1}, consider the mapping M_k defined by

$$M_k J = (1 - \lambda) T_{\mu_{k+1}} J_k + \lambda T_{\mu_{k+1}} J. \tag{2.39}$$

Assume that $T_{\mu_{k+1}}$ is a contraction mapping of modulus $\beta < 1$ with respect to some norm $\|\cdot\|$.

(a) The mapping M_k is a contraction mapping of modulus $\beta\lambda$ with respect to the norm $\|\cdot\|$.

(b) For any integer $m \geq 1$ and vector J, there holds

$$\begin{aligned} M_k^m J = {} & (1 - \lambda)\big(T_{\mu_{k+1}} J_k + \lambda T_{\mu_{k+1}}^2 J_k + \cdots + \lambda^{m-1} T_{\mu_{k+1}}^m J_k\big) \\ & + \lambda^m T_{\mu_{k+1}}^m J. \end{aligned} \tag{2.40}$$

(c) The vector J_{k+1} generated next by the λ-policy iteration method [cf. Eqs. (2.37)-(2.38)] is the unique fixed point of M_k. Furthermore,

$$J_{k+1} = (1-\lambda) \sum_{m=0}^{\infty} \lambda^m T_{\mu_{k+1}}^{m+1} J_k. \tag{2.41}$$

Proof: (a) For any two vectors J and $\overline{J}$, using the definition (2.39) of M_k, we have

$$\begin{aligned} \|M_k J - M_k \overline{J}\| &= \big\|\lambda(T_{\mu_{k+1}} J - T_{\mu_{k+1}} \overline{J})\big\| \\ &= \lambda \|T_{\mu_{k+1}} J - T_{\mu_{k+1}} \overline{J}\| \\ &\leq \beta\lambda \|J - \overline{J}\|. \end{aligned}$$

(b) The relation (2.40) holds for $m = 1$ by the definition (2.39) of M_k. It can be proved for all $m \geq 1$ by using a straightforward induction. In particular, we have

$$M_k^{m+1} J = M_k(M_k^m J) = (1-\lambda) T_{\mu_{k+1}} J_k + \lambda T_{\mu_{k+1}} (M_k^m J),$$

and after the expression (2.40) is used in the above equation, we obtain Eq. (2.40) with m replaced by $m+1$.

(c) From the definition (2.37)-(2.38) of J_{k+1}, we have

$$J_{k+1} - J_k = \overline{d}_k + \alpha\lambda P_{\mu_{k+1}} (J_{k+1} - J_k), \tag{2.42}$$

where $P_{\mu_{k+1}}$ is the transition probability matrix corresponding to the policy μ_{k+1}, and $\overline{d}_k$ is the vector of expected TD, with components given by

$$\overline{d}_k(i) = \sum_j p_{ij}\big(\mu_{k+1}(i)\big) d_k(i,j), \qquad \forall\ i.$$

Using the definition (2.36) of the TD, we have

$$\overline{d}_k = T_{\mu_{k+1}} J_k - J_k, \tag{2.43}$$

so Eq. (2.42) yields

$$\begin{aligned} J_{k+1} &= T_{\mu_{k+1}} J_k + \alpha\lambda P_{\mu_{k+1}} (J_{k+1} - J_k) \\ &= T_{\mu_{k+1}} J_k + \lambda(T_{\mu_{k+1}} J_{k+1} - T_{\mu_{k+1}} J_k) \\ &= (1-\lambda) T_{\mu_{k+1}} J_k + \lambda T_{\mu_{k+1}} J_{k+1} \\ &= M_k J_{k+1}. \end{aligned}$$

Thus, J_{k+1} is the fixed point of M_k. The expression (2.41) follows from Eq. (2.40) by taking the limit as $m \to \infty$. **Q.E.D.**

The following proposition shows the validity of the λ-policy iteration method and provides its convergence rate.

Proposition 2.8: (*TD-Based Policy Iteration Convergence*) Assume that $\lambda \in [0, 1)$, and let (J_k, μ_k) be the sequence generated by the λ-policy iteration algorithm.

(a) If $\alpha < 1$, then J_k converges to J^*. Furthermore, for all k greater than some index $\bar{k}$, we have

$$\|J_{k+1} - J^*\|_\infty \leq \frac{\alpha(1-\lambda)}{1-\alpha\lambda}\|J_k - J^*\|_\infty. \tag{2.44}$$

(b) If $\alpha = 1$, Assumptions 2.1 and 2.2 hold, and $TJ_0 \leq J_0$, then J_k converges to J^*.

Proof: (a) Let us first assume that $TJ_0 \leq J_0$. We show by induction that for all k, we have

$$J^* \leq TJ_{k+1} \leq J_{k+1} \leq TJ_k \leq J_k. \tag{2.45}$$

To this end, we fix k and we assume that $TJ_k \leq J_k$. We will show that $J^* \leq TJ_{k+1} \leq J_{k+1} \leq TJ_k$, and then Eq. (2.45) will follow from the hypothesis $TJ_0 \leq J_0$.

Using the fact $T_{\mu_{k+1}} J_k = TJ_k$ [cf. Eq. (2.35)] and the definition of M_k [cf. Eq. (2.39)], we have

$$M_k J_k = T_{\mu_{k+1}} J_k = TJ_k \leq J_k.$$

It follows from the monotonicity of $T_{\mu_{k+1}}$, which implies monotonicity of M_k, that for all positive integers m, we have $M_k^{m+1} J_k \leq M_k^m J_k \leq TJ_k \leq J_k$, so by taking the limit as $m \to \infty$, we obtain

$$J_{k+1} \leq TJ_k \leq J_k. \tag{2.46}$$

From the definition of M_k, we have

$$\begin{aligned} M_k J_{k+1} &= T_{\mu_{k+1}} J_k + \lambda(T_{\mu_{k+1}} J_{k+1} - T_{\mu_{k+1}} J_k) \\ &= T_{\mu_{k+1}} J_{k+1} + (1-\lambda)(T_{\mu_{k+1}} J_k - T_{\mu_{k+1}} J_{k+1}), \end{aligned}$$

Using the already shown relation $J_k - J_{k+1} \geq 0$ and the monotonicity of $T_{\mu_{k+1}}$, we obtain $T_{\mu_{k+1}} J_k - T_{\mu_{k+1}} J_{k+1} \geq 0$, so that

$$T_{\mu_{k+1}} J_{k+1} \leq M_k J_{k+1}.$$

Since $M_k J_{k+1} = J_{k+1}$, it follows that

$$TJ_{k+1} \leq T_{\mu_{k+1}} J_{k+1} \leq J_{k+1}. \tag{2.47}$$

Finally, the above relation and the monotonicity of $T_{\mu_{k+1}}$ imply that for all positive integers m, we have $T_{\mu_{k+1}}^m J_{k+1} \leq T_{\mu_{k+1}} J_{k+1}$, so by taking the limit as $m \to \infty$, we obtain

$$J^* \leq J^{\mu_{k+1}} \leq T_{\mu_{k+1}} J_{k+1}. \tag{2.48}$$

From Eqs. (2.46)-(2.48), we see that the inductive proof of Eq. (2.45) is complete.

From Eq. (2.45), it follows that the sequence J_k converges to some limit $\hat{J}$ with $J^* \leq \hat{J}$. Using the definition (2.39) of M_k, and the facts $J_{k+1} = M_k J_{k+1}$ and $T_{\mu_{k+1}} J_k = TJ_k$, we have

$$J_{k+1} = M_k J_{k+1} = TJ_k + \lambda(T_{\mu_{k+1}} J_{k+1} - T_{\mu_{k+1}} J_k),$$

so by taking the limit as $k \to \infty$ and by using the fact $J_{k+1} - J_k \to 0$, we obtain $\hat{J} = T\hat{J}$. Thus $\hat{J}$ is a solution of Bellman's equation, and it follows that $\hat{J} = J^*$.

To show the result without the assumption $TJ_0 \leq J_0$, note that we can replace J_0 by a vector $\hat{J}_0 = J_0 + se$, where $e = (1, \ldots, 1)$ and s is a scalar that is sufficiently large so that we have $T\hat{J}_0 \leq \hat{J}_0$; it can be seen that for any scalar $s \geq (1-\alpha)^{-1} \max_i \big(TJ_0(i) - J_0(i)\big)$, the relation $T\hat{J}_0 \leq \hat{J}_0$ holds. Consider the λ-policy iteration algorithm started with $(\hat{J}_0, \mu_0)$, and let $(\hat{J}_k, \hat{\mu}_k)$ be the generated sequence. Then it can be verified by induction that for all k we have

$$\hat{J}_k - J_k = \left(\frac{\alpha(1-\lambda)}{1-\alpha\lambda} \right)^k s, \qquad \hat{\mu}_k = \mu_k.$$

Hence $\hat{J}_k - J_k \to 0$. Since we have already shown that $\hat{J}_k \to J^*$, it follows that $J_k \to J^*$ as well.

Since $J_k \to J^*$, it follows that for all k larger than some index $\overline{k}$, μ_{k+1} is an optimal policy, so that $T_{\mu_{k+1}} J^* = TJ^* = J^*$. By using this fact, Eq. (2.41), and the linearity of $T_{\mu_{k+1}}$, we obtain for all $k \geq \overline{k}$,

$$\begin{aligned}
\|J_{k+1} - J^*\|_\infty &= \left\| (1-\lambda) \sum_{m=0}^{\infty} \lambda^m T_{\mu_{k+1}}^{m+1} J_k - J^* \right\|_\infty \\
&= (1-\lambda) \left\| \sum_{m=0}^{\infty} \lambda^m T_{\mu_{k+1}}^{m+1} (J_k - J^*) \right\|_\infty \\
&\leq (1-\lambda) \sum_{m=0}^{\infty} \lambda^m \alpha^{m+1} \|J_k - J^*\|_\infty \\
&= \frac{\alpha(1-\lambda)}{1-\alpha\lambda} \|J_k - J^*\|_\infty.
\end{aligned}$$

(b) The proof of this result is identical to the proof of the corresponding result of part (a). **Q.E.D.**

The convergence rate result of Prop. 2.8(a) can be generalized to cover the stochastic shortest path case where $\alpha = 1$. The proof of Eq. (2.44) can be adapted to show that if $\|\cdot\|$ is a norm with respect to which T_{μ^*} is a contraction of modulus β for all optimal policies μ^*, then

$$\|J_{k+1} - J^*\| \leq \frac{\beta(1-\lambda)}{1-\beta\lambda}\|J_k - J^*\|,$$

for all sufficiently large k. In the case of a stochastic shortest path problem where all policies are proper, the above relation holds with $\|\cdot\|$ being the weighted maximum norm with respect to which T and all T_μ are contractions (cf. Prop. 2.2).

2.4 PROBLEM FORMULATION AND EXAMPLES

We now provide some examples of infinite horizon DP problems. Some of these examples will form the basis for the case studies to be discussed in Ch. 8.

Example 2.1 (Reaching a Goal in Minimum Expected Time)

The case where

$$\alpha = 1, \qquad g(i,u,j) = 1, \qquad \forall\, i \neq 0, j,\ u \in U(i),$$

corresponds to a problem where the objective is to reach the termination state as fast as possible, on the average. The corresponding optimal cost-to-go $J^*(i)$ is the minimum expected time to termination starting from state i. Under our assumptions, the costs-to-go $J^*(i)$ uniquely solve Bellman's equation, which has the form

$$J^*(i) = \min_{u \in U(i)} \left[1 + \sum_{j=1}^{n} p_{ij}(u) J^*(j)\right], \qquad i = 1, \ldots, n.$$

In the special case where there is only one stationary policy, $J^*(i)$ represents the mean first passage time from i to 0 in the associated Markov chain (see Appendix B). These times, denoted by m_i, are the unique solution of the equations

$$m_i = 1 + \sum_{j=1}^{n} p_{ij} m_j, \qquad i = 1, \ldots, n.$$

Example 2.2 (Problems with Uncontrollable State Components)

Often in stochastic shortest path problems there is special structure that can be exploited to simplify the basic value and policy iteration algorithms, as discussed earlier. We describe here another type of simplification that will also be used in the three subsequent examples.

In many stochastic shortest path problems of interest the state is a composite (i, y) of two components i and y, and the evolution of the main component i can be directly affected by the control u, but the evolution of the other component y cannot. In particular, we assume that given the state (i, y) and the control u, the next state (j, z) is determined as follows: j is generated according to transition probabilities $p_{ij}(u, y)$, and z is generated according to conditional probabilities $p(z \mid j)$ that depend on the main component j of the new state. The simplest case arises when z is generated according to a fixed distribution independently of j. For an example, consider the case where z represents the control of another decision maker whose decision is selected randomly. As another example, consider a service facility that sequentially services customers of different types; then the next state consists of pairs (j, z), where j describes the status of the facility following the service of a customer and z is the type of the next customer.

Let us assume for notational convenience that the cost of a transition from state (i, y) is of the form $g(i, y, u, j)$ and does not depend on the uncontrollable component z of the next state (j, z). If g depends on z it can be replaced by

$$\hat{g}(i, y, u, j) = \sum_z p(z \mid j) g(i, y, u, j, z)$$

in the analysis that follows.

Stochastic shortest path problems with this structure admit substantial simplification. In particular, the value and the policy iteration algorithms can be carried out over a smaller state space, the space of the controllable component i. To see this, note that the DP mapping is given for all i, y by

$$(TJ)(i, y) = \min_{u \in U(i,y)} \sum_{j=0}^{n} p_{ij}(u, y) \left(g(i, y, u, j) + \sum_z p(z \mid j) J(j, z) \right),$$

and the corresponding mapping for a stationary policy μ is given for all i, y by

$$(T_\mu J)(i, y) = \sum_{j=0}^{n} p_{ij}\big(\mu(i, y), y\big) \left(g\big(i, y, \mu(i, y), j\big) + \sum_z p(z \mid j) J(j, z) \right).$$

If we define for each J,

$$\hat{J}(j) = \sum_z p(z \mid j) J(j, z),$$

we see that to evaluate $(TJ)(i, y)$ or $(T_\mu J)(i, y)$, it is sufficient to know the vector $\hat{J}$. Furthermore, to calculate an optimal policy, it is not necessary to know the optimal cost-to-go $J^*(j,z)$ for all (j,z). It is instead sufficient to know the vector $\hat{J}^*$ given by

$$\hat{J}^*(j) = \sum_z p(z \mid j) J^*(j,z),$$

for all i. We refer to $\hat{J}^*$ as the *reduced optimal cost-to-go vector*.

This leads us to introduce simplified versions of the value iteration and policy iteration algorithms over the space of the controllable state component i. Let us consider the mapping $\hat{T}$ defined by

$$\begin{aligned}(\hat{T}\hat{J})(i) &= \sum_y p(y \mid i)(TJ)(i,y) \\ &= \sum_y p(y \mid i)\left(\min_{u\in U(i,y)} \sum_{j=0}^n p_{ij}(u,y)\big(g(i,y,u,j) + \hat{J}(j)\big)\right)\end{aligned}$$

and the corresponding mapping for a stationary policy μ,

$$\begin{aligned}(\hat{T}_\mu\hat{J})(i) &= \sum_y p(y \mid i)(T_\mu J)(i,y) \\ &= \sum_y p(y \mid i)\left(\sum_{j=0}^n p_{ij}\big(\mu(i,y),y\big)\Big(g\big(i,y,\mu(i,y),j\big) + \hat{J}(j)\Big)\right).\end{aligned}$$

The simplified value iteration algorithm starts with arbitrary $\hat{J}(i)$, $i = 1,\ldots,n$, and sequentially produces $\hat{T}\hat{J}$, $\hat{T}^2\hat{J}$, $\hat{T}^3\hat{J}$, etc. The simplified policy iteration algorithm at the typical iteration, given the current policy $\mu_k(i,y)$, consists of two steps:

(a) The policy evaluation step, which computes the unique $\hat{J}^{\mu_k}(i)$, $i = 1,\ldots,n$, that solve the linear system of equations $\hat{J}^{\mu_k} = \hat{T}_{\mu_k}\hat{J}^{\mu_k}$ or equivalently

$$\hat{J}^{\mu_k}(i) = \sum_y p(y \mid i)\left(\sum_{j=0}^n p_{ij}\big(\mu_k(i,y)\big)\Big(g\big(i,y,\mu_k(i,y),j\big) + \hat{J}^{\mu_k}(j)\Big)\right),$$

for all $i = 1,\ldots,n$.

(b) The policy improvement step, which computes the improved policy $\mu_{k+1}(i,y)$, from the equation $\hat{T}_{\mu_{k+1}}\hat{J}^{\mu_k} = \hat{T}\hat{J}^{\mu_k}$ or equivalently

$$\mu_{k+1}(i,y) = \arg\min_{u\in U(i,y)} \sum_{j=0}^n p_{ij}(u,y)\big(g(i,y,u,j) + \hat{J}^{\mu_k}(j)\big),$$

for all (i,y).

It is possible to show that the preceding modified algorithms have satisfactory convergence properties under the standard assumptions. In particular, the simplified value iteration algorithm converges to the reduced optimal cost-to-go vector $\hat{J}^*$, which satisfies the reduced form of Bellman's equation $\hat{J}^* = \hat{T}\hat{J}^*$. The simplified policy iteration algorithm converges finitely to an optimal policy.

Example 2.3 (Tetris)

This example is typical of a number of one-player game contexts that can be modeled as stochastic shortest path problems and can be addressed using NDP methods.

Tetris is a popular video game played on a two-dimensional grid. Each square in the grid can be full or empty, making up a "wall of bricks" with "holes." The squares fill up as objects of different shapes fall from the top of the grid and are added to the top of the wall, giving rise to a "jagged top"; see Fig. 2.3. Each falling object can be moved horizontally and can be rotated by the player in all possible ways, subject to the constraints imposed by the sides of the grid. There is a finite set of standard shapes for the falling objects. The game starts with an empty grid and ends when a square in the top row becomes full and the top of the wall reaches the top of the grid. However, when a row of full squares is created, this row is removed, the bricks lying above this row move one row downward, and the player scores a point. (More than one row can be created and removed in a single step, in which case the number of points scored by the player is equal to the number of rows removed.) The player's objective is to maximize the score attained (total number of rows removed) up to termination of the game.

Figure 2.3: Illustration of the two-dimensional grid of the tetris game.

Assuming that for every policy the game terminates with probability 1 (something that is not really known at present), we can model the problem of finding an optimal tetris playing strategy as a stochastic shortest path problem. The control, denoted by u, is the horizontal positioning and rotation applied to the falling object. The state consists of two components:

(1) The board position, that is, a binary description of the full/empty status of each square, denoted by i.

(2) The shape of the current falling object, denoted by y.

As soon as the most recent object has been placed, the new component y is generated according to a probability distribution $p(y)$, independently of the preceding history of the game. Therefore, as shown in Example 2.2, it is possible to use the reduced form of Bellman's equation involving a reward-to-go vector $\hat{J}$ that depends only on the component i of the state. This equation has the form

$$\hat{J}(i) = \sum_y p(y) \max_u \Big[g(i, y, u) + \hat{J}\big(f(i, y, u)\big) \Big], \qquad \forall\ i,$$

where $g(i, y, u)$ and $f(i, y, u)$ are the number of points scored (rows removed), and the next board position, respectively, when the state is (i, y) and control u is applied.

Unfortunately, the number of states in the tetris problem is extremely large. It is roughly equal to $m2^{hw}$, where m is the number of different shapes of falling objects, and h and w are the height and width of the grid, respectively. In particular, for the reasonable numbers $m = 7$, $h = 20$, and $w = 10$, we have over 10^{61} states. Thus it is essential to use approximations. The control selection methods of NDP use a scoring function, which can be viewed as an approximation to the optimal reward function. In particular, the NDP methods that we will discuss later, construct scoring functions that evaluate a tetris position based on some characteristic features of the position. Such features are easily recognizable by experienced players, and include the current height of the wall, the presence of "holes" and "glitches" (severe irregularities) in the first few rows, etc; see the discussion in Section 8.3.

Example 2.4 (Maintenance with Limited Resources)

This problem relates to generic situations where the issue is to maintain certain assets in the face of breakdowns or adverse conditions, using limited resources. Consider a repair shop that has a number of spare parts that can be used to maintain a given collection of machines of different types and cost over a given number of time periods. We assume that breakdowns occur independently across time periods. A broken down machine can be repaired with the use of a spare part or it can be discarded, in which case a cost C_t is incurred. We want to minimize the expected cost-to-go incurred over N time periods. The essential issue is to decide whether to use a spare part to repair a machine of low cost rather than save the spare part for repair of a high cost machine in the future.

The state here has two components. The first component is

$$i = (m_1, \ldots, m_T, s),$$

where T is the number of machine types, m_t is the number of working machines of type t, and s is the number of remaining spare parts. This is the major component of the state. However, there is a second component of the state, which is the current breakdown vector, given by

$$y = (y_1, \ldots, y_T),$$

where y_t is the number of breakdowns of machines of type t in the current period. We assume that the conditional probabilities $p(y \mid i)$ are known. The control/decision to be selected is the vector

$$u = (u_1, \ldots, u_T),$$

where u_t is the number of spare parts used to repair broken down machines of type t.

Because the probability distribution of the breakdown vector y only depends on i, Bellman's equation simplifies as described in Example 2.2. In particular, let $J_k^*(i, y)$ denote the optimal expected cost-to-go starting at state (i, y) with k periods remaining, and let

$$\hat{J}_k^*(i) = \sum_y p(y \mid i) J_k^*(i, y).$$

Then $\hat{J}_k^*$ satisfies the following reduced form of Bellman's equation

$$\hat{J}_{k+1}^*(i) = \sum_y p(y \mid i) \min_u \Big[g(y, u) + \hat{J}_k^*\big(f(i, y, u)\big) \Big],$$

where

$$g(y, u) = \sum_{t=1}^{T} C_t \big(y_t - u_t\big)$$

is the one-period cost (the cost of the machines discarded during the period), and $f(i, y, u)$ denotes the controllable component of the next state:

$$f(i, y, u) = \left(m_1 - y_1 + u_1, \ldots, m_T - y_T + u_T, s - \sum_{t=1}^{T} u_t \right).$$

Note that this is a finite horizon problem.

There are a number of variations of this problem depending on the context. More complex versions arise when there are multiple types of spare parts and breakdowns. Another possible variation involves a limitation on the number of machines that can be repaired within a given time period, possibly due to a limited number of repair personnel. Then, an appropriate additional constraint should be placed on the control. Furthermore, if machines left unrepaired may be repaired at a later time, the inventory of unrepaired machines of each type must be included in the state. In all of these cases, the number of states can be extremely large, and it may be essential to resort to approximations.

Example 2.5 (Channel Allocation in Cellular Systems)

In cellular communication systems, an important problem is to allocate the communication resource (bandwidth) so as to maximize the service provided to a set of mobile users whose demand for service changes stochastically. A given geographical area is divided into disjoint cells, and each cell serves the users that are within its boundaries. The total system bandwidth is divided into channels, with each channel centered around a frequency. Each channel can be used simultaneously at different cells, provided these cells are sufficiently separated spatially, so that there is no interference between them. Thus the collection of subsets of cells where a channel can be simultaneously used is constrained; this is called the *reuse constraint*. When a user requests service in a given cell either a free channel (one that can be used at the cell without violating the reuse constraint) may be assigned to the user, or else, e.g., if no free channel can be found, the user is blocked from the system. Also, when a mobile user crosses from one cell to another, the user is "handed off" to the cell of entry; that is, a free channel is provided to the user at the new cell. If no such channel is available, the user must be dropped/disconnected from the system. One objective of a channel allocation strategy is to allocate the available channels to users so that the number of blocked users is minimized. An additional objective is to minimize the number of mobile users that are dropped when they are handed off to a busy cell. These two objectives must be weighted appropriately to reflect their relative importance, since dropping existing users is generally much more undesirable than blocking new users.

Many cellular systems are based on a fixed channel allocation; that is, channels are permanently assigned to cells. A more efficient strategy is *dynamic channel allocation*, which makes every channel available to every cell as needed, unless the channel is used in a nearby cell and the reuse constraint is violated.

We can formulate the dynamic channel allocation problem using DP. State transitions occur when a channel becomes free due to a user departure, or when a user arrives at a given cell and requests a channel, or when there is a handoff, which can be viewed as a simultaneous user departure from one cell and a user arrival at another cell. The state at each time consists of two components:

(1) The list of occupied channels at each cell.

(2) The event that causes the state transition (arrival, departure, or handoff). This component of the state is uncontrollable.

The control applied at the time of a user departure is the rearrangement of the channels in use with the aim of creating a more favorable channel packing pattern among the cells (one that will leave more options free for future assignments). The control exercised at the time of a user arrival is the assignment of resources to a user requesting service, or the blocking of the user if resources to support the user are not currently available or if assigning available resources to the user is likely to cause a disproportionate amount of blocking of other users in the future. Finally, a cost is incurred each time a user is blocked or dropped from the system. The objective is to find a policy, that is, a choice of control for each possible channel occupancy situation and type of transition, so as to minimize an infinite horizon cost: a sum of expected costs for blocking or dropping users, appropriately discounted. If we assume that all events occur at equally spaced discrete times, we obtain a form of the discounted cost problem of Section 2.3. Another possibility is to adopt a continuous-time Markov chain formulation, which, however, may be reduced to an equivalent discrete-time formulation using a standard transformation (see e.g., [Ber95a], Vol. II, Ch. 5).

To illustrate the qualitative nature of the channel assignment decisions, suppose that there are two channels and three cells. Assume that a channel may be used simultaneously in cells 1 and 3, but may not be used in channel 2 if it is already used in cell 1 or in cell 3. Suppose that the system is serving one call in cell 1 and another call in cell 3. Then serving both users on the same channel results in a better channel usage pattern than serving them on different channels, since in the former case the other channel is free to use in cell 2. The purpose of the channel assignment and channel rearrangement strategy is, roughly speaking, to create such favorable usage patterns that minimize the likelihood of calls being blocked. To illustrate the purpose of the blocking strategy, assume that channel 1 is used in cells 1 and 3, and channel 2 is not used in any cell. Then if a new call arrives (in any cell), it may be preferable to block this call, because assigning it to the available channel 2 will leave no room to accommodate a handoff to cell 2 of any of the two existing calls in cells 1 and 3.

Unfortunately, the optimal solution of the problem is completely intractable because the number of states is enormous. Thus suboptimal methods of solution are called for, and we will discuss in Section 8.5 the use of NDP methods for this problem.

Example 2.6 (Discrete Optimization)

Let us consider a general type of combinatorial optimization problem, that includes as special cases problems such as assignment, scheduling, matching, etc. The problem is characterized by a set of data d, by a finite set U_d of feasible solutions, and by a cost function $g_d(u)$. (We parametrize the cost function and the constraint set by the problem data d for reasons that will become apparent later.) Each solution u has N components; that is, it has the form $u = (u_1, u_2, \ldots, u_N)$, where N is a positive integer. Given problem data d, we want to find a solution $u \in U_d$ that minimizes $g_d(u)$. To simplify our notation, we will assume that the constraint set U_d has the form

$$U_d = \big\{(u_1, \ldots, u_N) \mid u_m \in U_{d,m},\ m = 1, \ldots, N\big\},$$

where for every m, $U_{d,m}$ is a given finite set. There is no loss of generality in doing so because if U_d does not have the above form, it is typically easy to determine sets $U_{d,m}$ such that the set U_d is contained in the set

$$\hat{U}_d = \big\{(u_1, \ldots, u_N) \mid u_m \in U_{d,m},\ m = 1, \ldots, N\big\}.$$

We can then define the cost $g_d(u)$ to be equal to a very large number for all u that do not belong to U_d, and replace U_d with the set $\hat{U}_d$, without essentially changing the problem.

This type of problem arises in many contexts. An example of an important special case is the general $\{0,1\}$ integer programming problem. For another example that is particularly relevant to our purposes, suppose that we have a DP problem, say of the stochastic shortest path type discussed in Section 2.2. Assume that we have obtained by some means an approximate cost-to-go function $\tilde{J}(i)$, and that we want to implement a policy based on $\tilde{J}$; that is, we want to be able to calculate on-line the control $\mu(i)$ given by

$$\mu(i) = \arg \min_{u \in U(i)} \sum_{j=0}^{n} p_{ij}(u)\big(g(i,u,j) + \tilde{J}(j)\big), \qquad i = 1, \ldots, n, \tag{2.49}$$

for any state i. This is a discrete optimization problem, which depending on the number of elements of $U(i)$, may be quite difficult to solve.

Let us view the problem as a deterministic DP problem with N stages. We refer to an n-tuple $(u_{m_1}, \ldots, u_{m_n})$ consisting of n components of a solution as an *n-solution*, and we associate n-solutions with the nth stage. In particular, for $n = 1, \ldots, N$, the states of the nth stage of the DP problem are of the form $(d, m_1, u_{m_1}, \ldots, m_n, u_{m_n})$, where $(u_{m_1}, \ldots, u_{m_n})$ is an n-solution. The initial state is the problem data d. From this state we may select any index m_1 and move to any state (d, m_1, u_{m_1}), where $u_{m_1} \in U_{d,m_1}$. More generally, from a state of the form

$$(d, m_1, u_{m_1}, \ldots, m_{n-1}, u_{m_{n-1}}),$$

we may select any index

$$m_n \notin \{m_1, \ldots, m_{n-1}\},$$

and move to any one of the states

$$(d, m_1, u_{m_1}, \ldots, m_{n-1}, u_{m_{n-1}}, m_n, u_{m_n}) \quad \text{with } u_{m_n} \in U_{d,m_n}.$$

Thus, the controls available at state $(d, m_1, u_{m_1}, \ldots, m_{n-1}, u_{m_{n-1}})$ can be identified with the pairs (m_n, u_{m_n}), where $m_n \notin \{m_1, \ldots, m_{n-1}\}$ and $u_{m_n} \in U_{d,m_n}$. The terminal states correspond to the N-solutions $(u_1, \ldots, u_N)$, and the only nonzero cost is the terminal cost $g_d(u_1, \ldots, u_N)$.

Unfortunately, in most cases, an optimal solution to the preceding DP problem cannot be found without examining a large portion or even the entire set of solutions. However, there is the possibility of an approximate solution using the NDP methods of this book. Roughly speaking, in the discrete optimization context, the objective of NDP is to "learn" how to construct, component-by-component, a solution by using experience gathered from many example problems that are similar to the one whose solution is sought.

Note that the NDP approach makes sense when we are interested in solving an *entire class of similar problems* rather than just a single problem;

it is only within such a context that "learning from experience" can be helpful. As an illustration, consider a workshop that has to construct its production schedule on a daily basis so as to optimize a certain criterion. The daily scheduling problems depend on the production targets and the available production capacity, which may change from day to day; yet they may be fairly similar, since they likely involve similar products and a similar production environment. The objective of the NDP approach is to obtain a scheduling algorithm that can solve every possible daily scheduling problem that may appear. As another example, consider the minimization over the control constraint set in Eq. (2.49). To implement the policy μ, it is necessary to perform a large number of minimizations corresponding to different states. In many cases, these minimizations are similar.

In Sections 3.1.4 and 8.4, we will discuss a specific NDP method for solving discrete optimization problems. This method constructs a function approximation to the optimal cost-to-go of the corresponding DP problem by using a "training" process involving a large number of sample problems from the given class. Once the "training" has been completed, the NDP approach yields an algorithm that obtains an approximately optimal solution to any given problem from the class in N steps. In particular, at each step, this algorithm augments a partial solution with one more component, by using the optimal cost-to-go approximation to evaluate the various candidate components. This is similar to evaluating the possible moves in a chess position by using a "scoring function" (see the discussion in Section 1.1).

Example 2.7 (Discrete Optimization by Local Search)

Let us consider a discrete optimization problem similar to the one in the preceding example, and discuss an alternative DP formulation, which is amenable to solution by NDP methods. In contrast to the preceding example, here the states of the equivalent DP problem will be the complete solutions of the discrete optimization problem. We consequently use the suggestive symbol i to denote a solution. In particular, given problem data d, the problem is to minimize a cost function $g_d(i)$ subject to the constraint $i \in I_d$, where I_d is a given finite set. Note, that this formulation is flexible enough to allow the set I_d to contain some solutions that are "undesirable" or even infeasible in a practical sense, since we can assign a very high cost to such solutions by suitable choice of the cost function g_d.

Discrete optimization problems are often approached by local search methods such as genetic algorithms, tabu search, simulated annealing, and random search. These search methods are iterative; that is, they generate a sequence of solutions starting from some initial solution. The successor solutions are obtained according to some search rules, which may be probabilistic. Usually, the search rules are based on various ideas of "local perturbation" of the current solution.

For example, consider a scheduling problem, where solutions are various schedules for completing a set of tasks. Then a successor schedule may be obtained by interchanging two tasks of the current schedule, or by scheduling one currently unscheduled task, or by replacing a scheduled task in the current schedule by an unscheduled task. Similarly, consider an assignment problem where we want to assign persons to jobs, and a solution is a set of assigned person/job pairs. Then a successor assignment may be obtained from the current assignment by exchanging the jobs of two assigned persons, or by assigning some unassigned person to an unassigned job, or by switching the assignment of an assigned job from one person to another.

The final solution of a search method is usually obtained either when some criterion indicating that exact or approximate optimality has been met or else when further search is deemed unlikely to generate solutions of lower cost than the one achieved so far.

The success of search methods for discrete optimization crucially depends on how the successor solution is chosen from the set of all possible successor solutions. DP offers one possible framework for optimizing this process. In particular, assume that given a solution i we have two choices:

(a) Incur a positive cost c, select one of several search options u from a subset $U(i)$, and generate another solution j according to some given transition probabilities $p_{ij}(u)$. Here c can be viewed as a cost intended to discourage very long searches. Also, typically the transition probabilities $p_{ij}(u)$ are not available explicitly; instead, transitions can be simulated by running the search method.

(b) Accept the current solution i, incur the cost $g_d(i)$, and terminate the search.

We view this problem as a stochastic shortest path problem with the obvious identifications of states, controls, costs, and transition probabilities. Because c (the cost per search) is positive, the total cost of never terminating (which would be incurred under an improper policy) is ∞. Since there exists at least one proper policy (the one that terminates at every solution), the standing assumptions of Section 2.2 are satisfied. The cumulative cost incurred if the search is stopped at a solution i is $g_d(i)$, plus c times the number of successive solutions generated in the process of reaching i. Thus by choosing c sufficiently small, an optimal policy will terminate with a solution of minimal cost that can be reached from the starting solution.

Unfortunately, an optimal solution to the DP problem is impractical when the number of possible solutions is large. However, there is the possibility of approximate solution using the NDP methodology of this book. The objective of NDP is to "learn" how to search more intelligently by using experience gathered from problems that are similar to the one whose solution is sought. In particular, the objective of the NDP approach is to supplement the standard search methods with a "scoring function," which is used to evaluate the search options at each solution.

2.5 NOTES AND SOURCES

2.1. Treatments of DP can be found in a number of textbooks, including those by Bertsekas [Ber95a], Puterman [Put94], and Ross [Ros83].

These books contain many references to the standard DP literature.

2.2. The analysis of the stochastic shortest path problem (Props. 2.1 and 2.2) is due to Bertsekas and Tsitsiklis [BeT91a], and can also be found in the textbooks [BeT89] and [Ber95a]. The proof of Prop. 2.2 is new and has also been independently discovered by Littman [Lit96]. Asynchronous value iteration algorithms for DP were proposed and analyzed by Bertsekas [Ber82a]; see also Bertsekas and Tsitsiklis [BeT89] for a detailed discussion. The modified policy iteration algorithm was proposed and analyzed by van Nunen [Van76], and Puterman and Shin [PuS78]. The convergence of the asynchronous policy iteration algorithm was shown by Williams and Baird [WiB93].

2.3. The λ-policy iteration algorithm and its analysis are new (see Bertsekas and Ioffe [BeI96]).

2.4. The formulation of tetris as a DP problem and its solution by NDP techniques are described by Van Roy [Van95]. The application of NDP methods to a problem that is similar to the maintenance example is described in Logan et. al. [LBH96]. The formulation of combinatorial optimization problems as DP problems in a way that local search strategies can be learned by NDP techniques follows the ideas of Zhang and Dietterich [ZhD95], and their solution of some large scheduling problems (see Section 8.7).

Don't train for so long that you get into a local min ...
Don't worry, all local mins give about the same accuracy ...
(On training neural networks – Anonymous)

3

Neural Network Architectures and Training

Contents

In this chapter we develop methods for approximating functions through the use of neural networks. We use the term "neural networks" to refer broadly to parametric approximation structures (or architectures), and we consider the types of parametric approximation that are best suited for the dynamic programming context. We also discuss in detail the training of neural networks, that is, the methods used to set the free parameters so as to optimize the quality of the approximation. The training methods of this chapter involve primarily deterministic iterative algorithms. In some contexts, it is more appropriate to use stochastic algorithms for training. We develop and analyze algorithms of this type in the next chapter.

3.1 ARCHITECTURES FOR APPROXIMATION

As discussed in Ch. 1, our objective is to construct approximate representations of the optimal cost-to-go function or other functions of interest, because this is the only possible method for breaking the curse of dimensionality in the face of very large state spaces. While the discussion to follow is framed in terms of building approximations to the optimal cost-to-go function, everything applies verbatim to the approximation of Q-factors.

The first step in the development of an approximate representation is to choose an *approximation architecture*, that is, a certain functional form involving a number of free parameters. These parameters are to be tuned so as to provide a best fit of the function to be approximated. The process of tuning these parameters is often referred to as *training* or *learning*. A closely related issue is the choice of a representation or encoding of the input i.

There are two important preconditions for the development of an effective approximation. First, we need the approximation architecture to be rich enough so that it can provide an acceptably close approximation of the function that we are trying to approximate. In this respect, the choice of a suitable architecture requires some practical experience or theoretical analysis that provides some rough information on the shape of the function to be approximated. Second, we need effective algorithms for tuning the parameters of the approximation architecture. These two objectives are often conflicting. Having a rich set of approximating functions usually means that there is a large number of parameters to be tuned or that the dependence on the parameters is nonlinear. In either case, the computational complexity of the training problem increases.

In most applications of approximation theory (or neural networks), we are given training data pairs (x_i, y_i), and we wish to construct a function $y = f(x)$ that best explains these data pairs. In our context, since we are interested in approximating the optimal cost-to-go function J^*, an ideal set of training data would consist of pairs $\big(i, J^*(i)\big)$, where i ranges over some subset of the state space. However, the function J^* is neither known nor can it be measured experimentally, and such training data are unavailable. As a result, NDP is faced with an additional layer of difficulty, as compared to classical uses of neural networks. In essence, the approximation needs to be carried out in conjunction with an algorithm that tries to compute J^*.

The situation becomes easier if we are dealing with a single policy μ because the expected cost-to-go $J^\mu(i)$ under that policy can be estimated by setting the initial state to i and simulating the system under the policy μ. For this reason, some of the most popular NDP methods focus on a single policy at any one time, and try to estimate the cost-to-go function of this policy by using simulation. Then, on occasion, they switch to another policy that appears more promising on the basis of the simulation data obtained, similar to the policy iteration and the modified policy iteration methods discussed in Ch. 2.

3.1.1 An Overview of Approximation Architectures

The approximation problem can be framed as follows. Suppose that we are interested in approximating a function $J : S \mapsto \Re$, where S is some set; in our context, S will usually be the state space of a dynamic programming problem and J will be the optimal cost-to-go function of the problem or the cost-to-go function of some policy. We let r be a vector of parameters and we postulate the functional form $J(i) \approx \tilde{J}(i,r)$. Here, $\tilde{J}$ is a known function describing the architecture that we have chosen and which is easy to evaluate once the vector r is fixed. We then try to choose the parameter vector r so as to minimize some measure of the distance between the function J that we are trying to fit and the approximant $\tilde{J}$.

Approximation architectures can be broadly classified into two main categories: linear and nonlinear ones. A *linear architecture* is of the general form

$$\tilde{J}(i,r) = \sum_{k=0}^{K} r(k)\phi_k(i), \tag{3.1}$$

where $r(k)$, $k = 0, 1, \ldots, K$, are the components of the parameter vector r, and ϕ_k are fixed, easily computable functions. For example, if the state i ranges over the integers and if $\phi_k(i) = i^k$, $k = 0, 1, \ldots, K$, the approximant is a polynomial in i of degree K. We will refer to the functions ϕ_k as the *basis functions* even though they do not always form a basis of a vector space in the strict sense of the term. Note that it is common to use the constant function $\phi_0(i) \equiv 1$ as a basis function in linear architectures, so that Eq. (3.1) takes the form

$$\tilde{J}(i,r) = r(0) + \sum_{k=1}^{K} r(k)\phi_k(i). \tag{3.2}$$

This allows a tunable offset from 0 for the approximation $\tilde{J}(i,r)$, and enlarges the range of mappings that the linear architecture can effectively approximate.

If we have some training data pairs $\big(i, J(i)\big)$ that we wish to fit using a linear architecture, we may pose a least squares problem whereby we wish to minimize

$$\sum_i \left(J(i) - \sum_k r(k)\phi_k(i)\right)^2, \tag{3.3}$$

over all vectors r. This is a linear least squares problem that can be addressed using linear algebra techniques. Several well-tested algorithms are available for this problem and some of these algorithms are discussed later in this chapter.

With a *nonlinear architecture*, the dependence of $\tilde{J}(i,r)$ on r is nonlinear. In this case, the least squares problem of minimizing

$$\sum_i \big(J(i) - \tilde{J}(i,r)\big)^2 \tag{3.4}$$

cannot be reduced to a linear algebraic problem and must be solved by means of nonlinear programming methods, some of which are discussed in Section 3.2.

Multilayer Perceptrons

A common nonlinear architecture is the *multilayer perceptron* or *feedforward neural network* with a *single hidden layer*; see Fig. 3.1(a). Under this architecture, the state i is encoded as a vector x with components $x_\ell(i)$, $\ell = 1, \ldots, L$, which is then transformed linearly through a "linear layer," involving the coefficients $r(k,\ell)$, to give the K scalars

$$\sum_{\ell=1}^{L} r(k,\ell)x_\ell(i), \qquad k = 1, \ldots, K.$$

Each of these scalars becomes the input to a function $\sigma(\cdot)$, called a *sigmoidal function*, which is differentiable, monotonically increasing and has the property

$$-\infty < \lim_{\xi \to -\infty} \sigma(\xi) < \lim_{\xi \to \infty} \sigma(\xi) < \infty;$$

see Fig. 3.2. Some common choices are the *hyperbolic tangent* function

$$\sigma(\xi) = \tanh(\xi) = \frac{e^{\xi} - e^{-\xi}}{e^{\xi} + e^{-\xi}},$$

and the *logistic* function

$$\sigma(\xi) = \frac{1}{1 + e^{-\xi}}.$$

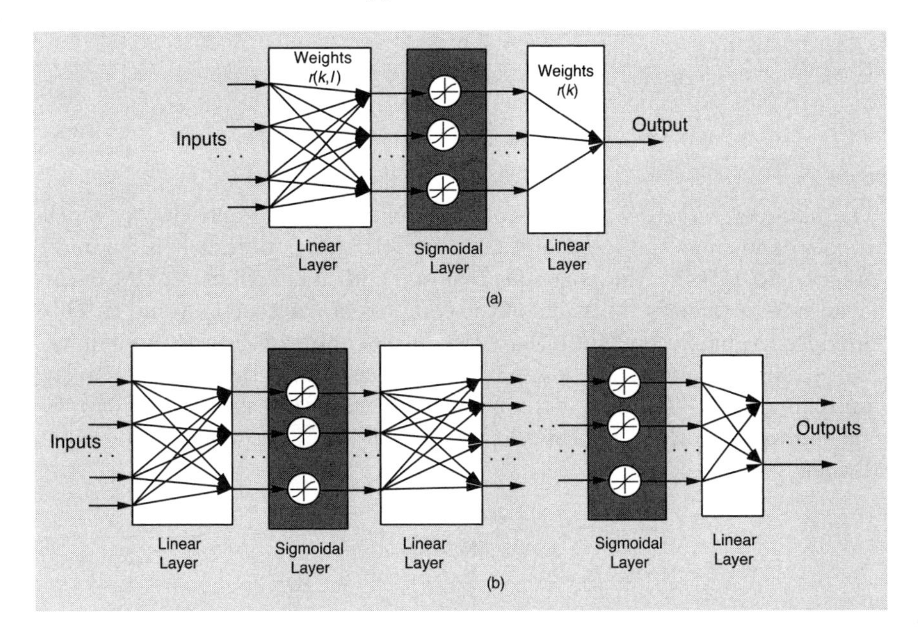

Figure 3.1: (a) A feedforward neural network with a single hidden layer and a single output. The inputs are $x_\ell(i)$. The outputs of the first linear layer are

$$\sum_\ell r(k,\ell)x_\ell(i).$$

The outputs of the sigmoidal units are

$$\sigma\left(\sum_\ell r(k,\ell)x_\ell(i)\right).$$

The final output is

$$\sum_k r(k)\sigma\left(\sum_\ell r(k,\ell)x_\ell(i)\right).$$

The term "hidden" refers to the portion of the architecture that is not directly connected with neither the inputs nor the outputs of the architecture.

(b) A feedforward neural network with multiple hidden layers and multiple outputs.

At the outputs of the sigmoidal functions, the scalars

$$\sigma\left(\sum_{\ell=1}^{L} r(k,\ell)x_\ell(i)\right), \qquad k=1,\ldots,K,$$

are obtained. These scalars are linearly combined using coefficients $r(k)$ to produce the final output

$$\tilde{J}(i,r) = \sum_{k=1}^{K} r(k)\sigma\left(\sum_{\ell=1}^{L} r(k,\ell)x_\ell(i)\right). \tag{3.5}$$

The parameter vector r consists of the coefficients $r(k)$ and $r(k,\ell)$, which are also known as the *weights* of the network. We note that it is common practice to provide the constant 1 as an additional input to the linear layer, or equivalently to fix one of the components of x at the value 1. This provides a tunable constant bias to each of the inputs of the sigmoidal units, and expands the range of mappings that the architecture can effectively approximate. For example, without such a bias, in the case where $\sigma(0) = 0$, the approximation $\tilde{J}(i,r)$ provided by Eq. (3.5) would be forced to the value 0 when $x = 0$.

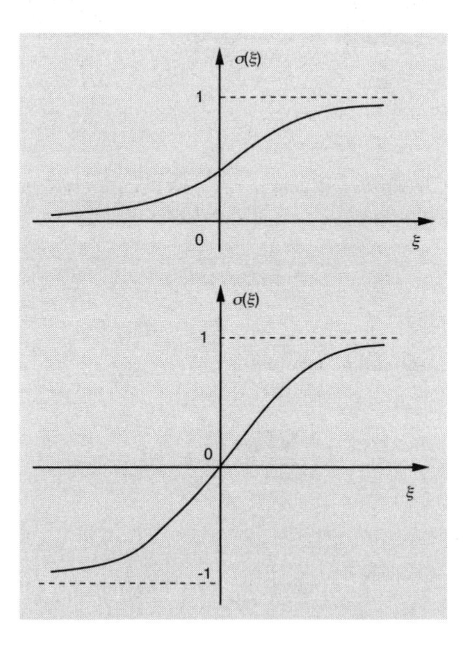

Figure 3.2: Some examples of sigmoidal functions.

There are also generalizations of the single hidden layer perceptron architecture that involve alternation of multiple layers of linear and sigmoidal functions; see Fig. 3.1(b). Multilayer perceptrons can be represented compactly by introducing certain mappings to describe the linear and the sigmoidal layers. In particular, let $L_1, \ldots, L_{m+1}$ denote the matrices representing the linear layers; that is, at the output of the 1st linear layer we obtain the vector $L_1 x$ and at the output of the kth linear layer ($k > 1$) we obtain $L_k\xi$, where ξ is the output of the preceding sigmoidal layer. Similarly, let $\Sigma_1, \ldots, \Sigma_m$ denote the mappings representing the sigmoidal layers; that is, when the input of the kth sigmoidal layer ($k > 1$) is the vector y with components $y(j)$, we obtain at the output the vector $\Sigma_k y$ with components $\sigma\big(y(j)\big)$. The output of the multilayer perceptron is

$$F(L_1, \ldots, L_{m+1}, x) = L_{m+1}\Sigma_m L_m \cdots \Sigma_1 L_1 x.$$

The special nature of this formula has an important computational consequence: the gradient (with respect to the weights) of the squared error between the output and a desired output y,

$$E(L_1, \ldots, L_{m+1}) = \frac{1}{2}\big(y - F(L_1, \ldots, L_{m+1}, x)\big)^2, \tag{3.6}$$

can be efficiently calculated using a special procedure known as *backpropagation*, which is just an intelligent way of using the chain rule.† In particular, the partial derivative of the cost function $E(L_1, \ldots, L_{m+1})$ with respect to $L_k(i,j)$, the ijth component of the matrix L_k, is given by

$$\frac{\partial E(L_1, \ldots, L_{m+1})}{\partial L_k(i,j)} = -e' L_{m+1}\overline{\Sigma}_m L_m \cdots L_{k+1}\overline{\Sigma}_k I_{ij}\Sigma_{k-1}L_{k-1} \cdots \Sigma_1 L_1 x, \tag{3.7}$$

where e is the error vector

$$e = y - F(L_1, \ldots, L_{m+1}, x),$$

$\overline{\Sigma}_n$, $n = 1, \ldots, m$, is the diagonal matrix with diagonal terms equal to the derivatives of the sigmoidal functions σ of the nth hidden layer evaluated at the appropriate points, and I_{ij} is the matrix obtained from L_k by setting all of its components to 0 except for the ijth component which is set to 1. This formula can be used to obtain efficiently all of the terms needed in the partial derivatives (3.7) of E using a two-step calculation:

(a) Use a forward pass through the network to calculate sequentially the outputs of the linear layers $L_1 x, L_2\Sigma_1 L_1 x, \ldots, L_{m+1}\Sigma_m L_m \cdots \Sigma_1 L_1 x$. This is needed in order to obtain the points at which the derivatives in the matrices $\overline{\Sigma}_n$ are evaluated, and also in order to obtain $e = y - F(L_1, \ldots, L_{m+1}, x)$.

(b) Use a backward pass through the network to calculate sequentially the terms $e' L_{m+1}\overline{\Sigma}_m L_m \cdots L_{k+1}\overline{\Sigma}_k$ in the derivative formulas (3.7), starting with $e' L_{m+1}\overline{\Sigma}_m$, proceeding to $e' L_{m+1}\overline{\Sigma}_m L_m \overline{\Sigma}_{m-1}$, and continuing to $e' L_{m+1}\overline{\Sigma}_m \cdots L_2\overline{\Sigma}_1$.

---

† The name *backpropagation* is used in several different ways in the neural networks literature. For example feedforward neural networks of the type shown in Fig. 3.1 are sometimes referred to as *backpropagation networks*.

Note that for the usual case where there is only one sigmoidal layer, the gradient of E can also be efficiently calculated starting from first principles.

There is an important result that relates to the approximation capability of the multilayer perceptron. It can be shown that this network, with a good choice of the weights, can approximate arbitrarily closely any function $J : S \mapsto \Re$ that is continuous over a closed and bounded set S, provided the number of sigmoidal functions used in the hidden layers is sufficiently large (see Cybenko [Cyb89], Funahashi [Fun89], and Hornik, Stinchcombe, and White [HSW89]). For this, a single hidden layer is sufficient. Intuitive explanations of this fundamental approximation property of multilayer perceptrons are given in Bishop ([Bis95], pp. 129-130) and Jones [Jon90]. In practice, the number of hidden sigmoidal layers is usually one or two, and almost never more than three.

3.1.2 Features

A general approximation architecture $\tilde{J}(i,r)$ can be visualized as in Fig. 3.3(a). It is often the case that the function J to be approximated is a highly complicated nonlinear map and it is sensible to try to break this complexity into smaller, less complex pieces. This is often done by selecting a set of *features* of the state and feeding them to an approximation architecture. These features are usually handcrafted, based on whatever human intelligence, insight, or experience is available, and are meant to capture the most important aspects of the current state.

Formally, a feature can be defined as a mapping $f_k : S \mapsto \Re$. Given a collection $f_1, \ldots, f_K$ of features, we form the feature vector $f(i) = \big(f_1(i), \ldots, f_K(i)\big)$. We may then consider approximation architectures of the form $\tilde{J}\big(f(i), r\big)$, as in Fig. 3.3(b), or, more generally, of the form $\tilde{J}\big(i, f(i), r\big)$, as in Fig. 3.3(c). We note that for the case of a feature-based architecture [cf. Fig. 3.3(b)], as long as the architecture is linear, the resulting approximation problem is again a linear least squares problem, even if the feature extraction mapping $f(\cdot)$ is nonlinear. This is because we are interested in minimizing

$$\sum_i \left(J(i) - \sum_k r(k)\phi_k\big(f(i)\big) \right)^2,$$

and the crucial factor is the linear dependence on the free parameters $r(k)$.

We observe that the architecture shown in Fig. 3.3(c) involves a redundant encoding of the state i because the feature vector $f(i)$ does not carry any information that is not already contained in i. Also, in Fig. 3.3(b), the transformation from i to $f(i)$ results in some loss of information. However, when dealing with features, the information content of an encoding of i is not the main issue. A more appropriate point of view is

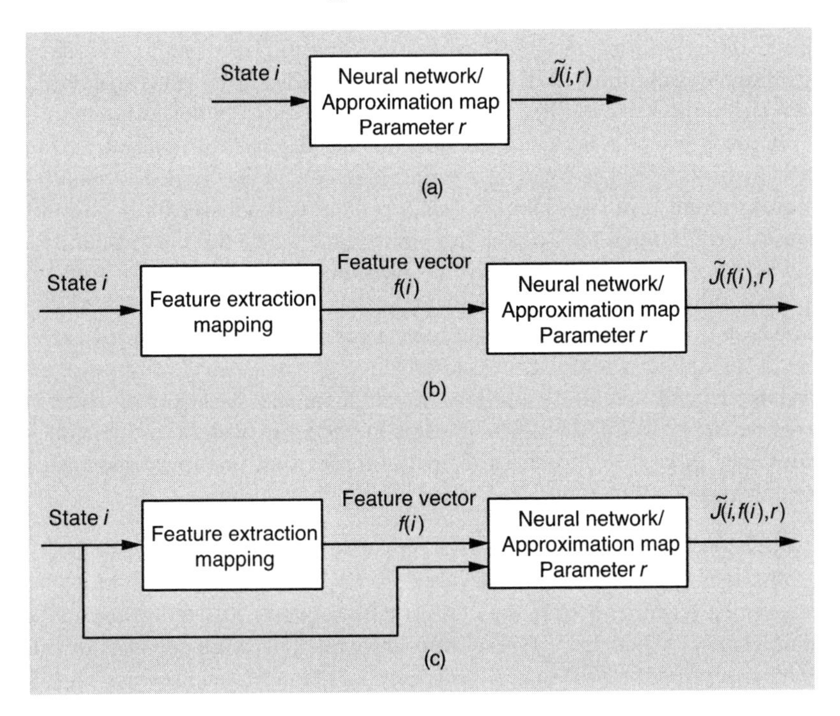

Figure 3.3: (a) A general approximation architecture. (b) A feature-based approximation architecture. (c) An architecture that uses both a raw encoding of the input as well as a feature vector.

to regard the mapping from states into features as a "nonlinear coordinate transformation" that results into a hopefully more structured problem.

An interesting possibility arises in the context of NDP where we want to approximate the optimal cost-to-go function J^*. Then, if we know a fairly good suboptimal policy μ (based for example on some heuristic), and we have a reasonable approximation $\tilde{J}^\mu$ of the corresponding cost-to-go function J^μ, then $\tilde{J}^\mu$ may be an effective feature for approximating J^*. The idea here is that $\tilde{J}^\mu$ may capture some important aspects of the functional form of J^*.

It is also possible to consider a *feature iteration approach*, whereby a set of features is selected, some computation is done using these features, and based on the results of the computation, some new features are introduced. Examples of this approach will be given later.

The process of feature extraction is intimately related to the choice of a representation of the data. For example, in the multilayer perceptron shown in Fig. 3.1, the state i is encoded as a vector $\big(x_1(i), \ldots, x_L(i)\big)$. In some cases there is a "natural" encoding of the data that is adopted; quite often, however, the choice of an encoding is a conscious or subconscious process of feature extraction. For example, suppose that we were given sample data points generated using the function $\sin(\cos x)$. By simply eyeballing data points of the form $(x, \sin(\cos x))$, it is quite difficult to guess the right form of the functional dependence. On the other hand, if we choose $y = \cos x$ as a feature, we are faced with the task of fitting a function to data of the form $(y, \sin y)$ which can be easily done visually or computationally. In this example, the problem was substantially simplified because we "happened" to use the right feature. Generally, without any knowledge of how the data to be fit have been generated, we should not expect to hit upon the most relevant features but, to the extent that we have some qualitative knowledge about a problem, we should try to use it.

It would be highly desirable to have a theory or a computational mechanism that could be used as a guide in the selection of the most useful features. While there exist some general purpose methods that try to discover interesting features from the data in an unsupervised fashion, often there is no substitute for sound engineering judgment.

Feature-Based Aggregation

The use of a feature vector $f(i)$ to represent the state i in an approximation architecture of the form $\tilde{J}\big(f(i), r\big)$ implicitly involves *state aggregation*, that is, the grouping of states into subsets. In particular, let us assume that the feature vector can take only a finite number of values, and let us define for each possible value v, the subset of states S_v whose feature vector is equal to v:

$$S_v = \big\{i \mid f(i) = v\big\}.$$

We refer to the sets S_v as the *aggregate states* induced by the feature vector. These sets form a partition of the state space. An approximate cost-to-go function of the form $\tilde{J}\big(f(i), r\big)$ is piecewise constant with respect to this partition; that is, it assigns the same cost-to-go value $\tilde{J}(v, r)$ to all states in the set S_v.

An often useful approach to deal with problem complexity in DP is to introduce an "aggregate" DP problem, whose states are some suitably defined feature vectors $f(i)$ of the original problem. The precise form of the aggregate problem may depend on intuition and/or heuristic reasoning, based on our understanding of the original problem. Suppose now that the aggregate problem is simple enough to be solved exactly by DP, and let $\hat{J}(v)$ be its optimal cost-to-go when the initial value of the feature vector is v. Then $\hat{J}\big(f(i)\big)$ provides an approximation architecture for the original problem, i.e., the architecture that assigns to state i the (exactly) optimal cost-to-go $\hat{J}\big(f(i)\big)$ of the feature vector $f(i)$ in the aggregate problem. There is considerable freedom on how one formulates and on how one solves aggregate problems. In particular, one may apply the DP methods of value and policy iteration discussed in Ch. 2, but one may also use special simulation-oriented methods of the type to be developed in Ch. 5 and in Section 6.7.

Features Viewed as State Measurements

A feature extraction mapping that is used for approximation of the cost-to-go function of a DP problem can be viewed as a measurement device: it produces a feature vector $f(i)$ that provides partial information about the state i. In general, we cannot reconstruct the state from its feature vector, so the feature vectors from preceding periods may provide genuinely useful additional information about the current state. This can be exploited by using approximation architectures that depend on several past values of the feature vector. For example, it may be fruitful to introduce a dependence on a running average of the past values of some of the features.

In this connection, given a DP problem, it is conceptually useful to consider a corresponding problem of *imperfect state information*, where at time t we are allowed to have access only to the feature vector $f(i_t)$ in place of the system state i_t. According to the theory of imperfect state information DP (see e.g., [Ber95a]), the optimal cost-to-go function of this problem at time t depends on the feature vectors of all preceding periods. It has the general form $\hat{J}_t\big(f(i_0), f(i_1), \ldots, f(i_t)\big)$, where $\hat{J}$ is some function, and it provides an upper bound on the optimal cost-to-go function $J^*(i_t)$ of the original DP problem and a lower bound on the performance of any policy that is based on an approximation architecture of the form $\tilde{J}\big(f(i_t), r\big)$. However, the performance gap can be narrowed by considering an approximation architecture of the form $\tilde{J}_t\big(f(i_0), f(i_1), \ldots, f(i_t), r\big)$ or an architecture of the form $\tilde{J}\big(f(i_{t-k}), f(i_{t-k+1}), \ldots, f(i_t), r\big)$, which involves the current feature vector and the feature vectors of the preceding k time periods.

Features Viewed as Approximate Sufficient Statistics

Feature extraction mappings take a special meaning when dealing with DP problems with imperfect state information. In such problems, the control is chosen on the basis of noise-corrupted measurements of the state, rather than the state itself. These measurements accumulate over time, and in order to deal with their expanding number, one often searches for *sufficient statistics*, that is, for quantities which summarize the essential content of the measurements as far as optimal control is concerned (see [Ber95a], Ch. 5). For example, under favorable circumstances, an estimator that provides some form of state reconstruction may be such a sufficient statistic. Since a sufficient statistic is, for optimization purposes, equivalent to the entire available information, it may be viewed as an "exact" feature extraction mapping. This leads to the conjecture that good feature extraction mappings are provided by "approximate sufficient statistics," that is, mappings which in a heuristic sense are close to being a sufficient statistic for the given problem. For example, in some problems, a suboptimal state estimator may serve as a fairly good feature extraction mapping. For another useful context, suppose that "most" of the relevant information for the purposes of future optimization is provided by a small subset of the available measurements, say those of the last k periods, where k is a small integer. Then these measurements may constitute an effective set of features for an approximation architecture.

3.1.3 Partitioning

A simple method to construct complex and sophisticated approximation architectures, is to partition the state space into several subsets and construct a separate approximation in each subset. For example, by using a linear or quadratic polynomial approximation in each subset of the partition, one can construct piecewise linear or piecewise quadratic approximations over the entire state space. An important issue here is the choice of the method for partitioning the state space. Regular partitions (e.g., grid partitions) may be used, but they often lead to a large number of subsets and very time-consuming computations. Generally speaking, each subset of the partition should contain "similar" states so that the variation of the optimal cost-to-go over the states of the subset is relatively smooth and can be approximated with smooth functions. An interesting possibility is to use features as the basis for partition. In particular, one may use a more or less regular discretization of the space of features, which induces a possibly irregular partition of the original state space (see Fig. 3.4). In this way, each subset of the irregular partition contains states with "similar features."

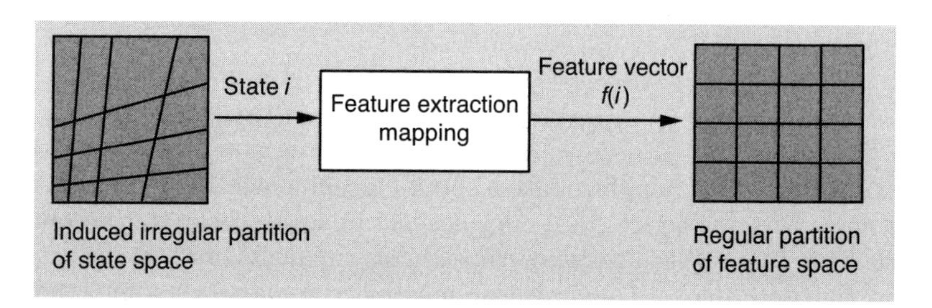

Figure 3.4: Feature-based partitioning of the state space. A regular partition of the space of features defines an irregular partition of the state space.

Partitioning and "local" approximations can also be used to enhance the quality of approximation in parts of the space where the mapping we are trying to approximate has some special character. For example, suppose that the state space S is partitioned in subsets $S_1, \ldots, S_M$ and consider approximations of the form

$$\tilde{J}(i,r) = \hat{J}(i,\hat{r}) + \sum_{m=1}^{M} \sum_{k=1}^{K_m} r_m(k)\phi_{k,m}(i), \tag{3.8}$$

where each $\phi_{k,m}(i)$ is a basis function which contributes to the approximation only on the set S_m; that is, it takes the value 0 for $i \notin S_m$. Here $\hat{J}(i, \hat{r})$ is an architecture of the type discussed earlier, such as a multilayer perceptron, and the parameter vector r consists of $\hat{r}$ and the coefficients $r_m(k)$ of the basis functions. Thus the portion $\hat{J}(i, \hat{r})$ of the architecture is used to capture "global" aspects of the mapping we are trying to approximate, while each portion $\sum_{k=1}^{K_m} r_m(k)\phi_{k,m}(i)$ is used to capture aspects of the mapping that are "local" to the subset S_m.

It should be noted that the solution of least squares problems involving the "local-global" architecture (3.8) can be facilitated with the use of decomposition. In particular, minimizing

$$\sum_i \left(J(i) - \hat{J}(i, \hat{r}) - \sum_{m=1}^{M} \sum_{k=1}^{K_m} r_m(k)\phi_{k,m}(i) \right)^2, \tag{3.9}$$

over the parameters $\hat{r}$ and $r_m = \big(r_m(1), \ldots, r_m(K_m)\big)$, $m = 1, \ldots, M$, can be done in two stages: first solving for the optimal local approximation with the global parameter $\hat{r}$ held fixed, and then minimizing over all $\hat{r}$. In particular, we first fix $\hat{r}$ and we minimize for each m separately the cost function (3.9) over all r_m; for each m, this is a linear least squares problem that can be solved in closed form (see Section 3.2.2), and that involves only the squared terms corresponding to $i \in S_m$. The optimal values of r_m are linear functions of $J(i) - \hat{J}(i, \hat{r})$, $i \in S_m$, and when they are substituted in the cost function (3.9), they yield a "reduced cost function" that depends exclusively on $\hat{r}$. This reduced cost function [which incidentally is quadratic if $\hat{J}(i, \hat{r})$ is linear in $\hat{r}$] can then be minimized with respect to $\hat{r}$; see also our discussion of linear least squares problems in Section 3.2.2.

Still another possibility for partitioning and decomposition arises when dealing with DP problems that have special structure. As an example, consider a stochastic shortest path problem that has the sequential decomposition structure discussed in Section 2.2.2. In particular, assume that there is a sequence of subsets of states $S_1, S_2, \ldots, S_M$ such that each of the nontermination states belongs to one and only one of these subsets, and the following property holds:

> For all $m = 1, \ldots, M$ and states $i \in S_m$, and under any policy, the successor state j is either the termination state or else belongs to one of the subsets $S_m, S_{m-1}, \ldots, S_1$.

Then, the solution of the problem decomposes into the solution of M stochastic shortest path problems, each involving the states in a subset S_m. If the solutions of these problems are approximated using NDP methods, each cost-to-go function approximation needs only to be done over the corresponding subset S_m. Furthermore, one may consider solving exactly the smaller ones of these problems, so that when the larger problems are solved approximately, we can use optimal cost-to-go values for the states of the smaller problems.

A potential difficulty with partitioned architectures is that there is discontinuity of the approximation along the boundaries of the partition. For this reason, a variant, called *soft partitioning*, is sometimes employed, whereby the subsets of the partition are allowed to overlap and the discontinuity is smoothed out over their intersection. In particular, once a function approximation is obtained in each subset, the approximate cost-to-go in the overlapping regions is taken to be a smoothly varying linear combination of the function approximations of the corresponding subsets.

3.1.4 Using Heuristic Policies to Construct Features

Suppose that we have a difficult DP problem for which we know some reasonable heuristic policies, say $\mu_1, \ldots, \mu_K$. The method by which these policies have been obtained is immaterial; for example, appropriate heuristic policies could be guessed, they could be obtained by solving an analytically tractable variant of the original problem, or they could be generated by some type of policy iteration. What is important for our purposes is that for each state i and policy μ_k, we can calculate fairly easily an approximate cost-to-go $\tilde{J}^{\mu_k}(i)$, perhaps through a closed-form expression, through some fast and convenient algorithm, or through some already trained neural network architecture.

Now suppose that for the states i of interest, at least some of the evaluations $\tilde{J}^{\mu_k}(i)$ are fairly close to the optimal cost-to-go $J^*(i)$. Then it appears sensible to use $\tilde{J}^{\mu_k}(i)$ as features to be appropriately weighted in an architecture that aims at approximating J^*. In particular, let us consider the functions $\tilde{J}^{\mu_k}(i)$ as features in an architecture of the form

$$\tilde{J}(i,r) = w_0(i,r_0) + \sum_{k=1}^{K} w_k(i,r_k)\tilde{J}^{\mu_k}(i), \tag{3.10}$$

where each $w_k(\cdot,r_k)$ is a tunable coefficient that depends on the state i as well as on a parameter vector r_k, and $r = (r_0, r_1, \ldots, r_K)$ is the overall parameter vector of the architecture.

The idea behind this architecture is that by using different weights $w_k(i,r_k)$ at different states i, we may be able to learn how to identify the most promising heuristic policy to use at different portions of the state space. Therefore, with proper training of the parameter vector r, one may hope to obtain a policy that performs better than all of the heuristic policies. Note that the architecture is very general, since by taking $w_k(i,r_k)$ to be identically 0 for $k = 1, \ldots, K$, it reduces to the generic form $w_0(i,r_0)$. On the other hand, the architecture may allow very effective approximation of the optimal cost function J^* by using simple parametrizations of the weights $w_k(i,r_k)$, since the functions $\tilde{J}^{\mu_k}$ may have already captured the most important characteristics of J^*.

In particular, suppose that we have an M-dimensional feature vector

$$f(i) = \big(f_1(i), \ldots, f_M(i)\big).$$

Then we may consider a linear parametrization of $w_k(i, r_k)$ of the form

$$w_k(i, r_k) = r_k(0) + \sum_{m=1}^{M} r_k(m) f_m(i), \tag{3.11}$$

where $r_k = \big(r_k(0), r_k(1), \ldots, r_k(M)\big)$ is the parameter vector. With this parametrization, the architecture (3.10) is written as

$$\begin{aligned} \tilde{J}(i, r) = r_0(0) &+ \sum_{m=1}^{M} r_0(m) f_m(i) + \sum_{k=1}^{K} r_k(0) \tilde{J}^{\mu_k}(i) \\ &+ \sum_{m=1}^{M} \sum_{k=1}^{K} r_k(m) f_m(i) \tilde{J}^{\mu_k}(i), \end{aligned} \tag{3.12}$$

so it is equivalent to a feature-based architecture where the features are $f_m(i)$, $\tilde{J}^{\mu_k}(i)$, and the products $f_m(i)\tilde{J}^{\mu_k}(i)$.

Alternatively, let the space of feature vectors $f(i)$ be partitioned into subsets $S_1, \ldots, S_L$. Then we can consider functions $w_k(i, r_k)$ that are piecewise constant on the (irregular) partition induced by the subsets $S_1, \ldots, S_L$ (cf. Section 3.1.3); that is, functions $w_k(i, r_k)$ of the form

$$w_k(i, r_k) = \sum_{l=1}^{L} r_k(l) \delta\big(f(i) \mid S_l\big), \tag{3.13}$$

where $\delta(\cdot \mid S_l)$ is the indicator function of the set S_l,

$$\delta\big(f(i) \mid S_l\big) = \begin{cases} 1, & \text{if } f(i) \in S_l, \\ 0, & \text{if } f(i) \notin S_l, \end{cases}$$

and $r_k = \big(r_k(1), r_k(2), \ldots, r_k(L)\big)$ is the parameter vector. Note that with the choice (3.13), the architecture (3.10) is written as

$$\tilde{J}(i, r) = \sum_{l=1}^{L} \left(r_0(l) + \sum_{k=1}^{K} r_k(l) \tilde{J}^{\mu_k}(i) \right) \delta\big(f(i) \mid S_l\big), \tag{3.14}$$

so it amounts to weighting differently the evaluations of the heuristic policies $\tilde{J}^{\mu_k}(i)$ on each subset $\{i \mid f(i) \in S_l\}$.

Note that here we are using features in two different ways. On one hand, in the architecture (3.10), we use the $\tilde{J}^{\mu_k}(i)$ as features to construct a cost-to-go approximation $\tilde{J}$. On the other hand, we use a different set of features $f_m(i)$ to represent the weighting coefficients $w_k(i, r_k)$ of the first set of features $\tilde{J}^{\mu_k}(i)$. Note also that both architectures (3.12) and (3.14) are linear, and can be trained using linear least squares methods.

The approach of using heuristic algorithms to construct features is illustrated through a simple example in Section 8.4. We close this section by specializing the approach to an important combinatorial optimization context.

Application to Discrete Optimization Problems

Let us consider the general discrete optimization problem of Example 2.6 in Section 2.4. The problem is characterized by a set of data d, by a cost function $g_d(u)$, and by a finite set U_d of N-component solutions $u = (u_1, \ldots, u_N)$, where N is a positive integer. As in Example 2.6, we view the problem as a deterministic DP problem with N stages. For $n = 1, \ldots, N$, the states of the nth stage of the DP problem are of the form $(d, m_1, u_{m_1}, \ldots, m_n, u_{m_n})$, where $(u_{m_1}, \ldots, u_{m_n})$ is an n-solution (a partial solution involving just n components). The initial state is the problem data d. From a state of the form

$$(d, m_1, u_{m_1}, \ldots, m_{n-1}, u_{m_{n-1}}),$$

we may select any index

$$m_n \notin \{m_1, \ldots, m_{n-1}\},$$

and move to any one of the states

$$(d, m_1, u_{m_1}, \ldots, m_{n-1}, u_{m_{n-1}}, m_n, u_{m_n}) \quad \text{with } u_{m_n} \in U_{d,m_n}.$$

The controls at a state $(d, m_1, u_{m_1}, \ldots, m_{n-1}, u_{m_{n-1}})$ can be identified with the pairs (m_n, u_{m_n}), where $m_n \notin \{m_1, \ldots, m_{n-1}\}$ and $u_{m_n} \in U_{d,m_n}$. The terminal states correspond to the N-solutions $(u_1, \ldots, u_N)$, and the only nonzero cost is the terminal cost $g_d(u_1, \ldots, u_N)$.

Let us assume that we have K heuristic algorithms, the kth of which, given problem data d and an n-solution $(u_{m_1}, \ldots, u_{m_n})$, produces an N-solution whose cost is denoted by $H_{k,d}(m_1, u_{m_1}, \ldots, m_n, u_{m_n})$. Thus, starting from any partial solution, the heuristic algorithms can generate complete N-component solutions, although there are no guarantees on the quality of these solutions. Consistently with our earlier discussion, we can use the heuristic algorithms as feature extraction mappings in an approximation architecture. In particular, we may use for each n, an architecture of the form

$$\begin{aligned}\tilde{J}(d, m_1, u_{m_1}, \ldots, m_n, u_{m_n}, r) &= w_0(d, m_1, u_{m_1}, \ldots, m_n, u_{m_n}, r_0) \\ &+ \sum_{k=1}^{K} w_k(d, m_1, u_{m_1}, \ldots, m_n, u_{m_n}, r_k) H_{k,d}(m_1, u_{m_1}, \ldots, m_n, u_{m_n}),\end{aligned}$$

where w_k are tunable coefficients that depend on the parameter vector $r = (r_0, r_1, \ldots, r_K)$. The functional form of the coefficients w_k may be feature-based, that is, it may include a dependence on features of the problem data d or the n-solution $(u_{m_1}, \ldots, u_{m_n})$, as discussed earlier. Furthermore, the parameter vector r may include a separate component for each stage n.

Such an architecture can be used to generate, for any problem data d, a complete N-component solution in N steps as follows. Given d, at the first step we generate u_{m_1} by solving the problem

$$\min_{\substack{m_1 \in \{1,\ldots,N\} \\ u_{m_1} \in U_{d,m_1}}} \tilde{J}(d, m_1, u_{m_1}, r). \tag{3.15}$$

At the nth step, given the preceding selections $m_1, u_{m_1}, \ldots, m_{n-1}, u_{m_{n-1}}$, we generate u_{m_n} by solving the problem

$$\min_{\substack{m_n \notin \{m_1,\ldots,m_{n-1}\} \\ u_{m_n} \in U_{d,m_n}}} \tilde{J}(d, m_1, u_{m_1}, \ldots, m_{n-1}, u_{m_{n-1}}, m_n, u_{m_n}, r). \tag{3.16}$$

Note that at each step of this process, several heuristic algorithms are executed and their corresponding complete N-solutions and costs are generated. These solutions may be compared with the solution obtained via the N-step (approximate) DP process (3.15)-(3.16) in terms of their costs, and the best solution may be used as the final answer.

Let us now mention a few variants of the preceding formulation of the discrete optimization problem as an N-stage DP problem. A simple and effective variation is to introduce additional controls/decisions that correspond to an immediate transition from the current state to a complete N-solution provided by one of the heuristic algorithms starting from that state. In particular, from a state $(d, m_1, u_{m_1}, \ldots, m_{n-1}, u_{m_{n-1}})$ we may move to any one of the states $(d, m_1, u_{m_1}, \ldots, m_{n-1}, u_{m_{n-1}}, m_n, u_{m_n})$, where $m_n \notin \{m_1, \ldots, m_{n-1}\}$ and $u_{m_n} \in U_{d,m_n}$, *or* to the complete N-solution produced by any one of the heuristic algorithms starting from $(d, m_1, u_{m_1}, \ldots, m_{n-1}, u_{m_{n-1}})$.

Another, more complex possibility involves the use of a *local search algorithm*, which given an n-solution $(u_{m_1}, \ldots, u_{m_n})$, produces a subset S of n-solutions that includes $(u_{m_1}, \ldots, u_{m_n})$ as well as some "neighboring" n-solutions to $(u_{m_1}, \ldots, u_{m_n})$ (compare with Example 2.7 in Section 2.4). The nth stage decision, at state $(d, m_1, u_{m_1}, \ldots, m_{n-1}, u_{m_{n-1}})$, would then consist of selecting $m_n \notin \{m_1, \ldots, m_{n-1}\}$ and the value of the component u_{m_n} from the set U_{d,m_n}, *and* then selecting an n-solution from the corresponding subset S. Examples of suitable local search algorithms abound in the combinatorial optimization literature; they include interchanging the order of a few cities in a traveling salesman tour, exchanging the job assignments of a few persons in a person-to-job assignment problem, etc.

Note that while all of the preceding DP formulations are equivalent to the original discrete optimization problem, they give rise to different approximation architectures. Thus, when these architectures are trained by NDP techniques, they will generate solutions of the original optimization problem that differ accordingly. Generally, one may hope that a more complex DP formulation will result in a more powerful architecture that is capable of yielding better solutions of the original problem. On the other hand, a more powerful architecture requires more time-consuming computations for training, as well as for generating problem solutions after training has been completed.

3.2 NEURAL NETWORK TRAINING

Neural network training problems are optimization problems where we want to find the set of parameters or weights of an approximation architecture that provide the best fit between the input/output map realized by the architecture and a given set of input/output data pairs. Typically, these problems are of the least squares form

$$\begin{aligned} &\text{minimize} \quad \frac{1}{2}\sum_{i=1}^{m} \|g_i(r)\|^2 \\ &\text{subject to} \quad r \in \Re^n, \end{aligned}$$

where g_i is for each i a continuously differentiable vector-valued function. Throughout this section, $\|\cdot\|$ will stand for the Euclidean norm, given by $\|x\| = \sqrt{x'x}$.

As discussed in Section 3.1, least squares problems arise when we want to approximate a DP cost-to-go function using sample pairs $\big(i, J(i)\big)$ of states i and corresponding (perhaps noisy) cost-to-go estimates $J(i)$ that have been obtained in some way (for example by simulation). In this context, depending on the type of architecture used, we will be minimizing the linear least squares cost function

$$\sum_i \Big(J(i) - \sum_k r(k)\phi_k(i)\Big)^2,$$

or its nonlinear version

$$\sum_i \big(J(i) - \tilde{J}(i,r)\big)^2,$$

[cf. Eqs. (3.3) and (3.4)]. To provide some further orientation, let us discuss some other related examples of least squares problems.

Example 3.1 (Model Construction – Curve Fitting)

Suppose that we want to estimate n parameters of a mathematical model so that it fits well a physical system, based on a set of measurements. In particular, we hypothesize an approximate relation of the form

$$y = h(x, r),$$

where h is a known function representing the model and

$r \in \Re^n$ is a vector of unknown parameters,
$y \in \Re^r$ is the model's output,
$x \in \Re^p$ is the model's input.

Given a set of m input-output data pairs $(x_1, y_1), \ldots, (x_m, y_m)$ from the physical system that we are trying to model, we want to find the vector of parameters r that best matches the data in the sense that it minimizes the sum of squared errors

$$\sum_{i=1}^{m} \|y_i - h(x_i, r)\|^2.$$

For example, to fit the data pairs by a cubic polynomial approximation, we would choose

$$h(x, r) = r(0) + r(1)x + r(2)x^2 + r(3)x^3,$$

where $r = \big(r(0), r(1), r(2), r(3)\big)$ is the vector of unknown coefficients of the cubic polynomial.

Example 3.2 (Least Squares Parameter Estimation)

Consider two jointly distributed random vectors x and y. We view x as a measurement that provides some information about y. We are interested in obtaining a function $f(\cdot)$, called an *estimator*, where $f(x)$ is the estimate of y given x. A commonly used estimator is the *least squares estimator*, which is obtained by minimization of the following expected squared error over all $f(\cdot)$:

$$E\big[\|y - f(x)\|^2\big].$$

Since

$$E\big[\|y - f(x)\|^2\big] = E\Big[E\big[\|y - f(x)\|^2 \mid x\big]\Big],$$

it is clear that $f^*(\cdot)$ is a least squares estimator if $f^*(x)$ minimizes the conditional expectation in the right-hand side above for every x. We have for every fixed vector z,

$$E\big[\|y - z\|^2 \mid x\big] = E\big[\|y\|^2 \mid x\big] - 2z'E[y \mid x] + \|z\|^2.$$

By setting to zero the gradient with respect to z, we see that the above expression is minimized by $z = E[y \mid x]$. Thus the least squares estimator of y given x is $E[y \mid x]$.

Suppose now that $E[y \mid x]$ is very complicated or otherwise difficult to calculate. It would then make sense to consider instead minimization over r of

$$E\big[\|y - h(x, r)\|^2\big]$$

over a class of estimators from a given architecture $h(\cdot, r)$ that depends on a parameter vector r. For example, $h(x, r)$ could be provided by some neural network. Assume further that we have sample pairs $(x_1, y_1), \ldots, (x_m, y_m)$. We can then approximate the preceding expected value by

$$\frac{1}{m} \sum_{i=1}^{m} \|y_i - h(x_i, r)\|^2.$$

Then, if r^* minimizes the above sum of squares, $h(x, r^*)$ is an approximation to the least squares estimator $E[y \mid x]$.

3.2.1 Optimality Conditions

In order to discuss computational methods for least squares problems, it is most economical to embed them within the general class of unconstrained minimization problems

$$\begin{aligned} &\text{minimize} \quad f(r) \\ &\text{subject to} \quad r \in \Re^n. \end{aligned}$$

A vector r^* is a *local minimum* of f if it is no worse than its neighbors; that is, if there exists an $\epsilon > 0$ such that

$$f(r^*) \leq f(r), \qquad \forall\ r \text{ with } \|r - r^*\| < \epsilon.$$

A vector r^* is a *global minimum* of f if it is no worse than all other vectors; that is,

$$f(r^*) \leq f(r), \qquad \forall\ r \in \Re^n.$$

The local or global minimum r^* is said to be *strict* if the corresponding inequality above is strict for $r \neq r^*$. Local and global maxima are similarly defined. In particular, r^* is a local (global) maximum of f, if r^* is a local (global) minimum of the function $-f$.

The main optimality conditions are given in the following two classical propositions.

Proposition 3.1: (*Necessary Optimality Conditions*) Let r^* be a local minimum of $f : \Re^n \mapsto \Re$, and assume that f is continuously differentiable. Then,

$$\nabla f(r^*) = 0. \qquad \text{(First Order Necessary Condition)}$$

If in addition f is twice continuously differentiable, then,

$\nabla^2 f(r^*)$ is positive semidefinite. (Second Order Necessary Condition)

Proof: Fix some $s \in \Re^n$. Then

$$s'\nabla f(r^*) = \lim_{\gamma \downarrow 0} \frac{f(r^* + \gamma s) - f(r^*)}{\gamma} \geq 0,$$

where the inequality follows from the assumption that r^* is a local minimum. Since s is arbitrary, the same inequality holds with s replaced by $-s$. Therefore, $s'\nabla f(r^*) = 0$ for all $s \in \Re^n$, which shows that $\nabla f(r^*) = 0$.

Assume that f is twice continuously differentiable, and let s be any vector in $\Re^n$. For all $\gamma \in \Re$, the second order Taylor series expansion yields

$$f(r^* + \gamma s) - f(r^*) = \gamma \nabla f(r^*)'s + \frac{\gamma^2}{2} s'\nabla^2 f(r^*)s + o(\gamma^2).$$

Using the condition $\nabla f(r^*) = 0$ and the local optimality of r^*, we see that

$$0 \leq \frac{f(r^* + \gamma s) - f(r^*)}{\gamma^2} = \frac{1}{2} s'\nabla^2 f(r^*)s + \frac{o(\gamma^2)}{\gamma^2},$$

for sufficiently small $\gamma \neq 0$. Taking the limit as $\gamma \to 0$ and using the fact

$$\lim_{\gamma \to 0} \frac{o(\gamma^2)}{\gamma^2} = 0,$$

we obtain $s'\nabla^2 f(r^*)s \geq 0$ for all $s \in \Re^n$, showing that $\nabla^2 f(r^*)$ is positive semidefinite. **Q.E.D.**

In what follows, we refer to a vector r^* satisfying the condition $\nabla f(r^*) = 0$ as a *stationary point*.

Proposition 3.2: (*Second Order Sufficient Optimality Conditions*) Let $f : \Re^n \mapsto \Re$ be twice continuously differentiable. Suppose that a vector r^* satisfies the conditions

$$\nabla f(r^*) = 0, \qquad \nabla^2 f(r^*) \text{ is positive definite.}$$

Then, r^* is a strict local minimum of f. In particular, there exist scalars $\alpha > 0$ and $\epsilon > 0$ such that

$$f(r) \geq f(r^*) + \alpha \|r - r^*\|^2, \qquad \forall\ r \text{ with } \|r - r^*\| < \epsilon. \tag{3.17}$$

Proof: Let λ be the smallest eigenvalue of $\nabla^2 f(r^*)$. Then λ is positive since $\nabla^2 f(r^*)$ is positive definite. Furthermore, $s'\nabla^2 f(r^*)s \geq \lambda \|s\|^2$ for all

$s \in \Re^n$. Using this relation, the hypothesis $\nabla f(r^*) = 0$, and the second order Taylor series expansion, we have for all s

$$\begin{aligned} f(r^* + s) - f(r^*) &= \nabla f(r^*)' s + \frac{1}{2} s' \nabla^2 f(r^*) s + o(\|s\|^2) \\ &\geq \frac{\lambda}{2}\|s\|^2 + o(\|s\|^2) \\ &= \left(\frac{\lambda}{2} + \frac{o(\|s\|^2)}{\|s\|^2}\right)\|s\|^2. \end{aligned}$$

It is seen that Eq. (3.17) is satisfied for any $\epsilon > 0$ and $\alpha > 0$ such that

$$\frac{\lambda}{2} + \frac{o(\|s\|^2)}{\|s\|^2} \geq \alpha, \qquad \forall\ s \ \text{with} \ \|s\| < \epsilon.$$

Q.E.D.

The above conditions can be substantially strengthened when the cost function f is convex. Appendix A gives the definition and some of the properties of convex functions. In particular, if f is twice differentiable, then f is convex if and only if for every r the Hessian matrix $\nabla^2 f(r)$ is positive semidefinite. Furthermore, for convex functions there is no distinction between local and global minima; it can be shown that every local minimum is also global. In addition, the first order necessary condition $\nabla f(r^*) = 0$ is also sufficient for optimality if f is convex. This last property is an immediate consequence of the relation $f(r) \geq f(r^*) + \nabla f(r^*)'(r - r^*)$, which holds for all vectors r and r^* when f is convex and differentiable.

Example 3.3 (Quadratic Minimization Problems)

Consider the quadratic function

$$f(r) = \frac{1}{2} r' Q r - b' r,$$

where Q is a symmetric $n \times n$ matrix and b is a vector in $\Re^n$. If r^* is a local minimum of f, we must have, by the necessary optimality conditions,

$$\nabla f(r^*) = Qr^* - b = 0, \qquad \nabla^2 f(r^*) = Q : \text{positive semidefinite}.$$

There are two cases to consider:

(a) Q is not positive semidefinite. Then f can have no local minima because the second order necessary condition is violated.

(b) Q is positive semidefinite. Then f is convex (because its Hessian matrix is positive semidefinite) and any vector r^* satisfying the first order condition $\nabla f(r^*) = Qr^* - b = 0$ is a global minimum of f. If Q is positive definite (and hence invertible), the equation $Qr^* - b = 0$

has a unique solution and the vector $r^* = Q^{-1}b$ is the unique global minimum. However, if Q is positive semidefinite and singular, there are two possibilities:

(1) The vector b is in the range of the matrix Q, in which case there exist infinitely many solutions [if $Q\overline{r} = b$ and d is any vector satisfying $Qd = 0$, then $Q(\overline{r} + \epsilon d) = b$ for all scalars ϵ]. All of these solutions are global minima.

(2) The vector b is not in the range of the matrix Q, in which case the minimal value of the cost function $f(r)$ is $-\infty$ and no optimal solution exists. To see this, take any nonzero vector $\overline{r}$ in the nullspace of Q that is not orthogonal to b (such a vector exists because b is not in the range of Q, and is therefore not orthogonal to the nullspace of Q). Consider the cost value $f(\alpha\overline{r})$ of vectors $\alpha\overline{r}$ where α is a scalar. Since $Q\overline{r} = 0$, we have $f(\alpha\overline{r}) = -\alpha b'\overline{r}$, and because $b'\overline{r} \neq 0$, the cost $f(\alpha\overline{r})$ can be made arbitrarily small by suitable choice of α.

3.2.2 Linear Least Squares Methods

Consider the least squares problem for the case where the functions $g_i(r)$ are linear of the form

$$g_i(r) = y_i - X_i r,$$

where $y_i \in \Re^{m_i}$ are given vectors and X_i are given $m_i \times n$ matrices. In other words, we are trying to fit a linear model to the set of input-output pairs $(y_1, X_1) \ldots, (y_m, X_m)$. We refer to each such pair as a *data block*. The cost function has the form

$$f(r) = \frac{1}{2} \sum_{i=1}^{m} \|y_i - X_i r\|^2,$$

and is quadratic as well as convex (cf. Example 3.3). Its gradient and Hessian matrix are

$$\nabla f(r) = -\sum_{i=1}^{m} X_i'(y_i - X_i r), \qquad \nabla^2 f(r) = \sum_{i=1}^{m} X_i' X_i.$$

Thus, if the matrix

$$\sum_{i=1}^{m} X_i' X_i$$

is invertible, the problem has a unique solution r^*, obtained by setting the gradient to zero,

$$r^* = \left(\sum_{i=1}^{m} X_i' X_i \right)^{-1} \sum_{i=1}^{m} X_i' y_i. \tag{3.18}$$

If the matrix $\sum_{i=1}^m X_i'X_i$ is not invertible, the problem has an infinite number of solutions. To see this, note that a convex quadratic problem that has no optimal solution must have an optimal value that is $-\infty$, as seen in Example 3.3, while in least squares problems, the optimal value is bounded from below by 0. Thus at least one solution must exist, and in fact the set of solutions is the linear manifold that is parallel to the nullspace of $\sum_{i=1}^m X_i'X_i$ and passes through any one of the solutions of the least squares problem.

There are a number of methods for computing the solution when the solution is unique. We discuss two of the most popular: the singular value decomposition method and the Kalman filtering algorithm. These methods can also be appropriately modified to deal with the case where there is an infinite number of solutions.

Singular Value Decomposition

The singular value decomposition (SVD) method first performs the factorization

$$\sum_{i=1}^m X_i'X_i = E\Lambda E', \tag{3.19}$$

where the matrices E and Λ have the form

$$E = [\, e_1 \quad e_2 \quad \ldots \quad e_k \,], \qquad \Lambda = \begin{bmatrix} \lambda_1 & 0 & \ldots & 0 \\ 0 & \lambda_2 & \ldots & 0 \\ \ldots & \ldots & \ldots & \ldots \\ 0 & 0 & \ldots & \lambda_k \end{bmatrix}.$$

Here the scalars λ_i and the vectors e_i are the nonzero eigenvalues and corresponding orthonormal eigenvectors, respectively, of the matrix $\sum_{i=1}^m X_i'X_i$.

In the case where $\sum_{i=1}^m X_i'X_i$ is invertible, the number of its nonzero eigenvalues is n and the matrix E is a square orthonormal matrix, that is, EE' is the identity matrix. The inverse of E is E', so that

$$(E\Lambda E')^{-1} = E\Lambda^{-1}E',$$

and using Eqs. (3.18) and (3.19), we see that the unique solution of the least squares problem has the form

$$r^* = E\Lambda^{-1}E' \sum_{i=1}^m X_i'y_i; \tag{3.20}$$

see Fig. 3.5(a).

In the case where $\sum_{i=1}^m X_i'X_i$ is not invertible, the vector r^* given by the formula (3.20) is still well defined. It can be shown that out of the infinite set of optimal solutions of the linear least squares problem, r^* is the optimal solution that has minimum Euclidean norm [see Fig. 3.5(b)]. The computation of r^* and the SVD can be performed with well-established algorithms that can be found in most linear algebra software packages (see e.g., the textbooks by Hager [Hag88] or Strang [Str76]).

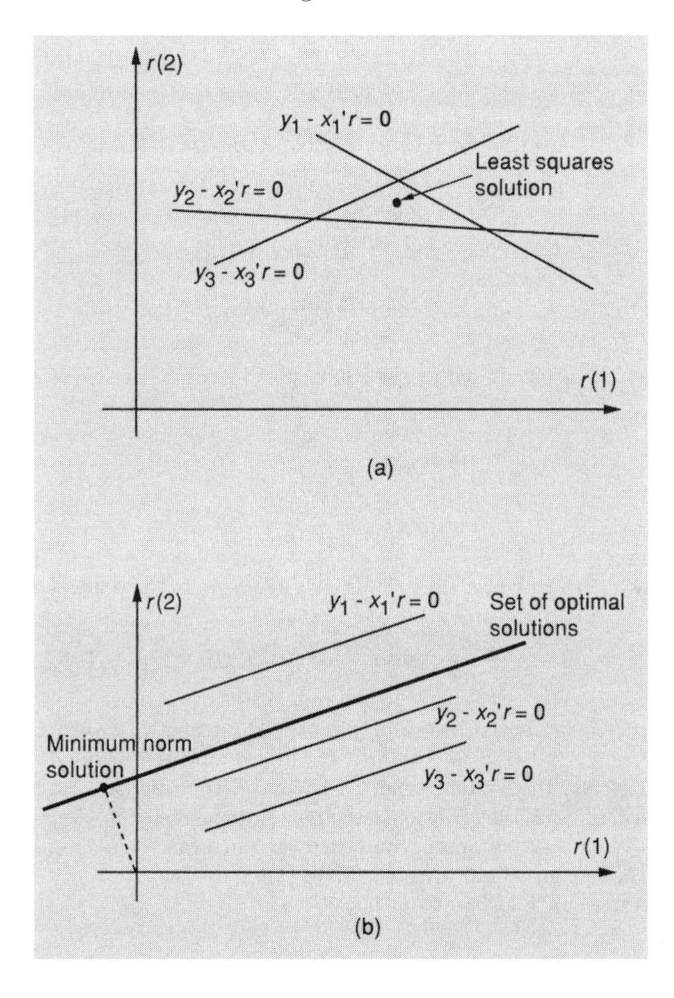

Figure 3.5: (a) Solution of the linear least squares problem when the rows of the matrices X_i (the vectors x_1', x_2', and x_3' in the figure) span the entire space and the matrix $\sum_{i=1}^m X_i'X_i$ is invertible. (b) Illustration of the case where the matrix $\sum_{i=1}^m X_i'X_i$ is not invertible. Then the set of vectors that are orthogonal to all the rows of the matrices X_i is a nontrivial subspace. The set of all optimal solutions of the linear least squares problem is a manifold parallel to this subspace. The SVD formula (3.20) gives the solution of minimum norm on this manifold.

Iterative Methods for Linear Least Squares – The Kalman Filter

The linear least squares estimate can also be computed recursively using the *Kalman filter*. This algorithm has many important applications in control and communication systems, and has been studied extensively. An important characteristic of the algorithm is that it processes the least squares terms $\|y_i - X_i r\|^2$ incrementally, one at a time. To get a sense of how this can be done, consider the following simple example:

Example 3.4 (Recursive Mean Estimation of a Random Variable)

Suppose that we want to calculate the mean of a random variable y using sample values $y_1, y_2, \ldots, y_m$. The sample mean is

$$r^* = \frac{y_1 + y_2 + \cdots + y_m}{m},$$

and can also be calculated as the solution of the problem

$$\begin{aligned} &\text{minimize} \quad \frac{1}{2}\sum_{i=1}^{m}(y_i - r)^2 \\ &\text{subject to} \quad r \in \Re. \end{aligned}$$

It is possible to generate the sample mean by using the recursive formula

$$r_{t+1} = r_t + \frac{1}{t+1}(y_{t+1} - r_t), \qquad t = 0, \ldots, m-1, \tag{3.21}$$

starting from the initial condition $r_0 = 0$. The scalar r_t is the sample mean $(y_1 + \cdots + y_t)/t$ based on the first t samples, and the optimal solution r^* is equal to r_m, the sample mean based on all the samples. Each iteration using formula (3.21) involves only the new sample y_{t+1}; the effect of the past samples $y_1, \ldots, y_t$ is summarized in their sample mean. Note that the formula (3.21) can be generalized to process more than one sample, say s samples, at each iteration as follows:

$$r_{t+s} = r_t + \frac{s}{t+s}\left(\frac{\sum_{i=1}^{s} y_{t+i}}{s} - r_t\right). \tag{3.22}$$

The iterative formulas (3.21) and (3.22) are special cases of the Kalman filtering algorithm.

Consider the general case of the linear least squares problem where

$$g_i(r) = y_i - X_i r.$$

The Kalman filter sequentially generates the vectors that solve the *partial* least squares subproblems involving only the first i squared terms, where $i = 1, \ldots, m$:

$$\psi_i = \arg\min_{r \in \Re^n} \sum_{j=1}^{i} \|y_j - X_j r\|^2, \qquad i = 1, \ldots, m.$$

Thus the algorithm first finds ψ_1 by minimizing the first term $\|y_1 - X_1 r\|^2$, then finds ψ_2 by minimizing the sum of the first two terms $\|y_1 - X_1 r\|^2 + \|y_2 - X_2 r\|^2$, etc. The solution of the original least squares problem is obtained at the last step as

$$r^* = \psi_m.$$

The method can be conveniently implemented so that the solution of each new subproblem is easy given the solution of the earlier subproblems, as will be shown in the following proposition. We state this proposition for the more general case where the least squares problem has the form

$$\begin{aligned} &\text{minimize} && \sum_{j=1}^{m} \lambda^{m-j} \|y_j - X_j r\|^2, \\ &\text{subject to} && r \in \Re^n, \end{aligned}$$

where λ is a scalar with

$$0 < \lambda \leq 1,$$

sometimes called the *fading factor* or the *exponential forgetting factor*. When $\lambda = 1$, we obtain the usual linear least squares problem. A value $\lambda < 1$ deemphasizes the effect of the initial terms $\|y_j - X_j r\|^2$. This is sometimes useful in situations where the mechanism that generates the data changes slowly with time.

Proposition 3.3: (*Kalman Filter*) Assuming that the matrix $X_1' X_1$ is positive definite, the least squares estimates

$$\psi_i = \arg \min_{r \in \Re^n} \sum_{j=1}^{i} \lambda^{i-j} \|y_j - X_j r\|^2, \qquad i = 1, \ldots, m,$$

can be generated by the algorithm

$$\psi_i = \psi_{i-1} + H_i^{-1} X_i' (y_i - X_i \psi_{i-1}), \qquad i = 1, \ldots, m, \tag{3.23}$$

where ψ_0 is an arbitrary vector, and the positive definite matrices H_i are generated by

$$H_i = \lambda H_{i-1} + X_i' X_i, \qquad i = 1, \ldots, m, \tag{3.24}$$

with

$$H_0 = 0.$$

More generally, for all $\bar{i} < i$ we have

$$\psi_i = \psi_{\bar{i}} + H_i^{-1} \sum_{j=\bar{i}+1}^{i} \lambda^{i-j} X_j'(y_j - X_j \psi_{\bar{i}}), \qquad i = 1, \ldots, m. \quad (3.25)$$

Proof: We first establish the result for the case of two squared terms in the following lemma:

Lemma 3.1: Let y_1, y_2 be given vectors, and X_1, X_2 be given matrices such that $X_1'X_1$ is positive definite. Then the vectors

$$\psi_1 = \arg \min_{r \in \Re^n} \|y_1 - X_1 r\|^2, \quad (3.26)$$

and

$$\psi_2 = \arg \min_{r \in \Re^n} \{\|y_1 - X_1 r\|^2 + \|y_2 - X_2 r\|^2\}, \quad (3.27)$$

are also given by

$$\psi_1 = \psi_0 + (X_1'X_1)^{-1} X_1'(y_1 - X_1\psi_0), \quad (3.28)$$

and

$$\psi_2 = \psi_1 + (X_1'X_1 + X_2'X_2)^{-1} X_2'(y_2 - X_2\psi_1), \quad (3.29)$$

where ψ_0 is an arbitrary vector.

Proof: By carrying out the minimization in Eq. (3.26), we obtain

$$\psi_1 = \left(X_1'X_1\right)^{-1} X_1'y_1, \quad (3.30)$$

yielding for any ψ_0,

$$\psi_1 = \psi_0 - \left(X_1'X_1\right)^{-1} X_1'X_1\psi_0 + \left(X_1'X_1\right)^{-1} X_1'y_1,$$

from which the desired Eq. (3.28) follows.

Also, by carrying out the minimization in Eq. (3.27), we obtain

$$\psi_2 = \left(X_1'X_1 + X_2'X_2\right)^{-1} \left(X_1'y_1 + X_2'y_2\right),$$

or equivalently, using also Eq. (3.30),

$$\begin{aligned}
(X_1'X_1 + X_2'X_2)\psi_2 &= X_1'y_1 + X_2'y_2 \\
&= X_1'X_1\psi_1 + X_2'y_2 \\
&= (X_1'X_1 + X_2'X_2)\psi_1 - X_2'X_2\psi_1 + X_2'y_2,
\end{aligned}$$

from which, by multiplying both sides with $(X_1'X_1 + X_2'X_2)^{-1}$, the desired Eq. (3.29) follows. **Q.E.D.**

Proof of Prop. 3.3: Equation (3.25) follows by applying Lemma 3.1 with the correspondences $\psi_0 \sim \psi_0$, $\psi_1 \sim \psi_{\bar{i}}$, $\psi_2 \sim \psi_i$, and

$$y_1 \sim \begin{pmatrix} \sqrt{\lambda^{i-1}} y_1 \\ \vdots \\ \sqrt{\lambda^{i-\bar{i}}} y_{\bar{i}} \end{pmatrix}, \qquad X_1 \sim \begin{pmatrix} \sqrt{\lambda^{i-1}} X_1 \\ \vdots \\ \sqrt{\lambda^{i-\bar{i}}} X_{\bar{i}} \end{pmatrix},$$

$$y_2 \sim \begin{pmatrix} \sqrt{\lambda^{i-\bar{i}-1}} y_{\bar{i}+1} \\ \vdots \\ y_i \end{pmatrix}, \qquad X_2 \sim \begin{pmatrix} \sqrt{\lambda^{i-\bar{i}-1}} X_{\bar{i}+1} \\ \vdots \\ X_i \end{pmatrix},$$

and by carrying out the straightforward algebra. Equation (3.23) is the special case of Eq. (3.25) corresponding to $\bar{i} = i - 1$. **Q.E.D.**

Note that the positive definiteness assumption on $X_1'X_1$ in Prop. 3.3 is needed to guarantee that the first matrix H_1 is positive definite and hence invertible; then the positive definiteness of the subsequent matrices $H_2, \ldots, H_m$ follows from Eq. (3.24). As a practical matter, it is usually possible to guarantee the positive definiteness of $X_1'X_1$ by lumping a sufficient number of measurements into the first data block (X_1 should contain n linearly independent columns). An alternative is to redefine ψ_i as

$$\psi_i = \arg\min_{r \in \Re^n} \left\{ \delta\lambda^i \|r - \psi_0\|^2 + \sum_{j=1}^{i} \lambda^{i-j} \|y_j - X_j r\|^2 \right\}, \qquad i = 1, \ldots, m,$$

where δ is a small positive scalar. Then it can be seen from the proof of Prop. 3.3 that ψ_i is generated by the same equations (3.23) and (3.24), except that the initial condition $H_0 = 0$ is replaced by

$$H_0 = \delta I,$$

where I is the identity matrix, so that $H_1 = \lambda\delta I + X_1'X_1$ is positive definite even if $X_1'X_1$ is not. Note, however, that in this case, the last estimate ψ_m is only approximately equal to the least squares estimate r^* (the approximation error depends on the size of δ).

Decomposition of Linear Least Squares Problems

Some linear least squares problems have special structure that lends itself to more efficient solution. As an example, consider a least squares problem of the form

$$\min_{r, r_1, \ldots, r_M} \sum_{m=1}^{M} \sum_{i \in I_m} \big(y_i - r'\phi(i) - r_m'\phi_m(i)\big)^2, \tag{3.31}$$

where r and r_m, $m = 1, \ldots, M$, are the unknown parameter vectors, $I_1, \ldots, I_M$ are subsets of indices, and $\phi(i)$ and $\phi_m(i)$, $m = 1, \ldots, M$, are known vectors for each i. Note that the "local-global" architecture (3.8) leads to least squares problems of this type, where $\phi(i)$ represents the "global" feature extraction mapping and $\phi_m(i)$, $m = 1, \ldots, M$, represent the "local" feature extraction mappings. If the number M of subsets is large, the dimension of the least squares problem will be proportionally large. However, thanks to its special structure, problem (3.31) can be decomposed into problems of lower dimension, as we now describe.

The minimization in Eq. (3.31) can be broken down in two parts: first minimizing with respect to r_m for each m, while keeping r fixed, and then minimizing with respect to r. The minimization with respect to r_m has the form

$$\min_{r_m} \sum_{i \in I_m} \big(y_i - r'\phi(i) - r_m'\phi_m(i)\big)^2, \tag{3.32}$$

and yields an optimal parameter vector r_m^* which is a linear combination of some vectors v_i weighted by the scalars $y_i - r'\phi(i)$, $i \in I_m$,

$$r_m^* = \sum_{i \in I_m} \big(y_i - r'\phi(i)\big) v_i,$$

[cf. the general linear least squares solution formula (3.18); $y_i - r'\phi(i)$ and v_i above correspond to y_i and $\left(\sum_{j=1}^m X_j' X_j\right)^{-1} X_i'$, respectively, in that formula]. The vectors v_i can be calculated from the corresponding formula (3.18). Thus, r_m^* can be expressed as

$$r_m^* = a_m + B_m r,$$

where a_m and B_m are the vector and matrix given by

$$a_m = \sum_{i \in I_m} y_i v_i, \qquad B_m = -\sum_{i \in I_m} v_i \phi(i)'.$$

Substituting these expressions into the least squares problem (3.31), we see that this problem is equivalent to

$$\min_{r} \sum_{m=1}^{M} \sum_{i \in I_m} \Big(y_i - a_m'\phi_m(i) - r'\big(\phi(i) + B_m'\phi_m(i)\big)\Big)^2, \tag{3.33}$$

which involves only the vector r. Thus the decomposition method consists of the solution of the linear least squares problems (3.32) and (3.33), which are of much smaller dimension than the original.

3.2.3 Gradient Methods

Most of the interesting algorithms for unconstrained minimization of a continuously differentiable function $f : \Re^n \mapsto \Re$ rely on an important idea, called *iterative gradient descent*, that works as follows. We start at some point r_0 (an initial guess) and we successively generate vectors $r_1, r_2, \ldots$, according to the algorithm

$$r_{t+1} = r_t + \gamma_t s_t, \qquad t = 0, 1, \ldots$$

where γ_t is a positive stepsize, and the direction s_t is a *descent direction*, that is,

$$\nabla f(r_t)' s_t < 0,$$

assuming that $\nabla f(r_t) \neq 0$. If $\nabla f(r_t) = 0$, then the method stops, that is, $r_{t+1} = r_t$ (equivalently we choose $s_t = 0$). We call algorithms of this type *gradient methods*.

There is a large variety of possibilities for choosing the direction s_t in a gradient method. Here are some examples, which are discussed in more detail in references such as [Lue84] and [Ber95b].

Steepest descent method

Here

$$s_t = -\nabla f(r_t).$$

It can be shown that among all directions s with $\|s\| = 1$, the steepest, i.e. the one that minimizes $\nabla f(r_t)' s$, the first order rate of change of f along the direction s, is $-\nabla f(r_t)/\|\nabla f(r_t)\|$. Steepest descent is simple, but suffers from slow convergence, particularly for cost functions involving "long and narrow" level sets.

Newton's method

Here

$$s_t = -\left(\nabla^2 f(r_t)\right)^{-1} \nabla f(r_t),$$

provided $\nabla^2 f(r_t)$ is positive definite. If $\nabla^2 f(r_t)$ is not positive definite, some modification is necessary (see e.g., [Ber95b]). The idea in Newton's method is to minimize at each iteration the quadratic approximation of f around the current point r_t given by

$$\tilde{f}_t(r) = f(r_t) + \nabla f(r_t)'(r - r_t) + \frac{1}{2}(r - r_t)' \nabla^2 f(r_t)(r - r_t).$$

By setting the derivative of $\tilde{f}_t(r)$ to zero,

$$\nabla f(r_t) + \nabla^2 f(r_t)(r - r_t) = 0,$$

we obtain the minimizing point

$$r_{t+1} = r_t - \left(\nabla^2 f(r_t)\right)^{-1} \nabla f(r_t).$$

This is the Newton iteration corresponding to a stepsize $\gamma_t = 1$. It follows that Newton's method finds the global minimum of a positive definite quadratic function in a single iteration (assuming $\gamma_t = 1$). More generally, Newton's method typically converges very fast asymptotically. The price for the gain in speed of convergence is the overhead required to calculate the Hessian matrix and to solve the linear system of equations $\nabla^2 f(r_t)s_t = -\nabla f(r_t)$ in order to find the Newton direction.

Quasi-Newton methods

Here

$$s_t = -D_t \nabla f(r_t),$$

where D_t is a positive definite symmetric matrix. Then s_t is a descent direction since, by the positive definiteness of D_t, we have $\nabla f(r_t)' s_t = -\nabla f(r_t)' D_t \nabla f(r_t) < 0$, assuming that $\nabla f(r_t) \neq 0$. In some of the most successful methods of this type, the matrix D_t is chosen to be an approximation of the inverse Hessian $\left(\nabla^2 f(r_t)\right)^{-1}$. Diagonal approximations that simply discard the off-diagonal second derivatives in the Hessian matrix are a common and simple possibility (see the subsequent discussion on scaling). More sophisticated approaches use gradient information that is collected in the course of the algorithm to construct inverse Hessian approximations along increasingly larger subspaces. Examples are the DFP (Davidon-Fletcher-Powell) method and the BFGS (Broyden-Fletcher-Goldfarb-Shanno) method (see e.g., [Ber95b] or [Lue84]). These two methods can also be viewed as special cases of an important class of methods, the class of conjugate direction methods, which have an interesting property: they require at most n iterations to minimize an n-dimensional positive definite quadratic function (as opposed to an infinite number of iterations for steepest descent and only one iteration for Newton's method). Generally, for nonquadratic functions, Quasi-Newton methods tend to achieve in part the fast convergence rate of Newton's method without incurring the extra overhead for calculating the Newton direction.

Gauss-Newton method

This is a very popular method for least squares problems of the form

$$\text{minimize} \quad f(r) = \frac{1}{2}\|g(r)\|^2 = \frac{1}{2}\sum_{i=1}^{m} \|g_i(r)\|^2$$

$$\text{subject to} \quad r \in \Re^n,$$

where g_i is for each i a given function from $\Re^n$ to a Euclidean space $\Re^{m_i}$ and $g = (g_1, \ldots, g_m)$. Given r_t, the pure form of the Gauss-Newton iteration is based on linearizing g around r_t to obtain the function

$$\tilde{g}(r, r_t) = g(r_t) + \nabla g(r_t)'(r - r_t)$$

and then minimizing the norm of the linearized function $\tilde{g}$, that is,

$$\begin{aligned} r_{t+1} &= \arg \min_{r \in \Re^n} \frac{1}{2} \|\tilde{g}(r, r_t)\|^2 \\ &= \arg \min_{r \in \Re^n} \frac{1}{2} \big(\|g(r_t)\|^2 + 2(r - r_t)'\nabla g(r_t) g(r_t) \\ &\qquad\qquad + (r - r_t)'\nabla g(r_t) \nabla g(r_t)'(r - r_t) \big). \end{aligned}$$

Assuming that the $n \times n$ matrix $\nabla g(r_t) \nabla g(r_t)'$ is invertible, the above quadratic minimization yields

$$r_{t+1} = r_t - \big(\nabla g(r_t) \nabla g(r_t)'\big)^{-1} \nabla g(r_t) g(r_t). \tag{3.34}$$

Note that if g is already a linear function, we have $\|g(r)\|^2 = \|\tilde{g}(r, r_t)\|^2$, and the method converges in a single iteration.

To ensure descent, and to deal with a singular matrix $\nabla g(r_t) \nabla g(r_t)'$ (and also to enhance convergence when this matrix is nearly singular), the method is often implemented in the modified form

$$r_{t+1} = r_t - \gamma_t \big(\nabla g(r_t) \nabla g(r_t)' + \Delta_t\big)^{-1} \nabla g(r_t) g(r_t), \tag{3.35}$$

where γ_t is a positive stepsize and Δ_t is a diagonal matrix such that

$$\nabla g(r_t) \nabla g(r_t)' + \Delta_t \text{ is positive definite.}$$

An early proposal, known as the *Levenberg-Marquardt method*, is to choose Δ_t to be a positive multiple of the identity matrix.

The Gauss-Newton method bears a close relation to Newton's method. In particular, assuming that each g_i is a scalar function, the Hessian of the cost of the least squares problem is

$$\nabla^2 f(r_t) = \nabla g(r_t) \nabla g(r_t)' + \sum_{i=1}^{m} \nabla^2 g_i(r_t) g_i(r_t),$$

so it is seen that the Gauss-Newton iteration (3.34) is an approximation of Newton's method, where the second order term

$$\sum_{i=1}^{m} \nabla^2 g_i(r_t) g_i(r_t)$$

is neglected. Thus, in the Gauss-Newton method, we save the computation of this term at the expense of some deterioration in the convergence rate. If, however, the neglected second order term is relatively small near a solution, the convergence rate of the Gauss-Newton method is quite fast. This is often true in many applications such as for example when the components $g_i(r)$ are nearly linear or when they are fairly small near the solution.

Stepsize Rules

There are also a number of rules for choosing the stepsize γ_t in a gradient method. We list some that are used widely in practice:

Minimization and Limited Minimization Rule

Here γ_t is such that the cost function is minimized along the direction s_t; that is, γ_t satisfies

$$f(r_t + \gamma_t s_t) = \min_{\gamma \geq 0} f(r_t + \gamma s_t).$$

The limited minimization rule is a version of the minimization rule, which is more easily implemented in many cases. A fixed scalar $\overline{\gamma} > 0$ is selected and γ_t is chosen to yield the greatest cost reduction over all stepsizes in the interval $[0, \overline{\gamma}]$, i.e.,

$$f(r_t + \gamma_t s_t) = \min_{\gamma \in [0, \overline{\gamma}]} f(r_t + \gamma s_t).$$

The minimization and limited minimization rules must be implemented with the aid of a one-dimensional line search method. In general, the minimizing stepsize cannot be computed exactly, and in practice, the line search is stopped once a stepsize γ_t satisfying some termination criterion is obtained. Still, the one-dimensional line search method is typically a fairly complicated iterative algorithm, and for this reason, the minimization rules are not used very frequently in neural network training problems.

Constant Stepsize

Here a fixed stepsize $\gamma > 0$ is selected and

$$\gamma_t = \gamma, \qquad t = 0, 1, \ldots$$

The constant stepsize rule is very simple and does not require any cost function evaluations for its implementation. However, if the stepsize is too large, divergence will occur, while if the stepsize is too small, the rate of convergence may be very slow.

Diminishing Stepsize

Here the stepsize converges to zero according to some more or less predetermined formula:

$$\gamma_t \to 0.$$

This stepsize rule shares with the constant stepsize rule the advantage of simplicity. One difficulty with a diminishing stepsize is that it may diminish so fast that substantial progress cannot be maintained, even when far from a stationary point. For this reason, we require that

$$\sum_{t=0}^{\infty} \gamma_t = \infty.$$

This condition guarantees that r_t does not converge to a nonstationary point. Indeed, if $r_t \to \bar{r}$, then for any large indexes m and n $(m > n)$ we have

$$r_m \approx r_n \approx \bar{r}, \qquad r_m \approx r_n - \left(\sum_{t=n}^{m-1} \gamma_t \right) \nabla f(\bar{r}),$$

which is a contradiction when $\bar{r}$ is nonstationary and $\sum_{t=n}^{m-1} \gamma_t$ can be made arbitrarily large. Generally, the diminishing stepsize rule has good theoretical convergence properties. The associated convergence rate tends to be slow, so this stepsize rule is used primarily in situations where slow convergence is inevitable, such as in singular problems or when the gradient is calculated with error (see the subsequent discussion on gradient methods with errors in this section).

Convergence Issues

Given a gradient method, ideally we would like the generated sequence r_t to converge to a global minimum. Unfortunately, this is too much to expect, at least when f is not convex, because of the presence of local minima that are not global. Indeed a gradient method is guided downhill by the form of f near the current iterate, while being oblivious to the global structure of f, and thus, can easily get attracted to any type of minimum, global or not. Furthermore, if a gradient method starts or lands at any stationary point, including a local maximum, it stops at that point. Thus, for nonconvex problems, dealing with local minima is one of the fundamental weaknesses of gradient methods, and the most we can expect as a guarantee is that the method converges to a stationary point.

Generally, depending on the nature of the cost function f, the sequence r_t generated by a gradient method need not have a limit point; in fact r_t is typically unbounded if f has no local minima. Even if r_t has a limit point, convergence to a single limit point may not be easy to guarantee in general. However, there is analysis that shows roughly speaking that local minima which are isolated stationary points (they are the unique stationary points within some open sphere), tend to attract most types of gradient methods; that is, once a gradient method gets sufficiently close to such a local minimum, it converges to it (see [Ber95b], Prop. 1.2.5 for a proof). Generally, if there is a connected set of multiple local minima, it is theoretically possible for r_t to have multiple limit points, but the occurrence of such a phenomenon has never been documented in practice.

We now address the question whether each limit point of a sequence r_t generated by a gradient method is a stationary point. From the first order Taylor expansion

$$f(r_{t+1}) = f(r_t) + \gamma_t \nabla f(r_t)' s_t + o(\gamma_t),$$

we see that if the slope of f at r_t along the direction s_t, which is $\nabla f(r_t)' s_t$, has "substantial" magnitude, the rate of progress of the method will also tend to be substantial. If on the other hand, the directions s_t tend to become asymptotically orthogonal to the gradient direction as r_t approaches a nonstationary point, the slope $\nabla f(r_t)' s_t$ will tend to zero. As a result, there is a chance that the method will get "stuck" near that point. To ensure that this does not happen, we consider a technical condition on the directions s_t, which is either naturally satisfied or can be easily enforced in most algorithms of interest. This condition is that for all t, we have

$$c_1 \|\nabla f(r_t)\|^2 \le -\nabla f(r_t)' s_t, \qquad \|s_t\| \le c_2 \|\nabla f(r_t)\|, \tag{3.36}$$

where c_1 and c_2 are some positive scalars. This condition guarantees that the vectors s_t and $\nabla f(r_t)$ will not become asymptotically orthogonal near a nonstationary point, as well as that their norms will be comparable (within a constant of each other).

We will provide convergence results for the cases of a constant and of a diminishing stepsize. A basic idea in the proof is that if the rate of change of the gradient of f is limited, then one can construct a quadratic function $\tilde{f}_t$ that majorizes f [satisfies $\tilde{f}_t(r_t + \gamma s_t) \ge f(r_t + \gamma s_t)$ for all $\gamma \ge 0$]; see Fig. 3.6.

Proposition 3.4: (*Convergence for a Constant Stepsize*) Let r_t be a sequence generated by a gradient method $r_{t+1} = r_t + \gamma s_t$, where s_t satisfies condition (3.36). Assume that for some constant $L > 0$, we have

$$\|\nabla f(r) - \nabla f(\bar{r})\| \le L \|r - \bar{r}\|, \qquad \forall\, r, \bar{r} \in \Re^n, \tag{3.37}$$

and

$$0 < \gamma < \frac{2c_1}{L c_2^2}. \tag{3.38}$$

Then either $f(r_t) \to -\infty$ or else $f(r_t)$ converges to a finite value and $\lim_{t\to\infty} \nabla f(r_t) = 0$. Furthermore, every limit point of r_t is a stationary point of f.

Proof: Fix two vectors r and z, let ξ be a scalar parameter, and let $g(\xi) = f(r + \xi z)$. The chain rule yields $(dg/d\xi)(\xi) = z'\nabla f(r + \xi z)$. We have

$$\begin{aligned} f(r+z) - f(r) = g(1) - g(0) &= \int_0^1 \frac{dg}{d\xi}(\xi)\,d\xi = \int_0^1 z'\nabla f(r+\xi z)\,d\xi \\ &\leq \int_0^1 z'\nabla f(r)\,d\xi + \left| \int_0^1 z'\big(\nabla f(r+\xi z) - \nabla f(r)\big)\,d\xi \right| \\ &\leq z'\nabla f(r) + \int_0^1 \|z\| \cdot \|\nabla f(r+\xi z) - \nabla f(r)\| d\xi \\ &\leq z'\nabla f(r) + \|z\| \int_0^1 L\xi \|z\|\,d\xi \\ &= z'\nabla f(r) + \frac{L}{2}\|z\|^2. \end{aligned} \tag{3.39}$$

Applying this relation with $r = r_t$ and $z = \gamma s_t$, we obtain

$$f(r_t + \gamma s_t) - f(r_t) \leq \gamma \nabla f(r_t)' s_t + \frac{1}{2}\gamma^2 L \|s_t\|^2,$$

from which by using the conditions (3.36), we have

$$\begin{aligned} f(r_t) - f(r_{t+1}) &\geq c_1 \gamma \|\nabla f(r_t)\|^2 - \frac{1}{2}\gamma^2 L c_2^2 \|\nabla f(r_t)\|^2 \\ &= \frac{\gamma L c_2^2}{2}\left(\frac{2c_1}{Lc_2^2} - \gamma\right)\|\nabla f(r_t)\|^2. \end{aligned} \tag{3.40}$$

In view of the assumption (3.38), the above relation implies that the value of the cost function $f(r_t)$ is monotonically nonincreasing. Thus either $f(r_t) \to -\infty$ or else $f(r_t)$ converges to a finite value. In the latter case, we have $f(r_t) - f(r_{t+1}) \to 0$, so Eq. (3.40) implies that $\nabla f(r_t) \to 0$. Furthermore, if a subsequence of r_t converges to $\overline{r}$, $f(r_t)$ must converge to the finite value $f(\overline{r})$. Thus we have $\nabla f(r_t) \to 0$, implying that $\nabla f(\overline{r}) = 0$. **Q.E.D.**

A condition of the form

$$\|\nabla f(r) - \nabla f(\overline{r})\| \leq L\|r - \overline{r}\|, \qquad \forall\, r, \overline{r} \in \Re^n,$$

[cf. Eq. (3.37)] is called a *Lipschitz continuity* condition on ∇f, and requires roughly that the "curvature" of f is no more than L in all directions. It is possible to show that this condition is satisfied if f is twice differentiable and the Hessian $\nabla^2 f$ is bounded over $\Re^n$. Actually for Prop. 3.4 to hold, it is sufficient to have a weaker form of the Lipschitz condition; it is sufficient that it holds within the set $\{r \mid f(r) \leq f(r_0)\}$ rather than the entire space $\Re^n$, as can be verified by the reader. Unfortunately, however, it is generally difficult to obtain an estimate of the constant L, so in most cases the

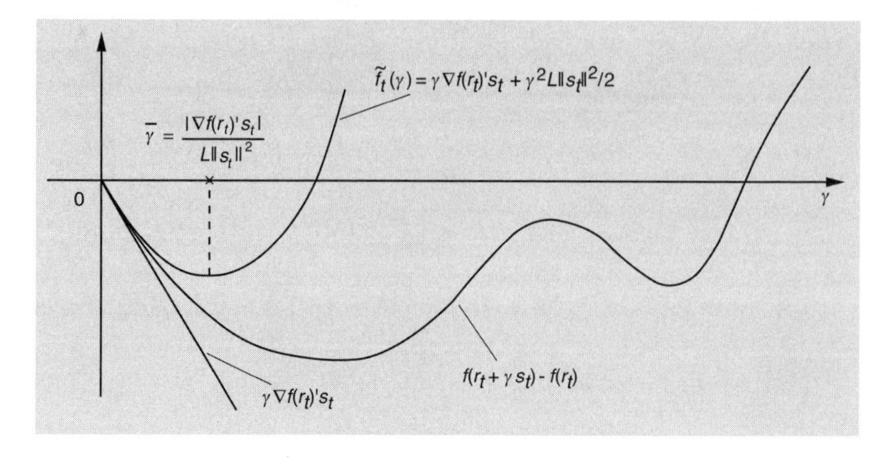

Figure 3.6: The idea of the proof of Prop. 3.4. Given r_t and the descent direction s_t, the cost difference $f(r_t + \gamma s_t) - f(r_t)$ is majorized by the quadratic function

$$\tilde{f}_t(\gamma) = \gamma \nabla f(r_t)'s_t + \frac{1}{2}\gamma^2 L\|s_t\|^2$$

[based on the Lipschitz assumption (3.37)]. Minimization of this function over γ yields the stepsize

$$\bar{\gamma} = \frac{|\nabla f(r_t)'s_t|}{L\|s_t\|^2}.$$

This stepsize reduces the cost function f as well by an amount that is proportional to $\|\nabla f(r_t)\|^2$ [cf. Eq. (3.40)].

interval of stepsizes that guarantee convergence [cf. Eq. (3.38)] is unknown. Thus, experimentation may be necessary to obtain an appropriate range of stepsizes.

The convergence proof of the following proposition, for the case of a diminishing stepsize, bears similarity to the one for a constant stepsize. We use again the construction of Fig. 3.6 to show that when the stepsize γ_t becomes sufficiently small, the cost function is decreased by an amount that is proportional to $\gamma_t\|\nabla f(r_t)\|^2$. We then use the hypothesis $\sum_{t=0}^{\infty} \gamma_t = \infty$ to conclude that if $f(r_t)$ is bounded below, then $\liminf_{t\to\infty} \|\nabla f(r_t)\| = 0$. With a somewhat technical argument, we strengthen this conclusion and show that $\lim_{t\to\infty} \nabla f(r_t) = 0$, and that every limit point of r_t is a stationary point of f.

Proposition 3.5: (*Convergence for a Diminishing Stepsize*) Let r_t be a sequence generated by a gradient method $r_{t+1} = r_t + \gamma_t s_t$, where s_t satisfies condition (3.36). Assume that for some constant $L > 0$, we have

$$\|\nabla f(r) - \nabla f(\overline{r})\| \leq L\|r - \overline{r}\|, \qquad \forall\, r, \overline{r} \in \Re^n, \tag{3.41}$$

and that

$$\gamma_t \to 0, \qquad \sum_{t=0}^{\infty} \gamma_t = \infty.$$

Then either $f(r_t) \to -\infty$ or else $f(r_t)$ converges to a finite value and $\lim_{t\to\infty} \nabla f(r_t) = 0$. Furthermore, every limit point of r_t is a stationary point of f.

Proof: Applying Eq. (3.39) with $z = \gamma_t s_t$ and using also Eq. (3.36), we have

$$\begin{aligned} f(r_{t+1}) &\leq f(r_t) + \gamma_t \left(\frac{1}{2}\gamma_t L \|s_t\|^2 - |\nabla f(r_t)' s_t| \right) \\ &\leq f(r_t) - \gamma_t \left(c_1 - \frac{1}{2}\gamma_t c_2^2 L \right) \|\nabla f(r_t)\|^2. \end{aligned}$$

Since $\gamma_t \to 0$, we have for some positive constant c and all t greater than some index $\overline{t}$,

$$f(r_{t+1}) \leq f(r_t) - \gamma_t c \|\nabla f(r_t)\|^2. \tag{3.42}$$

From this relation, we see that for $t \geq \overline{t}$, $f(r_t)$ is monotonically nonincreasing, so either $f(r_t) \to -\infty$ or $f(r_t)$ converges to a finite value. If the former case holds we are done, so assume the latter case. By adding Eq. (3.42) over all $t \geq \overline{t}$, we obtain

$$c \sum_{t=\overline{t}}^{\infty} \gamma_t \|\nabla f(r_t)\|^2 \leq f(r_{\overline{t}}) - \lim_{t\to\infty} f(r_t) < \infty.$$

We see that there cannot exist an $\epsilon > 0$ such that $\|\nabla f(r_t)\|^2 > \epsilon$ for all t greater than some $\hat{t}$, since this would contradict the assumption $\sum_{t=0}^{\infty} \gamma_t = \infty$. Therefore, we must have $\liminf_{t\to\infty} \|\nabla f(r_t)\| = 0$.

To show that $\lim_{t\to\infty} \nabla f(r_t) = 0$, assume the contrary; that is, $\limsup_{t\to\infty} \|\nabla f(r_t)\| > 0$. Then there exists an $\epsilon > 0$ such that $\|\nabla f(r_t)\| < \epsilon/2$ for infinitely many t and also $\|\nabla f(r_t)\| > \epsilon$ for infinitely many t. Therefore, there is an infinite subset of integers $\mathcal{T}$ such that for each $t \in \mathcal{T}$, there exists an integer $i(t) > t$ such that

$$\begin{aligned} &\|\nabla f(r_t)\| < \epsilon/2, \qquad \|\nabla f(r_{i(t)})\| > \epsilon, \\ &\epsilon/2 \leq \|\nabla f(r_i)\| \leq \epsilon, \qquad \text{if } t < i < i(t). \end{aligned}$$

Since

$$\begin{aligned} \|\nabla f(r_{t+1})\| - \|\nabla f(r_t)\| &\leq \|\nabla f(r_{t+1}) - \nabla f(r_t)\| \\ &\leq L\|r_{t+1} - r_t\| = \gamma_t L \|s_t\| \leq \gamma_t L c_2 \|\nabla f(r_t)\|, \end{aligned}$$

it follows that for all $t \in \mathcal{T}$ that are sufficiently large so that $\gamma_t L c_2 < 1$, we have

$$\epsilon/4 \leq \|\nabla f(r_t)\|;$$

otherwise, the condition $\epsilon/2 \leq \|\nabla f(r_{t+1})\|$ would be violated. Without loss of generality, we assume that the above relations hold for all $t \in \mathcal{T}$.

We have for all $t \in \mathcal{T}$, using the condition $\|s_t\| \leq c_2 \|\nabla f(r_t)\|$ and the Lipschitz condition (3.41),

$$\begin{aligned}
\frac{\epsilon}{2} &\leq \|\nabla f(r_{i(t)})\| - \|\nabla f(r_t)\| \leq \|\nabla f(r_{i(t)}) - \nabla f(r_t)\| \\
&\leq L\|r_{i(t)} - r_t\| \leq L \sum_{i=t}^{i(t)-1} \gamma_i \|s_i\| \\
&\leq L c_2 \sum_{i=t}^{i(t)-1} \gamma_i \|\nabla f(r_i)\| \leq L c_2 \epsilon \sum_{i=t}^{i(t)-1} \gamma_i,
\end{aligned}$$

and finally

$$\frac{1}{2Lc_2} \leq \sum_{i=t}^{i(t)-1} \gamma_i. \tag{3.43}$$

Using Eq. (3.42) for sufficiently large $t \in \mathcal{T}$, and the relation $\|\nabla f(r_i)\| \geq \epsilon/4$ for $i = t, t+1, \ldots, i(t) - 1$, we have

$$f(r_{i(t)}) \leq f(r_t) - c\left(\frac{\epsilon}{4}\right)^2 \sum_{i=t}^{i(t)-1} \gamma_i, \qquad \forall\, t \in \mathcal{T}.$$

Since $f(r_t)$ converges to a finite value, the preceding relation implies that

$$\lim_{t\to\infty,\, t\in\mathcal{T}} \sum_{i=t}^{i(t)-1} \gamma_i = 0,$$

contradicting Eq. (3.43). Thus, $\lim_{t\to\infty} \nabla f(r_t) = 0$. Finally, if $\overline{r}$ is a limit point of r_t, then $f(r_t)$ converges to the finite value $f(\overline{r})$. Thus we have $\nabla f(r_t) \to 0$, implying that $\nabla f(\overline{r}) = 0$. **Q.E.D.**

A condition that guarantees the existence of at least one limit point in Props. 3.4 and 3.5 is that the level sets $\{r \mid f(r) \leq \beta\}$ are bounded for all scalars β. Since the sequence $f(r_t)$ is bounded above, this condition implies that the sequence r_t is bounded, so that it must have a convergent subsequence. Another condition that guarantees boundedness of r_t is that f is bounded from below and that the set $\{r \mid \|\nabla f(r_t)\| \leq \epsilon\}$ is bounded for some $\epsilon > 0$. This follows from the conclusion $\|\nabla f(r_t)\| \to 0$, which holds when $\lim_{t\to\infty} f(r_t) > -\infty$.

Rate of Convergence of Steepest Descent

We now discuss the rate (or speed) of convergence characteristics of gradient methods. Many of these characteristics can be understood by analyzing the special case where the cost function is quadratic. To see why, assume that a gradient method is applied to the minimization of a twice continuously differentiable function function $f : \Re^n \mapsto \Re$, and that it generates a sequence r_t converging to a local minimum r^* with positive definite $\nabla^2 f(r^*)$. Since $\nabla f(r^*) = 0$, by Taylor's theorem we have

$$f(r) = f(r^*) + \frac{1}{2}(r - r^*)'\nabla^2 f(r^*)(r - r^*) + o\big(\|r - r^*\|^2\big).$$

Therefore, since $\nabla^2 f(r^*)$ is positive definite, f can be accurately approximated near r^* by the quadratic function

$$f(r^*) + \frac{1}{2}(r - r^*)'\nabla^2 f(r^*)(r - r^*).$$

[Note that if $\nabla^2 f(r^*)$ were singular, this would not necessarily be true, since the term $o\big(\|r - r^*\|^2\big)$ could be significant.] We thus expect that asymptotic convergence rate results obtained for the quadratic cost case have direct analogs for the general case. This conjecture can indeed be established by rigorous analysis and has been substantiated by extensive numerical experimentation. For this reason, we take the positive definite quadratic case as our point of departure. We subsequently discuss what happens when $\nabla^2 f(r^*)$ is not positive definite, in which case an analysis based on a quadratic model is inadequate.

Suppose that the cost function f is quadratic with positive definite Hessian Q. We may assume without loss of generality that f is minimized at $r^* = 0$ and that $f(r^*) = 0$ [otherwise we can use the change of variables $r \leftarrow r - r^*$ and subtract the constant $f(r^*)$ from $f(r)$]. Thus we have

$$f(r) = \frac{1}{2}r'Qr, \qquad \nabla f(r) = Qr, \qquad \nabla^2 f(r) = Q.$$

The steepest descent method takes the form

$$r_{t+1} = r_t - \gamma_t \nabla f(r_t) = (I - \gamma_t Q) r_t.$$

Therefore, we have

$$\|r_{t+1}\|^2 = r_t'(I - \gamma_t Q)^2 r_t.$$

Using the properties of eigenvalues of symmetric matrices, we have

$$\max_{\|r\|=1} r'(I - \gamma_t Q)^2 r = \text{maximum eigenvalue of } (I - \gamma_t Q)^2,$$

so we obtain

$$\|r_{t+1}\|^2 \leq \Big(\text{maximum eigenvalue of } (I - \gamma_t Q)^2\Big) \|r_t\|^2.$$

It can be seen that the eigenvalues of $(I - \gamma_t Q)^2$ are equal to $(1 - \gamma_t \lambda_i)^2$, where λ_i are the eigenvalues of Q. Therefore,

$$\text{maximum eigenvalue of } (I - \gamma_t Q)^2 = \max\big\{(1 - \gamma_t m)^2, (1 - \gamma_t M)^2\big\},$$

where

$$m: \text{ smallest eigenvalue of } Q,$$

$$M: \text{ largest eigenvalue of } Q.$$

It follows that for $r_t \neq 0$, we have

$$\frac{\|r_{t+1}\|}{\|r_t\|} \leq \max\big\{|1 - \gamma_t m|, |1 - \gamma_t M|\big\}. \tag{3.44}$$

It can be shown that the value of γ_t that minimizes the above bound is

$$\gamma^* = \frac{2}{M + m},$$

in which case

$$\frac{\|r_{t+1}\|}{\|r_t\|} \leq \frac{\frac{M}{m} - 1}{\frac{M}{m} + 1}.$$

This is the best convergence rate bound for steepest descent with constant stepsize. The bound approaches 1 as the ratio M/m increases, indicating slow convergence.

The ratio M/m is called the *condition number* of Q, and problems where M/m is large are referred to as *ill-conditioned*. Such problems are characterized by very elongated elliptical level sets. The steepest descent method converges slowly for these problems.

The estimate (3.44) shows that for a constant stepsize $\gamma_t = \gamma$ in the range $(0, 2/M)$, the sequence $\|r_t\|$ converges to zero at least as fast as a geometric progression; this is known as *linear* or *geometric* convergence. On the other hand, for a diminishing stepsize, the estimate (3.44) indicates that we may have $\|r_{t+1}\|/\|r_t\| \to 1$; this is known as *sublinear* convergence. Indeed it can be seen that since r_{t+1} differs from r_t by $\gamma_t Q r_t$, we will typically have $\|r_{t+1}\|/\|r_t\| \to 1$ whenever $\gamma_t \to 0$. Thus, asymptotically, a diminishing stepsize leads to a much slower speed of convergence than a constant stepsize.

Diagonal Scaling

Many practical problems are ill-conditioned because of poor relative scaling of the optimization variables. By this we mean that the units in which the variables are expressed are incongruent in the sense that single unit changes of different variables have disproportionate effects on the cost.

The ill-conditioning in such problems can be significantly alleviated by changing the units in which the optimization variables are expressed, which amounts to diagonal scaling of the variables. In particular, we work with a vector y that is related to r by a transformation,

$$r = Sy,$$

where S is a diagonal matrix. In lack of further information, a reasonable choice of S is one that makes all the diagonal elements of the Hessian

$$S\nabla^2 f(r)S$$

of the cost $h(y) = f(Sy)$ approximately equal to 1, that is,

$$s(i) \approx \left(\frac{\partial^2 f(r)}{\big(\partial r(i)\big)^2} \right)^{-1/2},$$

where $s(i)$ is the ith diagonal element of S. With this choice, and for a positive definite quadratic cost, equal deviations from the optimal solution along any one coordinate produce equal cost changes. It can be verified that steepest descent in the y-coordinate system, when translated to the r-coordinate system, yields the *diagonally scaled steepest descent method*

$$r_{t+1} = r_t - \gamma_t D_t \nabla f(r_t),$$

where D_t is a diagonal matrix with diagonal entries

$$D_t(i) \approx \left(\frac{\partial^2 f(r_t)}{\big(\partial r(i)\big)^2} \right)^{-1}.$$

The resulting method can be viewed as an approximation to Newton's method where the off-diagonal elements of the Hessian matrix have been discarded. This scaling method is often surprisingly effective, and also provides a side benefit: because the direction $D_t \nabla f(r_t)$ approximates the Newton direction, a constant stepsize around 1 often works well.

Some least squares problems are special because the variables can be scaled automatically by scaling some of the coefficients in the data blocks. As an example consider the linear least squares problem

$$\begin{aligned} &\text{minimize} \quad \frac{1}{2} \sum_{i=1}^{m} (y_i - x_i' r)^2 \\ &\text{subject to} \quad r \in \Re^n \end{aligned}$$

where y_i are given scalars and x_i are given vectors in $\Re^n$ with coordinates x_{ij}, $j = 1, \ldots, n$. The jth diagonal element of the Hessian matrix is

$$\sum_{i=1}^{m} x_{ij}^2.$$

Thus reasonable scaling of the variables is obtained by multiplying for each j, all coefficients x_{ij} with a common scalar β_j so that they all lie in the range $[-1, 1]$.

We finally note that the method

$$r_{t+1} = r_t - \gamma_t D_t \nabla f(r_t)$$

can be viewed as a scaled version of steepest descent even in the case where D_t is a possibly nondiagonal positive definite symmetric matrix. In particular, it can be viewed as steepest descent for the function $h_t(y) = f(S_t y)$, where S_t is an invertible matrix that satisfies $D_t = S_t S_t'$ and defines a transformation of variables $r = S_t y$. The convergence rate of this method is governed by the condition number of $\nabla^2 h_t(y) = S_t' \nabla^2 f(r) S_t$. Thus, with an appropriate choice of S_t, a favorable condition number and corresponding convergence rate can be achieved. In particular, when D_t is chosen to be equal to $\big(\nabla^2 f(r_t)\big)^{-1}$, which corresponds to Newton's method, the matrix $S_t' \nabla^2 f(r_t) S_t$ is equal to the identity matrix, and the corresponding condition number has the smallest possible value. Unfortunately, finding suitable nondiagonal scaling matrices is often difficult, particularly when second derivatives are unavailable.

Difficult Problems

We now turn to problems where the Hessian matrix either does not exist or is (nearly) singular near local minima of interest. Expressed mathematically, there are local minima r^* and directions s such that the slope of f along s, which is $\nabla f(r^* + \gamma s)'d$, changes very slowly or very rapidly with γ, that is, either

$$\lim_{\gamma \to 0} \frac{\nabla f(r^* + \gamma s)'s - \nabla f(r^*)'s}{\gamma} = \infty, \tag{3.45}$$

or

$$\lim_{\gamma \to 0} \frac{\nabla f(r^* + \gamma s)'s - \nabla f(r^*)'s}{\gamma} = 0. \tag{3.46}$$

In the case of Eq. (3.45) the cost rises steeply along the direction s. In the case of Eq. (3.46) the reverse is true; the cost is nearly flat along s and large excursions from r^* produce small changes in cost. This situation occurs often in approximation problems that are overparametrized.

As an example, consider the function

$$f\big(r(1), r(2)\big) = |r(1)|^4 + |r(2)|^{3/2},$$

where for the minimum $r^* = (0,0)$, Eq. (3.46) holds along the direction $s = (1,0)$ and Eq. (3.45) holds along the direction $s = (0,1)$. Gradient methods that use directions with norm that is comparable to the norm of the gradient may require very small stepsizes in the case of Eq. (3.45), and very large stepsizes in the case of Eq. (3.46). This suggests potential difficulties in the implementation of a good stepsize rule. For example a constant stepsize may lead to an oscillation in the case where Eq. (3.45) holds, and may lead to very slow convergence in the case where Eq. (3.46) holds. From the point of view of speed of convergence one may view the cases of Eqs. (3.45) and (3.46) as corresponding to an "infinite condition number," thereby suggesting slow convergence (see [Ber95b] or [Pol87] for further analysis).

Problems with singular Hessian matrices near local minima are not the only ones for which gradient methods may converge slowly. There are problems where a given method may have excellent asymptotic rate of convergence, but its progress when far from the eventual limit can be very slow because of singularity and/or indefiniteness of the Hessian matrix, giving rise to complex nonlinearities. This situation is common in the training of sigmoidal neural networks. For example, consider the training problem for a neural network with a single hidden layer involving the minimization of the cost [cf. Eqs. (3.4) and (3.5)]

$$\sum_i \left(J(i) - \sum_k r(k)\sigma\left(\sum_\ell r(k,\ell)x_\ell(i)\right)\right)^2,$$

over the weights $r(k)$ and $r(k,\ell)$. For values of the weights that are large, the sigmoidal units operate in their flat saturation regions, where the first and second derivatives σ' and σ'' of σ are nearly zero, giving rise to near-singularity of the Hessian matrix. In fact, as illustrated in Fig. 3.7, the cost function tends to a constant as the weight vector is changed along lines of the form $r + \gamma s$, where $\gamma > 0$ and s is a direction along which the weights $r(k,\ell)$ [but not the weights $r(k)$] are changed.

Generally, there is a tendency to think that difficult problems should be addressed with sophisticated methods, such as Newton-like methods. This is often true, particularly for problems with a Hessian that is positive definite but poorly conditioned. However, it is important to realize that *often the reverse is true*, namely that for difficult problems it is best to use simple methods such as (perhaps diagonally scaled) steepest descent with simple stepsize rules such as a constant or a diminishing stepsize. The reason is that methods that use sophisticated descent directions and stepsize rules often rely on assumptions that are likely to be violated in difficult problems.

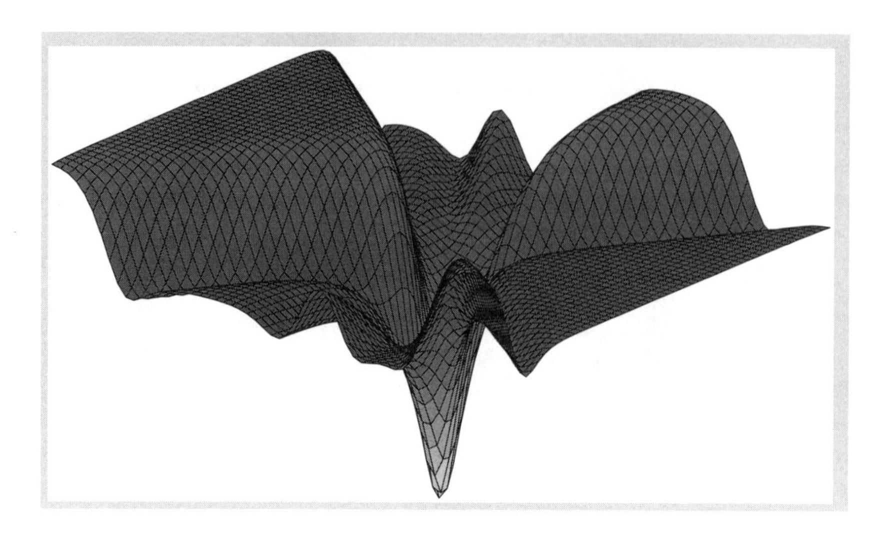

Figure 3.7: Three-dimensional plot of a least squares cost function

$$\frac{1}{2}\sum_{i=1}^{5}\Big(y_i - r(1)\sigma\big(r(1,0) + x_i r(1,1)\big)\Big)^2,$$

as a function of the weights $r(1,0)$ and $r(1,1)$, with the third weight $r(1)$ held fixed, for a simple neural network training problem. Here there is a single input x (plus a constant bias), a single output y, and only one hyperbolic tangent sigmoidal unit. The weights of the first linear layer are $r(1,0)$ and $r(1,1)$. The cost function tends to a constant as $r(1,0)$ and $r(1,1)$ are increased to infinity at a fixed ratio. For r near the origin the cost function can be quite complicated alternately involving flat and steep regions. This characteristic is common in neural network training problems involving multilayer perceptrons.

Steepest Descent With Momentum

There is variation of the steepest descent method (called the *heavy ball method* by its originator, Poljak [Pol64]) that is often used for the training of neural networks in place of steepest descent. This method has the form

$$r_{t+1} = r_t - \gamma\nabla f(r_t) + \beta(r_t - r_{t-1}), \qquad t = 1,2,\ldots,$$

where γ is a constant positive stepsize and β is a scalar with $0 \le \beta < 1$. The term $\beta(r_t - r_{t-1})$ is known as the *momentum term*. If we write the method as

$$r_{t+1} = r_t - \gamma\sum_{k=0}^{t}\beta^{t-k}\nabla f(r_k),$$

we see that the effect of the momentum term is to accelerate the progress of the method over a sequence of iterations where the gradients $\nabla f(r_t)$ are roughly aligned along the same direction, and to restrict the incremental changes of the method when successive gradients are roughly opposite to each other. Furthermore, when inside a "narrow valley," the momentum term has the effect of reinforcing the "general direction" of the valley and "filtering out" the corresponding perpendicular directions. It is generally conjectured that in comparison to steepest descent, the method tends to behave better for difficult problems where the cost function is alternatively very flat and very steep. This type of cost function is common in neural network training problems (see Fig. 3.7).

The convergence properties of the method are similar to those of steepest descent, and one can show corresponding variants of Props. 3.4 and 3.5. It also turns out, that the convergence rate is more favorable than the one for steepest descent. In particular, when the heavy ball method is applied to the positive definite quadratic function $f(r) = r'Qr/2$, it can be shown that with optimal choices of γ and β, we have

$$\frac{\|r_{t+1}\|}{\|r_t\|} \leq \frac{\sqrt{M} - \sqrt{m}}{\sqrt{M} + \sqrt{m}},$$

where m and M are the smallest and the largest eigenvalues of Q, respectively. The proof is based on writing the method as

$$\begin{bmatrix} r_{t+1} \\ r_t \end{bmatrix} = \begin{bmatrix} (1+\beta)I - \gamma Q & -\beta I \\ I & 0 \end{bmatrix} \begin{bmatrix} r_t \\ r_{t-1} \end{bmatrix}$$

and then showing that v is an eigenvalue of the matrix in the above equation if and only if $v + \beta/v$ is equal to $1 + \beta - \gamma\lambda$ where λ is an eigenvalue of Q.

Gradient Methods with Random and Nonrandom Errors

Frequently in optimization problems, the gradient $\nabla f(r_t)$ is not computed exactly. Instead, one has available

$$\nabla f(r_t) + w_t,$$

where w_t is an uncontrollable error vector. There are several potential sources of error; roundoff error and discretization error due to finite difference approximations to the gradient are two possibilities, but there are other situations of particular relevance to neural network training problems, which will be discussed later (see the analysis of the incremental gradient method in the next subsection). Let us consider the method of steepest descent with errors, where

$$s_t = -\big(\nabla f(r_t) + w_t\big),$$

and discuss several qualitatively different cases:

(a) w_t **is small relative to the gradient**, that is,

$$\|w_t\| < \|\nabla f(r_t)\|, \qquad \forall\, t.$$

Then, assuming $\nabla f(r_t) \neq 0$, s_t is a direction of descent, that is, $\nabla f(r_t)' s_t < 0$. This is illustrated in Fig. 3.8, and is verified by the calculation

$$\begin{aligned} \nabla f(r_t)' s_t &= -\|\nabla f(r_t)\|^2 - \nabla f(r_t)' w_t \\ &\leq -\|\nabla f(r_t)\|^2 + \|\nabla f(r_t)\| \cdot \|w_t\| \\ &= \|\nabla f(r_t)\| \big(\|w_t\| - \|\nabla f(r_t)\|\big) \\ &< 0. \end{aligned} \tag{3.47}$$

In this case, convergence results that are analogous to Props. 3.4 and 3.5 can be shown.

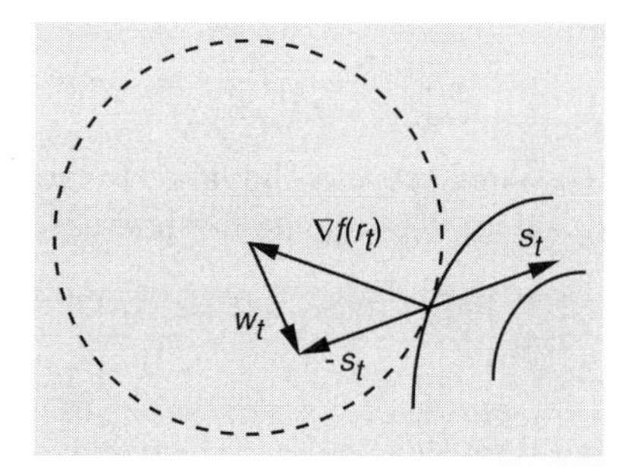

Figure 3.8: Illustration of the descent property of the direction $s_t = -(\nabla f(r_t) + w_t)$. If the error w_t has smaller norm than the gradient $\nabla f(r_t)$, then $-s_t$ lies strictly within the sphere centered at $\nabla f(r_t)$ with radius $\|\nabla f(r_t)\|$, and thus makes an angle of less than 90 degrees with $\nabla f(r_t)$.

(b) w_t **is bounded**, that is,

$$\|w_t\| \leq \delta, \qquad \forall\, t,$$

where δ is some scalar. Then by the preceding calculation (3.47), the method operates like a descent method within the region

$$\big\{r \mid \|\nabla f(r)\| > \delta\big\}.$$

In the complementary region where $\|\nabla f(r)\| \leq \delta$, the method can behave quite unpredictably. For example, if the errors w_t are constant, say $w_t \equiv w$, then since $s_t = -(\nabla f(r_t) + w)$, the method will essentially be trying to minimize $f(r) + w'r$ and will typically converge to a point $\overline{r}$ with $\nabla f(\overline{r}) = -w$. If the errors w_t vary substantially, the method will tend to oscillate within the region where $\|\nabla f(r)\| \leq \delta$. The precise behavior will depend on the precise nature of the errors, and also on whether a constant or a diminishing stepsize is used (see also the following cases).

(c) w_t **is proportional to the stepsize**, that is,

$$\|w_t\| \leq q\gamma_t, \qquad \forall\ t,$$

where q is some scalar. If the stepsize is constant, we come under case (b), while if the stepsize is diminishing, the behavior described in case (b) applies, but with $\delta \to 0$, so the method will tend to converge to a stationary point of f. In fact, if the stepsize is diminishing, by using the insight from case (a) above, it can be seen that for convergence it is sufficient to have for some positive scalars p and q,

$$\|w_t\| \leq \gamma_t\big(q + p\|\nabla f(r_t)\|\big), \qquad \forall\ t.$$

(d) w_t **are independent zero mean random vectors with finite variance**. An important special case where such errors arise is when f is of the form

$$f(r) = E_v\big[F(r, v)\big],$$

where $F : \Re^{m+n} \mapsto \Re$ is some function, v is a random vector in $\Re^m$, and $E_v[\cdot]$ denotes expected value. Under mild assumptions it can be shown that if F is continuously differentiable, the same is true of f and

$$\nabla f(r) = E_v\big[\nabla_r F(r, v)\big].$$

Often an approximation s_t to $\nabla f(r_t)$ is computed by simulation or by using a limited number of samples of $\nabla_r F(r_t, v)$, with potentially substantial error resulting. In the extreme case, we have

$$s_t = -\nabla_r F(r_t, v_t),$$

where v_t is a single sample value. Then the error

$$w_t = \nabla_r F(r_t, v_t) - \nabla f(r_t) = \nabla_r F(r_t, v_t) - E_v\big[\nabla_r F(r_t, v)\big]$$

need not diminish with $\|\nabla f(r_t)\|$, but has zero mean, and under appropriate conditions, its effects are "averaged out." What is happening here is that the descent condition $\nabla f(r_t)' s_t < 0$ holds *on the average* at nonstationary points r_t. It is still possible that for some sample values of v_t, the direction s_t is "bad," but with a diminishing stepsize, the occasional use of a bad direction cannot deteriorate the cost enough for the method to oscillate, given that on the average the method uses "good" directions. The analysis of gradient methods with random errors will be given in Ch. 4.

3.2.4 Incremental Gradient Methods for Least Squares

Let us return to the least squares problem

$$\text{minimize} \quad f(r) = \frac{1}{2}\|g(r)\|^2 = \frac{1}{2}\sum_{i=1}^{m}\|g_i(r)\|^2$$
$$\text{subject to} \quad r \in \Re^n,$$

where g is a continuously differentiable function with component functions $g_1, \ldots, g_m$, where $g_i : \Re^n \mapsto \Re^{m_i}$. Motivated by the problems of primary interest to us, we will view each component g_i as a *data block*, and we will refer to the entire function $g = (g_1, \ldots, g_m)$ as the *data set*.

In situations where there are many data blocks, gradient methods, including Newton's method and the Gauss-Newton method, may be ineffective because the size of the data set makes each iteration very costly. For such problems it may be more attractive to use an incremental method that does not wait to process the entire data set before updating r; instead, the method cycles through the data blocks in sequence and updates the estimate of r after each data block is processed. For example, given r_t, we may obtain r_{t+1} as

$$r_{t+1} = \psi_m,$$

where ψ_m is obtained at the last step of the following algorithm

$$\psi_i = \psi_{i-1} - \gamma_t h_i, \qquad i = 1, \ldots, m, \tag{3.48}$$

where

$$\psi_0 = r_t, \tag{3.49}$$

$\gamma_t > 0$ is a stepsize, and the direction h_i is the gradient of the cost associated with the ith data block,

$$h_i = \nabla g_i(\psi_{i-1}) g_i(\psi_{i-1}). \tag{3.50}$$

(Note that the vectors ψ_i depend on t, but for simplicity, we suppress this dependence in our notation.) This method can be written as

$$r_{t+1} = r_t - \gamma_t \sum_{i=1}^{m} \nabla g_i(\psi_{i-1}) g_i(\psi_{i-1}), \tag{3.51}$$

and is used extensively in the training of neural networks. It will be referred to as the *incremental gradient method*, and it should be compared with the steepest descent method (also referred to as the *batch* gradient method), which is

$$r_{t+1} = r_t - \gamma_t \nabla f(r_t) = r_t - \gamma_t \sum_{i=1}^{m} \nabla g_i(r_t) g_i(r_t). \tag{3.52}$$

Thus, we see that a cycle of the incremental gradient method through the data set differs from a pure steepest descent iteration only in that the evaluation of g_i and ∇g_i is done at the corresponding current estimates ψ_{i-1} rather than at the estimate r_t available at the start of the cycle.

One potential advantage to the incremental approach is that estimates of r become available as data are accumulated, making the approach suitable for real-time operation. Another advantage is that for a very large data set, it may converge much faster than the corresponding steepest descent method, particularly when far from the eventual limit. This type of behavior is most vividly illustrated in the case where the data blocks are linear and the vector r is one-dimensional, as in the following example.

Example 3.5

Assume that r is a scalar, and that the least squares problem has the form

$$\begin{aligned} &\text{minimize} \quad f(r) = \frac{1}{2}\sum_{i=1}^{m}(a_i r - b_i)^2 \\ &\text{subject to} \quad r \in \Re, \end{aligned}$$

where a_i and b_i are given scalars with $a_i \neq 0$ for all i. The minimum of each of the data blocks

$$f_i(r) = \frac{1}{2}(a_i r - b_i)^2$$

is

$$r_i^* = \frac{b_i}{a_i},$$

while the minimum of the least squares cost function f is

$$r^* = \frac{\sum_{i=1}^m a_i b_i}{\sum_{i=1}^m a_i^2}.$$

It can be seen that r^* lies within the range of the data block minima

$$R = \left[\min_i r_i^*,\ \max_i r_i^*\right],$$

and that for all r *outside* the range R, the gradient

$$\nabla f_i(r) = a_i(a_i r - b_i)$$

has the same sign as $\nabla f(r)$ (see Fig. 3.9). As a result, the incremental gradient method

$$\psi_i = \psi_{i-1} - \gamma_t a_i(a_i\psi_{i-1} - b_i)$$

approaches r^* at each step [cf. Eq. (3.48)], provided the stepsize γ_t is small enough. In fact it is sufficient that

$$\gamma_t < \min_i \frac{1}{a_i^2}.$$

However, for r *inside* the region R, the ith step of a cycle of the incremental gradient method need not make progress. It will approach r^* (for small enough stepsize γ_t) only if r_i^* and r^* lie on the same side of the current point ψ_{i-1}. This induces an oscillatory behavior within the region R, and as a result, the incremental gradient method will typically not converge to r^* unless $\gamma_t \to 0$. By contrast, it can be shown that the steepest descent method, which takes the form

$$r_{t+1} = r_t - \gamma_t \sum_{i=1}^{m} a_i(a_i r_t - b_i),$$

converges to r^* for any constant stepsize satisfying

$$\gamma_t < \frac{2}{\sum_{i=1}^{m} a_i^2}.$$

However, unless the stepsize choice is particularly favorable, for r outside the region R, a full iteration of steepest descent need not make more progress towards the solution than a single step of the incremental gradient method. In other words, for this one-dimensional example, *far from the solution (outside R), a single pass through the entire data set by incremental gradient is roughly as effective as m passes through the data set by steepest descent.*

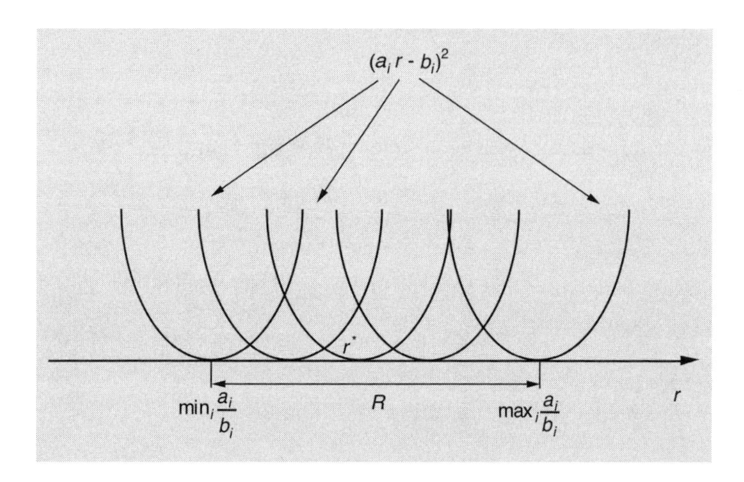

Figure 3.9: Illustrating the advantage of incrementalism when far from the optimal solution (cf. Example 3.5). The ith step in an incremental gradient cycle is a gradient step for minimizing $(a_i r - b_i)^2$, so if r lies outside the region of data block minima

$$R = \left[\min_i r_i^*, \max_i r_i^*\right],$$

and the stepsize is small enough, progress towards the solution r^* is made.

The preceding example relies on r being one-dimensional, but in many multidimensional problems the same qualitative behavior can be observed. In particular, a pass through the ith data block g_i by the incremental gradient method can make progress towards the solution in the region where the data block gradient $\nabla g_i(\psi_{i-1})g_i(\psi_{i-1})$ makes an angle less than 90 degrees with the cost function gradient $\nabla f(\psi_{i-1})$. If the data blocks g_i are not "too dissimilar," this is likely to happen in a region of points that are not too close to the optimal solution set.

There is a popular incremental version of the steepest descent method with momentum discussed earlier. This method uses the update

$$\psi_i = \psi_{i-1} - \gamma_t h_i + \beta(\psi_{i-1} - \psi_{i-2}), \qquad i = 1, \ldots, m, \tag{3.53}$$

where $\beta \in [0, 1)$, in place of the update $\psi_i = \psi_{i-1} - \gamma_t h_i$. The momentum term $\psi_{i-1} - \psi_{i-2}$ is often helpful in dealing with the peculiar features of the cost functions of neural network training problems. In addition it helps to "filter" out the fluctuations of h_i caused by dissimilarities of the different data blocks.

Another popular technique for incremental methods is to reshuffle randomly the order of the data blocks after each cycle through the data set. A variation of this method is to pick randomly a data block from the data set at each iteration rather than to pick each data block exactly once in each cycle according to a randomized order. If we take the view that an incremental method is basically a gradient method with errors, we see that randomization of the order of the data blocks tends to randomize the size of the errors, and it appears that this tends to improve the convergence properties of the method. This is further discussed in Ch. 4.

Stepsize Selection

The choice of the stepsize γ_t plays an important role in the performance of incremental gradient methods. On close examination, it turns out that the direction used by the method differs from the gradient direction by an error that is proportional to the stepsize, and for this reason a diminishing stepsize is essential for convergence to a stationary point of f (see the following discussion and Prop. 3.8 in particular). However, it turns out that a peculiar form of convergence also typically occurs for a constant but sufficiently small stepsize. In this case, the iterates converge to a "limit cycle," whereby the ith iterates ψ_i within the cycles converge to a different limit than the jth iterates ψ_j for $i \neq j$. The sequence r_t that consists of the iterates obtained at the end of cycles converges, except that the limit obtained *need not* be a stationary point. The limit tends to be close to a stationary point when the constant stepsize is small; see the following example and Prop. 3.6.

Example 3.6 (Limit Cycling When the Stepsize is Constant)

Consider the case where there are two data blocks, $g_1(r) = y_1 - r$ and $g_2(r) = y_2 - r$, where y_1 and y_2 are given scalars. Consider the incremental gradient method with a constant stepsize γ. At the first step of cycle t, we generate

$$\psi_{1,t} = r_t - \gamma(r_t - y_1),$$

while at the second step, we obtain the next iterate r_{t+1} by

$$r_{t+1} = \psi_{1,t} - \gamma(\psi_{1,t} - y_2).$$

It can be seen that if $0 < \gamma < 2$, the sequences $\psi_{1,t}$ and r_t converge. By taking the limits in the preceding equations, we can verify that the limit of the sequence r_t is

$$\frac{(1-\gamma)y_1 + y_2}{2-\gamma},$$

while the limit of the sequence $\psi_{1,t}$ is

$$\frac{y_1 + (1-\gamma)y_2}{2-\gamma}.$$

We see therefore that within each cycle, there is an oscillation around the minimum $(y_1 + y_2)/2$ of $(y_1 - r)^2 + (y_2 - r)^2$. The size of the oscillation diminishes as γ approaches 0.

The preceding example suggests that the "average" iterate within a cycle, which is

$$\frac{1}{m}\sum_{i=1}^{m} \psi_i,$$

may be a better estimate of the optimum than the last iterate, which is ψ_m. Indeed, this is often the case and there is some theoretical support for various forms of averaging (see the discussion of Section 4.2.2). Still, however, when the stepsize γ_t does not diminish to zero, the "average" iterate will typically not converge to an optimal solution.

In practice, it is common to use a constant stepsize for a (possibly prespecified) number of iterations, then decrease the stepsize by a certain factor, and repeat, up to the point where the stepsize reaches a prespecified minimum. The stepsize levels and the numbers of iterations between stepsize reductions are problem dependent and are typically chosen by trial and error. An alternative possibility is to use a stepsize rule of the form

$$\gamma_t = \min\left\{\gamma, \frac{\alpha}{t+\beta}\right\},$$

where α, β, and γ are some positive scalars, usually chosen by trial and error.

Incremental Gradient Methods with Constant Stepsize

There are variants of the incremental gradient method that use a constant stepsize throughout, but involve a diminishing degree of incrementalism as the method progresses. A simple approach to reduce the incremental nature of the method, called *batching*, is to lump several data blocks into bigger data blocks and process them simultaneously. The number of data blocks lumped together may increase as the method progresses. In the extreme case where all the data blocks are eventually lumped into one, the method reverts to the steepest descent method, which is convergent for a constant but sufficiently small stepsize (cf. Prop. 3.4).

There is another approach (due to Bertsekas [Ber95e]) that uses a constant stepsize, but also involves a time-varying parameter $\eta_t \geq 0$ to control the degree of incrementalism and to gradually switch from the incremental gradient method to the steepest descent method. Consider the method which given r_t, generates r_{t+1} according to

$$r_{t+1} = \psi_m,$$

where ψ_m is generated at the last step of the algorithm

$$\psi_i = r_t - \gamma_t h_i, \qquad i = 1, \ldots, m, \tag{3.54}$$

and the vectors h_i are defined as follows:

$$h_i = \sum_{j=1}^{i} w_{ij}(\eta_t) \nabla g_j(\psi_{j-1}) g_j(\psi_{j-1}), \tag{3.55}$$

where

$$\psi_0 = r_t,$$

and

$$w_{ij}(\eta_t) = \frac{1 + \eta_t + \cdots + \eta_t^{i-j}}{1 + \eta_t + \cdots + \eta_t^{m-j}}, \qquad i = 1, \ldots, m, \ 1 \leq j \leq i.$$

It can be verified using induction that the vectors h_i can be generated recursively using the formulas

$$h_i = \eta_t h_{i-1} + \sum_{j=1}^{i} \xi_j(\eta_t) \nabla g_j(\psi_{j-1}) g_j(\psi_{j-1}), \qquad i = 1, \ldots, m, \tag{3.56}$$

where $h_0 = 0$ and

$$\xi_i(\eta_t) = \frac{1}{1 + \eta_t + \cdots + \eta_t^{m-i}}, \qquad i = 1, \ldots, m.$$

Thus the computation of h_i using Eq. (3.56) requires no more storage or overhead per iteration that either the steepest descent method or the incremental gradient method.

Since $w_{mj}(\eta_t) = 1$ for all j, it follows from Eqs. (3.54)-(3.55) that the vector ψ_m obtained at the end of a pass through all the data blocks is

$$\psi_m = r_{t+1} = r_t - \gamma_t h_m = r_t - \gamma_t \sum_{j=1}^{m} \nabla g_j(\psi_{j-1}) g_j(\psi_{j-1}).$$

In the special case where $\eta_t = 0$, we have $\xi_i(\eta_t) = w_{ij}(\eta_t) = 1$ for all i and j, and it is seen from Eqs. (3.54)-(3.55)that the method coincides with the incremental gradient method. In the case where $\eta_t \to \infty$, it is seen from Eqs. (3.54)-(3.55) that we have $w_{ij}(\eta_t) \to 0$, $h_i \to 0$, and $\psi_i \to r_t$ for $i = 0, 1, \ldots, m-1$, so the method approaches the steepest descent method.

Generally, it can be seen that as η_t increases, the method becomes "less incremental." Geometric convergence of the method can be proved for the case of a constant but sufficiently small stepsize γ_t under some mild technical conditions, which require among other things that $\eta_t \to \infty$ so that the method asymptotically approaches the steepest descent method (see [Ber95e]). In practice, it may be better to change γ_t with η_t. In particular, when η_t is near zero and the method is similar to the incremental gradient method, the stepsize γ_t should be larger, while when η_t is large, the stepsize should tend to a constant and be of comparable magnitude to a stepsize that is appropriate for steepest descent. We refer to [Ber95e] for specific suggestions on how to update η_t and γ_t, and for further implementation details.

Scaling

We mentioned earlier that scaling can have a significant effect on the performance of gradient methods, and that in linear least squares problems, a reasonable form of diagonal scaling can be achieved by *input scaling*. In particular, for the linear least squares problem

$$\begin{aligned} &\text{minimize} \quad \frac{1}{2} \sum_{i=1}^{m} (y_i - x_i' r)^2 \\ &\text{subject to} \quad r \in \Re^n \end{aligned}$$

where y_i are given scalars and x_i are given vectors in $\Re^n$ with components x_{ij}, $j = 1, \ldots, n$, we may multiply for each j, all coefficients x_{ij} with a common scalar β_j so that they all lie in the range $[-1, 1]$.

A more general scaling possibility is to preprocess the input vector through a linear transformation. In particular, we may replace the vectors x_i in the preceding linear least squares problem with Ax_i, where A

is a square invertible matrix. The problem then is transformed into the equivalent version

$$\begin{aligned} &\text{minimize} \quad \frac{1}{2}\sum_{i=1}^{m}(y_i - x_i'A'z)^2 \\ &\text{subject to} \quad z \in \Re^n \end{aligned}$$

This transformation is equivalent to scaling the vector r in the manner discussed earlier in Section 3.2.3, that is, letting $r = A'z$. The Hessian matrix of the above cost function is

$$A\left(\sum_{i=1}^{m} x_i x_i'\right)A',$$

and may have a more favorable condition number than the Hessian matrix $\sum_{i=1}^{m} x_i x_i'$ of the original problem. One possibility is to use a matrix A that makes the diagonal elements of the above Hessian matrix roughly equal to each other.

One may similarly use diagonal or nondiagonal scaling for nonlinear least squares problems, involving for example multilayer perceptrons or other nonlinear architectures. Unfortunately, however, finding appropriate scaling transformations within this context may not be easy.

Initialization

The choice of the starting vector r_0 can be very important for the success of a gradient-like method. Clearly, one should choose this vector to be as close as possible to the eventual limit of the method. It is also sometimes necessary to use multiple initial starting points in order to reduce the risk of convergence to a poor local minimum. In training problems involving multilayer perceptrons, it is common to use initial weights that are small enough so that no sigmoidal unit operates in the saturation region, where the first and second derivatives of the corresponding sigmoidal function are nearly zero. A typical method is to choose the initial weights randomly from a small interval centered at zero.

Generally, it may be said that while incremental methods are used widely in practical neural network training problems, their effective use often requires skill, insight into the problem's structure, and trial and error.

3.2.5 Convergence Analysis of Incremental Gradient Methods

We now discuss the convergence properties of the incremental gradient method

$$r_{t+1} = r_t - \gamma_t \sum_{i=1}^{m} \nabla g_i(\psi_{i-1}) g_i(\psi_{i-1})$$

[cf. Eqs. (3.48)-(3.51)]. In Prop. 3.6, we consider the case where the data blocks are linear, while in Prop. 3.8, we take up the nonlinear case. The main idea of the convergence proofs is that the incremental gradient method can be viewed as the regular steepest descent iteration where the gradient is perturbed by an error term that is proportional to the stepsize. In particular, if we compare the above incremental gradient method with the steepest descent method

$$r_{t+1} = r_t - \gamma_t \sum_{i=1}^{m} \nabla g_i(r_t) g_i(r_t),$$

we see that the error term in the gradient direction is

$$\sum_{i=1}^{m} \big(\nabla g_i(\psi_{i-1}) g_i(\psi_{i-1}) - \nabla g_i(r_t) g_i(r_t) \big).$$

The norm of this term is proportional to the norms of the differences $\psi_{i-1} - r_t$, which are in turn proportional to the stepsize γ_t [cf. Eqs. (3.48)-(3.50)].

The following lemma will be needed for the proof of Prop. 3.6.

Lemma 3.3: Suppose that e_t and δ_t are nonnegative sequences, and c is a positive constant such that

$$e_{t+1} \leq (1 - \delta_t) e_t + c\delta_t^2, \qquad \delta_t \leq 1, \qquad \forall\, t = 0, 1, \ldots,$$

and

$$\delta_t \to 0, \qquad \sum_{t=0}^{\infty} \delta_t = \infty.$$

Then $e_t \to 0$.

Proof: Given any $\epsilon > 0$, we claim that $e_t < \epsilon$ for infinitely many t. Indeed, assuming this were not so and letting $\bar{t}$ be such that $e_t \geq \epsilon$ and $c\delta_t \leq \epsilon/2$ for all $t \geq \bar{t}$, we would have for all $t \geq \bar{t}$

$$e_{t+1} \leq e_t - \delta_t e_t + c\delta_t^2 \leq e_t - \delta_t \epsilon + \delta_t \epsilon/2 = e_t - \delta_t \epsilon/2.$$

Therefore, for all $m \geq \bar{t}$ we would have

$$e_{m+1} \leq e_{\bar{t}} - \frac{\epsilon}{2} \sum_{t=\bar{t}}^{m} \delta_t.$$

Since $0 \leq e_{m+1}$, this contradicts the assumption $\sum_{t=0}^{\infty} \delta_t = \infty$.

Thus, given any $\epsilon > 0$, there exists $\bar{t}$ such that $c\delta_t < \epsilon$ for all $t \geq \bar{t}$ and $e_{\bar{t}} < \epsilon$. We then have

$$e_{\bar{t}+1} \leq (1 - \delta_t)e_{\bar{t}} + c\delta_{\bar{t}}^2 < (1 - \delta_t)\epsilon + \delta_t\epsilon = \epsilon.$$

By repeating this argument, we obtain $e_t < \epsilon$ for all $t \geq \bar{t}$. Since ϵ can be arbitrarily small, it follows that $e_t \to 0$. **Q.E.D.**

Proposition 3.6: (*Convergence of the Incremental Gradient Method - Linear Least Squares Case*) Consider a linear least squares problem, where,

$$g_i(r) = y_i - X_i r, \qquad i = 1, \ldots, m, \tag{3.57}$$

and the incremental gradient method

$$r_{t+1} = r_t + \gamma_t \sum_{i=1}^{m} X_i'(y_i - X_i \psi_{i-1}),$$

where $\psi_0 = r_t$ and

$$\psi_i = \psi_{i-1} + \gamma_t X_i'(y_i - X_i \psi_{i-1}), \qquad i = 1, \ldots, m. \tag{3.58}$$

Assume that $\sum_{i=1}^{m} X_i' X_i$ is a positive definite matrix and let r^* be the optimal solution of the corresponding least squares problem. Then:

(a) There exists $\bar{\gamma} > 0$ such that if γ_t is equal to some constant $\gamma \in (0, \bar{\gamma}]$ for all t, r_t converges to some vector $r(\gamma)$ that depends on γ. Furthermore, we have $\lim_{\gamma \to 0} r(\gamma) = r^*$.

(b) If $\gamma_t > 0$ for all t, and

$$\gamma_t \to 0, \qquad \sum_{t=0}^{\infty} \gamma_t = \infty,$$

then r_t converges to r^*.

Proof: (a) Let γ_t be equal to some $\gamma > 0$ for all t. The idea of the proof is to write the iteration as $r_{t+1} = A(\gamma)r_t + b(\gamma)$, where $A(\gamma)$ and $b(\gamma)$ are some matrix and vector, respectively, that depend on γ. We then show that for sufficiently small γ, the eigenvalues of $A(\gamma)$ are strictly within the unit circle, which guarantees convergence.

We first show by induction that for all $i = 1, \ldots, m$, the vectors ψ_i of

Eq. (3.58) have the form

$$\psi_i = r_t + \gamma \sum_{j=1}^{i} X_j'(y_j - X_j r_t) + \gamma^2 L_i(r_t, \gamma), \tag{3.59}$$

where $L_i(r_t, \gamma)$ is a vector function of r_t and γ that is independent of t, and has the form

$$L_1(r_t, \gamma) = 0, \qquad L_i(r_t, \gamma) = \sum_{j=1}^{i-1} \gamma^{j-1}(\Phi_{ij} r_t + \phi_{ij}), \quad i = 2, \ldots, m, \tag{3.60}$$

where Φ_{ij} and ϕ_{ij} are some matrices and vectors, respectively. Indeed, this relation holds by definition for $i = 1$, with $L_1(r_t, \gamma) = 0$. Suppose that it holds for $i = p$. We have, using the induction hypothesis,

$$\begin{aligned}
\psi_{p+1} &= \psi_p + \gamma X_{p+1}'(y_{p+1} - X_{p+1}\psi_p) \\
&= r_t + \gamma \sum_{j=1}^{p} X_j'(y_j - X_j r_t) + \gamma^2 L_p(r_t, \gamma) \\
&\quad + \gamma X_{p+1}'(y_{p+1} - X_{p+1} r_t) - \gamma X_{p+1}' X_{p+1}(\psi_p - r_t) \\
&= r_t + \gamma \sum_{j=1}^{p+1} X_j'(y_j - X_j r_t) + \gamma^2 L_p(r_t, \gamma) \\
&\quad - \gamma X_{p+1}' X_{p+1} \left(\gamma \sum_{j=1}^{p} X_j'(y_j - X_j r_t) + \gamma^2 L_p(r_t, \gamma) \right),
\end{aligned}$$

which, by defining

$$L_{p+1}(r_t, \gamma) = L_p(r_t, \gamma) - X_{p+1}' X_{p+1} \left(\sum_{j=1}^{p} X_j'(y_j - X_j r_t) + \gamma L_p(r_t, \gamma) \right),$$

can be seen to be of the form (3.59)-(3.60), thereby completing the induction.

For $i = m$, Eqs. (3.59) and (3.60) yield

$$r_{t+1} = A(\gamma) r_t + b(\gamma), \tag{3.61}$$

with

$$A(\gamma) = I - \gamma \sum_{j=1}^{m} X_j' X_j + \gamma^2 \sum_{j=1}^{m-1} \gamma^{j-1} \Phi_{mj}, \tag{3.62}$$

$$b(\gamma) = \gamma \sum_{j=1}^{m} X_j' y_j + \gamma^2 \sum_{j=1}^{m-1} \gamma^{j-1} \phi_{mj}, \tag{3.63}$$

where Φ_{mj} and ϕ_{mj} are some matrices and vectors, respectively. Let us choose γ small enough so that the eigenvalues of $A(\gamma)$ are all strictly within the unit circle; this is possible since $\sum_{j=1}^{m} X_j' X_j$ is assumed positive definite and the last term in Eq. (3.62) involves powers of γ that are greater than 1. Define

$$r(\gamma) = \big(I - A(\gamma)\big)^{-1} b(\gamma). \tag{3.64}$$

Then $b(\gamma) = \big(I - A(\gamma)\big) r(\gamma)$, and by substituting this expression in Eq. (3.61), it can be seen that

$$r_{t+1} - r(\gamma) = A(\gamma)\big(r_t - r(\gamma)\big),$$

from which we obtain

$$r_{t+1} - r(\gamma) = A(\gamma)^t \big(r_0 - r(\gamma)\big), \qquad \forall\ t. \tag{3.65}$$

Since all the eigenvalues of $A(\gamma)$ are strictly within the unit circle, we have $A(\gamma)^t \to 0$, so $r_t \to r(\gamma)$.

To prove that $\lim_{\gamma \to 0} r(\gamma) = r^*$, we first note that r^* is given by

$$r^* = \left(\sum_{i=1}^{m} X_i' X_i \right)^{-1} \sum_{i=1}^{m} X_i' y_i; \tag{3.66}$$

[cf. Eq. (3.18)]. Then, we use Eq. (3.64) to write

$$r(\gamma) = \left(\frac{I}{\gamma} - \frac{A(\gamma)}{\gamma} \right)^{-1} \frac{b(\gamma)}{\gamma},$$

and we see from Eqs. (3.62) and (3.63) that

$$\lim_{\gamma \to 0} r(\gamma) = \left(\sum_{i=1}^{m} X_i' X_i \right)^{-1} \sum_{i=1}^{m} X_i' y_i = r^*.$$

(b) For $\gamma = \gamma_t$ and $i = m$, Eqs. (3.59) and (3.60) yield

$$r_{t+1} = r_t + \gamma_t \sum_{j=1}^{m} X_j'(y_j - X_j r_t) + \gamma_t^2 E_t (r_t - r^*) + \gamma_t^2 e_t, \tag{3.67}$$

where

$$E_t = \sum_{j=1}^{m-1} \gamma_t^{j-1} \Phi_{mj}, \tag{3.68}$$

$$e_t = \sum_{j=1}^{m-1} \gamma_t^{j-1} (\Phi_{mj} r^* + \phi_{mj}). \tag{3.69}$$

Using also the expression (3.66) for r^*, we can write Eq. (3.67) as

$$r_{t+1} - r^* = \left(I - \gamma_t \sum_{j=1}^{m} X_j' X_j + \gamma_t^2 E_t \right) (r_t - r^*) + \gamma_t^2 e_t. \tag{3.70}$$

For large enough t, the eigenvalues of $\gamma_t \sum_{j=1}^{m} X_j' X_j$ are bounded from above by 1, and hence the matrix $I - \gamma_t \sum_{j=1}^{m} X_j' X_j$ is positive definite. Without loss of generality, we assume that this is so for all t. Then we have

$$\left\| \left(I - \gamma_t \sum_{j=1}^{m} X_j' X_j \right) (r_t - r^*) \right\| \leq (1 - \gamma_t A) \|r_t - r^*\|, \tag{3.71}$$

where A is the smallest eigenvalue of $\sum_{j=1}^{m} X_j' X_j$. Let also B and δ be positive scalars such that for all t we have

$$\big\| E_t (r_t - r^*) \big\| \leq B \|r_t - r^*\|, \qquad \|e_t\| \leq \delta. \tag{3.72}$$

Combining Eqs. (3.70)-(3.72), we have

$$\begin{aligned} \|r_{t+1} - r^*\| &\leq \left\| \left(I - \gamma_t \sum_{j=1}^{m} X_j' X_j \right) (r_t - r^*) \right\| + \gamma_t^2 \big\| E_t (r_t - r^*) \big\| + \gamma_t^2 \|e_t\| \\ &\leq (1 - \gamma_t A + \gamma_t^2 B) \|r_t - r^*\| + \gamma_t^2 \delta. \end{aligned} \tag{3.73}$$

Let $\bar{t}$ be such that $\gamma_t B \leq A/2$ for all $t \geq \bar{t}$. Then from Eq. (3.73) we obtain

$$\|r_{t+1} - r^*\| \leq (1 - \gamma_t A/2) \|r_t - r^*\| + \gamma_t^2 \delta, \qquad \forall\, t \geq \bar{t},$$

and Lemma 3.3 can be used to show that $\|r_t - r^*\| \to 0$. **Q.E.D.**

In the case where the data blocks are nonlinear, the qualitative convergence behavior of the incremental gradient method is similar to the one shown in the above proposition for the linear case. Again, if the stepsize is constant, oscillatory behavior within data cycles can be expected. The size of the oscillation becomes smaller if the constant stepsize is chosen smaller. Finally, for a diminishing stepsize, convergence to a stationary point can be generally expected. To address this case, we first generalize the convergence result for gradient methods with diminishing stepsize (cf. Prop. 3.5) for the case where the descent direction involves an error that is proportional to the stepsize. Then we specialize this result to the case of the incremental gradient method with nonlinear data blocks. We will need the following lemma.

Lemma 3.4: Let Y_t, W_t, and Z_t be three sequences such that W_t and Z_t are nonnegative for all t. Assume that

$$Y_{t+1} \leq Y_t - W_t + Z_t, \qquad t = 0, 1, \ldots,$$

and that $\sum_{t=0}^{\infty} Z_t < \infty$. Then either $Y_t \to -\infty$, or else Y_t converges to a finite value and $\sum_{t=0}^{\infty} W_t < \infty$.

Proof: Let $\bar{t}$ be any nonnegative integer. By adding the relation $Y_{t+1} \leq Y_t + Z_t$ over all $t \geq \bar{t}$ and by taking the limit superior as $t \to \infty$, we obtain

$$\limsup_{t\to\infty} Y_t \leq Y_{\bar{t}} + \sum_{t=\bar{t}}^{\infty} Z_t < \infty.$$

By taking the limit inferior of the right-hand side as $\bar{t} \to \infty$ and by using the fact $\lim_{\bar{t}\to\infty} \sum_{t=\bar{t}}^{\infty} Z_t = 0$, we obtain

$$\limsup_{t\to\infty} Y_t \leq \liminf_{\bar{t}\to\infty} Y_{\bar{t}} < \infty.$$

This implies that either $Y_t \to -\infty$ or else Y_t converges to a finite value. In the latter case, by adding the relation $Y_{i+1} \leq Y_i - W_i + Z_i$ from $i = 0$ to $i = t$, we obtain

$$\sum_{i=0}^{t} W_i \leq Y_0 + \sum_{i=0}^{t} Z_i - Y_{t+1}, \qquad t = 0, 1, \ldots,$$

which implies that $\sum_{i=0}^{\infty} W_i \leq Y_0 + \sum_{i=0}^{\infty} Z_i - \lim_{t\to\infty} Y_t < \infty$. **Q.E.D.**

Proposition 3.7: (*Convergence for a Gradient Method with Errors*) Consider the problem of unconstrained minimization of a continuously differentiable function $f : \Re^n \mapsto \Re$. Let r_t be a sequence generated by the method

$$r_{t+1} = r_t + \gamma_t(s_t + w_t),$$

where γ_t is a positive stepsize, s_t is a descent direction satisfying for some positive scalars c_1 and c_2, and all t,

$$c_1 \|\nabla f(r_t)\|^2 \leq -\nabla f(r_t)' s_t, \qquad \|s_t\| \leq c_2 \|\nabla f(r_t)\|, \tag{3.74}$$

and w_t is an error vector satisfying for some positive scalars p and q, and all t,

$$\|w_t\| \leq \gamma_t\big(q + p\|\nabla f(r_t)\|\big).$$

Assume that for some constant $L > 0$, we have

$$\|\nabla f(r) - \nabla f(\overline{r})\| \leq L\|r - \overline{r}\|, \qquad \forall\, r, \overline{r} \in \Re^n,$$

and that

$$\sum_{t=0}^{\infty} \gamma_t = \infty, \qquad \sum_{t=0}^{\infty} \gamma_t^2 < \infty.$$

Then either $f(r_t) \to -\infty$ or else $f(r_t)$ converges to a finite value and $\lim_{t\to\infty} \nabla f(r_t) = 0$. Furthermore, every limit point of r_t is a stationary point of f.

Proof: The proof is similar to the convergence proof for gradient methods with a diminishing stepsize (Prop. 3.5), with the appropriate modifications to deal with the error vectors w_t. We apply Eq. (3.39) with $r = r_t$ and $z = \gamma_t(s_t + w_t)$. We obtain

$$f(r_{t+1}) \leq f(r_t) + \gamma_t \nabla f(r_t)'(s_t + w_t) + \frac{\gamma_t^2 L}{2}\|s_t + w_t\|^2.$$

Using our assumptions, we have

$$\begin{aligned}\nabla f(r_t)'(s_t + w_t) &\leq -c_1\|\nabla f(r_t)\|^2 + \|\nabla f(r_t)\|\,\|w_t\| \\ &\leq -c_1\|\nabla f(r_t)\|^2 + \gamma_t q\|\nabla f(r_t)\| + \gamma_t p\|\nabla f(r_t)\|^2,\end{aligned}$$

and

$$\begin{aligned}\|s_t + w_t\|^2 &\leq 2\|s_t\|^2 + 2\|w_t\|^2 \\ &\leq 2c_2^2\|\nabla f(r_t)\|^2 + 2\gamma_t^2 q^2 + 4\gamma_t^2 pq\|\nabla f(r_t)\| + 2\gamma_t^2 p^2\|\nabla f(r_t)\|^2.\end{aligned}$$

Combining the above relations, we have

$$\begin{aligned}f(r_{t+1}) \leq f(r_t) &- \gamma_t(c_1 - \gamma_t c_2^2 L - \gamma_t p - \gamma_t^3 p^2 L)\|\nabla f(r_t)\|^2 \\ &+ \gamma_t^2(q + 2\gamma_t^2 pqL)\|\nabla f(r_t)\| + \gamma_t^4 q^2 L.\end{aligned}$$

Since $\gamma_t \to 0$, we have for some positive constants c and d, and all t sufficiently large

$$f(r_{t+1}) \leq f(r_t) - \gamma_t c\|\nabla f(r_t)\|^2 + \gamma_t^2 d\|\nabla f(r_t)\| + \gamma_t^4 q^2 L.$$

Using the inequality $\|\nabla f(r_t)\| \leq 1 + \|\nabla f(r_t)\|^2$, the above relation yields for all t

$$f(r_{t+1}) \leq f(r_t) - \gamma_t(c - \gamma_t d)\|\nabla f(r_t)\|^2 + \gamma_t^2 d + \gamma_t^4 q^2 L. \tag{3.75}$$

Consider Eq. (3.75) for all t sufficiently large so that $c - \gamma_t d > 0$. By using Lemma 3.4 and the assumption $\sum_{t=0}^{\infty} \gamma_t^2 < \infty$, we see that either $f(r_t) \to -\infty$ or else $f(r_t)$ converges and

$$\sum_{t=0}^{\infty} \gamma_t \|\nabla f(r_t)\|^2 < \infty. \tag{3.76}$$

If there existed an $\epsilon > 0$ and an integer $\bar{t}$ such that $\|\nabla f(r_t)\| \geq \epsilon$ for all $t \geq \bar{t}$, we would have

$$\sum_{t=\bar{t}}^{\infty} \gamma_t \|\nabla f(r_t)\|^2 \geq \epsilon^2 \sum_{t=\bar{t}}^{\infty} \gamma_t = \infty,$$

which contradicts Eq. (3.76). Therefore, $\liminf_{t\to\infty} \|\nabla f(r_t)\| = 0$. From this point, proving that $\lim_{t\to\infty} \nabla f(r_t) = 0$ and that all limit points of r_t are stationary is very similar to the proof of Prop. 3.5 following Eq. (3.42) [we use Eq. (3.75) in place of Eq. (3.42) at the appropriate point in the proof]. **Q.E.D.**

Proposition 3.8: (*Convergence of the Incremental Gradient Method*) Consider a nonlinear least squares problem with cost function $f(r) = (1/2)\sum_{i=1}^{m} \|g_i(r)\|^2$, and let r_t be a sequence generated by the incremental gradient method. Assume that for some positive constants L, C, and D, and all $i = 1, \ldots, m$, we have

$$\|\nabla g_i(r) g_i(r) - \nabla g_i(\bar{r}) g_i(\bar{r})\| \leq L\|r - \bar{r}\|, \qquad \forall\, r, \bar{r} \in \Re^n,$$

and

$$\|\nabla g_i(r) g_i(r)\| \leq C + D\|\nabla f(r)\|, \qquad \forall\, r \in \Re^n. \tag{3.77}$$

Assume also that

$$\sum_{t=0}^{\infty} \gamma_t = \infty, \qquad \sum_{t=0}^{\infty} \gamma_t^2 < \infty.$$

Then $f(r_t)$ converges to a finite value and $\lim_{t\to\infty} \nabla f(r_t) = 0$. Furthermore, every limit point of r_t is a stationary point of f.

Proof: The idea of the proof is to formulate the incremental gradient method as a gradient method with errors that are proportional to the stepsize, and then to apply Prop. 3.7. For simplicity we will assume that there are only two data blocks; that is, $m = 2$. The proof is similar when $m > 2$. We have

$$\psi_1 = r_t - \gamma_t \nabla g_1(r_t) g_1(r_t),$$

$$r_{t+1} = \psi_1 - \gamma_t \nabla g_2(\psi_1) g_2(\psi_1).$$

By adding these two relations, we obtain

$$r_{t+1} = r_t + \gamma_t\big(-\nabla f(r_t) + w_t\big),$$

where

$$w_t = \nabla g_2(r_t) g_2(r_t) - \nabla g_2(\psi_1) g_2(\psi_1).$$

We have

$$\|w_t\| \leq L\|r_t - \psi_1\| = \gamma_t L \|\nabla g_1(r_t) g_1(r_t)\| \leq \gamma_t\big(LC + LD\|\nabla f(r_t)\|\big).$$

Thus Prop. 3.7 applies, and by using also the nonnegativity of $f(r)$, the result follows. **Q.E.D.**

Since the gradient $\nabla f(r)$ is the sum of the m terms $\nabla g_i(r) g_i(r)$, the assumption (3.77) does not seem very restrictive. We note that there is a convergence result analogous to Prop. 3.8 for the incremental gradient method with a momentum term of Eq. (3.53) [see Mangasarian and Solodov [MaS94], who assume boundedness of the generated sequence r_t instead of Eq. (3.77)].

3.2.6 Extended Kalman Filtering

We now consider a generalization of the Kalman filter, known as the *extended Kalman filter* (EKF for short). This method may be viewed as an incremental version of the Gauss-Newton method. The method starts with some r_0, then updates r via a Gauss-Newton-like iteration aimed at minimizing

$$\|g_1(r)\|^2,$$

then updates r via a Gauss-Newton-like iteration aimed at minimizing

$$\|g_1(r)\|^2 + \|g_2(r)\|^2,$$

and similarly continues, with the ith step consisting of a Gauss-Newton-like iteration aimed at minimizing the partial sum

$$\sum_{j=1}^{i} \|g_j(r)\|^2.$$

Thus, a cycle through the data set of the EKF sequentially generates the vectors

$$\psi_i = \arg \min_{r \in \Re^n} \sum_{j=1}^{i} \|\tilde{g}_j(r, \psi_{j-1})\|^2, \qquad i = 1, \ldots, m, \tag{3.78}$$

where $\tilde{g}_j(r, \psi_{j-1})$ are the linearized functions

$$\tilde{g}_j(r, \psi_{j-1}) = g_j(\psi_{j-1}) + \nabla g_j(\psi_{j-1})'(r - \psi_{j-1}), \tag{3.79}$$

and $\psi_0 = r_0$ is the initial estimate of r. Using the formulas (3.23) and (3.24) of Prop. 3.3 with the identifications

$$y_i = g_i(\psi_{i-1}) - \nabla g_i(\psi_{i-1})'\psi_{i-1}, \qquad X_i = -\nabla g_i(\psi_{i-1})',$$

the algorithm can be written in the incremental form

$$\psi_i = \psi_{i-1} - H_i^{-1} \nabla g_i(\psi_{i-1}) g_i(\psi_{i-1}), \qquad i = 1, \ldots, m, \tag{3.80}$$

where the matrices H_i are generated by

$$H_i = H_{i-1} + \nabla g_i(\psi_{i-1}) \nabla g_i(\psi_{i-1})', \qquad i = 1, \ldots, m, \tag{3.81}$$

with

$$H_0 = 0. \tag{3.82}$$

To contrast the EKF with the pure form of the Gauss-Newton method (unit stepsize), note that the first iteration of the latter can be written as

$$r_1 = \arg \min_{r \in \Re^n} \sum_{i=1}^{m} \|\tilde{g}_i(r, r_0)\|^2. \tag{3.83}$$

Thus, by comparing Eq. (3.78) for $i = m$ and $r = r_0$ with Eq. (3.83), we see that a cycle of the EKF through the data set differs from a pure Gauss-Newton iteration only in that the linearization of the data blocks g_i is done at the corresponding current estimates ψ_{i-1} rather than at the estimate r_0 available at the start of the cycle. This is similar to the relation between the incremental gradient method and the steepest descent method.

Convergence Issues for the Extended Kalman Filter

We have considered so far a single cycle of the EKF. To obtain an algorithm that cycles through the data set multiple times, we can simply create a larger data set by concatenating multiple copies of the original data set, that is, by forming what we refer to as *the extended data set*

$$(g_1, g_2, \ldots, g_m, g_1, g_2, \ldots, g_m, g_1, g_2, \ldots). \tag{3.84}$$

The EKF when applied to the extended data set asymptotically resembles a gradient method with diminishing stepsize of the type described earlier. To get a sense of this, let us denote by r_t the iterate at the end of cycle t through the data set, that is,

$$r_t = \psi_{tm}, \qquad t = 1, 2, \ldots$$

Then by using Eq. (3.25) with $i = (t+1)m$ and $\bar{i} = tm$, we obtain

$$r_{t+1} = r_t - H_{(t+1)m}^{-1} \left(\sum_{i=1}^{m} \nabla g_i(\psi_{tm+i-1}) g_i(\psi_{tm+i-1}) \right). \tag{3.85}$$

Now $H_{(t+1)m}$ grows roughly in proportion to $t+1$ because, by Eq. (3.81), we have

$$H_{(t+1)m} = \sum_{j=0}^{t} \sum_{i=1}^{m} \nabla g_i(\psi_{jm+i-1}) \nabla g_i(\psi_{jm+i-1})'. \tag{3.86}$$

It is therefore reasonable to expect that the method tends to make slow progress when t is large, which means that the vectors ψ_{tm+i-1} in Eq. (3.85) are roughly equal to r_t. Thus for large t, the sum in the right-hand side of Eq. (3.85) is roughly equal to the gradient $\nabla g(r_t) g(r_t)$, while from Eq. (3.86), $H_{(t+1)m}$ is roughly equal to $(t+1)\nabla g(r_t) \nabla g(r_t)'$, where $g = (g_1, g_2, \ldots, g_m)$ is the original data set. It follows that for large t, the EKF iteration (3.85) can be written approximately as

$$r_{t+1} \approx r_t - \frac{1}{t+1} \big(\nabla g(r_t) \nabla g(r_t)' \big)^{-1} \nabla g(r_t) g(r_t), \tag{3.87}$$

that is, as an approximate Gauss-Newton iteration with diminishing stepsize.

A variation of the EKF is obtained by introducing a fading factor $\lambda < 1$ (cf. Prop. 3.3), which determines the influence of old data blocks on new estimates. To account for λ, we should modify the updating formula (3.81) as follows:

$$H_i = \lambda H_{i-1} + \nabla g_i(\psi_{i-1}) \nabla g_i(\psi_{i-1})', \qquad i = 1, 2, \ldots, \tag{3.88}$$

with $H_0 = 0$. Generally, as λ tends towards zero, the effect of old data blocks is discounted faster, and the successive estimates produced by the method tend to change more rapidly. Thus one may obtain a faster rate of progress when $\lambda < 1$. On the other hand, when $\lambda < 1$, the matrix H_i^{-1} generated by the preceding recursion will typically not diminish to zero. Furthermore, as the following example shows, a limit cycling phenomenon may occur, similar to the case of the incremental gradient method with a constant stepsize (cf. Example 3.6).

Example 3.7

Consider the case where there are two data blocks, $g_1(r) = y_1 - r$ and $g_2(r) = y_2 - r$, where y_1 and y_2 are given scalars. Each cycle of the EKF with a fading factor λ consists of two steps. At the second step of cycle t, we minimize

$$\sum_{i=1}^{t} \left(\lambda^{2i-1}(y_1 - r)^2 + \lambda^{2i-2}(y_2 - r)^2\right),$$

which is equal to a scalar multiple of the expression $\lambda(y_1 - r)^2 + (y_2 - r)^2$. Thus at the second step, we obtain the minimizer of this expression, which is

$$\psi_{2t} = \frac{\lambda y_1 + y_2}{\lambda + 1}.$$

At the first step of cycle t, we minimize

$$(y_1 - r)^2 + \lambda \sum_{i=1}^{t-1} \left(\lambda^{2i-1}(y_1 - r)^2 + \lambda^{2i-2}(y_2 - r)^2\right),$$

which is equal to the following scalar multiple of $(y_1 - r)^2 + \lambda(y_2 - r)^2$

$$(1 + \lambda^2 + \cdots + \lambda^{2t-4})\left((y_1 - r)^2 + \lambda(y_2 - r)^2\right),$$

plus the diminishing term $\lambda^{2t-2}(y_1 - r)^2$. Thus at the first step, we obtain approximately (for large t) the minimum of $(y_1 - r)^2 + \lambda(y_2 - r)^2$,

$$\psi_{2t-1} \approx \frac{y_1 + \lambda y_2}{1 + \lambda}.$$

It follows that within each cycle, there is an oscillation around the minimum $(y_1 + y_2)/2$ of $(y_1 - r)^2 + (y_2 - r)^2$. The size of the oscillation diminishes as λ approaches 1.

In practice, a hybrid method that uses a different value of λ within each cycle may work best. One may start with a relatively small λ to attain a fast initial rate of convergence, and then progressively increase λ towards 1 in order to attain high solution accuracy. The following proposition shows convergence for the case where λ is either identically equal to 1, or else tends to 1 at a sufficiently fast rate. Note that the proposition assumes boundedness of the generated sequence of iterates. It is plausible that this assumption can be replaced by an assumption such as Eq. (3.77) in Prop. 3.8, but the details have not been worked out.

Proposition 3.9: Assume that the matrix $\nabla g_1(r)\nabla g_1(r)'$ has rank n for all r, and that for some $L > 0$ and all $i = 1, \ldots, m$, we have

$$\|\nabla g_i(r)g_i(r) - \nabla g_i(\overline{r})g_i(\overline{r})\| \leq L\|r - \overline{r}\|, \qquad \forall\, r, \overline{r} \in \Re^n.$$

Assume also that there is a constant $c > 0$ such that the scalar λ used in the updating formula (3.88) within cycle t, call it λ_t, satisfies

$$0 \leq 1 - \lambda_t^m \leq \frac{c}{t}, \qquad \forall\, t = 1, 2, \ldots.$$

Then if the EKF applied to the extended data set (3.84) generates a bounded sequence of vectors ψ_i, each of the limit points of r_t is a stationary point of the least squares problem.

Proof: *(Abbreviated)* Using the Kalman filter recursion (3.25), r_t satisfies

$$r_{t+1} = r_t - H_{(t+1)m}^{-1}\left(\sum_{i=1}^{m} \lambda_t^{m-i} \nabla g_i(\psi_{tm+i-1}) g_i(\psi_{tm+i-1})\right).$$

It can be shown using the continuity of ∇g_i, the rank assumption on $\nabla g_1(r)\nabla g_1(r)'$, the growth assumption on λ_t, the boundedness of r_t, and the preceding analysis that the eigenvalues of the matrices H_{tm} are within an interval $[c_1 t, c_2 t]$, where c_1 and c_2 are some positive constants (see [Ber95d]). The proof then follows the line of argument of the convergence proof of gradient methods with diminishing stepsize (see Prop. 3.5). **Q.E.D.**

Note that because of our assumption $H_0 = 0$, the matrix $\nabla g_1(r)\nabla g_1(r)'$ must have rank n, at least for $r = r_0$, in order for the matrix H_1 to be invertible. Proposition 3.9 can also be proved when H_0 is any positive definite matrix rather than $H_0 = 0$. In this case it is unnecessary to assume that $\nabla g_1(r)\nabla g_1(r)'$ has rank n, as long as enough alternative assumptions are imposed to guarantee that the eigenvalues of the matrices H_{tm} are within an interval $[c_1 t, c_2 t]$, where c_1 and c_2 are some positive constants.

3.2.7 Comparison of Various Methods

A large variety of methods for neural network training was given in this chapter, so it is worth summarizing their relative strengths and weaknesses.

For linear architectures and the corresponding linear least squares problems, one must choose between the exact/finite computation methods such as the SVD and the Kalman filter, and the incremental gradient method (possibly with a momentum term), which is simpler but converges only asymptotically. Surprisingly, in practice, the incremental gradient method is often preferred because of its simplicity and its small overhead per iteration. On the other hand, the incremental gradient method can be very slow, so experimentation with scaling and stepsize selection is often essential for success.

For nonlinear architectures, one important choice is between batch and incremental methods. As we explained earlier, for large data sets, incremental methods are clearly preferable. Among incremental methods, the chief candidates are the incremental gradient method and its variants (momentum term, data block randomization, etc), and the EKF. The latter method requires more overhead per iteration, but may converge faster. As in the case of a linear architecture, the use of appropriate scaling and stepsize choice can be crucial for the success of the incremental gradient method.

Among batch methods, one may use sophisticated algorithms such as the Gauss-Newton method, Quasi-Newton methods, or Newton's method. However, despite the power of some of these algorithms, the attainable speed of convergence may still be poor, because neural network problems often involve singular local minima and regions where the Hessian matrix is not positive semidefinite.

Finally, one must strive to be convinced that the final answer obtained from the training is a reasonable approximation to a (global) optimum. There are no systematic methods for this, but a few heuristics often help. In particular, in least squares problems, a small cost function value is a strong indication of success. No method, simple or sophisticated, incremental or batch, can address adequately the issue of multiple local minima. The technique of using many starting points for the training method is often useful (but not always successful) in dealing with the difficulties caused by multiple local minima.

3.3 NOTES AND SOURCES

3.1. Discussions of various neural network architectures, and multilayer perceptrons in particular, can be found in many books on neural networks (see e.g., Haykin [Hay94] and Bishop [Bis95]). A comparative discussion of neural network and statistical methodologies can be found in the review paper by Cheng and Titterington [ChT94]. The quotation appearing in the beginning of this chapter is taken from L. Breiman's comments in this reference. The idea of feature-based partitioning in conjunction with a "local-global" architecture and least squares problem decomposition seems to be new. The idea of an architecture that uses multiple heuristic algorithms as feature-extraction mappings, and its use in solving combinatorial optimization problems is also new.

3.2. Detailed discussions of nonlinear optimization can be found in books such as Luenberger [Lue84], Bazaraa, Sherali, and Shetty [BSS93], and Bertsekas [Ber95b]. The line of analysis of gradient methods with diminishing stepsize (Prop. 3.5) and of gradient methods with errors (Prop. 3.7) is classical (see for example Poljak [Pol87]). However, our results improve on those found in the literature. In particular, we show that if $f(r_t)$ is bounded below then $\lim_{t\to\infty} \nabla f(r_t) = 0$, while the strongest earlier results under the same assumptions, show that $\liminf_{t\to\infty} \|\nabla f(r_t)\| = 0$, and require either convexity of f or boundedness of r_t in order to assert that $\nabla f(r_t) \to 0$.

Various practical aspects of neural network training problems and algorithms are discussed in Ch. 6 of Haykin's book [Hay94]. The ill-conditioned character of these problems is discussed in Saarinen, Bramley, and Cybenko [SBC93]. Incremental gradient methods for linear least squares problems are attributed to Widrow and Hoff [WiH60]. The application of these methods to nonlinear least squares problems received wide attention following the work of Rumelhart, Hinton, and Williams [RHW86]. Related ideas are found in Werbös [Wer74] and Le Cun [LeC85]. These works were based on informal gradient descent arguments, and did not contain a rigorous convergence analysis.

The convergence result of Prop. 3.6 is due to Luo [Luo91], and stems from an earlier result of Kohonen [Koh74]. The convergence result of Prop. 3.8 was shown by Mangasarian and Solodov [MaS94], under a boundedness assumption on the generated sequence r_t. The alternative result given here and its method of proof based on general convergence results for gradient methods with errors (cf. Prop. 3.7) are new. For related results and variations of incremental gradient methods, see Gaivoronski [Gai94], Grippo [Gri94], Luo and Tseng [LuT94], Mangasarian and Solodov [MaS94], and Bertsekas [Ber95e].

The incremental version of the Gauss-Newton method was proposed, without a convergence analysis, by Davidon [Dav76]. The development given here in terms of the EKF and the corresponding convergence result are due to Bertsekas [Ber95d]. An alternative analysis of the EKF is given by Ljung [Lju79], who assuming $\lambda = 1$, used a stochastic formulation (i.e., an infinite data set) and his ODE approach [Lju77] to prove satisfactory convergence properties for a version of the EKF that is closely related to the one considered here.

I of dice possess the science
and in numbers thus am skilled.
(From the Mahábarata)

4

Stochastic Iterative Algorithms

Contents

Optimization problems as well as systems of equations are often solved by means of iterative algorithms; for example, Bellman's equation can be solved using the value iteration algorithm, described in Ch. 2. In many situations, however, the information needed to carry out an iterative algorithm is not directly available and one has to work with information corrupted by noise. (As an example, consider the "noise" introduced by using the average of several sample values of a random variable, as an estimate of its expectation.) Stochastic iterative algorithms are variants of deterministic iterative algorithms that can operate in the presence of noise. In this chapter, we provide a general discussion of stochastic iterative algorithms and overview some available tools for studying their convergence. These algorithms can be used both for solving optimization problems and for solving systems of equations. They will be used in subsequent chapters to study the convergence of simulation-based methods for exact and approximate dynamic programming.

Suppose that we are interested in solving a system of equations of the form

$$Hr = r,$$

where H is a function from $\Re^n$ into itself. [Note that we are using the "operator" notation Hr, as opposed to the more conventional notation $H(r)$.] An interesting special case is obtained if we have

$$Hr = r - \nabla f(r),$$

for some cost function f. In that case, the system $Hr = r$ is of the form

$$\nabla f(r) = 0,$$

which is closely related to the problem of finding a minimum of the function f, as discussed in Ch. 3.

One possible algorithm for solving the system $Hr = r$ is provided by the iteration

$$r := Hr,$$

or its small stepsize version

$$r := (1 - \gamma)r + \gamma Hr, \tag{4.1}$$

where γ is a positive stepsize parameter, usually chosen to be smaller than 1. [For the case where $Hr = r - \nabla f(r)$, Eq. (4.1) reduces to the gradient method.] If this algorithm converges to a limit r^*, and if H is continuous at r^*, then the limit must satisfy $Hr^* = r^*$, and there are several conditions available under which such convergence is guaranteed to take place. Suppose now that the functional form of the function H is not known precisely or that an exact evaluation of Hr is difficult, but that we have access to a random variable s of the form $s = Hr + w$, where w is a random noise term. For example, s might be generated using simulation or by means of some experiment whose outcome is corrupted by measurement noise. It is then reasonable to use s in place of Hr in Eq. (4.1), to obtain

$$r := (1-\gamma)r + \gamma(Hr + w). \tag{4.2}$$

(Note that the gradient methods with errors studied in Section 3.2.3 are a special case.) The resulting algorithm is called a *stochastic iterative* or *stochastic approximation* algorithm. In an algorithm of this type, there is an incentive to use a small stepsize γ, because this reduces the sensitivity to the noise w. On the other hand, smaller values of γ generally lead to slower progress. This conflict can be resolved by using a variable stepsize that decreases towards zero as the algorithm progresses. (Recall also the discussion of deterministic methods with diminishing stepsize in Sections 3.2.3 and 3.2.5.)

A more concrete setting is obtained as follows. Let v be a random variable with a known probability distribution $p(v \mid r)$ that depends on r. Suppose that we are interested in solving for r an equation of the form

$$E\big[g(r,v)\big] = r,$$

where g is a known function and where the expectation is with respect to the conditional distribution $p(v \mid r)$ of the random variable v. One possibility is to use the deterministic algorithm

$$r := (1-\gamma)r + \gamma E\big[g(r,v)\big].$$

Note that if v is a vector-valued random variable, the expectation $E[g(r,v)]$ amounts to a multidimensional integral, which is generally hard to evaluate. On the other hand, if the distribution $p(v \mid r)$ is known or can be simulated, it is often straightforward to generate random samples $\tilde{v}$ of v and use them to estimate $E[g(r,v)]$. For example, we might obtain several random samples $\tilde{v}_1, \ldots, \tilde{v}_k$, and carry out an update using the sample mean

$$\frac{1}{k}\sum_{i=1}^{k} g(r, \tilde{v}_i)$$

in the place of $E[g(r,v)]$. As k becomes large, the sample mean converges to the true mean and we recover the deterministic algorithm $r := (1-\gamma)r + \gamma E[g(r,v)]$. At the other extreme, we may let $k = 1$ and base an update on a single sample $\tilde{v}$. The resulting algorithm, known as the *Robbins-Monro* stochastic approximation algorithm, is of the form

$$r := (1-\gamma)r + \gamma g(r, \tilde{v}),$$

where $\tilde{v}$ is a single random sample, generated according to the distribution $p(v \mid r)$. If we rewrite the Robbins-Monro algorithm as

$$r := (1-\gamma)r + \gamma\Big(E\big[g(r,v)\big] + g(r,\tilde{v}) - E\big[g(r,v)\big]\Big),$$

we see that we are dealing with the special case of the algorithm

$$r := (1-\gamma)r + \gamma(Hr + w)$$

[cf. Eq. (4.2)], where

$$Hr = E\big[g(r,v)\big],$$

and where

$$w = g(r,\tilde{v}) - E\big[g(r,v)\big]$$

is a zero mean noise term.

In the remainder of this chapter, we will consider stochastic approximation algorithms in some detail. After providing a formal model, we will continue with an overview of a few available convergence results and of the different methodologies that can be used to establish convergence.

4.1 THE BASIC MODEL

We now provide a precise model of the stochastic approximation algorithms we will be interested in. This model does not require all components of r to be updated at each iteration, and is similar in spirit to the asynchronous value iteration algorithm discussed in Ch. 2.

Consider an algorithm that performs noisy updates of a vector $r \in \Re^n$, for the purpose of solving a system of equations of the form

$$Hr = r,$$

where H is a mapping from $\Re^n$ into itself. A solution, that is, a vector r^* that satisfies $Hr^* = r^*$, will be called a *fixed point* of H. We use $(Hr)(1), \ldots, (Hr)(n)$ to denote the components of Hr, that is,

$$Hr = \big((Hr)(1), \ldots, (Hr)(n)\big), \qquad \forall\, r \in \Re^n.$$

We use an integer variable t to index the different iterations of the algorithm. This variable is simply a counter and need not have any relation with "real time." Let r_t be the value of the vector r at time t, and let $r_t(i)$ denote its ith component. Let T^i be an infinite set of integers indicating the set of times at which an update of $r(i)$ is performed. We assume that

$$r_{t+1}(i) = r_t(i), \qquad t \notin T^i, \tag{4.3}$$

and

$$r_{t+1}(i) = \big(1 - \gamma_t(i)\big)r_t(i) + \gamma_t(i)\big((Hr_t)(i) + w_t(i)\big), \qquad t \in T^i. \qquad (4.4)$$

Here, $\gamma_t(i)$ is a nonnegative stepsize parameter, $w_t(i)$ is a noise term, and

$$(Hr_t)(i) - r_t(i) + w_t(i)$$

is the step direction. Note that the stepsize $\gamma_t(i)$ is allowed to depend on the component i, which will provide us with some additional flexibility. In order to bring Eqs. (4.3) and (4.4) into a common format, it is convenient to assume that $\gamma_t(i)$ and $w_t(i)$ are defined for every i and t, but that $\gamma_t(i) = 0$ for $t \notin T^i$. This is indeed what we will do throughout this chapter, and we will henceforth assume that Eq. (4.4) applies for all times t. Let us also mention that the stepsizes $\gamma_t(i)$ often need to be viewed as random variables, for reasons to be discussed later.

Stepsize Choice

Let us now motivate and introduce some general assumptions on the stepsize choice. Consider a situation where the noise term $w_t(i)$ is statistically independent from r_t and has a constant variance σ^2. Suppose that the stepsize $\gamma_t(i)$ is kept equal to some positive constant γ. It follows from Eq. (4.4) that the variance of $r_{t+1}(i)$ is at least $\gamma^2\sigma^2$ for all $t \in T^i$. In particular, r_t cannot converge to any limit vector r. In the best of circumstances, r_t will reach a neighborhood of a solution r^* and start moving randomly in that neighborhood. The size of the neighborhood can be controlled and can be made smaller by using a smaller γ, but convergence to r^* cannot be obtained as long as γ is a positive constant. If on the other hand, we allow $\gamma_t(i)$ to decrease to zero, the effect of the noise $w_t(i)$ on the variance of $r_{t+1}(i)$ becomes vanishingly small and the possibility of convergence remains open. Of course, $\gamma_t(i)$ cannot be allowed to decrease too quickly. The reason is that [cf. Eq. (4.4)]

$$|r_t(i) - r_0(i)| \le \sum_{\tau=0}^{t-1} \gamma_\tau(i)\big|(Hr_\tau)(i) - r_\tau(i) + w_\tau(i)\big|.$$

In particular, if the steps $(Hr_\tau)(i) - r_\tau(i) + w_\tau(i)$ have bounded magnitude and if $\sum_{\tau=0}^{\infty} \gamma_\tau(i) \le A < \infty$, then the algorithm will be confined within a fixed radius from r_0; if the desired solution r^* happens to be outside that radius, the algorithm will never succeed in getting to r^*. This motivates part (a) of our first assumption.

Assumption 4.1: The stepsizes $\gamma_t(i)$ are nonnegative and $\gamma_t(i) = 0$ for $t \notin T^i$. Furthermore, the following hold with probability 1:

(a) For every i, we have

$$\sum_{t=0}^{\infty} \gamma_t(i) = \infty.$$

(b) For every i, we have

$$\sum_{t=0}^{\infty} \gamma_t^2(i) < \infty.$$

For the case where all components are updated at every time instance, a common (deterministic) choice that conforms to Assumption 4.1 is given by $\gamma_t(i) = c/t$, for $t > 0$, where c is a positive constant. More generally, if component i is updated less frequently, at times $t_1, t_2, \ldots$, we may let $\gamma_{t_k}(i) = \alpha/(k + \beta)$, where α and β are positive constants. In this case, if the order of updating the various components is random, the stepsizes $\gamma_t(i)$ are also random. Furthermore, Assumption 4.1 holds if and only if all components are updated infinitely often, with probability 1.

The following example, among other things, motivates the need for Assumption 4.1(b).

Example 4.1 (Estimation of an Unknown Mean)

Let v_t, $t = 1, 2, \ldots$, be independent identically distributed scalar random variables with mean μ and unit variance. Suppose that μ is an unknown constant that we wish to estimate using the random samples v_t. This is the same as wishing to solve the equation $E[v] = r$ for r. Starting with the algorithm $r := (1 - \gamma)r + \gamma E[v]$ and using a single sample estimate of $E[v]$, we obtain a Robbins-Monro stochastic approximation algorithm of the form

$$r_{t+1} = (1 - \gamma_t)r_t + \gamma_t v_t. \tag{4.5}$$

Let us assume that the sequence γ_t is deterministic. Let V_t be the variance of r_t. Since r_t and v_t are independent, the variance of r_{t+1} is the sum of the variances of $(1 - \gamma_t)r_t$ and of $\gamma_t v_t$, and we obtain

$$V_{t+1} = (1 - \gamma_t)^2 V_t + \gamma_t^2.$$

With the stepsizes being nonnegative, it can be shown that V_t converges to zero if and only if γ_t converges to zero and $\sum_{t=0}^{\infty} \gamma_t = \infty$ [cf. Assumption 4.1(a)]. (Actually, convergence to zero under this assumption follows from Lemma 3.3 in Section 3.2.4.)

Besides convergence of the variance to zero, one is usually interested in having r_t converge to μ, with probability 1. It is evident from Eq. (4.5) that for r_t to converge to μ, we need $\gamma_t(v_t - \mu)$ to converge to zero. As it turns out, Assumption 4.1(b) is useful in that respect. Under Assumption 4.1(b), we have

$$E\left[\sum_{t=1}^{\infty} \gamma_t^2 (v_t - \mu)^2\right] = \sum_{t=1}^{\infty} \gamma_t^2 E\big[(v_t - \mu)^2\big] = \sum_{t=1}^{\infty} \gamma_t^2 < \infty.$$

(The interchange of the infinite summation and the expectation can be justified by appealing to the monotone convergence theorem [Ash72].) It follows that the random variable $\sum_{t=1}^{\infty} \gamma_t^2 (v_t - \mu)^2$ must be finite with probability 1 (otherwise, its expectation would be infinite), and we conclude that $\gamma_t(v_t - \mu)$ converges to zero with probability 1. With more technical work, it can be shown that r_t converges to μ with probability 1, as desired. A more general result of this type is provided in the next section.

Let us now consider the special case where $\gamma_t = 1/t$ for all $t > 0$. Assuming that $r_1 = 0$, it is easily shown, by induction on t, that

$$r_{t+1} = \frac{\sum_{\tau=1}^{t} v_\tau}{t}, \qquad t = 1, 2, \ldots.$$

Since the sample mean of the measurements v_t is a sensible estimate of the true mean μ, this example suggests that $\gamma_t = 1/t$ is a sound choice for the stepsize parameter. (The convergence of r_t to μ for this case is known as the "strong law of large numbers.")

In practice, the condition $\sum_{t=0}^{\infty} \gamma_t^2(i) < \infty$ of Assumption 4.1(b) is not always adhered to. One reason is that once the stepsize becomes very small, the algorithm may stop making any noticeable progress. Another reason is that in certain on-line situations, the nature of the problem may slowly change with time. In order for the algorithm to be able to respond to problem changes, a nonnegligible adaptation rate must be present, and we need to prevent the stepsize parameter from becoming too small.

Some Conventions

For our purposes, it is convenient to model all of the variables entering the description of the algorithm as random variables. We wish to allow the choice of the components $r(i)$ to be updated at any given time t to depend on the results of previous computations. Since previous results are affected by the probabilistically described noise terms $w_t(i)$, we end up with an implicitly probabilistic mechanism for choosing which variables to update. Finally, the choice of whether a variable $r(i)$ is to be updated at time t may determine whether $\gamma_t(i)$ will be zero or not. To summarize, the variables $\gamma_t(i)$ will in general depend on past noise terms and should be viewed as random variables. Furthermore, all stated conditions on their values are implicitly assumed to hold with probability 1.

Let $\mathcal{F}_t$ stand for the entire history of the algorithm up to and including the point at which the stepsizes $\gamma_t(i)$ are selected, but just before the update direction is determined. More concretely, we may let

$$\mathcal{F}_t = \{r_0(i), \ldots, r_t(i), w_0(i), \ldots, w_{t-1}(i), \gamma_0(i), \ldots, \gamma_t(i),\ i = 1, \ldots, n\},$$

although, in some cases, we may wish to include in $\mathcal{F}_t$ additional information from the past of the algorithm that could be of relevance. (For example, the generation of the stepsizes or the noise terms might involve some auxiliary probabilistic mechanisms whose details may need to be included in $\mathcal{F}_t$.) Nevertheless, we will always assume that $\mathcal{F}_t$ is an increasing sequence in the sense that the history $\mathcal{F}_{t+1}$ contains the history $\mathcal{F}_t$.

Throughout this chapter, whenever we say that a sequence of random variables converges to another random variable, we mean that we have convergence with probability 1. In the same vein, many of the statements to be made are true subject to the qualification "with probability 1." However, we often refrain from making this qualification explicit, so as not to overburden the discussion.

Lines of Analysis and Results

Given a stochastic approximation algorithm with the structure that we have postulated, it can be quite difficult to determine whether it converges to the desired fixed point. The tools used in the analysis of such algorithms are often sophisticated. In the remaining sections of this chapter, we provide three different types of convergence results relating to the convergence of the iteration

$$r_{t+1}(i) = \big(1 - \gamma_t(i)\big)r_t(i) + \gamma_t(i)\big((Hr_t)(i) + w_t(i)\big).$$

(a) In Section 4.2, we derive convergence results for the case where the expected update directions at each iteration are descent directions with respect to a smooth potential (or Lyapunov) function. This includes the optimization case where $Hr = r - \nabla f(r)$ and the case where H is a contraction with respect to the Euclidean norm on $\Re^n$. The line of analysis here is quite similar to the arguments used in the convergence analysis of Section 3.2.

(b) In Section 4.3, we consider the case where the mapping H is a contraction with respect to a weighted maximum norm or satisfies certain monotonicity assumptions. These assumptions are intended to capture the properties of the dynamic programming operator T for different DP models, and the results are therefore useful in studying stochastic versions of DP methods such as value iteration.

(c) The results in Sections 4.2 and 4.3 more or less require the update noises $w_t(i)$ to be independent from one iteration to the next. In Section 4.4 we discuss some analytical tools that allow us to study convergence when such independence is absent. This case is of interest if, for example, the noise $w_t(i)$ is due to the randomness in some underlying Markov process, e.g., the Markov chain that we are interested in controlling.

4.2 CONVERGENCE BASED ON A SMOOTH POTENTIAL FUNCTION

In this section, we present a few useful convergence results that rely on the decrease of a smooth potential function. Rigorous proofs of these results are provided at the end of this section and can be skipped without loss of continuity.

4.2.1 A Convergence Result

A common method for establishing the convergence of a deterministic iterative algorithm to some r^* is to introduce a Lyapunov or potential function $f(r)$ that provides a measure of the distance from r^*. If the "distance" from r^* keeps decreasing, in the sense that $f(r_{t+1}) < f(r_t)$ whenever $r_t \neq r^*$, and under some minor additional technical conditions, convergence to r^* follows. An example of this approach was seen during our discussion of descent methods for deterministic optimization in Ch. 3, where the cost function of the optimization problem served as the potential function.

For stochastic algorithms, a similar approach can be used except that in the presence of noise, it would be overly restrictive to require that the condition $f(r_{t+1}) < f(r_t)$ hold at every step. Instead, it is more natural to assume that the *expected* direction of update is a direction of decrease of f.

In this section, we consider an algorithm of the form

$$r_{t+1} = r_t + \gamma_t s_t, \tag{4.6}$$

and we introduce an assumption that will allow us to establish convergence using the potential function method. Note that this is slightly more general than the model introduced in Section 4.1, in which we had

$$s_t = Hr_t - r_t + w_t.$$

For the purposes of this section, we will be using a single scalar stepsize parameter γ_t for every component that is being updated. In order to handle the general case, it suffices to divide every positive $\gamma_t(i)$ by a certain factor and multiply $s_t(i)$ by the same factor, so as to come back to the case where all positive $\gamma_t(i)$, $i = 1, \ldots, n$, are equal. Those $\gamma_t(i)$ that are zero can also be set to the common value γ_t if we redefine the corresponding component $s_t(i)$ to be zero.

The assumption that follows is closely related to the descent condition

$$c_1 \big\| \nabla f(r_t) \big\|^2 \leq -\nabla f(r_t)' s_t,$$

that was used in our analysis of gradient methods in deterministic optimization, properly adapted to the stochastic setting. Throughout this section, $\| \cdot \|$ will stand for the Euclidean norm defined by

$$\|r\|^2 = r'r.$$

Furthermore, we use $\mathcal{F}_t$ to denote the history of the algorithm until time t. For example, we may let

$$\mathcal{F}_t = \{r_0, \ldots, r_t, s_0, \ldots, s_{t-1}, \gamma_0, \ldots, \gamma_t\},$$

but additional information can also be included in $\mathcal{F}_t$, if necessary.

Assumption 4.2: We assume that there exists a function $f : \Re^n \mapsto \Re$ with the following properties:

(a) There holds $f(r) \geq 0$ for all $r \in \Re^n$.

(b) (*Lipschitz continuity of* ∇f) The function f is continuously differentiable and there exists some constant L such that

$$\big\| \nabla f(r) - \nabla f(\bar{r}) \big\| \leq L \|r - \bar{r}\|, \qquad \forall\, r, \bar{r} \in \Re^n.$$

(c) (*Pseudogradient property*) There exists a positive constant c such that

$$c \big\| \nabla f(r_t) \big\|^2 \leq -\nabla f(r_t)' E[s_t \mid \mathcal{F}_t], \qquad \forall\, t.$$

(d) There exist positive constants K_1, K_2 such that

$$E\big[\|s_t\|^2 \mid \mathcal{F}_t\big] \leq K_1 + K_2 \big\| \nabla f(r_t) \big\|^2, \qquad \forall\, t.$$

Part (b) of Assumption 4.2 is a smoothness condition on the function f and was used to establish convergence of deterministic gradient methods (cf. Props. 3.4 and 3.5 in Ch. 3). This condition is satisfied, in particular, if f is twice differentiable and all of its second derivatives are bounded (globally) by some constant. Unlike the case of deterministic optimization (cf. the discussion following Prop. 3.4), convergence cannot be guaranteed if the inequality $\|\nabla f(r) - \nabla f(\bar{r})\| \leq L\|r - \bar{r}\|$ is only imposed for r and $\bar{r}$ belonging to a certain level set of f. The reason is that even if we start within such a level set, the randomness in the algorithm could always take us outside. In fact, if the second derivatives of f are finite but unbounded, adding noise to a convergent deterministic gradient algorithm can result in divergence.

Part (c) is a key assumption and requires that the expected direction $E[s_t \mid \mathcal{F}_t]$ of update make an angle of more than ninety degrees with the gradient of f; thus, the expected update is in a direction of cost decrease of the function f. Finally, part (d) provides a bound on the mean square magnitude of the update; it plays a role similar to the condition

$$\|s_t\| \leq c_2\|\nabla f(r_t)\|$$

that we had imposed in the context of deterministic optimization. A notable difference, however, is that Assumption 4.2(d) allows s_t to be nonzero even if $\nabla f(r_t)$ is zero.

We now have the following convergence result, which is in many ways analogous to Prop. 3.5 in Section 3.2. The proof is given in Section 4.2.3.

Proposition 4.1: Consider the algorithm

$$r_{t+1} = r_t + \gamma_t s_t,$$

where the stepsizes γ_t are nonnegative and satisfy

$$\sum_{t=0}^{\infty} \gamma_t = \infty, \qquad \sum_{t=0}^{\infty} \gamma_t^2 < \infty.$$

Under Assumption 4.2, the following hold with probability 1:

(a) The sequence $f(r_t)$ converges.

(b) We have

$$\lim_{t \to \infty} \nabla f(r_t) = 0.$$

(c) Every limit point of r_t is a stationary point of f.

Proposition 4.1 says nothing about the convergence or boundedness of the sequence r_t. However, if the function f has bounded level sets, part (a) of the proposition implies that the sequence r_t is bounded. If, in addition, f has a unique stationary point r^*, part (c) implies that r^* is the only limit point of r_t and, therefore, r_t converges to r^*.

We now continue with some applications of Prop. 4.1.

Example 4.2 (Stochastic Gradient Algorithm)

The stochastic gradient algorithm for minimizing a cost function f is described by

$$r_{t+1} = r_t - \gamma_t\big(\nabla f(r_t) + w_t\big). \tag{4.7}$$

We observe that this is a special case of algorithm (4.6), with

$$s_t = -\nabla f(r_t) - w_t.$$

We assume that the stepsizes are nonnegative and satisfy

$$\sum_{t=0}^{\infty} \gamma_t = \infty, \qquad \sum_{t=0}^{\infty} \gamma_t^2 < \infty.$$

Furthermore, we assume that the function f is nonnegative and has a Lipschitz continuous gradient [Assumptions 4.2(a)-(b)]. Concerning the noise term w_t, we assume that

$$E[w_t \mid \mathcal{F}_t] = 0, \tag{4.8}$$

and

$$E\big[\|w_t\|^2 \mid \mathcal{F}_t\big] \le A + B\big\|\nabla f(r_t)\big\|^2, \tag{4.9}$$

for some constants A and B.

We now verify that parts (c) and (d) of Assumption 4.2 are satisfied. Indeed, using Eq. (4.8), we have

$$\nabla f(r_t)' E[s_t \mid \mathcal{F}_t] = \nabla f(r_t)'\big(-\nabla f(r_t) - E[w_t \mid \mathcal{F}_t]\big) = -\|\nabla f(r_t)\|^2,$$

and Assumption 4.2(c) holds with $c = 1$. Furthermore, using Eqs. (4.8) and (4.9),

$$\begin{aligned} E\big[\|s_t\|^2 \mid \mathcal{F}_t\big] &= \|\nabla f(r_t)\|^2 + E\big[\|w_t\|^2 \mid \mathcal{F}_t\big] + 2\nabla f(r_t)' E[w_t \mid \mathcal{F}_t] \\ &\le \|\nabla f(r_t)\|^2 + A + B\|\nabla f(r_t)\|^2 \\ &= A + (B+1)\|\nabla f(r_t)\|^2, \end{aligned}$$

and Assumption 4.2(d) is also satisfied. We have used here the fact that the value of $\nabla f(r_t)$ is completely determined by $\mathcal{F}_t$ and, therefore, can be pulled out of the conditional expectation.

Under the preceding assumptions, Prop. 4.1 applies and we conclude that $f(r_t)$ converges and that

$$\lim_{t\to\infty} \nabla f(r_t) = 0.$$

Furthermore, every limit point of the sequence r_t is a stationary point of f.

Our next example deals with the simplest possible application of the stochastic gradient algorithm and is intimately related to the problem of estimating the unknown mean of a random variable (cf. Example 4.1).

Example 4.3 (Estimation of an Unknown Mean, ctd.)

Let v_t be independent identically distributed random variables with unknown mean μ and finite variance. The Robbins-Monro stochastic approximation algorithm

$$r_{t+1} = (1 - \gamma_t) r_t + \gamma_t v_t,$$

considered in Example 4.1 [cf. Eq. (4.5)] can be written in the form

$$r_{t+1} = r_t - \gamma_t (r_t - \mu) + \gamma_t w_t,$$

where $w_t = v_t - \mu$. This is the stochastic gradient algorithm (4.7), applied to the potential function $(r - \mu)^2/2$. Assuming that the stepsizes satisfy Assumption 4.1, the gradient of the potential function converges to zero, which implies that r_t converges to μ.

For a related situation, let r_t be a scalar that is updated according to the recursion

$$r_{t+1} = (1 - \gamma_t) r_t + \gamma_t w_t.$$

We assume the usual conditions on the stepsizes γ_t and that the noise term w_t satisfies

$$E[w_t \mid \mathcal{F}_t] = 0,$$

and

$$E[w_t^2 \mid \mathcal{F}_t] \le A + B r_t^2.$$

This is the stochastic gradient algorithm applied to the potential function $f(r) = r^2/2$. By the discussion in Example 4.2, $\nabla f(r_t)$ converges to zero, and therefore r_t converges to zero.

Our next example provides an alternative view of the incremental gradient methods for deterministic optimization that were discussed in Section 3.2.4.

Example 4.4 (Incremental Gradient Methods – I)

Consider a nonnegative cost or potential function $f : \Re^n \mapsto \Re$ of the form

$$f(r) = \frac{1}{K} \sum_{k=1}^{K} f_k(r),$$

where each f_k is a nonnegative function from $\Re^n$ into $\Re$. One possible approach for minimizing the function f is to use the (deterministic) gradient algorithm, described by

$$r_{t+1} = r_t - \frac{\gamma_t}{K} \sum_{k=1}^{K} \nabla f_k(r_t).$$

In incremental gradient methods (cf. Section 3.2.4), instead of using the average of all gradients $\nabla f_k(r_t)$ at each update, one relies instead on the gradient

$\nabla f_k(r_t)$ of just one of these functions. In the most common implementation, the individual gradients $\nabla f_k(r_t)$ are used in a cyclical order. In an alternative that we consider here, each update involves a single gradient term $\nabla f_k(r_t)$ which is selected at random.

Let us be more specific. Let $k(t)$, $t = 1, 2, \ldots$, be a sequence of independent random variables, each distributed uniformly over the set $\{1, \ldots, K\}$. The algorithm under consideration is

$$r_{t+1} = r_t - \gamma_t \nabla f_{k(t)}(r_t), \tag{4.10}$$

where γ_t is a nonnegative scalar stepsize. We claim that this is a special case of the stochastic gradient algorithm. Indeed, the algorithm (4.10) can be rewritten as

$$r_{t+1} = r_t - \frac{\gamma_t}{K} \sum_{k=1}^{K} \nabla f_k(r_t) - \gamma_t \left(\nabla f_{k(t)}(r_t) - \frac{1}{K} \sum_{k=1}^{K} \nabla f_k(r_t) \right),$$

which is of the form

$$r_{t+1} = r_t - \gamma_t \nabla f(r_t) - \gamma_t w_t,$$

where

$$w_t = \nabla f_{k(t)}(r_t) - \frac{1}{K} \sum_{k=1}^{K} \nabla f_k(r_t).$$

We now verify that w_t satisfies the assumptions (4.8) and (4.9) that we had imposed during our discussion of the stochastic gradient algorithm. Due to the way that $k(t)$ is chosen, we have

$$E\big[\nabla f_{k(t)}(r_t) \mid \mathcal{F}_t\big] = \frac{1}{K} \sum_{k=1}^{K} \nabla f_k(r_t),$$

from which it follows that $E[w_t \mid \mathcal{F}_t] = 0$ and Eq. (4.8) is satisfied. We also have

$$\begin{aligned} E\big[\|w_t\|^2 \mid \mathcal{F}_t\big] &= E\Big[\big\|\nabla f_{k(t)}(r_t)\big\|^2 \mid \mathcal{F}_t\Big] - \Big\|E\big[\nabla f_{k(t)}(r_t) \mid \mathcal{F}_t\big]\Big\|^2 \\ &\leq E\Big[\big\|\nabla f_{k(t)}(r_t)\big\|^2 \mid \mathcal{F}_t\Big], \end{aligned}$$

which yields

$$E\big[\|w_t\|^2 \mid \mathcal{F}_t\big] \leq \max_k \big\|\nabla f_k(r_t)\big\|^2.$$

An inequality of this type is not enough to lead to a bound of the type (4.9); for this reason, we impose some additional conditions on the functions f_k. In particular, let us assume that there exist constants C and D such that

$$\big\|\nabla f_k(r)\big\|^2 \leq C + D\big\|\nabla f(r)\big\|^2, \qquad \forall\, k, r. \tag{4.11}$$

It follows that

$$E\left[\|w_t\|^2 \mid \mathcal{F}_t\right] \leq C + D\left\|\nabla f(r_t)\right\|^2,$$

condition (4.9) is satisfied, and convergence of $\nabla f(r_t)$ to zero follows, under the usual stepsize conditions.

Condition (4.11) is guaranteed to be satisfied if each f_k is of the form

$$f_k(r) = r'Q_k r + g_k' r + h_k,$$

where each Q_k is a positive semidefinite matrix, $\sum_{k=1}^{K} Q_k$ is positive definite, each g_k is a vector, and each h_k is a scalar. (This is the generic situation encountered in the training problem for linear approximation architectures.) Because of positive definiteness, there exists a unique vector r^* at which $\nabla f(r^*) = 0$, and this vector is the unique minimizer of f. The function f has bounded level sets and the convergence of $f(r_t)$ implies that the sequence r_t is bounded. Every limit point r of r_t must satisfy $\nabla f(r) = 0$, from which we conclude that r_t converges to r^*.

We close by noting that the incremental gradient algorithm (4.10) can be generalized to

$$r_{t+1} = r_t - \frac{\gamma_t}{|S(t)|} \sum_{k \in S(t)} \nabla f_k(r_t).$$

Here, $S(t)$ is a random subset of $\{1, \ldots, K\}$ and $|S(t)|$ is its cardinality. We assume that the cardinality of $S(t)$ is chosen at the same time that γ_t is chosen [that is, before any of the elements of $S(t)$ are sampled] and that every element of $\{1, \ldots, K\}$ has the same probability of belonging to $S(t)$. We can then repeat our earlier arguments and come to the same conclusions regarding convergence. This more general version of the algorithm, described as "batching" in Chapter 3, allows us to control the noise variance in the stochastic algorithm. In the beginning, we may use $|S(t)| = 1$, leading to high variance estimates of ∇f; as the algorithm approaches a minimum of f, more accurate estimates of ∇f can be used, by increasing the cardinality of $S(t)$. With such a modification, we can obtain convergence without the need for a diminishing stepsize, as long as in the limit we have $|S(t)| = K$.

Example 4.5 (Incremental Gradient Methods – II)

In Example 4.4, we dealt with a cost function that was the sum of finitely many terms. A similar development is also possible when the cost function is an integral of a continuum of cost functions.

Consider a nonnegative cost or potential function $f : \Re^n \mapsto \Re$ of the form

$$f(r) = \int f_\theta(r)\, dp(\theta) = E\left[f_\theta(r)\right],$$

where θ is a parameter ranging in some set, $p(\cdot)$ is a probability distribution over that set, and the expectation above is with respect to this probability distribution. Under some regularity assumptions that allow us to interchange expectation and differentiation, we have

$$\nabla f(r) = E\big[\nabla f_\theta(r)\big].$$

In this case, an alternative to the deterministic gradient method is obtained by replacing the above expectation by a single sample estimate. More precisely, we have the following algorithm. At each time t, we generate a sample value θ_t of θ, sampled according to the probability distribution $p(\cdot)$, and we perform the update

$$\begin{aligned} r_{t+1} &= r_t - \gamma_t \nabla f_{\theta_t}(r_t) \\ &= r_t - \gamma_t \nabla f(r_t) - \gamma_t w_t, \end{aligned}$$

where

$$w_t = \nabla f_{\theta_t}(r_t) - E\big[\nabla f_\theta(r_t) \mid \mathcal{F}_t\big].$$

We have $E[w_t \mid \mathcal{F}_t] = 0$ and

$$\begin{aligned} E\big[\|w_t\|^2 \mid \mathcal{F}_t\big] &= E\Big[\big\|\nabla f_\theta(r_t)\big\|^2 \mid \mathcal{F}_t\Big] - \Big\|E\big[\nabla f_\theta(r_t) \mid \mathcal{F}_t\big]\Big\|^2 \\ &\leq E\Big[\big\|\nabla f_\theta(r_t)\big\|^2 \mid \mathcal{F}_t\Big]. \end{aligned}$$

If we impose enough assumptions to guarantee that

$$E\Big[\big\|\nabla f_\theta(r_t)\big\|^2 \mid \mathcal{F}_t\Big] \leq C + D\big\|\nabla f(r_t)\big\|^2,$$

for some constants C and D, our convergence analysis of stochastic gradient algorithms (Example 4.2) applies.

Example 4.6 (Euclidean Norm Pseudo-Contractions)

In our last example for this section, we consider the Robbins-Monro stochastic approximation algorithm

$$r_{t+1} = (1 - \gamma_t) r_t + \gamma_t (H r_t + w_t),$$

where H is a mapping from $\Re^n$ into itself that satisfies

$$\|Hr - r^*\| \leq \beta \|r - r^*\|, \qquad \forall\, r \in \Re^n. \tag{4.12}$$

Here, β is constant with $0 \leq \beta < 1$ and r^* is a vector that satisfies $Hr^* = r^*$, that is, a fixed point of H. We say that the mapping H is a *pseudo-contraction* with respect to the Euclidean norm.

In order to study this algorithm, we introduce the potential function

$$f(r) = \frac{1}{2}\|r - r^*\|^2,$$

and note that $\nabla f(r) = r - r^*$. Assuming that $E[w_t \mid \mathcal{F}_t] = 0$, we obtain

$$E[s_t \mid \mathcal{F}_t] = Hr_t - r_t,$$

where

$$s_t = Hr_t - r_t + w_t$$

is the step direction. Using the Schwartz inequality and Eq. (4.12), we obtain

$$(Hr - r^*)'(r - r^*) \leq \|Hr - r^*\| \cdot \|r - r^*\| \leq \beta \|r - r^*\|^2.$$

We subtract $(r - r^*)'(r - r^*)$ from both sides, to obtain

$$(Hr - r)'(r - r^*) \leq -(1 - \beta)\|r - r^*\|^2.$$

With r equal to r_t, the latter inequality can be rewritten as

$$E[s_t \mid \mathcal{F}_t]'\nabla f(r_t) \leq -(1 - \beta)\big\|\nabla f(r_t)\big\|^2,$$

and the pseudogradient Assumption 4.2(c) is satisfied with $c = 1 - \beta$. If we now assume that $E\big[\|w_t\|^2 \mid \mathcal{F}_t\big] \leq A + B\|r_t - r^*\|^2$ for some A, B, and impose the usual stepsize conditions, we conclude that the gradient $r_t - r^*$ of the potential function converges to zero.

4.2.2 Two-Pass Methods

The analysis in this section was geared towards methods like the stochastic gradient algorithm. There are also stochastic approximation methods that use second derivative information, such as "stochastic Newton methods." Such second order methods tend to converge faster, at the expense of more complex calculations during each iteration, although of course not as fast as their deterministic counterparts. Surprisingly, however, second order methods are in a sense unnecessary in the stochastic context; namely, there is a variant of the stochastic gradient algorithm, involving a second pass through the sequence r_t, that combines the computational simplicity of first order methods and the desirable convergence properties of second order methods.

Consider the stochastic gradient algorithm and suppose that the stepsize γ is fixed at some small positive value. We then expect that r_t gets to the vicinity of a local minimum r^* and thereafter oscillates around r^*, with the magnitude of these oscillations depending on the size of γ (see the discussion at the end of Subsection 3.2.3). In that case, an estimate of r^* that is better than the current value of r_t can be constructed by forming the average of the past values of r_t. The estimate of r^* generated by such averaging has some remarkable properties. Not only does it converge to r^* (under suitable assumptions), but its mean squared error is in a certain asymptotic sense at least as good as the mean squared error of any other method, including methods that exploit second derivative information; see the notes at the end of the chapter.

In practical terms, in order for the averaged estimate to outperform the instantaneous estimate r_t, we need the algorithm to reach a "steady-state" and to be oscillating around r^*. In contrast, during the early stages of the algorithm, that is, when r_t is halfway between r_0 and the vicinity of r^*, there is no reason to expect that the averaged estimate will be of any particular use.

4.2.3 Convergence Proofs

This subsection contains the proof of Prop. 4.1 and can be skipped without loss of continuity. In many respects, the proof is similar to the convergence proofs for deterministic descent algorithms (e.g., Props. 3.4, 3.5, and 3.7); however, some deep results from probability theory are also required.

In all of the proofs in this section, we assume that we have already discarded a suitable set of measure zero, so that we do not need to keep repeating the qualification "with probability 1."

Martingales and Supermartingales

The supermartingale convergence theorem is a key result from probability theory that lies at the heart of many convergence proofs for stochastic approximation algorithms. It is a generalization to a probabilistic context of the fact that a bounded monotonic sequence converges, and also of Lemma 3.4 of Section 3.2. The proof of this theorem lies beyond the scope of this treatment and can be found in [Ash72] and [Nev75]. For our purposes, we need a slight generalization of the supermartingale convergence theorem (see Neveu [Nev75], p. 33).

Proposition 4.2: (*Supermartingale Convergence Theorem*) Let Y_t, X_t, and Z_t, $t = 0, 1, 2, \ldots$, be three sequences of random variables and let $\mathcal{F}_t$, $t = 0, 1, 2, \ldots$, be sets of random variables such that $\mathcal{F}_t \subset \mathcal{F}_{t+1}$ for all t. Suppose that:

(a) The random variables Y_t, X_t, and Z_t are nonnegative, and are functions of the random variables in $\mathcal{F}_t$.

(b) For each t, we have $E[Y_{t+1} \mid \mathcal{F}_t] \leq Y_t - X_t + Z_t$.

(c) There holds $\sum_{t=0}^{\infty} Z_t < \infty$.

Then, we have $\sum_{t=0}^{\infty} X_t < \infty$, and the sequence Y_t converges to a nonnegative random variable Y, with probability 1.

We also provide here a related result that will be of use later on [Nev75].

Proposition 4.3: (*Martingale Convergence Theorem*) Let X_t, $t = 0, 1, 2, \ldots$, be a sequence of random variables and let $\mathcal{F}_t$, $t = 0, 1, 2, \ldots$, be sets of random variables such that $\mathcal{F}_t \subset \mathcal{F}_{t+1}$ for all t. Suppose that:

(a) The random variable X_t is a function of the random variables in $\mathcal{F}_t$.

(b) For each t, we have $E[X_{t+1} \mid \mathcal{F}_t] = X_t$.

(c) There exists a constant M such that $E\big[|X_t|\big] \leq M$ for all t.

Then, the sequence X_t converges to a random variable X, with probability 1.

A sequence X_t that satisfies properties (a) and (b) in Prop. 4.3, together with the property $E\big[|X_t|\big] < \infty$ for all t, is called a *martingale*. We note that if the random variables X_t satisfy $E[X_t^2] \leq M$ for some constant M and for all t, we can use the inequality $|x| \leq 1 + x^2$, to obtain $E\big[|X_t|\big] \leq 1 + M$. In conclusion, if the second moment of a martingale X_t is bounded, the martingale convergence theorem can be applied.

Proof of Proposition 4.1

This proof is similar to the proof of Prop. 3.5 (convergence of deterministic gradient methods with diminishing stepsize) and Prop. 3.7 (convergence of deterministic gradient methods with errors), in Section 3.2.3. However, due to the presence of probabilistic noise, we need to work with suitable conditional expectations of $f(r_t)$. In particular, using a second order Taylor series expansion of f, we will show that $E\big[f(r_{t+1}) \mid \mathcal{F}_t\big] \leq f(r_t) + Z_t$, where Z_t satisfies $\sum_t Z_t < \infty$, and we will use the supermartingale convergence theorem to establish convergence of $f(r_t)$. In order to show that $\nabla f(r_t)$ converges to zero, we argue as in the proof of Prop. 3.5, except that some additional technical work is needed (Lemma 4.1), to bound the effects of the noise during the excursions of $\nabla f(r_t)$ away from zero.

By Assumption 4.2(b), the gradient of the potential function f is Lipschitz continuous and has the property

$$\big\|\nabla f(r) - \nabla f(\overline{r})\big\| \leq L\|r - \overline{r}\|, \qquad \forall\, r, \overline{r},$$

where L is some positive constant. As shown in the beginning of the proof of Prop. 3.4 in Section 3.2.3, we have

$$f(\overline{r}) \leq f(r) + \nabla f(r)'(\overline{r} - r) + \frac{L}{2}\|\overline{r} - r\|^2, \qquad \forall\, r, \overline{r}.$$

We apply this inequality with $r = r_t$ and $\overline{r} = r_{t+1} = r_t + \gamma_t s_t$ and obtain

$$f(r_{t+1}) \le f(r_t) + \gamma_t \nabla f(r_t)' s_t + \frac{L}{2}\gamma_t^2 \|s_t\|^2. \qquad (4.13)$$

We take expectations of both sides of Eq. (4.13), conditioned on $\mathcal{F}_t$, and use Assumption 4.2, to obtain

$$\begin{aligned} E\big[f(r_{t+1}) \mid \mathcal{F}_t\big] & \\ &\le f(r_t) + \gamma_t \nabla f(r_t)' E[s_t \mid \mathcal{F}_t] + \frac{L}{2}\gamma_t^2 \Big(K_1 + K_2 \big\|\nabla f(r_t)\big\|^2\Big) \\ &\le f(r_t) - \gamma_t \Big(c - \frac{LK_2\gamma_t}{2}\Big)\big\|\nabla f(r_t)\big\|^2 + \frac{LK_1\gamma_t^2}{2} \\ &= f(r_t) - X_t + Z_t, \end{aligned} \qquad (4.14)$$

where

$$X_t = \begin{cases} \gamma_t \Big(c - \frac{LK_2\gamma_t}{2}\Big)\big\|\nabla f(r_t)\big\|^2, & \text{if } LK_2\gamma_t \le 2c, \\ 0, & \text{otherwise,} \end{cases}$$

and

$$Z_t = \begin{cases} \frac{LK_1\gamma_t^2}{2}, & \text{if } LK_2\gamma_t \le 2c, \\ \frac{LK_1\gamma_t^2}{2} - \gamma_t \Big(c - \frac{LK_2\gamma_t}{2}\Big)\big\|\nabla f(r_t)\big\|^2, & \text{otherwise.} \end{cases}$$

Note that X_t and Z_t are nonnegative. Furthermore, X_t and Z_t are functions of the random variables in $\mathcal{F}_t$.

Using the assumption $\sum_{t=0}^{\infty} \gamma_t^2 < \infty$, γ_t converges to zero, and there exists some finite time after which $LK_2\gamma_t \le 2c$. It follows that after some finite time, we have $Z_t = LK_1\gamma_t^2/2$, and therefore $\sum_{t=0}^{\infty} Z_t < \infty$. Thus, the supermartingale convergence theorem (Prop. 4.2) applies and shows that $f(r_t)$ converges and also that $\sum_t X_t < \infty$. Since γ_t converges to zero, we have $LK_2\gamma_t \le c$ after some finite time, and

$$X_t = \gamma_t \Big(c - \frac{LK_2\gamma_t}{2}\Big)\big\|\nabla f(r_t)\big\|^2 \ge \frac{c}{2}\gamma_t \big\|\nabla f(r_t)\big\|^2.$$

Hence,

$$\sum_{t=0}^{\infty} \gamma_t \big\|\nabla f(r_t)\big\|^2 < \infty. \qquad (4.15)$$

Suppose that there exists some time t_0 and some $\delta > 0$ such that

$$\big\|\nabla f(r_t)\big\|^2 \ge \delta, \qquad \forall\, t \ge t_0.$$

Since $\sum_{t=0}^{\infty} \gamma_t = \infty$, this would provide a contradiction to Eq. (4.15). We conclude that $\big\|\nabla f(r_t)\big\|$ gets infinitely often arbitrarily close to zero, and

$$\liminf_{t\to\infty} \big\|\nabla f(r_t)\big\| = 0.$$

In order to prove that $\nabla f(r_t)$ converges to zero, we will be studying its excursions away from zero. To this effect, we fix a positive constant ϵ and we say that the time interval $\{t, t+1, \ldots, \bar{t}\}$ is an *upcrossing interval* (from $\epsilon/2$ to ϵ) if

$$\|\nabla f(r_t)\| < \frac{\epsilon}{2}, \qquad \|\nabla f(r_{\bar{t}})\| > \epsilon,$$

and

$$\frac{\epsilon}{2} \leq \|\nabla f(r_\tau)\| \leq \epsilon, \qquad t < \tau < \bar{t}.$$

We will eventually show that any sample path has a finite number of upcrossings from $\epsilon/2$ to ϵ. Towards this goal, we first develop some machinery (cf. Lemma 4.1 below) that will allow us to bound the effects of the noise terms w_t in the course of an upcrossing interval.

Let us define

$$\bar{s}_t = E[s_t \mid \mathcal{F}_t],$$

and

$$w_t = s_t - \bar{s}_t.$$

Using Assumption 4.2(d), we have

$$\|\bar{s}_t\|^2 + E\big[\|w_t\|^2 \mid \mathcal{F}_t\big] = E\big[\|s_t\|^2 \mid \mathcal{F}_t\big] \leq K_1 + K_2\big\|\nabla f(r_t)\big\|^2, \qquad \forall\, t. \tag{4.16}$$

We also define

$$\chi_t = \begin{cases} 1, & \text{if } \|\nabla f(r_t)\| \leq \epsilon, \\ 0, & \text{otherwise,} \end{cases}$$

and note that χ_t is a function of the random variables in $\mathcal{F}_t$.

Lemma 4.1: The sequence u_t defined by

$$u_t = \sum_{\tau=0}^{t-1} \chi_\tau \gamma_\tau w_\tau,$$

converges with probability 1.

Proof: We start by imposing the additional assumption that $\sum_{t=0}^{\infty} \gamma_t^2 \leq A$, where A is some deterministic constant. Later on, we will show how this assumption can be relaxed.

We first observe that u_t is a function of the random variables in $\mathcal{F}_t$, because the summation defining u_t runs up to $\tau = t-1$. We also note that

$$E[\chi_t \gamma_t w_t \mid \mathcal{F}_t] = \chi_t \gamma_t E[w_t \mid \mathcal{F}_t] = 0,$$

and therefore,

$$E[u_{t+1} \mid \mathcal{F}_t] = u_t.$$

If $\chi_t = 0$, then $E\big[\|u_{t+1}\|^2 \mid \mathcal{F}_t\big] = \|u_t\|^2$. If on the other hand, $\chi_t = 1$, we have

$$\begin{aligned} E\big[\|u_{t+1}\|^2 \mid \mathcal{F}_t\big] &= \|u_t\|^2 + 2u_t'\gamma_t E[w_t \mid \mathcal{F}_t] + \gamma_t^2 E\big[\|w_t\|^2 \mid \mathcal{F}_t\big] \\ &\leq \|u_t\|^2 + \gamma_t^2(K_1 + K_2\epsilon^2), \end{aligned}$$

where the last inequality is a consequence of Eq. (4.16). By taking unconditional expectations and summing over t, we obtain

$$E\big[\|u_t\|^2\big] \leq (K_1 + K_2\epsilon^2)E\Big[\sum_{\tau=0}^{\infty} \gamma_\tau^2\Big] \leq (K_1 + K_2\epsilon^2)A, \qquad \forall\, t,$$

and since $\|u_t\| \leq 1 + \|u_t\|^2$, we have

$$\sup_t E\big[\|u_t\|\big] < \infty.$$

We can now apply the martingale convergence theorem (Prop. 4.3) to (the components of) u_t and conclude that u_t converges, with probability 1.

Let us now consider the general case where $\sum_{t=0}^{\infty} \gamma_t^2$ is finite, but not necessarily bounded by a deterministic constant. Given any positive integer k, we consider the stopped process u_t^k which is equal to u_t as long as $\sum_{i=0}^{t-1} \gamma_i^2 \leq k$, and stays constant thereafter. Using the argument that we just gave, it follows that for any k, u_t^k converges with probability 1. Let Ω_k be the set of sample paths for which u_t^k does not converge. By discarding the zero probability set $\cup_{k=1}^{\infty} \Omega_k$, we can assume that u_t^k always converges, for every k. Since $\sum_{t=0}^{\infty} \gamma_t^2$ is finite, for every sample path there exists some k such that $\sum_{t=0}^{\infty} \gamma_t^2 \leq k$. Hence, for every sample path there exists some k such that $u_t = u_t^k$ for all t, and u_t converges. **Q.E.D.**

Let us now consider a sample path with an infinity of upcrossing intervals and let $\{t_k, \ldots, \bar{t}_k\}$ be the kth such interval. Using the definitions of χ_t and of upcrossing intervals, we have $\chi_t = 1$ for $t_k \leq t < \bar{t}_k$ and using Lemma 4.1, we obtain

$$\lim_{k\to\infty} \sum_{t=t_k}^{\bar{t}_k - 1} \gamma_t w_t = 0. \tag{4.17}$$

(If this were not the case, the sum of $\chi_t\gamma_t w_t$ could not converge.) For another useful consequence of Lemma 4.1, we observe that

$$\lim_{k\to\infty} \gamma_{t_k} w_{t_k} = 0. \tag{4.18}$$

We now have

$$\begin{aligned} \big\|\nabla f(r_{t_k+1})\big\| - \big\|\nabla f(r_{t_k})\big\| &\leq \big\|\nabla f(r_{t_k+1}) - \nabla f(r_{t_k})\big\| \\ &\leq L\|r_{t_k+1} - r_{t_k}\| \\ &= \gamma_{t_k} L\|\bar{s}_{t_k} + w_{t_k}\| \\ &\leq \gamma_{t_k} L\|\bar{s}_{t_k}\| + \gamma_{t_k} L\|w_{t_k}\|. \end{aligned} \tag{4.19}$$

Note that the right-hand side converges to zero as k tends to infinity, because $\|\bar{s}_{t_k}\|^2$ is bounded by $K_1 + K_2\epsilon^2$ [cf. Eq. (4.16)], γ_{t_k} goes to zero, and also because of Eq. (4.18). Since $\big\|\nabla f(r_{t_k+1})\big\| \geq \epsilon/2$, Eq. (4.19) implies that for all large enough k, we have

$$\big\|\nabla f(r_{t_k})\big\| \geq \frac{\epsilon}{4}. \tag{4.20}$$

For every k, we also have

$$\begin{aligned} \frac{\epsilon}{2} &\leq \big\|\nabla f(r_{\bar{t}_k})\big\| - \big\|\nabla f(r_{t_k})\big\| \\ &\leq \big\|\nabla f(r_{\bar{t}_k}) - \nabla f(r_{t_k})\big\| \\ &\leq L\|r_{\bar{t}_k} - r_{t_k}\| \\ &\leq L\sum_{t=t_k}^{\bar{t}_k-1} \gamma_t\|\bar{s}_t\| + L\bigg\|\sum_{t=t_k}^{\bar{t}_k-1} \gamma_t w_t\bigg\|. \end{aligned}$$

Note that the second term in the right-hand side converges to zero, by Eq. (4.17). Furthermore, for $t_k \leq t \leq \bar{t}_k - 1$, $\|\bar{s}_t\|^2$ is bounded by $K_1 + K_2\epsilon^2$, which implies that $\|\bar{s}_t\|$ is bounded by $d = 1 + K_1 + K_2\epsilon^2$ [cf. Eq. (4.16)]. Thus, by taking the limit inferior as k tends to infinity, we obtain

$$\liminf_{k\to\infty} \sum_{t=t_k}^{\bar{t}_k-1} \gamma_t \geq \frac{\epsilon}{2Ld}.$$

We now note that for $t_k \leq t < \bar{t}_k$, we have $\big\|\nabla f(r_t)\big\| \geq \epsilon/4$. [For $t = t_k$, this is a consequence of Eq. (4.20).] Thus,

$$\liminf_{k\to\infty} \sum_{t=t_k}^{\bar{t}_k-1} \gamma_t\big\|\nabla f(r_t)\big\|^2 \geq \frac{\epsilon}{2Ld}\cdot\frac{\epsilon^2}{16}.$$

By summing over all upcrossing intervals, we obtain

$$\sum_{t=0}^{\infty} \gamma_t\big\|\nabla f(r_t)\big\|^2 = \infty,$$

which contradicts Eq. (4.15). We conclude that the number of upcrossing intervals is finite.

Given that $\|\nabla f(r_t)\|$ comes infinitely often arbitrarily close to zero and since there are finitely many upcrossings, it follows that $\|\nabla f(r_t)\|$ can exceed ϵ only a finite number of times, and $\limsup_{t\to\infty} \|\nabla f(r_t)\| \le \epsilon$. Since ϵ was arbitrary, it follows that $\limsup_{t\to\infty} \|\nabla f(r_t)\| = 0$, and part (b) of the proposition has been proved. Finally, if r is a limit point of r_t, $\nabla f(r)$ is the limit of some subsequence of $\nabla f(r_t)$ and must be equal to 0, which establishes part (c). **Q.E.D.**

4.3 CONVERGENCE UNDER CONTRACTION OR MONOTONICITY ASSUMPTIONS

The results of the preceding section are very useful in a variety of contexts, but there are some important exceptions. For example, in the next chapter we will study stochastic approximation algorithms that build on the value iteration algorithm $J := TJ$. Unfortunately, it is unclear whether one can define a smooth potential function f such that the update of any component of J is along a descent direction with respect to f. The only properties of T that can be exploited are its monotonicity and, under further assumptions, the fact that it is a contraction with respect to a weighted maximum norm. In the latter case, the weighted maximum norm is a natural potential function and has been used, in a deterministic setting, to establish convergence; on the other hand, it is not differentiable and the results of the preceding section do not apply. In this section, we state and prove some convergence results that do not rely on a smooth potential function but only on certain contraction or monotonicity properties of the algorithm. The proofs are quite technical and are deferred to the end of the section so as not to disrupt the continuity of the presentation.

4.3.1 Algorithmic Model

For the purposes of this section, we assume that the ith component $r(i)$ of r is updated according to

$$r_{t+1}(i) = \big(1-\gamma_t(i)\big)r_t(i)+\gamma_t(i)\big((Hr_t)(i)+w_t(i)\big), \qquad t = 0,1,\ldots, \quad (4.21)$$

[cf. Eq. (4.4)], with the understanding that $\gamma_t(i) = 0$ if $r(i)$ is not updated at time t. Here, $w_t(i)$ is a random noise term. As before, we denote by $\mathcal{F}_t$ the history of the algorithm until time t, which can be defined as

$$\mathcal{F}_t = \big\{r_0(i),\ldots,r_t(i),w_0(i),\ldots,w_{t-1}(i),\gamma_0(i),\ldots,\gamma_t(i),\ i=1,\ldots,n\big\},$$

or may include some additional information.

We now introduce some assumptions on the noise statistics. Part (a) of the assumption below states that $w_t(i)$ has zero conditional mean and part (b) places a bound on its conditional variance.

Assumption 4.3:

(a) For every i and t, we have $E\big[w_t(i) \mid \mathcal{F}_t\big] = 0$.

(b) Given any norm $\|\cdot\|$ on $\Re^n$, there exist constants A and B such that

$$E\big[w_t^2(i) \mid \mathcal{F}_t\big] \leq A + B\|r_t\|^2, \qquad \forall\, i, t.$$

We continue with our assumptions on the iteration mapping H, and the corresponding convergence results. The assumptions we will impose correspond to the properties of the dynamic programming operator T for different dynamic programming problems.

4.3.2 Weighted Maximum Norm Contractions

We start by recalling some definitions from Ch. 2. Given any positive vector ξ, we define the *weighted maximum norm* $\|\cdot\|_\xi$ by

$$\|r\|_\xi = \max_i \frac{|r(i)|}{\xi(i)}.$$

When all components of ξ are equal to 1, the resulting norm is the *maximum norm*, denoted by $\|\cdot\|_\infty$. We say that a function $H : \Re^n \mapsto \Re^n$ is a *weighted maximum norm pseudo-contraction* if there exists some $r^* \in \Re^n$, a positive vector $\xi = \big(\xi(1), \ldots, \xi(n)\big) \in \Re^n$, and a constant $\beta \in [0, 1)$ such that

$$\|Hr - r^*\|_\xi \leq \beta\|r - r^*\|_\xi, \qquad \forall\, r.$$

In more detail, the pseudo-contraction condition can be written as

$$\frac{\big|(Hr)(i) - r^*(i)\big|}{\xi(i)} \leq \beta \max_j \frac{\big|r(j) - r^*(j)\big|}{\xi(j)}, \qquad \forall\, i, r. \tag{4.22}$$

The pseudo-contraction condition implies that r^* is a fixed point of H, that is, $Hr^* = r^*$. It is easily verified that H cannot have other fixed points. To see this, consider a vector r that satisfies $Hr = r$. We then have $\|r - r^*\|_\xi = \|Hr - r^*\|_\xi \leq \beta\|r - r^*\|_\xi$; since $\beta < 1$, we conclude that $\|r - r^*\|_\xi = 0$ and $r = r^*$.

It is well known that if H is a contraction mapping, that is, if it satisfies

$$\|Hr - H\overline{r}\|_\xi \leq \beta \|r - \overline{r}\|_\xi, \qquad \forall\, r, \overline{r},$$

then it is guaranteed to have a fixed point r^* (see e.g., [BeT89], p. 182, or [Lue69], p. 272). By letting $\overline{r} = r^*$, we see that H is automatically a pseudo-contraction.

We now have the following result.

Proposition 4.4: Let r_t be the sequence generated by the iteration (4.21). We assume the following.

(a) The stepsizes $\gamma_t(i)$ are nonnegative and satisfy

$$\sum_{t=0}^{\infty} \gamma_t(i) = \infty, \qquad \sum_{t=0}^{\infty} \gamma_t^2(i) = \infty.$$

(b) The noise terms $w_t(i)$ satisfy Assumption 4.3.

(c) The mapping H is a weighted maximum norm pseudo-contraction.

Then, r_t converges to r^*, with probability 1.

Proposition 4.4 will be used in Ch. 5 to establish the convergence of stochastic approximation methods based on the value iteration $J := TJ$ or related algorithms, for the case of discounted problems or stochastic shortest path problems in which all policies are proper.

We now provide some of the intuition behind Prop. 4.4, by considering a related but much simpler algorithm. In particular, let us assume that $r^* = 0$ (this is no loss of generality) and that H is a pseudo-contraction with respect to the maximum norm $\|\cdot\|_\infty$. Thus,

$$\big|Hr(i)\big| \leq \beta \max_i \big|r(i)\big|, \qquad \forall\, i,\ r.$$

Furthermore, let us assume that there is no noise in the algorithm, namely, $w_t(i) = 0$ for all i, t, and that the stepsizes $\gamma_t(i)$ are either unity [when $r(i)$ is updated] or zero. Suppose that we start with a vector r_0 and that $\big|r_0(i)\big| \leq c$ for some constant c. (We can always take $c = \|r_0\|_\infty$.) In words, r_0 lies within a "cube" of size $2c$ along each dimension, centered at the origin. Suppose that r_t also lies within that cube and that its ith component is updated, to $r_{t+1}(i) = (Hr_t)(i)$. Due to the pseudo-contraction condition, we have $\big|r_{t+1}(i)\big| \leq \beta c < c$, which shows that given the initial cube, any subsequent update of any component keeps us within that cube. Furthermore, any updated component is at most βc in magnitude, which means that when all components have been updated at least once, we find ourselves within a smaller cube, of size $2\beta c$ along each dimension. By continuing similarly, and assuming that each component is updated an infinite number of times, we get into progressively smaller cubes, and convergence to 0 follows. The proof of Prop. 4.4 follows the same lines, except that in the presence of a small stepsize it will take several updates before we get into the next smaller cube; in addition, we need to verify that the presence of the noise terms $w_t(i)$ cannot have an adverse affect on convergence.

4.3.3 Time-Dependent Maps and Additional Noise Terms

We provide here a generalization of Prop. 4.4 that encompasses two additional elements:

(a) We allow the iteration mapping H to change from one iteration to another and hence use the notation H_t. We will assume, however, that all of the mappings H_t are pseudo-contractions under the same weighted maximum norm, with the same fixed point, and with the same contraction factor β.

(b) We allow the algorithm to be driven by some additional noise terms, not necessarily zero mean. We will impose certain assumptions that guarantee that these additional terms are insignificant in the limit.

The algorithm we consider is of the form

$$r_{t+1}(i) = \big(1 - \gamma_t(i)\big) r_t(i) + \gamma_t(i)\big((H_t r_t)(i) + w_t(i) + u_t(i)\big), \qquad t = 0, 1, \ldots. \tag{4.23}$$

Here each H_t is assumed to belong to a family $\mathcal{H}$ of mappings. We assume that the selection at each step of a particular mapping H_t out of the family $\mathcal{H}$ is made without knowledge of the future. Formally, we require that the choice of H_t is only a function of the information contained in the past history $\mathcal{F}_t$ and, in particular, $H_t r_t$ is determined by $\mathcal{F}_t$.

Proposition 4.5: Let r_t be the sequence generated by the iteration (4.23). We assume the following.

(a) The stepsizes $\gamma_t(i)$ are nonnegative and satisfy

$$\sum_{t=0}^{\infty} \gamma_t(i) = \infty, \qquad \sum_{t=0}^{\infty} \gamma_t^2(i) < \infty.$$

(b) The noise terms $w_t(i)$ satisfy Assumption 4.3.

(c) There exists a vector r^*, a positive vector ξ, and a scalar $\beta \in [0, 1)$, such that

$$\|H_t r_t - r^*\|_\xi \leq \beta \|r_t - r^*\|_\xi, \qquad \forall\, t.$$

(d) There exists a nonnegative random sequence θ_t that converges to zero with probability 1, and is such that

$$|u_t(i)| \leq \theta_t\big(\|r_t\|_\xi + 1\big), \qquad \forall\, i, t.$$

Then, r_t converges to r^* with probability 1.

4.3.4 Convergence under Monotonicity Assumptions

For many stochastic shortest path problems, the dynamic programming operator T fails to be a weighted maximum norm pseudo-contraction. In such cases, convergence proofs for the deterministic value iteration algorithm $r := Tr$, as well as for its stochastic counterparts, need to rely on other properties of T, such as those captured by our next assumption.

Assumption 4.4:

(a) The mapping H is monotone; that is, if $r \leq \bar{r}$, then $Hr \leq H\bar{r}$.

(b) There exists a unique vector r^* satisfying $Hr^* = r^*$.

(c) If $e \in \Re^n$ is the vector with all components equal to 1, and if η is a positive scalar, then

$$Hr - \eta e \leq H(r - \eta e) \leq H(r + \eta e) \leq Hr + \eta e.$$

Parts (a) and (c) of Assumption 4.4 imply that the mapping H is continuous; to see this, note that for any $r, \bar{r} \in \Re^n$, we have $\bar{r} \leq r + \|\bar{r} - r\|_\infty e$, which implies that

$$H\bar{r} \leq H\big(r + \|\bar{r} - r\|_\infty e\big) \leq Hr + \|\bar{r} - r\|_\infty e.$$

A similar argument yields

$$Hr \leq H\bar{r} + \|\bar{r} - r\|_\infty e.$$

Therefore,

$$\|H\bar{r} - Hr\|_\infty \leq \|\bar{r} - r\|_\infty,$$

and H is continuous. Note that the last inequality is not strict, in general, and for this reason, H is not necessarily a maximum norm contraction.

Assumption 4.4 leads to a convergence result, which is somewhat weaker than our earlier result for weighted maximum norm pseudo-contractions; in particular, a separate boundedness assumption is imposed. It is not known whether the same result can be proved without this boundedness assumption.

Proposition 4.6: Let r_t be the sequence generated by the iteration (4.21). We assume the following:

(a) The stepsizes $\gamma_t(i)$ are nonnegative and satisfy

$$\sum_{t=0}^{\infty} \gamma_t(i) = \infty, \qquad \sum_{t=0}^{\infty} \gamma_t^2(i) < \infty.$$

(b) The noise terms $w_t(i)$ satisfy Assumption 4.3.

(c) The mapping H satisfies Assumption 4.4.

If the sequence r_t is bounded with probability 1, then r_t converges to r^* with probability 1.

4.3.5 Boundedness

Verifying the boundedness condition in Prop. 4.6 is often difficult and may require some special arguments tailored to the application at hand. We provide below a criterion which is sometimes useful.

Proposition 4.7: Let r_t be the sequence generated by the iteration (4.23). We assume the following:

(a) The stepsizes $\gamma_t(i)$ are nonnegative and satisfy

$$\sum_{t=0}^{\infty} \gamma_t(i) = \infty, \qquad \sum_{t=0}^{\infty} \gamma_t^2(i) < \infty.$$

(b) The noise terms $w_t(i)$ satisfy Assumption 4.3.

(c) There exists a positive vector ξ, a scalar $\beta \in [0,1)$, and a positive scalar D such that

$$\|H_t r_t\|_\xi \leq \beta \|r_t\|_\xi + D, \qquad \forall\ t.$$

(d) There exists a nonnegative random sequence θ_t that converges to zero with probability 1, and is such that

$$|u_t(i)| \leq \theta_t\big(\|r_t\|_\xi + 1\big), \qquad \forall\ i, t.$$

Then, the sequence r_t is bounded with probability 1.

Note that whenever H_t is a pseudo-contraction mapping with respect to a weighted maximum norm, the condition $\|H_t r_t\|_\xi \leq \beta\|r_t\|_\xi + D$ is automatically satisfied because

$$\begin{aligned} \|H_t r_t\|_\xi &\leq \|H_t r_t - r^*\|_\xi + \|r^*\|_\xi \\ &\leq \beta\|r_t - r^*\|_\xi + \|r^*\|_\xi \\ &\leq \beta\|r_t\|_\xi + (1+\beta)\|r^*\|_\xi. \end{aligned}$$

Therefore, in this case boundedness is guaranteed, consistently with Props. 4.4-4.5. Unfortunately, however, Prop. 4.7 does not always apply when H only satisfies the monotonicity Assumption 4.4.

We now discuss a general purpose approach for guaranteeing boundedness of stochastic approximation algorithms. The idea is to modify the algorithm so that r_t gets projected to a bounded set whenever r_t tends to become too large. Of course, the bounded set must be chosen large enough so that the desired fixed point r^* lies in that set. (This requires some prior knowledge on the structure of the problem at hand, but such prior knowledge is often available.) Once projections are introduced, the algorithm has been modified, and in order to establish convergence one needs to verify that the modified algorithm satisfies all required assumptions. For this reason, some additional technical work is usually required to complete a convergence proof. Fortunately, this is often feasible.

A mathematical description of one possible version of the projection method is as follows. The method uses two bounded sets, denoted by S and S_0, with $S_0 \subset S$. At any time t such that $r_t \notin S$, the next update is governed by

$$r_{t+1} = \pi(r_t) \in S_0,$$

instead of the original update equation. Here, the "projection" π is an arbitrary mapping from $\Re^n$ into S_0.

The convergence of stochastic approximation algorithms is usually established by showing that the algorithm tends to make some progress at each step, assuming that a suitable measure of progress has been defined. In order to establish convergence of the projection method, we should choose the sets S and S_0 so that whenever a projection is effected, progress is guaranteed. For example, if a potential function f is used, we should choose S and S_0 so that $f(r) < f(\overline{r})$ for every $r \in S_0$ and every $\overline{r} \notin S$.

4.3.6 Convergence Proofs

In this subsection we prove all of the results that have been stated in this section, except for Prop. 4.4, which is a special case of Prop. 4.5. Our first step is to develop a corollary of Prop. 4.1 from the previous section that will be needed in the proofs to follow.

As in Example 4.3, let r_t be a scalar and consider the recursion

$$r_{t+1} = (1 - \gamma_t)r_t + \gamma_t w_t. \tag{4.24}$$

We assume that the stepsizes satisfy $\gamma_t \geq 0$, $\sum_{t=0}^{\infty} \gamma_t = \infty$, and $\sum_{t=0}^{\infty} \gamma_t^2 < \infty$, and that the noise term w_t satisfies

$$E[w_t \mid \mathcal{F}_t] = 0$$

and

$$E[w_t^2 \mid \mathcal{F}_t] \leq A_t,$$

where A_t is a random variable which is a function of the random variables in $\mathcal{F}_t$. The only difference here from Example 4.3 and Assumption 4.2(d) is that the conditional variance of w_t is now bounded by a random variable A_t.

Corollary 4.1: Under the above assumptions, and if the sequence A_t is bounded with probability 1, the sequence r_t generated by Eq. (4.24) converges to zero, with probability 1.

Proof: We see that all of the assumptions of Prop. 4.1 are satisfied [with $f(r) = r^2$]. except that the deterministic constant K_1 in Assumption 4.2(d) is replaced by the random but bounded sequence A_t. With this substitution, convergence is established by repeating the proof of Prop. 4.1; in particular, we use the boundedness of A_t to ensure that $\sum_{t=0}^{\infty} Z_t < \infty$, and the rest of the proof remains unchanged. **Q.E.D.**

We continue with a proof of the boundedness result (Prop. 4.7), which will be needed for the proof of the weighted maximum norm pseudo-contraction convergence results (Prop. 4.5).

Proof of Proposition 4.7

Before developing a formal proof, we summarize the basic ideas. We first show that without loss of generality, we can restrict attention to the case where $\|\cdot\|_\xi$ is the maximum norm, to be denoted simply by $\|\cdot\|$. We then enclose r_0 in a "cube" of the form $\{r \mid \|r\| \leq G\}$, where G is chosen large enough so that $\|H_t r_t\| \leq G$, whenever $\|r_t\| \leq G$. In the absence of the noise terms u_t and w_t, this would imply that r_t stays forever in that cube. On the other hand, in the presence of noise, r_t may wander outside the cube. If it wanders far enough, so that $\|r_t\| > (1+\epsilon)G$ (here ϵ is a positive constant), we reset G and the corresponding cube so as to again enclose the current iterate r_t, and repeat our arguments. We use our assumptions to bound the effects of the noise terms (Lemma 4.2) and show that a reset of the bound G will take place only a finite number of times (Lemma 4.3), implying boundedness of r_t.

We now start the proof by showing that the case of a general positive weighting vector ξ can be reduced to the special case where $\xi = (1, \ldots, 1)$, by a suitable coordinate scaling. We define a new vector $\tilde{r}$ by

$$\tilde{r} = \Xi^{-1} r,$$

where Ξ is a diagonal matrix whose ith diagonal entry is equal to $\xi(i)$; in particular,

$$\tilde{r}(i) = \frac{r(i)}{\xi(i)}, \qquad i = 1, \ldots, n.$$

We then define a new mapping $\tilde{H}_t$ by

$$\tilde{H}_t \tilde{r} = \Xi^{-1} H_t(\Xi \tilde{r}),$$

and note that

$$\xi(i)(\tilde{H}_t \tilde{r})(i) = (H_t r)(i), \qquad i = 1, \ldots, n.$$

We use this relation in Eq. (4.23), and we also replace $r_t(i)$ by $\xi(i)\tilde{r}_t(i)$. We then divide by $\xi(i)$, to obtain

$$\tilde{r}_{t+1}(i) = \big(1 - \gamma_t(i)\big)\tilde{r}_t(i) + \gamma_t(i)\left((\tilde{H}_t \tilde{r}_t)(i) + \frac{w_t(i)}{\xi(i)} + \frac{u_t(i)}{\xi(i)}\right),$$

which is an iteration of the same form except that H_t has been replaced by $\tilde{H}_t$. Using assumption (c) in the statement of the proposition, we obtain

$$\big|(\tilde{H}_t \tilde{r})(i)\big| = \frac{\big|H_t r(i)\big|}{\xi(i)} \leq \beta \|r\|_\xi + D = \beta \|\tilde{r}\|_\infty + D,$$

where $\|\cdot\|_\infty$ is the maximum norm. We conclude that with this change of variables we have come to a situation where the norm $\|\cdot\|_\xi$ has been replaced by the maximum norm $\|\cdot\|_\infty$.

For the purposes of this proof, let us use the simpler notation $\|\cdot\|$ to denote the maximum norm. Given the preceding discussion, we can assume, without loss of generality, that there exist some $\beta \in [0, 1)$ and some $D > 0$ such that

$$\|H_t r_t\| \leq \beta \|r_t\| + D, \qquad \forall\, t. \tag{4.25}$$

Let us choose some $G \geq 1$ such that $\beta G + D < G$ and let η satisfy $\beta G + D = \eta G$. Note that $\beta < \eta < 1$. Let us also choose $\epsilon > 0$ so that

$$(1 + \epsilon)\eta = 1.$$

Our next step is to define some scaling factors that will be used to renormalize the noise terms so as to keep them bounded. We define a sequence G_t, recursively, as follows. We start with $G_0 = \max\{\|r_0\|, G\}$. Assuming that G_t has already been defined, let $G_{t+1} = G_t$ if $\|r_{t+1}\| \leq (1+\epsilon)G_t$. If $\|r_{t+1}\| > (1+\epsilon)G_t$, then let $G_{t+1} = G_0(1+\epsilon)^k$ where k is chosen so that

$$G_0(1+\epsilon)^{k-1} < \|r_{t+1}\| \leq G_0(1+\epsilon)^k = G_{t+1}.$$

What this definition has accomplished is that we have constructed a nondecreasing sequence $G_t \geq 1$ such that

$$\|r_t\| \leq (1+\epsilon)G_t, \qquad \forall\, t \geq 0, \tag{4.26}$$

and

$$\|r_t\| \leq G_t, \qquad \text{if } G_{t-1} < G_t. \tag{4.27}$$

We finally note that $\|r_t\|$ and G_t are functions of the random variables in $\mathcal{F}_t$ since they are completely determined by $r_0, r_1, \ldots, r_t$.

We observe that $\beta\epsilon + \eta < \eta\epsilon + \eta = 1$. Let θ^* be a small enough positive constant such that

$$\beta\epsilon + \eta + \theta^*(2 + \epsilon) \leq 1. \tag{4.28}$$

Finally, let t^* be a time such that

$$\theta_t \leq \theta^*, \qquad \forall\, t \geq t^*;$$

such a time exists since θ_t has been assumed to converge to zero.

We will now prove that

$$\|H_t r_t\| + \theta_t\big(\|r_t\| + 1\big) \leq G_t, \qquad \forall\, t \geq t^*. \tag{4.29}$$

Indeed, Eqs. (4.25), (4.26), and the definition $\beta G + D = \eta G$ of η yield

$$\begin{aligned}\|H_t r_t\| &\le \beta\|r_t\| + D \\ &\le \beta(1+\epsilon)G_t + D \\ &= \beta(1+\epsilon)G_t + (\eta - \beta)G \\ &\le \big(\beta(1+\epsilon) + \eta - \beta\big)G_t \\ &= (\beta\epsilon + \eta)G_t,\end{aligned}$$

where we have also used the property $G_t \ge G$. Hence, for $t \ge t^*$, we have

$$\begin{aligned}\|H_t r_t\| + \theta_t\big(\|r_t\| + 1\big) &\le (\beta\epsilon + \eta)G_t + \theta^*\big((1+\epsilon)G_t + 1\big) \\ &\le (\beta\epsilon + \eta)G_t + \theta^*(2+\epsilon)G_t \\ &\le G_t,\end{aligned}$$

as desired. [We have used here the property $G_t \ge 1$, Eq. (4.26), and Eq. (4.28).]

In order to motivate our next step, note that as long as there is a possibility that r_t is unbounded, $E\big[w_t^2(i) \mid \mathcal{F}_t\big]$ could also be unbounded [cf. Assumption 4.3(b)] and results such as Corollary 4.1 are inapplicable. To circumvent this difficulty, we will work with a suitably scaled version of $w_t(i)$ whose conditional variance is bounded.

We define

$$\tilde{w}_t(i) = \frac{w_t(i)}{G_t}, \qquad \forall\, t \ge 0.$$

Assumption 4.3 and the fact that G_t is a function of the random variables in $\mathcal{F}_t$ imply that

$$E\big[\tilde{w}_t(i) \mid \mathcal{F}_t\big] = \frac{E\big[w_t(i) \mid \mathcal{F}_t\big]}{G_t} = 0, \qquad \forall\, t \ge 0,$$

and

$$\begin{aligned}E\big[\tilde{w}_t^2(i) \mid \mathcal{F}_t\big] &= \frac{E\big[w_t^2(i) \mid \mathcal{F}_t\big]}{G_t^2} \\ &\le \frac{A + B\|r_t\|^2}{G_t^2} \\ &\le \frac{A + B(1+\epsilon)^2 G_t^2}{G_t^2} \\ &\le K, \qquad\qquad \forall\, t \ge 0,\end{aligned}$$

where K is a deterministic constant. [Here we made use of Eq. (4.26) to bound $\|r_t\|$, as well as the property $G_t \ge 1$.]

For any i and $t_0 \ge 0$, we define $\tilde{W}_{t_0;t_0}(i) = 0$ and

$$\tilde{W}_{t+1;t_0}(i) = \big(1 - \gamma_t(i)\big)\tilde{W}_{t;t_0}(i) + \gamma_t(i)\tilde{w}_t(i), \qquad t \ge t_0.$$

Lemma 4.2: For every $\delta > 0$, and with probability 1, there exists some t_0 such that $|\tilde{W}_{t;t_0}(i)| \leq \delta$, for all $t \geq t_0$.

Proof: From the convergence result in Example 4.3, we obtain

$$\lim_{t\to\infty} \tilde{W}_{t;0}(i) = 0.$$

For every $t \geq t_0$, we have, using linearity,

$$\tilde{W}_{t;0}(i) = \left[\prod_{\tau=t_0}^{t-1} \big(1 - \gamma_\tau(i)\big)\right] \tilde{W}_{t_0;0}(i) + \tilde{W}_{t;t_0}(i),$$

which implies that

$$\big|\tilde{W}_{t;t_0}(i)\big| \leq \big|\tilde{W}_{t;0}(i)\big| + \big|\tilde{W}_{t_0;0}(i)\big|.$$

The result follows by letting t_0 be large enough so that $|\tilde{W}_{t;0}(i)| \leq \delta/2$ for every $t \geq t_0$. **Q.E.D.**

We now assume that r_t is unbounded, in order to derive a contradiction. Then, Eq. (4.26) implies that G_t converges to infinity, and Eq. (4.27) implies that the inequality $\|r_t\| \leq G_t$ holds for infinitely many values of t. Using also Lemma 4.2, we conclude that there exists some t_0 such that $\|r_{t_0}\| \leq G_{t_0}$ and

$$\big|\tilde{W}_{t;t_0}(i)\big| \leq \epsilon, \qquad \forall\, t \geq t_0,\ \forall\, i. \tag{4.30}$$

Furthermore, by taking t_0 large enough we can also assume that

$$\gamma_t(i) \leq 1, \qquad \theta_t \leq \theta^*, \qquad \forall\, t \geq t_0,\ i. \tag{4.31}$$

The lemma that follows derives a contradiction to the unboundedness of G_t and concludes the proof of the proposition.

Lemma 4.3: Suppose that there exists some t_0 such that $\|r_{t_0}\| \leq G_{t_0}$ and such that Eqs. (4.30) and (4.31) hold. Then, for every $t \geq t_0$, we have $G_t = G_{t_0}$. Furthermore, for every i we have

$$\begin{aligned} -G_{t_0}(1+\epsilon) \leq -G_{t_0} + \tilde{W}_{t;t_0}(i)G_{t_0} &\leq r_t(i) \\ &\leq G_{t_0} + \tilde{W}_{t;t_0}(i)G_{t_0} \leq G_{t_0}(1+\epsilon). \end{aligned}$$

Proof: The proof proceeds by induction on t. For $t = t_0$, the result is obvious from $\|r_{t_0}\| \le G_{t_0}$ and $\tilde{W}_{t_0;t_0}(i) = 0$. Suppose that the result is true for some t and, in particular, $G_t = G_{t_0}$. We then use the induction hypothesis and Eq. (4.29) to obtain

$$\begin{aligned}
r_{t+1}(i) &= \big(1-\gamma_t(i)\big)r_t(i) + \gamma_t(i)(H_t r_t)(i) + \gamma_t(i)w_t(i) + \gamma_t(i)u_t(i) \\
&\le \big(1-\gamma_t(i)\big)\big(G_{t_0} + \tilde{W}_{t;t_0}(i)G_{t_0}\big) + \gamma_t(i)(H_t r_t)(i) + \gamma_t(i)\tilde{w}_t(i)G_{t_0} \\
&\qquad + \gamma_t(i)\theta_t\big(\|r_t\| + 1\big) \\
&\le \big(1-\gamma_t(i)\big)\big(G_{t_0} + \tilde{W}_{t;t_0}(i)G_{t_0}\big) + \gamma_t(i)G_{t_0} + \gamma_t(i)\tilde{w}_t(i)G_{t_0} \\
&= G_{t_0} + \Big(\big(1-\gamma_t(i)\big)\tilde{W}_{t;t_0}(i) + \gamma_t(i)\tilde{w}_t(i)\Big)G_{t_0} \\
&= G_{t_0} + \tilde{W}_{t+1;t_0}(i)G_{t_0}.
\end{aligned}$$

A symmetrical argument also yields $-G_{t_0} + \tilde{W}_{t+1;t_0}(i)G_{t_0} \le r_{t+1}(i)$. Using Eq. (4.30), we obtain $|r_{t+1}(i)| \le G_{t_0}(1+\epsilon)$, which also implies that $G_{t+1} = G_{t_0}$. **Q.E.D.**

Proof of Proposition 4.5

This proof follows the general argument outlined at the end of Section 4.3.2. According to that argument, and in the absence of noise, the iterates are guaranteed to eventually enter and stay inside progressively smaller cubes. Thus, we only need to show that the effects of the noise are small and cannot alter this type of behavior.

Without loss of generality, we assume that $r^* = 0$; this can be always accomplished by translating the origin of the coordinate system. Furthermore, as in the proof of Prop. 4.7, we assume that all components of the vector ξ are equal to 1 and use the notation $\|r\|$ instead of $\|r\|_\xi$. Notice that Prop. 4.7 applies and establishes that the sequence r_t is bounded.

Proposition 4.7 states that there exists some (generally random) D_0 such that $\|r_t\| \le D_0$, for all t. Fix some $\epsilon > 0$ such that $\beta + 2\epsilon < 1$. We define

$$D_{k+1} = (\beta + 2\epsilon)D_k, \qquad k \ge 0.$$

Clearly, D_k converges to zero.

To set up a proof by induction, suppose that there exists some time t_k such that

$$\|r_t\| \le D_k, \qquad \forall\, t \ge t_k.$$

Since $\gamma_t(i)$ converges to zero, we can also assume that $\gamma_t(i) \le 1$ for all i and all $t \ge t_k$. We will show that this implies that there exists some time t_{k+1} such that $\|r_t\| \le D_{k+1}$ for all $t \ge t_{k+1}$. This will complete the proof of convergence of r_t to zero.

Let $W_0(i) = 0$ and

$$W_{t+1}(i) = \big(1 - \gamma_t(i)\big)W_t(i) + \gamma_t(i)w_t(i). \tag{4.32}$$

Since r_t is bounded, so is the conditional variance of $w_t(i)$. Hence, Corollary 4.1 applies and we have $\lim_{t\to\infty} W_t(i) = 0$. For any time τ, we also define $W_{\tau;\tau}(i) = 0$ and

$$W_{t+1;\tau}(i) = \big(1 - \gamma_t(i)\big)W_{t;\tau}(i) + \gamma_t(i)w_t(i), \qquad t \geq \tau. \tag{4.33}$$

Using the same argument, Corollary 4.1 also implies that $\lim_{t\to\infty} W_{t;\tau}(i) = 0$ for all i and τ.

Note that

$$|u_t(i)| \leq \theta_t\big(\|r_t\| + 1\big) \leq \theta_t(D_k + 1), \qquad t \geq t_k.$$

Since θ_t converges to zero, so does $|u_t(i)|$. Let $\tau_k \geq t_k$ be such that $|u_t(i)| \leq \epsilon D_k$ for all $t \geq \tau_k$ and all i. We define $Y_{\tau_k}(i) = D_k$ and

$$Y_{t+1}(i) = \big(1 - \gamma_t(i)\big)Y_t(i) + \gamma_t(i)\beta D_k + \gamma_t(i)\epsilon D_k, \qquad t \geq \tau_k. \tag{4.34}$$

Lemma 4.4: For every i, we have

$$-Y_t(i) + W_{t;\tau_k}(i) \leq r_t(i) \leq Y_t(i) + W_{t;\tau_k}(i), \qquad \forall\, t \geq \tau_k. \tag{4.35}$$

Proof: We use induction on t. Since $Y_{\tau_k}(i) = D_k$ and $W_{\tau_k;\tau_k}(i) = 0$, the result is true for $t = \tau_k$. Suppose that Eq. (4.35) holds for some $t \geq \tau_k$. Note that $\big|(H_t r_t)(i)\big| \leq \beta\|r_t\| \leq \beta D_k$. We then have

$$\begin{aligned} r_{t+1}(i) &\leq \big(1 - \gamma_t(i)\big)\big(Y_t(i) + W_{t;\tau_k}(i)\big) + \gamma_t(i)(H_t r_t)(i) \\ &\qquad + \gamma_t(i)w_t(i) + \gamma_t(i)u_t(i) \\ &\leq \big(1 - \gamma_t(i)\big)\big(Y_t(i) + W_{t;\tau_k}(i)\big) + \gamma_t(i)\beta D_k + \gamma_t(i)w_t(i) + \gamma_t(i)\epsilon D_k \\ &= Y_{t+1}(i) + W_{t+1;\tau_k}(i). \end{aligned}$$

A symmetrical argument yields $-Y_{t+1}(i) + W_{t+1;\tau_k}(i) \leq r_{t+1}(i)$ and the inductive proof is complete. **Q.E.D.**

It is evident from Eq. (4.34) and the assumption $\sum_{t=0}^{\infty} \gamma_t(i) = \infty$ that $Y_t(i)$ converges to $\beta D_k + \epsilon D_k$ as $t \to \infty$. This fact, together with Eq. (4.35), and the fact that $\lim_{t\to\infty} W_{t;\tau_k}(i) = 0$, yield

$$\limsup_{t\to\infty} \|r_t\| \leq (\beta + \epsilon)D_k < D_{k+1}.$$

Therefore, there exists some time t_{k+1} such that $\|r_t\| \leq D_{k+1}$ for all $t \geq t_{k+1}$ and the induction is complete.

Proof of Proposition 4.6

At a high level, this proof is similar to the proof of Prop. 4.5, where we showed that r_t is guaranteed to enter and stay inside progressively smaller cubes. However, because we do not have a contraction property anymore, instead of working with cubes, we need to work with rectangular sets of the form $\{r \mid L^k \leq r \leq U^k\}$. The vectors L^k and U^k are constructed similar to the vectors J_k^- and J_k^+ used in the proof of asynchronous value iteration convergence (Prop. 2.3 in Section 2.2.2), and the entire proof can be viewed as an adaptation of that argument to account for the presence of noise.

Recall that e stands for the vector with all components equal to 1. Let η be a large enough positive scalar so that

$$r^* - \eta e \leq r_t \leq r^* + \eta e,$$

for all t. (Such a scalar exists by the boundedness assumption on r_t but is a random variable because $\sup_t \|r_t\|_\infty$ could be different for different sample paths.) Let

$$L^0 = \big(L^0(1), \ldots, L^0(n)\big) = r^* - \eta e$$

and

$$U^0 = \big(U^0(1), \ldots, U^0(n)\big) = r^* + \eta e.$$

Let us define two sequences U^k and L^k in terms of the recursions

$$U^{k+1} = \frac{U^k + HU^k}{2}, \qquad k \geq 0, \tag{4.36}$$

and

$$L^{k+1} = \frac{L^k + HL^k}{2}, \qquad k \geq 0. \tag{4.37}$$

Lemma 4.5: For every $k \geq 0$, we have

$$HU^k \leq U^{k+1} \leq U^k,$$

and

$$HL^k \geq L^{k+1} \geq L^k.$$

Proof: The proof is by induction on k. Notice that, by Assumption 4.4(c) and the fixed point property of r^*, we have $HU^0 = H(r^* + \eta e) \leq Hr^* + \eta e = r^* + \eta e = U^0$. Since U^1 is the average of U^0 and HU^0, we obtain $HU^0 \leq U^1 \leq U^0$. Suppose that the result is true for some k. The inequality $U^{k+1} \leq U^k$ and the monotonicity of H yield $HU^{k+1} \leq HU^k$. Equation

(4.36) then implies that $U^{k+2} \leq U^{k+1}$. Furthermore, since $HU^{k+1} \leq HU^k \leq U^{k+1}$ and since U^{k+2} is the average of HU^{k+1} and U^{k+1}, we also obtain $HU^{k+1} \leq U^{k+2}$. The inequalities for L^k follow by a symmetrical argument. **Q.E.D.**

Lemma 4.6: The sequences U^k and L^k converge to r^*.

Proof: We first prove, by induction, that $U^k \geq r^*$ for all k. This is true for U^0, by definition. Suppose that $U^k \geq r^*$. Then, by monotonicity, $HU^k \geq Hr^* = r^*$, from which the inequality $U^{k+1} \geq r^*$ follows. Therefore, the sequence U^k is bounded below. Since this sequence is monotonic (Lemma 4.5), it converges to some limit U. Using the continuity of H, we must have $U = (U + HU)/2$, which implies that $U = HU$. Since r^* was assumed to be the unique fixed point of H [Assumption 4.4(b)], it follows that $U = r^*$. Convergence of L^k to r^* follows from a symmetrical argument. **Q.E.D.**

We will now show that for every k, there exists some time t_k such that

$$L^k \leq r_t \leq U^k, \qquad \forall\, t \geq t_k. \tag{4.38}$$

(The value of t_k will not be the same for different sample paths and is therefore a random variable.) Once this is proved, the convergence of r_t to r^* follows from Lemma 4.6. For $k = 0$, Eq. (4.38) is certainly true, with $t_0 = 0$, because of the way that U^0 and L^0 were defined. We continue by induction on k. We fix some k and assume that there exists some t_k so that Eq. (4.38) holds. Furthermore, since $\gamma_t(i)$ converges to zero, we can assume that $\gamma_t(i) \leq 1$ for all i and all $t \geq t_k$.

Let $W_0(i) = 0$ and

$$W_{t+1}(i) = \big(1 - \gamma_t(i)\big)W_t(i) + \gamma_t(i)w_t(i).$$

Note that the bound $A + B\|r_t\|^2$ on the conditional variance of w_t (cf. Assumption 4.3) is a random variable. Since the sequence r_t has been assumed bounded, Corollary 4.1 applies and we obtain $\lim_{t\to\infty} W_t(i) = 0$. For any time τ, we also define $W_{\tau;\tau}(i) = 0$ and

$$W_{t+1;\tau}(i) = \big(1 - \gamma_t(i)\big)W_{t;\tau}(i) + \gamma_t(i)w_t(i), \qquad t \geq \tau. \tag{4.39}$$

Using Corollary 4.1 once more, we obtain $\lim_{t\to\infty} W_{t;\tau}(i) = 0$, for all τ.

We also define a sequence $X_t(i)$, $t \geq t_k$, by letting $X_{t_k}(i) = U^k(i)$ and

$$X_{t+1}(i) = \big(1 - \gamma_t(i)\big)X_t(i) + \gamma_t(i)(HU^k)(i), \qquad t \geq t_k. \tag{4.40}$$

Lemma 4.7: We have $r_t(i) \le X_t(i) + W_{t;t_k}(i)$, for all i and $t \ge t_k$.

Proof: The proof proceeds by induction on t. For $t = t_k$, the induction hypothesis (4.38) yields $r_{t_k}(i) \le U^k(i)$ and, by definition, we have $U^k(i) = X_{t_k}(i) + W_{t_k;t_k}(i)$. Suppose that the result is true for some t. Then, Eqs. (4.21), (4.38), (4.40), and (4.39) imply that

$$\begin{aligned} r_{t+1}(i) &= \big(1 - \gamma_t(i)\big) r_t(i) + \gamma_t(i)(Hr_t)(i) + \gamma_t(i) w_t(i) \\ &\le \big(1 - \gamma_t(i)\big)\big(X_t(i) + W_{t;t_k}(i)\big) + \gamma_t(i)(HU^k)(i) + \gamma_t(i) w_t(i) \\ &= X_{t+1}(i) + W_{t+1;t_k}(i). \end{aligned}$$

Q.E.D.

Let δ_k be equal to the minimum of $\big(U^k(i) - (HU^k)(i)\big)/4$, where the minimum is taken over all i for which $U^k(i) - (HU^k)(i)$ is positive. Since $HU^k \le U^k$, we see that δ_k is well-defined and positive unless $U^k = HU^k$. But in the latter case, we must have $U^k = r^* = U^{k+1}$, the inequality $r_t \le U^k$ implies that $r_t \le U^{k+1}$, and there is nothing more to be proved. We therefore assume that δ_k is well-defined and positive.

Let t'_k be such that $t'_k \ge t_k$,

$$\prod_{\tau = t_k}^{t'_k - 1} \big(1 - \gamma_\tau(i)\big) \le \frac{1}{4},$$

and

$$W_{t;t_k}(i) \le \delta_k,$$

for all $t \ge t'_k$ and all i. Such a t'_k exists because Assumption 4.1(a) implies that

$$\prod_{\tau = t_k}^{\infty} \big(1 - \gamma_\tau(i)\big) = 0$$

(this can be checked by taking logarithms and using a first order Taylor series expansion, or by appealing to Lemma 3.3), and because $W_{t;t_k}(i)$ converges to zero, as discussed earlier.

Lemma 4.8: We have $r_t(i) \le U^{k+1}(i)$, for all i and $t \ge t'_k$.

Proof: Fix some i. If $U^{k+1}(i) = U^k(i)$, the inequality $r_t(i) \le U^{k+1}(i)$ follows from the induction hypothesis (4.38). We therefore concentrate on the case where $U^{k+1}(i) < U^k(i)$. Equation (4.40) and the relation

$X_{t_k}(i) = U^k(i)$ imply that $X_t(i)$ is a convex combination of $U^k(i)$ and $(HU^k)(i)$, of the form

$$X_t(i) = \alpha_t(i)U^k(i) + \big(1 - \alpha_t(i)\big)(HU^k)(i),$$

where

$$\alpha_t(i) = \prod_{\tau=t_k}^{t-1} \big(1 - \gamma_\tau(i)\big) \leq \frac{1}{4},$$

for $t \geq t'_k$. Because $HU^k \leq U^k$, we obtain

$$\begin{aligned} X_t(i) &\leq \frac{1}{4}U^k(i) + \frac{3}{4}(HU^k)(i) \\ &= \frac{1}{2}U^k(i) + \frac{1}{2}(HU^k)(i) - \frac{1}{4}\big(U^k(i) - (HU^k)(i)\big) \\ &\leq U^{k+1}(i) - \delta_k. \end{aligned}$$

This inequality, together with the inequality $W_{t;t_k}(i) \leq \delta_k$ and Lemma 4.7, imply that $r_t(i) \leq U^{k+1}(i)$ for all $t \geq t'_k$. **Q.E.D.**

By an entirely symmetrical argument, we can also establish that $r_t(i) \geq L^{k+1}(i)$ for all t greater than some t''_k. By letting $t_{k+1} = \max\{t'_k, t''_k\}$, we see that we have proved Eq. (4.38), with k replaced by $k+1$, which concludes the induction, and completes the proof of the proposition.

4.4 THE ODE APPROACH

We discuss here another approach for studying the convergence of stochastic approximation algorithms. This method has become known as the ODE (ordinary differential equation) approach, because it leads to a deterministic differential equation that captures the aggregate behavior of the iterates r_t. This ODE approach can be developed in full rigor, but it is often used heuristically (without elaborating on technical conditions) in order to obtain a quick preliminary understanding of a given algorithm. In this section, we start along this heuristic avenue, and then conclude with a formal result.

Consider a stochastic approximation method of the form

$$r_{t+1} = (1 - \gamma_t)r_t + \gamma_t H r_t + \gamma_t w_t,$$

where γ_t is a nonnegative deterministic stepsize parameter. We assume that $\sum_t \gamma_t = \infty$ and $\sum_t \gamma_t^2 < \infty$. Let us define γ to be a small positive constant and let us choose an increasing sequence k_m of integers such that

$$\sum_{t=k_m}^{k_{m+1}-1} \gamma_t \approx \gamma.$$

We will monitor the progress of the algorithm by using a different "clock" under which the interval $[k_m, k_{m+1}-1]$ is treated as a unit interval.

Let us assume that the infinite sum $\sum_{t=0}^{\infty} \gamma_t w_t$ is convergent with probability 1. This assumption can be shown to hold if, for example, $E[w_t \mid \mathcal{F}_t] = 0$ and $E\big[\|w_t\|^2 \mid \mathcal{F}_t\big] \leq A$, where A is an absolute constant and $\mathcal{F}_t$ is the history of the algorithm until time t. (Under these conditions, the sequence $\sum_{t=0}^{T} \gamma_t w_t$, $T = 0, 1, \ldots$, is a martingale with bounded second moments and its convergence follows from the martingale convergence theorem.) Note that under our assumptions, $\sum_{t=k_m}^{k_{m+1}-1} \gamma_t w_t$ converges to zero as m tends to infinity.

Loosely speaking, we have $r_{t+1} = r_t + O(\gamma_t)$, which implies that

$$r_t = r_{k_m} + O(\gamma), \qquad \text{if } k_m \leq t \leq k_{m+1}. \tag{4.41}$$

We then have

$$\begin{aligned} r_{k_{m+1}} &= r_{k_m} + \sum_{t=k_m}^{k_{m+1}-1} \gamma_t (H r_t - r_t) + \sum_{t=k_m}^{k_{m+1}-1} \gamma_t w_t \\ &\approx r_{k_m} + \sum_{t=k_m}^{k_{m+1}-1} \gamma_t \big(H r_{k_m} - r_{k_m} + O(\gamma)\big) \\ &\approx r_{k_m} + \gamma (H r_{k_m} - r_{k_m}) + O(\gamma^2), \end{aligned}$$

where we have used the convergence of $\sum_{t=k_m}^{k_{m+1}-1} \gamma_t w_t$ to zero to eliminate that term. There is also an implicit assumption that

$$H r_t = H r_{k_m} + O\big(\|r_t - r_{k_m}\|\big) = H r_{k_m} + O(\gamma), \qquad k_m \leq t \leq k_{m+1}. \tag{4.42}$$

Thus, up to first order in γ, the algorithm is similar to the deterministic iteration

$$r_{k_{m+1}} = r_{k_m} + \gamma (H r_{k_m} - r_{k_m}),$$

or

$$r := r + \gamma (H r - r). \tag{4.43}$$

As mentioned earlier, we rescale the time axis so that $[k_m, k_{m+1}-1]$ becomes a unit interval. We divide by γ, and take the limit as γ tends to zero. (This is possible because the preceding argument was carried out for an arbitrary positive γ.) We then obtain that the algorithm has behavior similar to the deterministic differential equation

$$\frac{dr}{dt} = Hr - r. \tag{4.44}$$

In conclusion, a preliminary understanding of the long term behavior of the algorithm can be obtained by focusing on the deterministic iteration (4.43) or the differential equation (4.44).

The development above is far from rigorous. For example, the assumptions (4.41) and (4.42) need to be given precise meaning and must be justified under suitable conditions. Furthermore, the assumption that the sum $\sum_{t=0}^{\infty} \gamma_t w_t$ converges often requires the boundedness of the conditional variance of w_t which in turn may require r_t to be bounded. For this reason, rigorous developments of the ODE approach are often based on an assumption that r_t is bounded, something that needs to be independently verified. Nevertheless, it is usually the case that Eqs. (4.43) or (4.44) correctly capture the asymptotic behavior of the algorithm.

4.4.1 The Case of Markov Noise

Our development in Sections 4.2 and 4.3 rested on the assumption that the noise w_t has the property $E[w_t \mid \mathcal{F}_t] = 0$. In contrast, the ODE approach is more generally applicable; for example, w_t could be a Markov process. Note that when the stepsize γ_t is small, r_t changes very slowly, and can be viewed as a constant over fairly long time intervals. If the Markov process w_t reaches steady-state during such long intervals, then it is only the steady-state average value of w_t that matters; if this value is zero, the situation is similar to having imposed our earlier assumption $E[w_t \mid \mathcal{F}_t] = 0$. In this subsection, we derive a result that shows how a Markov process can be replaced by its mean value, and which will be used in later chapters. In fact, this result, as presented, does not involve an explicit ODE, but nevertheless relies on the informal ideas that we introduced earlier.

Let X_t, $t = 0, 1, \ldots$, be a time-homogeneous Markov process (that is, its transition probabilities do not depend on time) taking values in some set S, not necessarily finite. Let $A(\cdot)$ be a mapping that maps every $X \in S$ to an $n \times n$ matrix $A(X)$. Similarly, let $b(\cdot)$ be a mapping that maps every $X \in S$ to a vector $b(X)$ in $\Re^n$. We consider an algorithm of the form

$$r_{t+1} = r_t + \gamma_t \big(A(X_t) r_t + b(X_t)\big),$$

where γ_t is a nonnegative scalar stepsize.

For the remainder of this section, we let $\|\cdot\|$ be the Euclidean norm. In addition, for any matrix M, we define its Euclidean matrix norm by

$$\|M\| = \max_{\|x\| \neq 0} \frac{\|Mx\|}{\|x\|}.$$

In particular, we have $\|Mx\| \leq \|M\| \cdot \|x\|$, for every vector x. We introduce the following assumption.

Assumption 4.5:

(a) The stepsizes γ_t are nonnegative, deterministic, and satisfy $\sum_{t=0}^{\infty} \gamma_t = \infty$, $\sum_{t=0}^{\infty} \gamma_t^2 < \infty$.

(b) The Markov process X_t has an invariant (steady-state) distribution. Let $E_0[\,\cdot\,]$ stand for the expectation with respect to this invariant distribution.

(c) The matrix A defined by $A = E_0\big[A(X_t)\big]$ is negative definite.

(d) There exists a constant K such that $\big\|A(X)\big\| \leq K$ and $\big\|b(X)\big\| \leq K$, for all $X \in S$.

(e) There exist scalars C and ρ, with $0 \leq \rho < 1$, such that

$$\Big\|E\big[A(X_t) \mid X_0 = X\big] - A\Big\| \leq C\rho^t, \qquad \forall\, t \geq 0,\ X \in S,$$

and

$$\Big\|E\big[b(X_t) \mid X_0 = X\big] - b\Big\| \leq C\rho^t, \qquad \forall\, t \geq 0,\ X \in S,$$

where $b = E_0\big[b(X_t)\big]$.

In words, Assumption 4.5(e) requires that for any initial state X_0, the expectation of $A(X_t)$ and $b(X_t)$ converges exponentially fast to the steady-state expectation A and b, respectively. Thus, in a sense, we are assuming "rapid" convergence to steady-state.

The convergence result that follows is based on a rigorous version of the informal argument outlined in the beginning of this section. In particular, we show that the algorithm has essentially the same behavior as the deterministic iteration

$$r := r + \gamma(Ar + b),$$

which is convergent, for small γ, due to the negative definiteness assumption on A.

Proposition 4.8: Under Assumption 4.5, the sequence r_t converges with probability 1 to the unique solution r^* of the system $Ar^* + b = 0$.

Proof: To simplify notation, we provide the proof for the case where $b(X)$ is identically zero and, therefore, $r^* = 0$. The proof for the general case is entirely similar. Let us fix a positive constant γ that satisfies

$$e^{2\gamma K} \leq 2.$$

We choose an increasing sequence of integers k_m such that

$$\gamma \leq \sum_{t=k_m}^{k_{m+1}-1} \gamma_t \leq 2\gamma, \qquad \forall\, m.$$

We define

$$\overline{\gamma}_m = \sum_{t=k_m}^{k_{m+1}-1} \gamma_t,$$

and note that

$$\gamma \leq \overline{\gamma}_m \leq 2\gamma, \qquad \forall\, m.$$

Since γ_t converges to zero, we can assume without loss of generality that all γ_t are smaller than γ, and such a sequence k_m is guaranteed to exist. Furthermore, let us focus on m large enough so that $\gamma_t \leq \gamma^2$ for all $t \geq k_m$.

Let us define $q_m = r_{k_m}$. We have

$$\begin{aligned} q_{m+1} &= q_m + \sum_{t=k_m}^{k_{m+1}-1} \gamma_t A(X_t) r_t \\ &= q_m + \sum_{t=k_m}^{k_{m+1}-1} \gamma_t A q_m \\ &\quad + \sum_{t=k_m}^{k_{m+1}-1} \gamma_t \Big(E\big[A(X_t) \mid X_{k_m}\big] - A \Big) q_m \\ &\quad + \sum_{t=k_m}^{k_{m+1}-1} \gamma_t \Big(A(X_t) - E\big[A(X_t) \mid X_{k_m}\big] \Big) q_m \\ &\quad + \sum_{t=k_m}^{k_{m+1}-1} \gamma_t \big(A(X_t) r_t - A(X_t) q_m \big). \end{aligned} \tag{4.45}$$

We can write Eq. (4.45) in the form

$$q_{m+1} = q_m + g_{1,m} + g_{2,m} + g_{3,m} + g_{4,m}, \tag{4.46}$$

where $g_{1,m}, \ldots, g_{4,m}$ are the four sums that appear in the right-hand side of Eq. (4.45), in that order. We will now bound each one of these terms.

Since the matrix A is negative definite, there exists some constant $\beta > 0$ such that $q_m' A q_m \leq -\beta \|q_m\|^2$. Using this fact, together with the inequality $\|Aq_m\| \leq K\|q_m\|$, we obtain

$$\begin{aligned}
\|q_m + g_{1,m}\|^2 &= \left\| q_m + \sum_{t=k_m}^{k_{m+1}-1} \gamma_t A q_m \right\|^2 \\
&= q_m'(I + \overline{\gamma}_m A)'(I + \overline{\gamma}_m A) q_m \\
&= q_m' q_m + 2\overline{\gamma}_m q_m' A q_m + \overline{\gamma}_m^2 q_m' A' A q_m \\
&\leq \|q_m\|^2 - 2\beta \overline{\gamma}_m \|q_m\|^2 + \overline{\gamma}_m^2 K^2 \|q_m\|^2 \\
&= (1 - 2\beta\overline{\gamma}_m + K^2 \overline{\gamma}_m^2) \|q_m\|^2.
\end{aligned}$$

Let us assume that γ has been chosen small enough so that $2K^2\gamma \leq \beta$. Then, $K^2\overline{\gamma}_m^2 \leq 2K^2\gamma\overline{\gamma}_m \leq \beta\overline{\gamma}_m$. We then obtain

$$\|q_m + g_{1,m}\|^2 \leq (1 - \beta\overline{\gamma}_m)\|q_m\|^2 \leq (1 - \beta\gamma)\|q_m\|^2. \tag{4.47}$$

Regarding the second term $g_{2,m}$, we use Assumption 4.5(e), to obtain

$$\|g_{2,m}\| \leq \sum_{t=k_m}^{k_{m+1}-1} \gamma_t C \rho^{t-k_m} \|q_m\| \leq \gamma^2 \frac{C}{1-\rho} \|q_m\|. \tag{4.48}$$

(We have used here the assumption $\gamma_t \leq \gamma^2$, for $t \geq k_m$.)

Let $\mathcal{F}_m$ be the history of the algorithm up to and including the time that X_{k_m} is generated, that is,

$$\mathcal{F}_m = \{X_0, X_1, \ldots, X_{k_m}\}.$$

Using the Markov property of X_k, we obtain

$$E[g_{3,m} \mid \mathcal{F}_m] = E[g_{3,m} \mid X_{k_m}] = 0. \tag{4.49}$$

We also note that

$$\|g_{3,m}\| \leq 4\gamma K \|q_m\|. \tag{4.50}$$

In order to bound the magnitude of the fourth term, some more work is needed and we have the following lemma.

Lemma 4.9: For all t such that $k_m \leq t \leq k_{m+1}$, we have

$$\|r_t - q_m\| \leq 4\gamma K \|q_m\|.$$

Proof: Note that

$$\|r_{t+1}\| \leq \big\|I + \gamma_t A(X_t)\big\| \cdot \|r_t\| \leq (1 + \gamma_t K)\|r_t\|.$$

Hence, for all t satisfying $k_m \leq t \leq k_{m+1}$, we have

$$\|r_t\| \leq \|q_m\| \prod_{\tau=k_m}^{k_{m+1}-1} (1 + \gamma_\tau K) \leq e^{2\gamma K}\|q_m\| \leq 2\|q_m\|,$$

where we have used the inequality

$$\prod_{\tau=k_m}^{k_{m+1}-1} (1 + \gamma_\tau K) \leq \exp\left\{\sum_{\tau=k_m}^{k_{m+1}-1} \gamma_\tau K\right\} \leq e^{2\gamma K},$$

which is easily proved using the Taylor series expansion of the exponential function, and our assumption $e^{2\gamma K} \leq 2$.

We now have, for $k_m \leq t \leq k_{m+1}$,

$$\|r_t - q_m\| \leq \sum_{\tau=k_m}^{k_{m+1}-1} \gamma_\tau \big\|A(X_\tau)\big\| \cdot \|r_\tau\| \leq \sum_{\tau=k_m}^{k_{m+1}-1} \gamma_\tau K 2\|q_m\| \leq 4\gamma K\|q_m\|.$$

Q.E.D.

Lemma 4.9 yields

$$\|g_{4,m}\| \leq \sum_{t=k_m}^{k_{m+1}-1} \gamma_t \big\|A(X_t)\big\| \cdot \|r_t - q_m\| \leq (2\gamma)K\big(4K\gamma\|q_m\|\big) = 8K^2\gamma^2\|q_m\|. \tag{4.51}$$

We now use Eq. (4.46), to obtain

$$\begin{aligned} \|q_{m+1}\|^2 &= \|q_m + g_{1,m}\|^2 + \|g_{2,m} + g_{3,m} + g_{4,m}\|^2 \\ &\quad + 2(q_m + g_{1,m})'(g_{2,m} + g_{3,m} + g_{4,m}) \\ &\leq \|q_m + g_{1,m}\|^2 + \|g_{2,m} + g_{3,m} + g_{4,m}\|^2 \\ &\quad + 2(q_m + g_{1,m})'g_{3,m} + 2\|q_m + g_{1,m}\| \cdot \|g_{2,m} + g_{4,m}\|. \end{aligned}$$

We take expectations of both sides, conditioned on $\mathcal{F}_m$. Note that $q_m + g_{1,m}$ is completely determined by $\mathcal{F}_m$, and Eq. (4.49) yields

$$E\big[(q_m + g_{1,m})'g_{3,m} \mid \mathcal{F}_m\big] = 0.$$

We use Eq. (4.47) to bound the term $\|q_m + g_{1,m}\|^2$. We use Eqs. (4.48) and (4.51) to bound $\|g_{2,m} + g_{4,m}\|$ by $D\gamma^2\|q_m\|$, for some constant D. Finally, we use Eq. (4.50) to bound $\|g_{3,m}\|$ by $G\gamma\|q_m\|$, for some other constant

G. Putting everything together, we see that there exists a constant L such that

$$E\big[\|q_{m+1}\|^2 \mid \mathcal{F}_m\big] \leq (1 - \beta\gamma + L\gamma^2)\|q_m\|^2.$$

Suppose that γ was chosen small enough so that $L\gamma^2 \leq \beta\gamma/2$. Then,

$$E\big[\|q_{m+1}\|^2 \mid \mathcal{F}_m\big] \leq (1 - \beta\gamma/2)\|q_m\|^2.$$

Using the supermartingale convergence theorem (Prop. 4.2 in Section 4.2), with the identification $Y_m = \|q_m\|^2$, we see that $\|q_m\|^2$ converges and furthermore, $\sum_m \|q_m\|^2 < \infty$, which implies that q_m converges to zero. We then use Lemma 4.9 to conclude that r_t must also converge to zero. **Q.E.D.**

4.5 NOTES AND SOURCES

4.1. Some general references on stochastic approximation algorithms are the books by Kushner and Clark [KuC78], and by Benveniste, Metivier, and Priouret [BMP90].

4.2. Stochastic approximation algorithms under the pseudogradient assumption have been studied by Poljak and Tsypkin [PoT73] and part of the proof of Prop. 4.1 given here is adapted from that paper; see also [Pol87]. However, the convergence of $\nabla f(r_t)$ to zero is a new result, the strongest earliest result being $\liminf_{t\to\infty} \|\nabla f(r_t)\| = 0$. The connection of incremental gradient methods and the stochastic gradient algorithm has been made by White [Whi89]. Two-pass methods have been introduced and studied by Poljak and Juditsky [PoJ92], and have been further analyzed by Yin [Yin92], and by Kushner and Yang [KuY93]. A heuristic analysis is provided by Ljung [Lju94].

4.3 The results of this section are adapted from Tsitsiklis [Tsi94], and the proofs use ideas on asynchronous distributed algorithms developed in a deterministic context by Bertsekas [Ber82a] and expanded in Bertsekas and Tsitsiklis [BeT89]. An informal proof of Prop. 4.4 is also given by Jaakkola, Jordan, and Singh [JJS94].

The idea of projecting on a bounded set in order to ensure boundedness of a stochastic approximation algorithm has been discussed by Ljung [Lju77], and by Kushner and Clark [KuC78]. A formal proof of convergence for the case where a smooth potential function is employed is given by Chong and Ramadge [ChR92].

4.4 The ODE approach has been introduced by Ljung [Lju77]. A more formal treatment can be found in Kushner and Clark [KuC78]. The monograph by Benveniste et al. [BMP90] provides a comprehensive treatment and contains results much more general than Prop. 4.8.

You can observe an awful lot
by just watching.
(Yogi Berra)

5

Simulation Methods for a Lookup Table Representation

Contents

The computational methods for dynamic programming problems that were described in Ch. 2 apply when there is an explicit model of the cost structure and the transition probabilities of the system. In many problems, however, such a model is not available but instead, the system and the cost structure can be simulated. By this we mean that the state space and the control space are known and there is a computer program that simulates, for a given control u, the probabilistic transitions from any given state i to a successor state j according to the transition probabilities $p_{ij}(u)$, and also generates the corresponding transition cost $g(i, u, j)$. It is then of course possible to use repeated simulation to calculate (at least approximately) the transition probabilities of the system and the expected costs per stage by averaging, and then to apply the methods discussed in Ch. 2.

The methodology discussed in this chapter, however, is geared towards an alternative possibility, which is much more attractive when one contemplates approximations: the transition probabilities are not explicitly estimated, but instead the cost-to-go function of a given policy is progressively calculated by generating several sample system trajectories and associated costs. We consider several possibilities, which will be revisited in the next chapter in conjunction with approximation methods. More specifically, we discuss simulation-based methods for policy evaluation, including the method of so called *temporal differences.* We also present variants of the value iteration algorithm that use simulation so as to target the computations on the most important parts of the state space. We finally study the *Q-learning* algorithm, which is a simulation-based method for obtaining an optimal policy when a model of the system is unavailable.

The methods of this chapter involve a *lookup table representation* of the cost-to-go function J, in the sense that a separate variable $J(i)$ is kept in memory for each state i, in contrast with the *compact representations* of the next chapter where J is represented as a function of a smaller set of parameters. While the methods of this chapter cannot be used when the number of states is large, they are still of interest for several reasons:

(a) These methods are applicable whenever the problem is "easy" in the sense that the state space is of moderate size, but an exact model is unavailable.

(b) Some of the methods in the next chapter are extensions and modifications of the methods considered here. For this reason, the analysis of algorithms that use a lookup table representation provides a baseline against which the results of the next chapter are to be compared.

(c) Finally, some of the approximate methods studied in the next chapter can be viewed as lookup table methods (of the type considered here) for a suitable auxiliary (and usually much smaller) problem. Thus, the understanding gained here can be transferred to the study of some of the methods in the next chapter.

As a final comment, in this chapter we assume that the costs per stage $g(i, u, j)$ are a deterministic function of i, u, and j. All of the algorithms to be developed remain applicable if $g(i, u, j)$ is replaced by $g(i, u, j) + w$, where w is zero mean random noise with bounded variance. All of our convergence results generalize to cover this case, with minimal changes in the proofs. We have chosen, however, not to include this added noise term, so as to keep notation more manageable.

5.1 SOME ASPECTS OF MONTE CARLO SIMULATION

In this section, we consider some issues related to Monte Carlo simulation that will set the stage for our subsequent discussion of simulation-based methods for dynamic programming. Suppose that v is a random variable with an unknown mean m that we wish to estimate. The basic idea in Monte-Carlo simulation is to generate a number of samples $v_1, \ldots, v_N$ of the random variable v and then to estimate the mean of v by forming the *sample mean*

$$M_N = \frac{1}{N} \sum_{k=1}^{N} v_k.$$

Note that the sample mean can be computed recursively according to

$$M_{N+1} = M_N + \frac{1}{N+1}(v_{N+1} - M_N),$$

starting with $M_1 = v_1$.

The Case of i.i.d. Samples

In the simplest possible setting, the number of samples N is chosen ahead of time and the samples $v_1, v_2, \ldots, v_N$ are independent, identically distributed, with mean m. We then have

$$E[M_N] = \frac{1}{N} \sum_{k=1}^{N} E[v_k] = m, \tag{5.1}$$

and the estimator M_N is said to be *unbiased*. Furthermore, the variance of M_N is given by

$$\text{Var}(M_N) = \frac{1}{N^2} \sum_{k=1}^{N} \text{Var}(v_k) = \frac{\sigma^2}{N},$$

where σ^2 is the common variance of the random variables v_k. The variance of M_N converges to zero as $N \to \infty$ and we say that M_N converges to m

in the mean square sense. This can be used to show that for N large, M_N is close to the true mean m, with high probability. Another property is provided by the *strong law of large numbers* which asserts that the sequence M_N converges to m, with probability 1, and we say that the estimator is *consistent.* (The strong law of large numbers is actually a special case of the convergence result in Example 4.3 of Section 4.2, if we let the stepsize γ_t be $1/t$.)

The Case of Dependent Samples

Suppose now that the samples v_k are identically distributed, with common mean and variance m and σ^2, respectively, but that they are dependent. Equation (5.1) remains valid and we still have an unbiased estimator. On the other hand, the variance of M_N need not be equal to σ^2/N and its calculation requires knowledge of the covariance between the different samples v_k. In general, the dependence between the samples v_k can have either a favorable or an adverse effect on the variance of M_N, as shown by the examples that follow.

An interesting question that arises in this context has to do with the possibility of obtaining better estimators (with smaller variance) by discarding some of the available samples. In general, it may be preferable to use a weighted average of the samples v_k, of the form $\sum_k \zeta_k v_k$, where ζ_k are nonnegative weights that sum to one. On the other hand, except for certain singular situations, each sample contains some new and useful information and should not be entirely discarded; thus, typically all of the weights ζ_k should be nonzero.

Example 5.1

Suppose that $N = 3$, that v_1 and v_2 are independent identically distributed, with unit variance, and that $v_3 = v_2$. Then the sample mean is equal to

$$M_3 = \frac{v_1 + v_2 + v_3}{3} = \frac{1}{3}v_1 + \frac{2}{3}v_2.$$

Its variance is

$$\left(\frac{1}{3}\right)^2 + \left(\frac{2}{3}\right)^2 = \frac{5}{9}.$$

However, the variance of the estimator $(v_1 + v_2)/2$ is equal to

$$\left(\frac{1}{2}\right)^2 + \left(\frac{1}{2}\right)^2 = \frac{1}{2},$$

which is smaller. In fact, any estimator of the form $\zeta_1 v_1 + \zeta_2 v_2 + \zeta_3 v_3$ with $\zeta_1 = 1/2$ and $\zeta_2 + \zeta_3 = 1/2$ has variance equal to $1/2$, and it can be shown that this level of variance is the minimal that can be achieved using estimators of the form $\zeta_1 v_1 + \zeta_2 v_2 + \zeta_3 v_3$ that are constrained to be unbiased, that is, $\zeta_1 + \zeta_2 + \zeta_3 = 1$. In this example, v_3 does not contain any new information and can be discarded.

Example 5.2

Consider a situation where we have $2N$ samples $v_1, \ldots, v_N, \overline{v}_1, \ldots, \overline{v}_N$, all with the same mean and with unit variance. Suppose that the random pairs $(v_k, \overline{v}_k)$ are independent from each other and identically distributed for different k, but that the two samples v_k and $\overline{v}_k$ in the same pair are dependent. We may then ask whether we are better off using the sample mean

$$M_{2N} = \frac{1}{2N} \sum_{k=1}^{N} (v_k + \overline{v}_k),$$

or whether we should discard the dependent data and compute

$$M_N = \frac{1}{N} \sum_{k=1}^{N} v_k.$$

An easy computation shows that

$$\text{Var}(M_{2N}) = \frac{1+\rho}{2N},$$

where

$$\rho = E\big[(v_k - m)(\overline{v}_k - m)\big]$$

is the correlation coefficient between v_k and $\overline{v}_k$. Because $\rho \leq 1$, this is no worse than the variance $1/N$ of M_N, which means that we do not profit by discarding dependent data.

Regarding the convergence of M_N, it can be established that M_N converges to m, with probability 1, as long as the dependence of the random variables v_k is weak, e.g., if the distribution of v_k has a diminishing dependence on the random variables v_i realized in the remote past. For a concrete example, consider a finite state Markov chain i_t with a single recurrent class and let $\pi(i)$ be the steady-state probability of state i. Suppose that a cost of $g(i)$ is incurred whenever at state i, and let $v_t = g(i_t)$. Assuming that the Markov chain starts in steady-state, $E[v_t]$ is equal to the average cost per stage $\sum_i \pi(i)g(i)$, for all t. The random variables v_t are dependent but, nevertheless, it is well known that

$$\lim_{N\to\infty} \frac{1}{N} \sum_{t=1}^{N} v_t = \sum_i \pi(i)g(i),$$

with probability 1. More results of this type can be obtained using the ODE approach discussed in Section 4.4.

The Case of a Random Sample Size

The situation becomes more interesting and also more complex if the number of samples N is itself a random variable. In the simplest case, the mean of $v_1, \ldots, v_N$, conditioned on N, is the same as the unconditional mean m. We then have

$$E[M_N] = E\left[E\left[\frac{1}{N}\sum_{k=1}^{N} v_k \;\Big|\; N\right]\right] = E[m] = m,$$

and the sample mean is again an unbiased estimator. If, conditioned on N, the random variables v_k are independent identically distributed and their variance σ^2 does not depend on N, then the conditional variance of M_N given N is equal to σ^2/N, according to our earlier discussion. It follows that the unconditional variance of M_N is equal to $\sigma^2 E[1/N]$.

Suppose now that the conditional distribution of $v_1, \ldots, v_N$, conditioned on N, is different than the unconditional one. This would be the case, for example, if we draw a number of samples and we decide whether more samples should be drawn depending on the past sample values. We then have, in general,

$$E[M_N] \neq m,$$

and the sample mean is a biased estimator.

Example 5.3

Let v_i be independent identically distributed random variables that take either value 1 or -1, with equal probability. Let us choose the number of samples as follows. We first draw v_1. If $v_1 = 1$, we stop, else we draw one more sample. There are three possible events:

(a) If $v_1 = 1$ (probability 1/2), the sample mean is 1.

(b) If $v_1 = -1$ and $v_2 = 1$ (probability 1/4), the sample mean is 0.

(c) If $v_1 = -1$ and $v_2 = -1$ (probability 1/4), the sample mean is -1.

Thus, the expectation of the sample mean is

$$\frac{1}{2} \times 1 + \frac{1}{4} \times 0 + \frac{1}{4} \times (-1) = \frac{1}{4}.$$

Here, the sample mean is a biased estimator of the true mean (which is zero).

Fortunately, M_N is a consistent estimator, as long as the sequence $v_1, v_2, \ldots$ satisfies a strong law of large numbers. To turn this into a formal statement, suppose that the random number N of samples depends on a certain parameter K, and that $N_K \to \infty$ as K increases. We then have

$$\lim_{K\to\infty} \frac{1}{N_K}\sum_{k=1}^{N_K} v_k = \lim_{n\to\infty} \frac{1}{n}\sum_{k=1}^{n} v_k = m,$$

where the last equality follows from the assumed strong law of large numbers for the sequence $v_1, v_2, \ldots$

We close by developing an identity that is very useful when dealing with samples of a random size. Suppose that the random variables $v_1, v_2, \ldots$ have a common mean and

$$E[v_k \mid N \geq k] = E[v_1];$$

intuitively, the fact that we have decided to go ahead and obtain a kth sample does not alter our expectation for the value of that sample. We now claim that

$$E\left[\sum_{k=1}^{N} v_k\right] = E[v_1]\,E[N], \tag{5.2}$$

under certain conditions. Equation (5.2) is known as *Wald's identity.* For a simple proof, notice that

$$\begin{aligned} E\left[\sum_{k=1}^{N} v_k\right] &= \sum_{k=1}^{\infty} \mathrm{P}(N \geq k)\,E\big[v_k \mid N \geq k\big] \\ &= E[v_1] \sum_{k=1}^{\infty} \mathrm{P}(N \geq k) \\ &= E[v_1] \sum_{k=1}^{\infty} \sum_{n=k}^{\infty} \mathrm{P}(N = n) \\ &= E[v_1] \sum_{n=1}^{\infty} n\mathrm{P}(N = n) \\ &= E[v_1]\,E[N]. \end{aligned}$$

The interchange of the expectation and the summation in the first step of the proof can be justified by appealing to the dominated convergence theorem [Ash72], if we impose a condition of the form

$$E\big[|v_k| \mid N \geq k\big] \leq M, \qquad \forall\ k,$$

where M is some constant, and if we also assume that $E[N] < \infty$.

One possible corollary of Wald's identity is the following: while the expected value of the sample mean M_N has no simple relation with the unknown mean m, the sample sum $\sum_{k=1}^{N} v_k$ is a good indicator of the *sign* of $E[v_1]$, in the sense that the expected value of the sample sum has the same sign as $E[v_1]$.

5.2 POLICY EVALUATION BY MONTE CARLO SIMULATION

In this section, we assume that we have fixed a stationary policy μ and that we wish to calculate by simulation the corresponding cost-to-go vector J^μ. We consider the stochastic shortest path problem, as defined in Ch. 2, with state space $\{0, 1, \ldots, n\}$, where 0 is a cost-free absorbing state, and we focus on the case where the policy μ is proper. In order to simplify notation, we do not show the dependence of various quantities on the policy. In particular, the transition probability from i to j and the corresponding cost per stage are denoted by p_{ij} and $g(i,j)$, in place of $p_{ij}\big(\mu(i)\big)$ and $g\big(i, \mu(i), j\big)$, respectively.

One possibility is to generate, starting from each i, many sample state trajectories and average the corresponding costs to obtain an approximation to $J^\mu(i)$. While this can be done separately for each state i, a possible alternative is to use each trajectory to obtain cost samples for all states visited by the trajectory, by considering the cost of the trajectory portion that starts at each intermediate state.

To formalize the process, suppose that we perform a number of simulation runs, each ending at the termination state 0. Consider the mth time a given state i_0 is encountered, and let $(i_0, i_1, \ldots, i_N)$ be the remainder of the corresponding trajectory, where $i_N = 0$. Let $c(i_0, m)$ be the corresponding cumulative cost up to reaching state 0, that is,

$$c(i_0, m) = g(i_0, i_1) + \cdots + g(i_{N-1}, i_N).$$

We assume that different simulated trajectories are statistically independent and that each trajectory is generated according to the Markov process determined by the policy μ. In particular, for all states i and for all m, we have

$$J^\mu(i) = E\big[c(i,m)\big]. \tag{5.3}$$

We estimate $J^\mu(i)$ by forming the sample mean

$$J(i) = \frac{1}{K}\sum_{m=1}^{K} c(i,m), \tag{5.4}$$

subsequent to the Kth encounter with state i. We can iteratively calculate the sample means in Eq. (5.4) by using the update formula

$$J(i) := J(i) + \gamma_m\big(c(i,m) - J(i)\big), \qquad m = 1, 2, \ldots, \tag{5.5}$$

where

$$\gamma_m = \frac{1}{m}, \qquad m = 1, 2, \ldots,$$

starting with

$$J(i) = 0.$$

Consider a trajectory $(i_0, i_1, \ldots, i_N)$ and let k be an integer between 1 and N. We note that this trajectory contains the subtrajectory $(i_k, i_{k+1}, \ldots, i_N)$; this is a sample trajectory with initial state i_k and can therefore be used to update $J(i_k)$ according to Eq. (5.5). This leads to the following algorithm. At the end of a simulation run that generates the state trajectory $(i_0, i_1, \ldots, i_N)$, update the estimates $J(i_k)$ by using for each $k = 0, \ldots, N-1$, the formula

$$J(i_k) := J(i_k) + \gamma(i_k)\big(g(i_k, i_{k+1}) + g(i_{k+1}, i_{k+2}) + \cdots + g(i_{N-1}, i_N) - J(i_k)\big), \tag{5.6}$$

where the stepsize $\gamma(i_k)$ is allowed to change from one iteration to the next. There are many choices for the stepsizes $\gamma(i_k)$ under which the algorithm remains sound. In particular, using the results of Ch. 4, it can be shown that convergence of iteration (5.6) to the correct cost value $J^\mu(i_k)$ is obtained as long as $\gamma(i_k)$ diminishes at the rate of one over the number of visits to state i_k; such results will be established in the next section, within a more general setting.

While the preceding simulation approach is conceptually straightforward, there are some subtle issues relating to the correlation between different cost samples, which we proceed to discuss.

5.2.1 Multiple Visits to the Same State

It is important to note that if a state i is encountered multiple times within the same trajectory, Eq. (5.6) amounts to multiple updates of $J(i)$, with each update based on the cost of the subtrajectory that starts at the time of a different visit. (We call this the *every-visit* method.) These subtrajectories are dependent because they are all portions of the same trajectory $(i_0, i_1, \ldots, i_N)$. Consequently, the cost samples associated to the different visits are dependent. As discussed in Section 5.1, such dependencies usually have an effect on the variance of the sample mean. Furthermore, since the number of times a given state is encountered is random, the number of cost samples available for a given state is also random. According to the discussion in Section 5.1, the sample mean could be a biased estimator, and this is illustrated by the example that follows. Nevertheless, we will see later that the bias diminishes to zero as we increase the number of trajectories that are averaged.

Example 5.4

Consider the two-state Markov chain shown in Fig. 5.1. State 0 is a cost-free absorbing state but every transition out of state 1 carries a cost of 1. Let p be a shorthand for the probability p_{10} of moving from state 1 to state 0. Note that $J^\mu(1)$ satisfies

$$J^\mu(1) = 1 + (1-p)J^\mu(1),$$

and therefore,

$$J^\mu(1) = \frac{1}{p} = E[N],$$

where N is the number of transitions until the state becomes 0, starting from state 1.

A typical trajectory starting at state $i_0 = 1$ is of the form $(i_0, i_1, \ldots, i_{N-1}, i_N)$, with $i_0 = \cdots = i_{N-1} = 1$ and $i_N = 0$. Note that $(i_m, i_{m+1}, \ldots, i_N)$, for $m = 0, \ldots, N-1$ is a subtrajectory that starts at state 1. We therefore have a total of N subtrajectories and the cost sample $c(i, m)$ corresponding to the mth subtrajectory is equal to

$$c(i, m) = N - m, \qquad m = 0, \ldots, N-1.$$

The sample mean of all the subtrajectory costs is equal to

$$\frac{1}{N} \sum_{m=0}^{N-1} (N - m) = \frac{N+1}{2}.$$

The expectation of the sample mean is equal to

$$\frac{E[N+1]}{2} = \frac{1+p}{2p},$$

which is different than $J^\mu(1)$ and we therefore have a biased estimator.

More generally, if we generate K trajectories and if N_k is the time that the kth trajectory reaches state 0, a similar calculation shows that the sample mean of all subtrajectory costs is

$$\frac{\sum_{k=1}^K N_k(N_k+1)}{2\sum_{k=1}^K N_k},$$

and this turns out to be again a biased estimator.

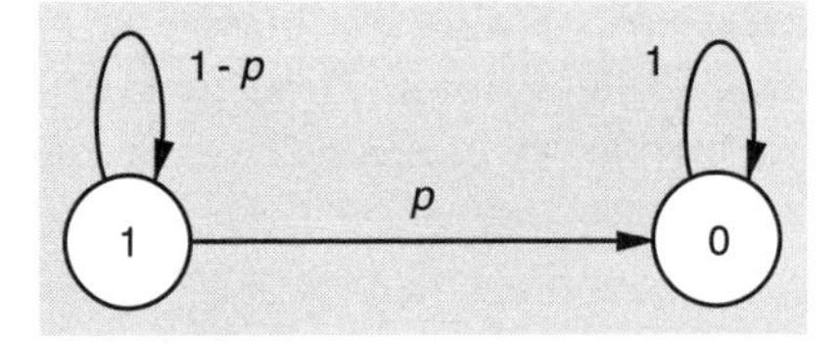

Figure 5.1: The Markov chain in Example 5.4.

Example 5.4 has shown that the sample mean of all available subtrajectory costs is a biased estimator of $J^\mu(i)$. Nevertheless, if we average over many independent trajectories the estimator is consistent as will be shown next.

Consistency of the Every-Visit Method

For the purposes of this discussion, let us focus on a particular state $i \neq 0$. Suppose that there is a total of K independently generated trajectories and that K_i of these trajectories visit state i. Suppose that the kth trajectory involves n_k visits to state i and generates n_k corresponding cost samples $c(i,1,k), c(i,2,k), \ldots, c(i,n_k,k)$. Here, $c(i,m,k)$ is the cost sample corresponding to the mth visit to state i during the kth trajectory. We note that conditioned on $n_k \geq 1$ (that is, if the kth trajectory visits state i at least once), the random variables n_k are independent and identically distributed. (The fact that they are identically distributed is a consequence of the Markov property; once the trajectory gets to state i, the statistics of the future trajectory and the number of revisits to state i only depend on i. The independence of the n_k is a consequence of the assumed independence of the different trajectories.) For similar reasons, the random variables $\sum_{m=1}^{n_k} c(i,m,k)$, conditioned on $n_k \geq 1$, are also independent and identically distributed for different k.

We now assume that as $K \to \infty$, so does K_i; that is, state i is encountered infinitely many times in the long run. We use the strong law of large numbers to see that the sample mean of all available cost samples $c(i,m,k)$ is given, asymptotically, by

$$\begin{aligned}\lim_{K\to\infty} \frac{\sum_{\{k \mid n_k \geq 1\}} \sum_{m=1}^{n_k} c(i,m,k)}{\sum_{\{k \mid n_k \geq 1\}} n_k} &= \lim_{K_i\to\infty} \frac{\frac{1}{K_i}\sum_{\{k \mid n_k \geq 1\}} \sum_{m=1}^{n_k} c(i,m,k)}{\frac{1}{K_i}\sum_{\{k \mid n_k \geq 1\}} n_k} \\ &= \frac{E\Big[\sum_{m=1}^{n_k} c(i,m,k) \mid n_k \geq 1\Big]}{E[n_k \mid n_k \geq 1]}.\end{aligned}$$

We now note that $E\big[c(i,m,k) \mid n_k \geq m\big] = J^\mu(i)$, which is a consequence of the Markov property: each time that state i is visited, the expected behavior of the remaining trajectory is the same as if state i were the initial state on the trajectory. Therefore, Wald's identity applies [cf. Eq. (5.2)] and we have

$$\frac{E\Big[\sum_{m=1}^{n_k} c(i,m,k) \mid n_k \geq 1\Big]}{E[n_k \mid n_k \geq 1]} = E\big[c(i,1,k) \mid n_k \geq 1\big] = J^\mu(i).$$

By combining these two relations, we see that for any state i that is visited infinitely often in the course of the simulation, the mean of all available cost samples converges to the correct value $J^\mu(i)$.

The First-Visit Method

The preceding argument has established the consistency of averaging all available cost samples $c(i,m,k)$, despite their dependence. On the other hand, the wisdom of doing so has been questioned, primarily because we end up with a biased estimator when the number of cost samples that are averaged is finite (cf. Example 5.4). In an alternative method, for any given state i and for each trajectory k, we only use the cost sample $c(i, 1, k)$ corresponding to the first visit to that state. Formally, this leads to the estimator

$$\frac{\sum_{\{k \mid n_k \geq 1\}} c(i, 1, k)}{K_i},$$

where K_i is the number of trajectories that visit state i. Let us assume that the total number K of trajectories and their starting states are fixed ahead of time. Then, the mean of the cost samples $c(i, 1, k)$, $k = 1, \ldots, K_i$, conditioned on K_i, is the same as the unconditional mean $J^\mu(i)$ and, according to the discussion in Section 5.1, we have an unbiased estimator.

Example 5.5

Consider the same system as in Example 5.4. Suppose that we generate a total of K independent trajectories $(i_0, \ldots, i_{N_k})$, $k = 1, \ldots, K$, all of them starting at state $i_0 = 1$. Here N_k is the time that the kth trajectory reaches state 0. Then, the first cost sample $c(i, 1, k)$ obtained from the kth trajectory is equal to N_k. The sample mean computed by the first-visit method is

$$\frac{1}{K} \sum_{k=1}^{K} N_k.$$

Its expectation is equal to $E[N_k]$, which is the same as $J^\mu(1)$, and we have an unbiased estimator.

We have seen so far that the first-visit method leads to unbiased estimates whereas the every-visit method can lead to biased estimates. This observation in itself does not necessarily mean that the first-visit method is preferable. For a meaningful comparison, we need to compute and compare the mean squared error that results from the two alternative methods.

Example 5.6

We consider again the same system as in Examples 5.4 and 5.5. Suppose that we generate a single trajectory and state 0 is reached at time N. As shown in Example 5.4, the every-visit method leads to the estimate

$$\hat{J}_E(1) = \frac{N+1}{2},$$

whereas the first-visit method leads to the estimate

$$\hat{J}_F(1) = N.$$

The random variable N has a geometric distribution with mean $1/p$ and

$$E[N^2] = \frac{2}{p^2} - \frac{1}{p}.$$

Thus,

$$\mathrm{Var}\big(\hat{J}_F(1)\big) = \mathrm{Var}(N) = E[N^2] - E[N]^2 = \frac{1}{p^2} - \frac{1}{p}.$$

On the other hand the mean squared error of the every-visit method is equal to

$$E\left[\left(\frac{N+1}{2} - \frac{1}{p}\right)^2\right] = \frac{1}{2p^2} - \frac{3}{4p} + \frac{1}{4}.$$

It is easily checked that the every-visit method has smaller mean squared error for every $p < 1$, even though it is biased.

Let us now consider an arbitrary Markov chain and focus on multiple visits to a fixed state, say state i. If we look at this Markov chain only at those times that the state is either i or 0, we obtain a reduced Markov chain whose state space is $\{0, i\}$. A transition from state i to itself in the reduced Markov chain corresponds to a sequence of transitions (a segment of a trajectory) in the original Markov chain that starts at i and eventually returns to i. Let the cost of such a transition in the reduced Markov chain be random and have the same distribution as the cost of a trajectory segment in the original chain that starts at i and returns to i. Transitions from state i to state 0 in the reduced Markov chain are modeled similarly. It can be seen that the reduced Markov chain contains all the information needed to analyze the behavior of the every-visit and first-visit methods. Thus, Examples 5.4-5.6 capture the essence of these methods for arbitrary Markov chains; the only difference is that in the reduced Markov chain that we have just constructed, the transition costs are random variables, whereas in Examples 5.4-5.6 they are equal to unity. The calculations in Examples 5.4-5.6 can be repeated with this randomness taken into account, and the same conclusions are reached; in particular, for the case of a single trajectory, the every-visit method outperforms the first-visit method, as far as the mean squared error is concerned [SiS96].

In practice, we are mostly interested in the case of a large number of simulated trajectories. It can be shown that the bias of the every-visit method converges to zero, but there is strong evidence, including experimental results, indicating that its mean squared error eventually (as the number of trajectories increases) becomes larger than the mean squared error of the first-visit method [SiS96]. Even though this may seem paradoxical, the first-visit method appears to be preferable.

Note, however, that the significance of the comparison of the every-visit and first-visit methods should not be overemphasized. For problems with a large state space, the likelihood of a trajectory visiting the same state twice is usually quite small and so the two methods essentially coincide.

5.2.2 Q-Factors and Policy Iteration

Policy evaluation procedures such as the one discussed in this section or the temporal difference methods to be discussed in the next, can be embedded within a simulation-based policy iteration approach for obtaining an optimal policy. Let us introduce the notion of the *Q-factor* of a state-control pair (i, u) and a stationary policy μ, defined as

$$Q^\mu(i, u) = \sum_{j=0}^{n} p_{ij}(u)\big(g(i, u, j) + J^\mu(j)\big), \tag{5.7}$$

which is the expected cost if we start at state i, use control u at the first stage, and use policy μ at the second and subsequent stages.

The Q-factors can be evaluated by first evaluating J^μ as discussed earlier, and then using further simulation and averaging (if necessary) to compute the right-hand side of Eq. (5.7) for all pairs (i, u). [Of course, if a model is available, the right-hand side of Eq. (5.7) can be computed directly.] Once this is done, one can execute a policy improvement step using the equation

$$\overline{\mu}(i) = \arg\min_{u \in U(i)} Q^\mu(i, u), \qquad i = 1, \ldots, n. \tag{5.8}$$

It is thus possible to implement a version of the policy iteration algorithm that alternates between a policy and Q-factor evaluation step using simulation, and a policy improvement step using Eq. (5.8).

Strictly speaking, we are dealing here with an approximate version of policy iteration because policy evaluation is based on simulation rather than exact computation. We therefore need to verify that inexact policy evaluation does not have too adverse an effect on the performance of the policy iteration algorithm. This issue will be addressed in greater generality in Section 6.2.

Finally, there are some important practical issues related to the way that the initial states of the simulated trajectories are chosen. Any given policy may have a tendency to steer the state to a restricted region R in the state space. We then obtain many cost samples for states within R and fewer cost samples for states outside R. Accordingly, the quality of our simulation-based estimates of $J^\mu(i)$ can be poor for states outside R. It is now possible that a policy update results in a new policy that steers the state to some region $\overline{R}$, disjoint from R. In that case, the actions of the new policy at the states that matter most (states in $\overline{R}$) have been designed on the basis of the poor estimates of $J^\mu(i)$ for that region. For this reason, the new policy can perform much worse than the previous one, in sharp contrast with the improvement guarantees of exact policy iteration. These difficulties suggest that great care must be exercised in choosing the initial states of simulated trajectories.

Issues of this type are important not just for approximate policy iteration but for any simulation-based method. The best way of addressing them usually depends on the nature of the particular problem at hand. One option, which we call *iterative resampling*, is the following. Suppose that under the old policy μ we have high quality estimates for states in a region R and poor estimates outside. If a policy update leads us to a new region $\overline{R}$ outside R, we can postpone the policy update, go back, and perform further simulations of μ with starting states chosen in $\overline{R}$.

5.3 TEMPORAL DIFFERENCE METHODS

In this section, we discuss an implementation of the Monte Carlo policy evaluation algorithm that incrementally updates the cost-to-go estimates $J(i)$, following each transition. We then provide a generalization of this method whereby later transitions are given less weight. This leads us to a family of policy evaluation algorithms, known as temporal difference methods. We discuss some related issues and close with an analysis of convergence properties.

Throughout this section, we adopt the following notational convention: given any trajectory $i_0, i_1, \ldots, i_N$, with $i_N = 0$, we define $i_k = 0$ for all $k > N$ and accordingly set $g(i_k, i_{k+1}) = 0$ for $k \geq N$. In addition, for any cost vector $J(\cdot)$ to be considered, we will always assume that $J(0)$ has been fixed to zero. Finally, the assumption that the policy under consideration is proper remains in effect.

5.3.1 Monte Carlo Simulation Using Temporal Differences

We recall [cf. Eq. (5.6)] that once a trajectory $(i_0, i_1, \ldots, i_N)$ is generated, our cost estimates $J(i_k)$, $k = 0, \ldots, N-1$, are updated according to

$$J(i_k) := J(i_k) + \gamma\big(g(i_k, i_{k+1}) + g(i_{k+1}, i_{k+2}) + \cdots + g(i_{N-1}, i_N) - J(i_k)\big).$$

This update formula can be rewritten in the form

$$\begin{aligned} J(i_k) := J(i_k) + \gamma\Big(&\big(g(i_k, i_{k+1}) + J(i_{k+1}) - J(i_k)\big) \\ &+ \big(g(i_{k+1}, i_{k+2}) + J(i_{k+2}) - J(i_{k+1})\big) \\ &+ \cdots \\ &+ \big(g(i_{N-1}, i_N) + J(i_N) - J(i_{N-1})\big)\Big), \end{aligned}$$

where we have made use of the property $J(i_N) = 0$. Equivalently,

$$J(i_k) := J(i_k) + \gamma(d_k + d_{k+1} + \cdots + d_{N-1}), \tag{5.9}$$

where the quantities d_k, which are called *temporal differences*, are defined by

$$d_k = g(i_k, i_{k+1}) + J(i_{k+1}) - J(i_k). \tag{5.10}$$

The temporal difference d_k represents the difference between an estimate

$$g(i_k, i_{k+1}) + J(i_{k+1})$$

of the cost-to-go based on the simulated outcome of the current stage, and the current estimate $J(i_k)$. In this sense, the temporal difference provides an indication as to whether our current estimates $J(i)$ should be raised or lowered.

Note that the ℓth temporal difference d_ℓ becomes known as soon as the transition from i_ℓ to $i_{\ell+1}$ is simulated. This raises the possibility of carrying out the update (5.9) incrementally, that is, by setting

$$J(i_k) := J(i_k) + \gamma d_\ell, \qquad \ell = k, \ldots, N-1,$$

as soon as d_ℓ becomes available.

We observe that the temporal difference d_ℓ appears in the update formula for $J(i_k)$ for every $k \leq \ell$. Thus, the final form of the algorithm is to let

$$\begin{aligned} J(i_k) &:= J(i_k) + \gamma d_\ell \\ &= J(i_k) + \gamma\big(g(i_\ell, i_{\ell+1}) + J(i_{\ell+1}) - J(i_\ell)\big), \qquad k = 0, \ldots, \ell, \end{aligned} \tag{5.11}$$

as soon as the transition to $i_{\ell+1}$ has been simulated. The stepsize γ can be different for every state i_k and can also be different at every iteration. One possibility is to let the stepsize used for updating $J(i_k)$ be roughly inversely proportional to the total number of visits to state i_k during past trajectories.

In full analogy with the discussion in the previous section, we can distinguish between the every-visit and the first-visit variant of the algorithm. For example, in the every-visit variant, the update Eq. (5.11) is carried out for every k, even if two different ks correspond to the same state i_k. In contrast, with the first-visit method, we can have at most one update of $J(i)$ for the same state i.

If the sample trajectory involves at most one visit to each state, the algorithm (5.11) is mathematically equivalent to the original update rule (5.6) or (5.9). Otherwise, there is a small difference between the update rules (5.9) and (5.11), for the following reason. Suppose that k and ℓ are such that $k < \ell$ and $i_k = i_\ell = i$. Then, under the rule (5.11), the update of $J(i)$ effected after the kth transition causes a change in the value of the temporal difference d_ℓ generated later. The nature of this discrepancy will be examined in more detail, and in a more general context, towards the end of the next subsection.

5.3.2 TD(λ)

In this subsection, we motivate the TD(λ) algorithm as a stochastic approximation method for solving a suitably reformulated Bellman equation, and then proceed to discuss a few of its variants.

The Monte Carlo policy evaluation algorithm of Section 5.2 and Eq. (5.6) in particular, can be viewed as a Robbins-Monro stochastic approximation method for solving the equations

$$J^\mu(i_k) = E\left[\sum_{m=0}^{\infty} g(i_{k+m}, i_{k+m+1})\right], \tag{5.12}$$

for the unknowns $J^\mu(i_k)$, as i_k ranges over the states in the state space. (Of course, this would be a pretty trivial system of equations if the expectation could be explicitly evaluated.) One can generate a range of algorithms with a similar flavor by starting from other systems of equations involving J^μ and then replacing expectations by single sample estimates. For example, if we start from Bellman's equation

$$J^\mu(i_k) = E\big[g(i_k, i_{k+1}) + J^\mu(i_{k+1})\big], \tag{5.13}$$

the resulting stochastic approximation method takes the form

$$J(i_k) := J(i_k) + \gamma\big(g(i_k, i_{k+1}) + J(i_{k+1}) - J(i_k)\big), \tag{5.14}$$

with such an update being executed each time that state i_k is visited.

Equation (5.12) relies on the sum of the costs over the entire trajectory whereas Eq. (5.13) only takes into account the immediate cost. A middle ground is obtained by fixing a nonnegative integer ℓ and taking into consideration the cost of the first $\ell+1$ transitions. For example, a stochastic approximation algorithm could be based on the $(\ell+1)$-step Bellman equation

$$J^\mu(i_k) = E\left[\sum_{m=0}^{\ell} g(i_{k+m}, i_{k+m+1}) + J^\mu(i_{k+\ell+1})\right]. \tag{5.15}$$

In the absence of any special knowledge that could make us prefer one value of ℓ over another, we may consider forming a weighted average of all possible multistep Bellman equations. More precisely, let us fix some $\lambda < 1$, multiply Eq. (5.15) by $(1-\lambda)\lambda^\ell$, and sum over all nonnegative ℓ. We then obtain

$$J^\mu(i_k) = (1-\lambda)E\left[\sum_{\ell=0}^{\infty}\lambda^\ell\left(\sum_{m=0}^{\ell} g(i_{k+m}, i_{k+m+1}) + J^\mu(i_{k+\ell+1})\right)\right]. \tag{5.16}$$

We interchange the order of the two summations in Eq. (5.16) and use the fact $(1-\lambda)\sum_{\ell=m}^{\infty}\lambda^\ell = \lambda^m$, to obtain

$$
\begin{aligned}
J^\mu(i_k) \\
&= E\left[(1-\lambda)\sum_{m=0}^{\infty} g(i_{k+m}, i_{k+m+1}) \sum_{\ell=m}^{\infty}\lambda^\ell + \sum_{\ell=0}^{\infty} J^\mu(i_{k+\ell+1})(\lambda^\ell - \lambda^{\ell+1})\right] \\
&= E\left[\sum_{m=0}^{\infty}\lambda^m\Big(g(i_{k+m}, i_{k+m+1}) + J^\mu(i_{k+m+1}) - J^\mu(i_{k+m})\Big)\right] + J^\mu(i_k).
\end{aligned}
\tag{5.17}
$$

[Recall here our convention that $i_k = 0$, $g(i_k, i_{k+1}) = 0$, and $J(i_k) = 0$ for k larger than or equal to the termination time N. Thus, the infinite sum in Eq. (5.17) is in reality a finite sum, albeit with a random number of terms.] In terms of the temporal differences d_m, given by

$$
d_m = g(i_m, i_{m+1}) + J^\mu(i_{m+1}) - J^\mu(i_m),
\tag{5.18}
$$

Eq. (5.17) can be rewritten as

$$
J^\mu(i_k) = E\left[\sum_{m=k}^{\infty}\lambda^{m-k} d_m\right] + J^\mu(i_k),
\tag{5.19}
$$

which is hardly surprising since from Bellman's equation [cf. Eq. (5.13)], we have $E[d_m] = 0$ for all m.

The Robbins-Monro stochastic approximation method based on Eq. (5.16), or the equivalent Eq. (5.19), is

$$
J(i_k) := J(i_k) + \gamma \sum_{m=k}^{\infty}\lambda^{m-k} d_m,
\tag{5.20}
$$

where γ is a stepsize parameter, possibly changing from one iteration to another. The above equation provides us with a family of algorithms, one for each choice of λ, and is known as TD(λ), where TD stands for "temporal differences." The use of a value of λ less than 1 tends to discount the effect of the temporal differences of state transitions in the far future on the cost estimate of the current state. (However, this is not to be confused with having a discount factor in the cost function of the dynamic programming problem.)

Note that if we let $\lambda = 1$ in Eq. (5.20), the resulting TD(1) method is the Monte Carlo policy evaluation method of the preceding subsection [cf. Eq. (5.9)]. Another limiting case is obtained if we let $\lambda = 0$; using the convention $0^0 = 1$, the resulting TD(0) algorithm is

$$
J(i_k) := J(i_k) + \gamma\big(g(i_k, i_{k+1}) + J(i_{k+1}) - J(i_k)\big),
$$

and coincides with the algorithm (5.14) that was based on the one-step Bellman equation. Note that this update can be carried out by picking an arbitrary state i_k and only simulating the transition to a next state i_{k+1} rather than an entire trajectory. Thus, TD(0) can be implemented by carrying out updates at an arbitrary sequence of states, much the same as in the asynchronous value iteration algorithm of Ch. 2. Nevertheless, we will reserve the name TD(0) for the variant of the algorithm that generates complete trajectories and updates $J(i)$ for each state i visited by the trajectory.

Every-Visit and First-Visit Variants

With the TD(λ) algorithm, once a sample trajectory is generated, the update (5.20) is to be carried out for every state i_k visited by the trajectory. If a state is visited more than once by the same trajectory, then under the every-visit variant, the update (5.20) is to be carried out more than once, with each update involving the summation of $\lambda^{m-k} d_m$ over the portion of the trajectory that follows the visit under consideration. More formally, fix some state i, and suppose that during a single trajectory, state i is visited a total of M times, namely, at times $m_1, m_2, \ldots, m_M$. (Note that M is a random variable.) Then, the total update of $J(i)$ is given by

$$J(i) := J(i) + \gamma \sum_{j=1}^{M} \sum_{m=m_j}^{\infty} \lambda^{m-m_j} d_m. \tag{5.21}$$

In essence, this amounts to updating $J(i_k)$ using several samples of the right-hand side of Eq. (5.19). These samples are in general correlated, because the same d_m may appear in the sum $\sum_{m=m_j}^{\infty} \lambda^{m-m_j} d_m$ for more than one j. Nevertheless, the soundness of the method is not affected for reasons similar to those discussed in Section 5.2.1.

The first-visit variant of the algorithm can be similarly described by the update formula

$$J(i) := J(i) + \gamma \sum_{m=m_1}^{\infty} \lambda^{m-m_1} d_m, \tag{5.22}$$

where m_1 is the time of the first visit to state i.

For $\lambda = 1$, the two variants of TD(1) are equivalent to the two corresponding variants of Monte Carlo policy evaluation that were discussed in Section 5.2. Both methods are guaranteed to converge and for a large enough number of simulated trajectories, the first-visit method is preferable, in the sense that its mean squared error is smaller.

For $\lambda < 1$, however, the situation is different. Consider, for example, the first-visit method for the case $\lambda = 0$. Then, the temporal difference corresponding to a transition from i_k to i_{k+1} leads to an update only if this is the first time during the trajectory that state i_k was visited. In particular, at most n temporal differences will be used by the update equations where n is the number of nonterminal states. Thus, if we are dealing with a problem for which the typical time to termination N is much larger than n, most of the information obtained from simulated transitions gets discarded. This suggests that first-visit TD(0) and, by extension, first-visit TD(λ) for small λ, may not be the best choice. Other, possibly more promising, alternatives will be introduced later in this section.

Off-Line and On-Line Variants

In the most straightforward implementation of TD(λ), all of the updates are carried out simultaneously, according to Eq. (5.21), after the entire trajectory has been simulated. This is called the *off-line* version of the algorithm. In an alternative implementation, called the *on-line* version of the algorithm, the running sums in Eq. (5.21) are evaluated one term at a time, following each transition, as shown below:

$$J(i_0) := J(i_0) + \gamma d_0, \qquad \text{following the transition } (i_0, i_1),$$

$$\begin{cases} J(i_0) := J(i_0) + \gamma\lambda d_1, \\ J(i_1) := J(i_1) + \gamma d_1, \end{cases} \qquad \text{following the transition } (i_1, i_2)$$

and more generally for $k = 0, \ldots, N-1$,

$$\begin{cases} J(i_0) := J(i_0) + \gamma\lambda^k d_k, \\ J(i_1) := J(i_1) + \gamma\lambda^{k-1} d_k, \\ \quad\vdots \qquad\qquad \vdots \\ J(i_k) := J(i_k) + \gamma d_k, \end{cases} \qquad \text{following the transition } (i_k, i_{k+1}).$$

If the same state is visited more than once by the same trajectory, then the two implementations are slightly different. The example that follows is meant to clarify the nature of the updates in the off-line and on-line algorithm. In particular, we shall see that the difference between the two variants is of second order in the stepsize γ and is therefore inconsequential as the stepsize diminishes to zero.

Example 5.7

Consider a trajectory that starts at state 1 and then visits states 2, 1, 0. We first consider the off-line version of every-visit TD(λ), initialized with some $J_0(1)$ and $J_0(2)$. The results of one iteration of the algorithm are denoted by $J_f(1)$ and $J_f(2)$. We have

$$\begin{aligned} J_f(1) &= J_0(1) + \gamma(d_0 + \lambda d_1 + \lambda^2 d_2 + d_2) \\ &= J_0(1) + \gamma\Big(\big(g(1,2) + J_0(2) - J_0(1)\big) + \lambda\big(g(2,1) + J_0(1) - J_0(2)\big) \\ &\qquad + \lambda^2\big(g(1,0) - J_0(1)\big) + \big(g(1,0) - J_0(1)\big)\Big), \end{aligned}$$

where the last temporal difference term d_2 is due to the subtrajectory that starts with the second visit at state 1. We also have

$$J_f(2) = J_0(2) + \gamma\Big(\big(g(2,1) + J_0(1) - J_0(2)\big) + \lambda\big(g(1,0) - J_0(1)\big)\Big).$$

We now consider the on-line version of the algorithm. Following the first transition, from state 1 to 2, we set

$$J_1(1) = J_0(1) + \gamma\big(g(1,2) + J_0(2) - J_0(1)\big),$$
$$J_1(2) = J_0(2).$$

Following the second transition, from state 2 back to state 1, we set

$$\begin{aligned} J_2(1) &= J_1(1) + \gamma\lambda\big(g(2,1) + J_1(1) - J_1(2)\big), \\ J_2(2) &= J_1(2) + \gamma\big(g(2,1) + J_1(1) - J_1(2)\big). \end{aligned} \tag{5.23}$$

Following the last transition, from state 1 back to the final state 0, we set

$$\begin{aligned} J_3(1) &= J_2(1) + \gamma\Big(\lambda^2\big(g(1,0) - J_2(1)\big) + \big(g(1,0) - J_2(1)\big)\Big), \\ J_3(2) &= J_2(2) + \gamma\lambda\big(g(1,0) + J_2(1)\big). \end{aligned} \tag{5.24}$$

If in the right-hand side of Eqs. (5.23) and (5.24), J_1 and J_2 were replaced by J_0, then J_f and J_3 would be the same. The difference of J_1 and J_0 is of the order of γ, which implies that the difference of J_2 and J_0 is also of the order of γ. It follows that the difference between J_f and J_3 is of the order of γ^2.

Preview of Convergence Results

It can be shown that the values $J(i)$ generated by the TD(λ) algorithm are guaranteed to converge to $J^\mu(i)$, with probability 1, provided that each state is visited by infinitely many trajectories and the stepsizes diminish towards zero at a suitable rate. A complete proof is provided later in this section, in a more general setting. Nevertheless, we provide here a summary of the proof as it would apply to off-line first-visit TD(λ).

The first step is to use the properness of the policy μ to conclude that there exists a positive vector ξ and some $\beta \in [0,1)$ such that $\|PJ\|_\xi \leq \beta\|J\|_\xi$, where P is the $n \times n$ matrix with the transition probabilities p_{ij}, $i,j = 1,\ldots,n$, under the given policy (cf. Prop. 2.2 in Section 2.2.1). Equation (5.16) can be written in vector notation as

$$J^\mu = G + (1-\lambda)\sum_{\ell=0}^{\infty} \lambda^\ell P^{\ell+1} J^\mu, \tag{5.25}$$

where G is a constant vector. Using the contraction property of P, it is seen that

$$\begin{aligned} \left\| (1-\lambda) \sum_{\ell=0}^{\infty} \lambda^\ell P^{\ell+1} J \right\|_\xi &\leq (1-\lambda) \sum_{\ell=0}^{\infty} \lambda^\ell \| P^{\ell+1} J \|_\xi \\ &\leq \beta (1-\lambda) \sum_{\ell=0}^{\infty} \lambda^\ell \| J \|_\xi \\ &= \beta \| J \|_\xi, \qquad \forall\ J. \end{aligned}$$

Thus, Eq. (5.25) is of the form $J^\mu = HJ^\mu$ where H is a contraction with respect to a weighted maximum norm. Recall now that TD(λ) is the Robbins-Monro stochastic approximation method based on Eq. (5.25). Convergence of the algorithm to the fixed point J^μ of Eq. (5.25) can be then obtained by applying our general results for Robbins-Monro stochastic approximation methods for solving a system $J = HJ$, when the mapping H is a weighted maximum norm contraction (Prop. 4.4 in Section 4.3.2).

Tradeoffs in Choosing λ

Now that we have available a whole family of algorithms, parametrized by λ, it is natural to inquire as to the best choice of λ. A systematic study has been carried out by Singh and Dayan [SiD96] who developed analytical formulas that can be used to predict the mean squared error of different methods, when applied to problems of small size. The general conclusion is that intermediate values of λ (neither 1 nor 0) seem to work best, but there is a very intricate interplay with the way that the stepsize is chosen.

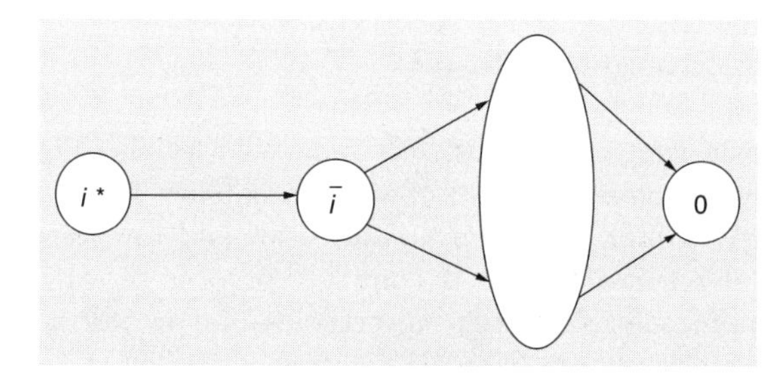

Figure 5.2: A Markov chain. Once at state i^*, we always move to state $\bar{i}$, at unit cost. When TD(1) is applied, the information in the cost samples with initial state $\bar{i}$ is not used to improve the estimate of $J(i^*)$. On the other hand, when TD(0) is applied, an accurate estimate of $J(\bar{i})$ leads to an accurate estimate of $J(i^*)$.

We now discuss a simple example that is meant to provide some insight. Consider the Markov chain shown in Fig. 5.2 and suppose that every simulated trajectory is initialized with $i_0 = i^*$. Let $c(i^*, m)$ be the cumulative cost in the mth trajectory. Since i^* is visited only once by each trajectory, the average of the sample trajectory costs $c(i^*, m)$ is an unbiased estimate of $J^\mu(i^*)$ and, furthermore, it is known to be a best estimate in a precise sense. [For example, it can be shown that the sample mean minimizes the error variance among all unbiased estimates of $J^\mu(i^*)$ that depend linearly on the simulated transition costs $g(i_k, i_{k+1})$.] Since the sample mean is the same as the result of TD(1) with $\gamma_t = 1/t$, it follows that $\lambda = 1$ should be the preferred choice.

The situation becomes more complex if there is some additional available information that could be of help in estimating $J^\mu(i^*)$. For example, suppose that an estimate of $J^\mu(\bar{i})$ is available, possibly because of additional simulations with $\bar{i}$ as the initial state. Under TD(1), this additional information would not be used in the estimation of $J^\mu(i^*)$. In contrast, TD(0) uses such information very efficiently because for $\gamma = 1$, it sets $J(i^*)$ to $J(\bar{i}) + 1$; if $J(\bar{i})$ is a very accurate estimate of $J^\mu(\bar{i})$, an equally accurate estimate of $J^\mu(i^*)$ is obtained in a single step.

In the example we have just discussed, the move from TD(1) to TD(0) amounts to replacing the simulated cost-to-go $c(\bar{i}, m)$ from state $\bar{i}$, which might have a high variance, with $J(\bar{i})$. Whether TD(1) or TD(0) will be better involves a comparison of the variance of $c(\bar{i}, m)$ with the inaccuracy (bias) of our estimate of $J^\mu(\bar{i})$; this is commonly referred to as the *bias-variance tradeoff.*

As long as we have a convergent algorithm, the quality of the estimates $J(i)$ should improve as the algorithm progresses. As the values of $J(i)$ get closer to $J^\mu(i)$, it seems plausible that we could rely more and more on the values of J at the nearest downstream states, and we should use a smaller λ. This suggests a strategy whereby we start with a large value of λ and then, as learning progresses, reduce the value of λ to zero. The convergence proofs to be given later on, cover the case where λ is allowed to vary from one iteration to another and, therefore, there are no theoretical obstacles to changing λ in the course of the algorithm. However, a mathematical argument substantiating the possible advantages of such a strategy is not available at present.

5.3.3 General Temporal Difference Methods

We will now introduce a generalization of TD(λ) that will provide us with additional flexibility and will allow us to develop a single convergence proof that covers all variants of interest.

Suppose that a sample trajectory $i_0, i_1, \ldots$ has been generated and let d_m be the corresponding temporal differences. We then postulate an update equation of the form

$$J(i) := J(i) + \gamma \sum_{m=0}^{\infty} z_m(i) d_m, \tag{5.26}$$

where the $z_m(i)$ are nonnegative scalars, called the *eligibility coefficients.* We will introduce shortly a number of assumptions on the eligibility coefficients but in order to motivate them, we will first discuss a number of special cases.

Let us concentrate on a particular state i and let $m_1, \ldots, m_M$ be the different times that the trajectory is at state i. With M being the total number of such visits, we also use the convention $m_{M+1} = \infty$. Recall that in TD(λ), a temporal difference d_m may lead to an update of $J(i)$ only if i has already been visited by time m. For this reason, in all of our examples, we assume that $z_m(i) = 0$ for $m < m_1$.

(a) If we let

$$z_m(i) = \lambda^{m-m_1}, \qquad m \geq m_1,$$

we recover the first-visit TD(λ) method [cf. Eq. (5.22)].

(b) If we let

$$z_m(i) = \sum_{\{j \mid m_j \leq m\}} \lambda^{m-m_j},$$

we recover the every-visit TD(λ) method [cf. Eq. (5.21)].

(c) Consider the choice

$$z_m(i) = \lambda^{m-m_j}, \qquad m_j \leq m < m_{j+1}, \ \forall\, j.$$

With this choice, subsequent to the first visit to i, we start forming the terms $\lambda^{m-m_1} d_m$, as in ordinary TD(λ). Once state i is visited for a second time, at time m_2, we essentially reset the eligibility coefficients $z_m(i)$ and proceed with TD(λ) updates as if we were faced with a new trajectory that started at time m_2 at state i. We call this the *restart* variant of TD(λ). Note that for $\lambda = 1$, the restart variant coincides with the first-visit method (the eligibility coefficients stay equal to 1), whereas for $\lambda = 0$ it coincides with the every-visit method. The available computational experience suggests that the restart variant outperforms the every-visit method over the entire range of values of λ [SiS96], and seems to be a better alternative.

(d) Let us define a *stopping time* as a random variable τ such that the event $\{\tau \leq k\}$ is completely determined by the history of our simulations up to and including the point that the state i_k is generated. Intuitively, a stopping time is a rule for making stopping decisions, that is not allowed to foresee the future: the decision whether or not to stop at state i_k must be made before generating subsequent states in the simulated trajectory. Some examples will be discussed shortly. Given a stopping time τ, we let

$$z_m(i) = \lambda^{m-m_1}, \qquad 0 \leq m < \tau,$$

and $z_m(i) = 0$ for $m \geq \tau$. Let us focus, for example, on the case where $\lambda = 1$. Then, the update equation for $J(i)$ is

$$\begin{aligned} J(i) &:= J(i) + \gamma \sum_{m=m_1}^{\tau-1} d_m \\ &= J(i) + \gamma \sum_{m=m_1}^{\tau-1} \big(g(i_m, i_{m+1}) + J(i_{m+1}) - J(i_m)\big) \qquad (5.27) \\ &= (1-\gamma)J(i) + \gamma \sum_{m=m_1}^{\tau-1} g(i_m, i_{m+1}) + \gamma J(i_\tau). \end{aligned}$$

This update rule is the stochastic approximation algorithm corresponding to a multistep Bellman equation, but with a random number of steps, namely,

$$J(i_0) = E\left[\sum_{m=0}^{\tau-1} g(i_m, i_{m+1}) + J(i_\tau)\right].$$

(e) If S is a subset of the state space, we can consider the stopping time τ defined by $\tau = \min\{k \geq 1 \mid i_k \in S\}$, that is, the first time that S is reached. The resulting method essentially treats the states in S the same way as the termination state 0. This method makes most sense if there is a subset S of the state space for which we are highly confident about the quality of our estimates $J(i)$, while on its complement we are less confident. Our confidence on $J(i)$, $i \in S$, means that once we reach a state $i \in S$, the expected cost-to-go for the remaining trajectory is readily available and need not be simulated, and this is exactly what our method does.

(f) Let $\lambda = 1$ and consider a trajectory that starts at state $i_0 = i$. Let the stopping time τ be the first positive time that the state either comes back to state i or reaches the termination state. For any given trajectory, there are two possibilities. If the trajectory reaches state 0 without returning to state i, then the resulting update is the same as for TD(1). If on the other hand, the trajectory returns to state i at time τ, we have $J(i_0) = J(i_\tau)$ and the update rule (5.27) becomes

$$J(i) := J(i) + \gamma \sum_{m=0}^{\tau-1} g(i_m, i_{m+1}).$$

Out of the preceding list of possible variants, the one involving stopping when high-confidence states are reached [variant (e)] is quite appealing. However, it has not been systematically investigated, one of the reasons being that there are no convenient and general methods for measuring the confidence on the quality of an estimate $J(i)$.

We now present a list of conditions on the eligibility coefficients $z_m(i)$ that are satisfied by all of the special cases we have discussed.

(a) $z_m(i) \geq 0$.

(b) $z_{-1}(i) = 0$. This is simply a convention to guarantee that the coefficients are initially zero.

(c) $z_m(i) \leq z_{m-1}(i)$ if $i_m \neq i$. Due to this condition, together with (b), the eligibility coefficients remain at zero until the first time that state i is visited.

(d) $z_m(i) \leq z_{m-1}(i) + 1$ if $i_m = i$. This condition allows the eligibility coefficients to increase (by at most 1) with each revisit to state i. In particular, this allows us to capture multiple-visit variants of TD(λ).

(e) $z_m(i)$ is completely determined by $i_0, \ldots, i_m$, together possibly with other information collected before the simulation of the current trajectory began. The essence of this condition is that the weight $z_m(i)$ given to the temporal difference d_m should be chosen before this temporal difference is generated. It is not hard to see that this condition is satisfied by all of the examples that were discussed earlier.

We finally note that the coefficients $z_m(i)$ play a role analogous to the stepsize parameter γ. For this reason, some additional assumption is required to guarantee that the terms $\gamma z_m(i)$ sum to infinity over the entire duration of the algorithm. A precise assumption of this form will be introduced in Section 5.3.5 where convergence of general temporal difference methods to J^μ will be established.

The On-Line Variant

We close by noting that the method described here has an on-line variant whereby instead of executing the update (5.26) at the end of a trajectory, we let

$$J(i) := J(i) + \gamma z_m(i) d_m,$$

as soon as the mth temporal difference is generated. This on-line variant is also convergent, as will be shown in Section 5.3.6.

5.3.4 Discounted Problems

Discounted problems can be handled in two different ways, leading to somewhat different types of simulation experiments. We present these two alternatives and discuss their relative merits.

The first approach is to convert a discounted problem into an equivalent stochastic shortest path problem, as discussed in Ch. 2, and then apply TD(λ) to the new problem. Recall that this conversion amounts to viewing the discount factor α as a continuation probability. At each stage, a transition is simulated and we then decide to terminate the process with probability $1-\alpha$. With this approach, every simulated trajectory is of finite duration, with probability 1. Its duration is random and can be arbitrarily large; in fact it gets larger, on the average, the closer the discount factor is to 1. The following example illustrates the variance of the sample trajectory costs associated with this approach.

Example 5.8

Consider a Markov chain with a single state, namely $i = 1$. In a small departure from our standard framework, we let the transition cost be a random variable g, which is independent from one transition to the next. Suppose, for simplicity, that g has mean zero and variance σ^2. Let the discount factor be $\alpha < 1$. Clearly, $J^\mu(0) = 0$.

In the equivalent stochastic shortest path problem, we have two states, namely $i = 0$ and $i = 1$. State 0 is a zero-cost termination state. Whenever at state 1, we move back to state 1 with probability α or to state 0 with probability $1-\alpha$; every such transition carries a random cost g. The off-line first-visit TD(1) method estimates $J^\mu(1)$ as the average of the available cost samples. Each cost sample is of the form $g_1 + \cdots + g_N$, where g_i is the cost incurred during the ith transition, and N is the number of transitions that take us back to state 1. The variance of this cost sample is computed to be

$$\begin{aligned} E\big[(g_1 + \cdots + g_N)^2\big] &= \sum_{k=1}^{\infty} E\big[(g_1 + \cdots + g_k)^2\big]\, \mathrm{P}(N = k) \\ &= \sum_{k=1}^{\infty} k\sigma^2 \mathrm{P}(N = k) \\ &= \sigma^2 E[N] \\ &= \frac{\sigma^2}{1-\alpha}. \end{aligned}$$

As seen in the preceding example, the algorithm treats the known discount factor α as an unknown probability which is essentially estimated via simulation. This is a wasteful use of simulation and therefore the method appears to be inefficient. A more sensible alternative is to develop a method that does not use the equivalent stochastic shortest path problem and treats α as a known quantity. This is indeed possible as will be shown next.

The development of a temporal differences method for discounted problems is essentially a repetition of our earlier derivation of TD(λ). We start with the $(\ell+1)$-step Bellman equation for discounted problems which is

$$J^\mu(i_k) = E\left[\sum_{m=0}^{\ell} \alpha^m g(i_{k+m}, i_{k+m+1}) + \alpha^{\ell+1} J^\mu(i_{k+\ell+1})\right]. \tag{5.28}$$

Fix some $\lambda < 1$, multiply Eq. (5.28) by $(1-\lambda)\lambda^\ell$, and sum over all nonnegative ℓ. We then obtain

$$J^\mu(i_k) = (1-\lambda)E\left[\sum_{\ell=0}^{\infty} \lambda^\ell \left(\sum_{m=0}^{\ell} \alpha^m g(i_{k+m}, i_{k+m+1}) + \alpha^{\ell+1} J^\mu(i_{k+\ell+1})\right)\right]. \tag{5.29}$$

We interchange the order of the two summations in Eq. (5.29) and use the fact $(1-\lambda)\sum_{\ell=m}^{\infty} \lambda^\ell = \lambda^m$, to obtain

$$\begin{aligned} J^\mu(i_k) &= E\Bigg[(1-\lambda)\sum_{m=0}^{\infty} \alpha^m g(i_{k+m}, i_{k+m+1}) \sum_{\ell=m}^{\infty} \lambda^\ell \\ &\qquad + \sum_{\ell=0}^{\infty} \alpha^{\ell+1} J^\mu(i_{k+\ell+1})(\lambda^\ell - \lambda^{\ell+1})\Bigg] \\ &= E\Bigg[\sum_{m=0}^{\infty} \lambda^m \big(\alpha^m g(i_{k+m}, i_{k+m+1}) \\ &\qquad + \alpha^{m+1} J^\mu(i_{k+m+1}) - \alpha^m J^\mu(i_{k+m})\big)\Bigg] + J^\mu(i_k). \end{aligned} \tag{5.30}$$

In terms of the temporal differences d_m, defined by

$$d_m = g(i_m, i_{m+1}) + \alpha J(i_{m+1}) - J(i_m), \tag{5.31}$$

Eq. (5.30) can be rewritten as

$$J^\mu(i_k) = E\left[\sum_{m=k}^{\infty} (\alpha\lambda)^{m-k} d_m\right] + J^\mu(i_k), \tag{5.32}$$

which is again not surprising since from Bellman's equation, we have $E[d_m] = 0$ for all m. From here on, the development is entirely similar to the development for the undiscounted case. The only difference is that α enters in the definition of the temporal differences [cf. Eq. (5.31)] and that λ is replaced by $\alpha\lambda$. In particular, Eq. (5.20) is replaced by

$$J(i_k) := J(i_k) + \gamma \sum_{m=k}^{\infty} (\alpha\lambda)^{m-k} d_m.$$

Example 5.9

Let us consider the same problem as in Example 5.8, and apply the off-line first-visit TD(1) method directly, without a conversion to a stochastic shortest path problem. We obtain cost samples by simulating infinitely long trajectories, and by accumulating the properly discounted costs. Thus, a typical cost sample is of the form $g_1 + \alpha g_2 + \alpha^2 g_3 + \cdots$, where g_k is the cost of the kth transition. The variance of such a cost sample is computed to be

$$E\left[\left(\sum_{k=0}^{\infty} \alpha^k g_{k+1}\right)^2\right] = \sigma^2 \sum_{k=0}^{\infty} \alpha^{2k} = \frac{\sigma^2}{1-\alpha^2} = \frac{\sigma^2}{(1-\alpha)(1+\alpha)}.$$

For α close to 1, this is better than the variance of the cost samples considered in Example 5.8, by a factor of about 2. In conclusion, there is an advantage in dealing with the problem directly, rather than through a conversion to a stochastic shortest path problem. On the other hand, each iteration of the method described here involves a never ending, infinitely long trajectory, which would have to be truncated with some bias resulting.

For discounted problems, the assumption that we eventually reach a zero-cost absorbing state is not usually imposed. (Indeed, in Example 5.9, there was no zero-cost state.) This implies that a trajectory may never end and if we use an off-line variant of TD(λ), we may have to wait infinitely long before a complete trajectory is obtained. This is not practical and for this reason, some changes are necessary. We must either truncate trajectories so that their length is finite or we must use an on-line variant. We review these two options in some more detail.

(a) As discussed in the preceding subsection, we can introduce a stopping time τ and collect temporal differences only up to that time. This stopping time could be deterministic (e.g., stop after 100 transitions) or random (e.g., stop as soon a certain subset of the state space is reached). As long as this stopping time is finite with probability 1, each trajectory would have to be simulated only for a finite number of steps and the off-line TD(λ) method becomes again practical. In the context of Example 5.9 this would amount to replacing the infinite sum $g_1 + \alpha g_2 + \alpha^2 g_3 + \cdots$ with

$$g_1 + \alpha g_2 + \cdots + \alpha^{\tau-1} g_\tau + \alpha^\tau J(i_\tau),$$

where the term $\alpha^\tau J(i_\tau)$ originates from the temporal difference generated by the τth transition, which is, $g_\tau + \alpha J(i_\tau) - J(i_{\tau-1})$ [compare with Eq. (5.28)].

(b) We can let the entire algorithm consist of a single infinitely long trajectory. Since we cannot afford to wait until the end of the trajectory in order to carry out an update, the on-line version of the algorithm must be used. Furthermore, it is essential that the stepsize γ be gradually reduced to zero in the course of this infinitely long simulation. The update equation for this case becomes

$$J_{m+1}(i) = J_m(i) + \gamma_m(i) z_m(i) d_m(i), \qquad \forall\ i,$$

where the $\gamma_m(i)$ are nonnegative stepsize coefficients. The only temporal differences methods that seem sensible in this case are the following:

(i) The every-visit TD(λ) method whereby the eligibility coefficients are determined by

$$z_m(i) = \begin{cases} \alpha\lambda z_{m-1}(i), & \text{if } i_m \neq i, \\ \alpha\lambda z_{m-1}(i) + 1, & \text{if } i_m = i. \end{cases} \tag{5.33}$$

(ii) The restart TD(λ) method whereby the eligibility coefficients are determined by

$$z_m(i) = \begin{cases} \alpha\lambda z_{m-1}(i), & \text{if } i_m \neq i, \\ 1, & \text{if } i_m = i. \end{cases} \tag{5.34}$$

Convergence of the algorithm to J^μ is guaranteed for all variants introduced in this subsection, under suitable assumptions, as will be discussed in more detail in Section 5.3.7.

5.3.5 Convergence of Off-Line Temporal Difference Methods

In this subsection, we consider the most general form of (off-line) temporal difference methods for undiscounted problems, as described in Section 5.3.3, and we prove convergence to the correct vector J^μ, with probability 1. The proof is rather technical and the reader may wish to read only the formal description of the algorithm and the statement of the convergence result. We start with a precise description of the method.

We use a discrete variable t to index the simulated trajectories that are generated by the algorithm. Let $\mathcal{F}_t$ represent the history of the algorithm up to the point at which the simulation of the tth trajectory is to commence and let J_t be the estimate of the cost-to-go vector available at that time.

Based on $\mathcal{F}_t$, we choose the initial state i_0^t of the tth trajectory and the stepsizes $\gamma_t(i)$, $i = 1, \ldots, n$, that will be used for updating each $J(i)$. We generate a trajectory $i_0^t, i_1^t, \ldots, i_{N_t}^t$, where N_t is the first time that the trajectory reaches state 0. We then update J_t by letting

$$J_{t+1}(i) = J_t(i) + \gamma_t(i) \sum_{m=0}^{N_t-1} z_m^t(i) d_{m,t}, \tag{5.35}$$

where the temporal differences $d_{m,t}$ are defined by

$$d_{m,t} = g(i_m^t, i_{m+1}^t) + J_t(i_{m+1}^t) - J_t(i_m^t),$$

and where $z_m^t(i)$ are the eligibility coefficients which are assumed to have the following properties (these are the same properties that were introduced in Section 5.3.3):

Assumption 5.1: For all m and t, we have:

(a) $z_m^t(i) \geq 0$.

(b) $z_{-1}^t(i) = 0$.

(c) $z_m^t(i) \leq z_{m-1}^t(i)$, if $i_m^t \neq i$.

(d) $z_m^t(i) \leq z_{m-1}^t(i) + 1$, if $i_m^t = i$.

(e) $z_m^t(i)$ is completely determined by $\mathcal{F}_t$ and $i_0^t, \ldots, i_m^t$.

We allow the possibility that no update of $J(i)$ is carried out even if a trajectory visits state i. However, for $J(i)$ to converge to the correct value, there should be enough trajectories that lead to a nontrivial update of $J(i)$. For this reason, an additional assumption is needed. To this effect, we define

$$q_t(i) = \mathrm{P}\Big(\text{there exists } m \text{ such that } z_m^t(i) > 0 \mid \mathcal{F}_t\Big).$$

Note that $q_t(i)$ is a function of the past history $\mathcal{F}_t$. We define

$$T^i = \{t \mid q_t(i) > 0\},$$

which corresponds to the set of trajectories that have a chance of leading to a nonzero update of $J(i)$. Observe that whether t belongs to T^i or not is only a function of the past history $\mathcal{F}_t$. We now introduce the following assumption.

Assumption 5.2:

(a) For any fixed i and t, $z_m^t(i)$ must be equal to 1 the first time that it becomes positive.

(b) There exists a deterministic constant $\delta > 0$ such that $q_t(i) \geq \delta$ for all $t \in T^i$ and all i.

(c) We have $\gamma_t(i) \geq 0$ for all $t \in T^i$, and $\gamma_t(i) = 0$ for $t \notin T^i$.

(d) $\sum_{t \in T^i} \gamma_t(i) = \infty$, for all i.

(e) $\sum_{t \in T^i} \gamma_t^2(i) < \infty$, for all i.

Assumption 5.2(a) is satisfied by all of the special cases discussed in Section 5.3.3 and is therefore not a major restriction. Assumption 5.2(b) is not necessary for the convergence result to be given below, but it is imposed because it allows for a simpler proof. In fact, Assumption 5.2(b) is only a minor restriction, as illustrated by the following example, which is representative of practical implementations of the method.

Example 5.10

Suppose that before starting the tth trajectory we choose two subsets $R_{1,t}$ and $R_{2,t}$ of the state space. The trajectory will be simulated and temporal differences will be generated only up to the time that the state enters the set $R_{1,t}$. Furthermore, TD(λ) updates of $J(i)$ will be carried out for all states $i \in R_{2,t}$ that are visited by the trajectory before the set $R_{1,t}$ is entered. Note that $z^t_m(i)$ becomes positive for some m if and only if $i \in R_{2,t}$ and state i is visited before the trajectory enters $R_{1,t}$. The probability that this happens, which we have defined as $q_t(i)$, only depends on $R_{1,t}$, $R_{2,t}$, and the starting state i^t_0, and can be written as $\delta(R_{1,t}, R_{2,t}, i^t_0)$. For all t such that $q_t(i)$ is positive, we have $q_t(i) \geq \min_{R_1,R_2,i_0} \delta(R_1, R_2, i_0)$, where the minimum is taken over all R_1, R_2, i_0 such that $\delta(R_1, R_2, i_0)$ is positive. Since there are only finitely many states and subsets of the state space, it follows that whenever $q_t(i)$ is positive, it is bounded below by a positive constant δ.

Our main result follows.

Proposition 5.1: Consider the off-line temporal differences algorithm, as described by Eq. (5.35), and let Assumptions 5.1 and 5.2 hold. Assume that the policy under consideration is proper. Then, $J_t(i)$ converges to $J^\mu(i)$ for all i, with probability 1.

Proof: In this proof, we will first express the off-line temporal difference method in the same form as the stochastic approximation algorithms studied in Ch. 4, and we will then apply the convergence results for the case where the iteration mapping is a weighted maximum norm pseudo-contraction. Given that the method for setting the eligibility coefficients can change from one trajectory to the next, we will see that the iteration mapping must be taken to be time-dependent. Furthermore, because the eligibility coefficients affect the magnitude of each update, they will have to be incorporated in the stepsize. As it turns out, the argument outlined towards the end of Section 5.3.2 cannot be used to verify the desired pseudo-contraction property, and a different approach will be taken, involving Lemma 5.3 below.

We start the proof by recalling that an update of $J(i)$ can take place only when $t \in T^i$. For $t \in T^i$, we define

$$I_t(i) = \{m \mid i_m^t = i\},$$

which is the set of times that the tth trajectory visits state i. We also define

$$\delta_t(i) = E\left[\sum_{m \in I_t(i)} z_m^t(i) \;\middle|\; \mathcal{F}_t\right], \qquad t \in T^i.$$

Note that for every $t \in T^i$, there is probability $q_t(i)$ that $z_m^t(i)$ is positive for some m, and that $z_m^t(i) \geq 1$ for some m [see Assumption 5.2(a)]. Furthermore, by Assumption 5.2(b), $q_t(i)$ is at least δ. We conclude that

$$\delta_t(i) \geq \delta, \qquad \forall\, t \in T^i.$$

For $t \notin T^i$, we have $J_{t+1}(i) = J_t(i)$. For $t \in T^i$, the algorithm (5.35) can be rewritten in the form

$$\begin{aligned} J_{t+1}(i) = {} & J_t(i)\big(1 - \gamma_t(i)\delta_t(i)\big) \\ & + \gamma_t(i)\delta_t(i)\left(\frac{E\Big[\sum_{m=0}^{N_t-1} z_m^t(i) d_{m,t} \mid \mathcal{F}_t\Big]}{\delta_t(i)} + J_t(i)\right) \\ & + \gamma_t(i)\delta_t(i)\frac{\sum_{m=0}^{N_t-1} z_m^t(i) d_{m,t} - E\Big[\sum_{m=0}^{N_t-1} z_m^t(i) d_{m,t} \mid \mathcal{F}_t\Big]}{\delta_t(i)}. \end{aligned} \tag{5.36}$$

We define a mapping $H_t : \Re^n \mapsto \Re^n$ by letting, for any vector $J \in \Re^n$,

$$(H_t J)(i) = \frac{E\Big[\sum_{m=0}^{N_t-1} z_m^t(i)\big(g(i_m^t, i_{m+1}^t) + J(i_{m+1}^t) - J(i_m^t)\big) \mid \mathcal{F}_t\Big]}{\delta_t(i)} + J(i).$$

and note that

$$(H_t J_t)(i) = \frac{E\Big[\sum_{m=0}^{N_t-1} z_m^t(i) d_{m,t} \mid \mathcal{F}_t\Big]}{\delta_t(i)} + J_t(i).$$

Thus, the algorithm (5.36) is a stochastic approximation algorithm of the form

$$J_{t+1}(i) = \big(1 - \hat{\gamma}_t(i)\big) J_t(i) + \hat{\gamma}_t(i)(H_t J_t)(i) + \hat{\gamma}_t(i) w_t(i),$$

where

$$\hat{\gamma}_t(i) = \gamma_t(i)\delta_t(i) = \gamma_t(i) E\left[\sum_{m \in I_t(i)} z_m^t(i) \;\middle|\; \mathcal{F}_t\right], \qquad t \in T^i,$$

$$\hat{\gamma}_t(i) = 0, \qquad t \notin T^i,$$

and

$$w_t(i) = \frac{\sum_{m=0}^{N_t-1} z_m^t(i) d_{m,t} - E\left[\sum_{m=0}^{N_t-1} z_m^t(i) d_{m,t} \mid \mathcal{F}_t\right]}{\delta_t(i)}, \qquad t \in T^i.$$

We plan to apply Prop. 4.5 from Section 4.3.3, which deals with the convergence of stochastic approximation methods based on weighted maximum norm pseudo-contractions. To this effect, we need to verify all the required assumptions. In particular, we need to verify that the standard stepsize conditions hold, establish a bound on the variance of the noise term $w_t(i)$, and finally establish that the mappings H_t are contractions with respect to a common weighted maximum norm. Note that the term $u_t(i)$ in Prop. 4.5 is absent in the present context.

We start by discussing the stepsize conditions. Note that by assumption, $\gamma_t(i)$ is a function of the history $\mathcal{F}_t$. The same is true for $\delta_t(i)$, since it is a conditional expectation given $\mathcal{F}_t$. It follows that the new stepsize $\hat{\gamma}_t(i)$ is a function of the history $\mathcal{F}_t$, as required. Next we verify that the new stepsizes $\hat{\gamma}_t(i)$ satisfy the usual conditions.

Lemma 5.1: The new stepsizes $\hat{\gamma}_t(i)$ sum to infinity and the expectation of the sum of their squares is finite.

Proof: Recall that $\delta_t(i) \geq \delta > 0$ for all $t \in T^i$. Using Assumption 5.2(d) it follows that $\sum_{t=0}^{\infty} \hat{\gamma}_t(i) = \sum_{t \in T^i} \hat{\gamma}_t(i) \geq \delta \sum_{t \in T^i} \gamma_t(i) = \infty$.

Next, we derive an upper bound for $\delta_t(i)$. Recall that N_t is the number of transitions in the tth trajectory until the termination state is reached. Since $z_m^t(i)$ can increase by at most 1 with each transition [Assumptions 5.1(c)-(d)], we have $z_m^t(i) \leq N_t$ for all m. We therefore have

$$\delta_t(i) = E\left[\sum_{m=0}^{N_t-1} z_m^t(i) \mid \mathcal{F}_t\right] \leq E\left[N_t^2 \mid \mathcal{F}_t\right].$$

Because the policy under consideration is proper, the tail of the probability distribution of N_t is bounded by a decaying exponential. Hence, all moments of N_t are finite and only depend on the initial state. Since there are only finitely many states, there exists a deterministic constant K such that

$$\delta_t(i) \leq K, \qquad \forall\ t,\ i.$$

Thus, $\hat{\gamma}_t(i) \leq K\gamma_t(i)$, and

$$\sum_{t=0}^{\infty} \hat{\gamma}_t^2(i) = \sum_{t \in T^i} \hat{\gamma}_i^2(t) \leq K^2 \sum_{t \in T^i} \gamma_i^2(t) < \infty,$$

where the last step follows from Assumption 5.2(e). **Q.E.D.**

Next we consider the noise term and show that it satisfies Assumption 4.3 of Section 4.3.1. The condition $E[w_t(i) \mid \mathcal{F}_t] = 0$ is a straightforward consequence of the definition of $w_t(i)$. The lemma that follows deals with the variance of $w_t(i)$ and shows that Assumption 4.3(b) is satisfied.

Lemma 5.2: Let $\|\cdot\|$ be any norm on $\Re^n$. Then, there exist constants A and B such that

$$E\big[w_t^2(i) \mid \mathcal{F}_t\big] \le A + B\|J_t\|^2, \qquad \forall\, t \in T^i, \ \forall\, i.$$

Proof: Recall that the denominator $\delta_t(i)$ in the definition of $w_t(i)$ is bounded below by the positive constant δ, for $t \in T^i$. We therefore, only need to concentrate on the numerator. Since the variance of any random variable is no larger than its mean square, the expectation of the numerator square is bounded above by

$$E\left[\left(\sum_{m=0}^{N_t-1} z_m^t(i) d_{m,t}\right)^2 \;\Bigg|\; \mathcal{F}_t\right]. \tag{5.37}$$

Let G be an upper bound for $|g(i,j)|$. Then, $|d_{m,t}| \le 2\|J_t\| + G$, and the expression (5.37) is bounded above by

$$\big(2\|J_t\| + G\big)^2 E\left[\left(\sum_{m=0}^{N_t-1} z_m^t(i)\right)^2 \;\Bigg|\; \mathcal{F}_t\right]. \tag{5.38}$$

As argued earlier, $z_m^t(i)$ is bounded above by the length N_t of the tth trajectory and

$$\sum_{m=0}^{N_t-1} z_m^t(i) \le N_t^2.$$

It follows that the conditional expectation in Eq. (5.38) is bounded above by $E[N_t^4 \mid \mathcal{F}_t]$. The latter quantity is only a function of the initial state of the trajectory and is bounded above by a deterministic constant, due to the properness of the policy. We also use the inequality $\big(2\|J_t\| + G\big)^2 \le 8\|J_t\|^2 + 2G^2$ in Eq. (5.38), and the desired result follows. **Q.E.D.**

We now turn to the heart of the proof, which deals with the contraction property of H_t. Our first step is to define a suitable norm.

Since we are dealing with a proper policy, the $n \times n$ matrix with entries p_{ij}, for $i, j = 1, \ldots, n$, is a contraction with respect to some weighted maximum norm (Prop. 2.2 in Section 2.2.1). In particular, there exist positive coefficients $\xi(j)$ and a constant $\beta \in [0, 1)$ such that

$$\sum_{j=1}^{n} p_{kj}\xi(j) \leq \beta\xi(k), \qquad k = 1, \ldots, n. \tag{5.39}$$

We define the corresponding weighted maximum norm $\|\cdot\|_\xi$ by letting

$$\|J\|_\xi = \max_{j=1,\ldots,n} \frac{|J(j)|}{\xi(j)}.$$

In order to apply Prop. 4.5 from Section 4.3.3, it suffices to show that

$$\|H_t J - J^\mu\|_\xi \leq \beta \|J - J^\mu\|_\xi, \qquad \forall\ J,\ t.$$

We start with the observation that $H_t J$ is an affine mapping of J. In particular,

$$(H_t J)(i) = (A_t J)(i) + b_t(i),$$

where A_t is the linear mapping defined by

$$(A_t J)(i) = \frac{E\Big[\sum_{m=0}^{N_t-1} z_m^t(i)\big(J(i_{m+1}^t) - J(i_m^t)\big) \mid \mathcal{F}_t\Big]}{\delta_t(i)} + J(i),$$

and

$$b_t(i) = \frac{E\Big[\sum_{m=0}^{N_t-1} z_m^t(i) g(i_m^t, i_{m+1}^t) \mid \mathcal{F}_t\Big]}{\delta_t(i)}.$$

We plan to show that

$$\|A_t J\|_\xi \leq \beta \|J\|_\xi, \qquad \forall\ J,\ t, \tag{5.40}$$

and

$$H_t J^\mu = J^\mu, \qquad \forall\ t. \tag{5.41}$$

Once these two properties are shown, the desired contraction property for H_t follows because

$$\|H_t J - J^\mu\|_\xi = \|H_t J - H_t J^\mu\|_\xi = \|A_t(J - J^\mu)\|_\xi \leq \beta\|J - J^\mu\|_\xi.$$

We start by showing Eq. (5.41). Let $\mathcal{F}_{m,t}$ stand for the history of the algorithm up to the point that i_m^t is generated and the coefficients $z_m^t(i)$

are chosen, but just before the next state i_{m+1}^t is generated. Because the state sequence is a Markov chain, we have

$$
\begin{aligned}
E\Big[g(i_m^t, i_{m+1}^t) + J^\mu(i_{m+1}^t) - J^\mu(i_m^t) \mid \mathcal{F}_{m,t}\Big] \\
&= E\Big[g(i_m^t, i_{m+1}^t) + J^\mu(i_{m+1}^t) - J^\mu(i_m^t) \mid i_t^m\Big] \\
&= 0,
\end{aligned}
\tag{5.42}
$$

where the last equality is simply Bellman's equation for the case of a fixed policy.

Conditioned on $\mathcal{F}_{m,t}$, $z_m^t(i)$ is a deterministic constant [cf. Assumption 5.1(e)] and Eq. (5.42) yields

$$
E\Big[z_m^t(i)\big(g(i_m^t, i_{m+1}^t) + J^\mu(i_{m+1}^t) - J^\mu(i_m^t)\big) \mid \mathcal{F}_{m,t}\Big] = 0.
$$

Taking the expectation conditioned on $\mathcal{F}_t$, we obtain

$$
E\Big[z_m^t(i)\big(g(i_m^t, i_{m+1}^t) + J^\mu(i_{m+1}^t) - J^\mu(i_m^t)\big) \mid \mathcal{F}_t\Big] = 0.
$$

If we add the latter equality for all values of m, and interchange the summation and the expectation (which can be justified by appealing to the dominated convergence theorem [Ash72]), we obtain

$$
E\left[\sum_{m=0}^{N_t-1} z_m^t(i)\big(g(i_m^t, i_{m+1}^t) + J^\mu(i_{m+1}^t) - J^\mu(i_m^t)\big) \mid \mathcal{F}_t\right] = 0.
$$

It follows that

$$
(H_t J^\mu)(i) = J^\mu(i),
$$

and Eq. (5.41) has been established.

We now turn to the proof of Eq. (5.40). By comparing the definition of $(H_t J)(i)$ and $(A_t J)(i)$, we see that it is sufficient to prove that $\|H_t J\|_\xi \leq \beta \|J\|_\xi$ for the special case where the transition costs $g(i,j)$ are identically zero, which we will henceforth assume.

Consider some vector $J \in \Re^n$ We define a new vector $\overline{J}$ by letting

$$
\overline{J}(i) = \xi(i)\|J\|_\xi.
$$

Note that

$$
\|\overline{J}\|_\xi = \|J\|_\xi.
$$

Also, since $|J(i)| \leq \xi(i)\|J\|_\xi$ for all i, we see that

$$
|J(i)| \leq \overline{J}(i), \qquad \forall\, i.
$$

Lemma 5.3: For every $i \neq 0$ and $t \in T^i$, we have

$$E\left[\sum_{m=0}^{N_t-1} z_m^t(i)\big(J(i_{m+1}^t) - J(i_m^t)\big) + \sum_{m \in I_t(i)} z_m^t(i) J(i) \;\Big|\; \mathcal{F}_t\right]$$

$$\leq \beta E\left[\sum_{m \in I_t(i)} z_m^t(i) \overline{J}(i) \;\Big|\; \mathcal{F}_t\right].$$

Proof: Let us fix some state $i \neq 0$ and some $t \in T^i$. For the purposes of this proof only, we suppress t from our notation, and we do not show explicitly the conditioning on $\mathcal{F}_t$. Furthermore, having fixed i and t, we use the notation z_m to indicate $z_m^t(i)$. Similarly, we use the notation I and N instead of $I_t(i)$ and N_t, respectively.

The left-hand side of the inequality in the statement of the lemma is equal to

$$E\left[\sum_{m=0}^{N-1} z_m\big(J(i_{m+1}) - J(i_m)\big) + \sum_{m \in I} z_m J(i)\right]$$

$$= E\left[\sum_{m=0}^{N-1} (z_{m-1} - z_m) J(i_m) + \sum_{m \in I} z_m J(i)\right], \qquad (5.43)$$

where we have made use of the properties $z_{-1} = 0$ [cf. Assumption 5.1(b)] and $J(i_N) = J(0) = 0$. Note that whenever $i_m \neq i$, the coefficient $z_{m-1} - z_m$ multiplying $J(i_m)$ is nonnegative [cf. Assumption 5.1(c)]. Furthermore, whenever $i_m = i$, the coefficient $z_{m-1} - z_m + z_m$ multiplying $J(i)$ is also nonnegative. Thus, the right-hand side in Eq. (5.43) can be upper bounded by

$$E\left[\sum_{m=0}^{N-1} (z_{m-1} - z_m) \overline{J}(i_m) + \sum_{m \in I} z_m \overline{J}(i)\right],$$

which is the same as

$$E\left[\sum_{m=0}^{N-1} z_m\big(\overline{J}(i_{m+1}) - \overline{J}(i_m)\big) + \sum_{m \in I} z_m \overline{J}(i)\right].$$

Hence, in order to prove the lemma, it suffices to show that

$$E\left[\sum_{m=0}^{N-1} z_m\big(\overline{J}(i_{m+1}) - \overline{J}(i_m)\big) + (1 - \beta) \sum_{m \in I} z_m \overline{J}(i)\right] \leq 0. \qquad (5.44)$$

Since the vector $\overline{J}$ is proportional to ξ, Eq. (5.39) yields

$$\sum_{j=1}^{n} p_{kj}\overline{J}(j) \le \beta\overline{J}(k), \qquad \forall\ k,$$

which implies that

$$\begin{aligned} E\big[\overline{J}(i_{m+1}) - \overline{J}(i_m) \mid i_m\big] &\le \beta\overline{J}(i_m) - \overline{J}(i_m) \\ &= -(1-\beta)\overline{J}(i_m). \end{aligned}$$

Hence,

$$E\Big[z_m\big(\overline{J}(i_{m+1}) - \overline{J}(i_m)\big)\Big] \le -(1-\beta)E\big[z_m\overline{J}(i_m)\big].$$

Thus, the left-hand side of Eq. (5.44) is upper bounded by

$$E\left[-(1-\beta)\sum_{m=0}^{N-1} z_m\overline{J}(i_m) + (1-\beta)\sum_{m\in I} z_m\overline{J}(i_m)\right],$$

which is clearly nonpositive and the proof of lemma is complete. (We are using here the fact that for $m \in I$, we have $i_m = i$.) **Q.E.D.**

Let us now take the result of Lemma 5.3 and divide both sides by $E\big[\sum_{m\in I_t(i)} z_m^t(i) \mid \mathcal{F}_t\big]$, which is the same as $\delta_t(i)$. We then obtain

$$(A_tJ)(i) \le \beta\overline{J}(i).$$

By an argument entirely symmetrical to the proof of Lemma 5.3, we obtain

$$-\beta\overline{J}(i) \le (A_tJ)(i).$$

Thus,

$$\big|(A_tJ)(i)\big| \le \beta\overline{J}(i) = \beta\xi(i)\|J\|_\xi.$$

We divide both sides by $\xi(i)$ and take the maximum over all i, to conclude that

$$\|A_tJ\|_\xi \le \beta\|J\|_\xi,$$

as desired. The proof is now complete because all of the conditions of Prop. 4.5 are satisfied, and we get convergence with probability 1 to the common fixed point J^μ of the mappings H_t. **Q.E.D.**

Remarks on Stepsize Selection

Assumptions 5.2(d)-(e) on the stepsizes are rather natural, but are they easily enforced? We discuss a few alternatives.

(a) If we let $\gamma_t(i) = 1/t$, we have $\sum_t \gamma_t(i) = \infty$. Still, the desired condition $\sum_{t \in T^i} \gamma_t(i) = \infty$ does not follow unless we introduce additional assumptions such as a requirement that $t \in T^i$ at least once every K time steps, where K is a constant. In many cases, this latter condition is natural.

(b) We may let $\gamma_t(i) = 1/k$ if t is the kth element of T^i. As long as T^i is an infinite set, this rule will satisfy Assumptions 5.2(d)-(e), but is not easy to implement, because at each time t we would have to determine whether t belongs to T^i or not, which may be nontrivial.

(c) A last and most natural alternative is to let $\gamma_t(i) = 1/(k+1)$ if there have been exactly k past trajectories during which $J(i)$ was updated; that is, if there have been exactly k past trajectories during which $z(i)$ became positive. This stepsize rule does satisfy Assumptions 5.2(d)-(e), but this is not entirely obvious, and we provide a proof.

Let us concentrate on some state i. We use γ_t and q_t to denote $\gamma_t(i)$ and $q_t(i)$, respectively. Let χ_t be equal to 1 if $z(i)$ becomes positive at some point during the tth trajectory; let χ_t be zero otherwise. Note that $E[\chi_t \mid \mathcal{F}_t] = q_t$ which is either zero, if $t \notin T^i$, or larger than or equal to δ, if $t \in T^i$. We assume that $\sum_{t=0}^{\infty} \chi_t = \infty$, that is, that $J(i)$ is updated an infinite number of times. (This assumption is clearly necessary in order to prove convergence.)

Let us look at the sequence $\gamma_t \chi_t$. We see that $\gamma_t \chi_t$ is nonzero only if $\chi_t = 1$. Furthermore, if this is the kth time that $\chi_t = 1$, then $\gamma_t \chi_t = 1/k$. We conclude that the sequence $\gamma_t \chi_t$ is the same as the sequence $1/k$, with zeroes interleaved between the elements $1/k$. In particular, we obtain $\sum_{t=0}^{\infty} \gamma_t \chi_t = \infty$ and $\sum_{t=0}^{\infty} \gamma_t^2 \chi_t < \infty$. The first relation implies that $\sum_{t \in T^i} \gamma_t = \infty$ and the validity of Assumption 5.2(d) has been established.

Using the definition $q_t = E[\chi_t \mid \mathcal{F}_t]$ and the fact that γ_t is a function of $\mathcal{F}_t$, we obtain $\gamma_t^2 q_t = E[\gamma_t^2 \chi_t \mid \mathcal{F}_t]$. Taking unconditional expectations, we obtain $E[\gamma_t^2 q_t] = E[\gamma_t^2 \chi_t]$. Therefore,

$$E\left[\sum_{t=0}^{\infty} \gamma_t^2 q_t\right] = E\left[\sum_{t=0}^{\infty} \gamma_t^2 \chi_t\right] = \sum_{t=0}^{\infty} \frac{1}{t^2} < \infty.$$

From this, it follows that $\sum_{t=0}^{\infty} \gamma_t^2 q_t < \infty$, with probability 1. For $t \in T^i$, we have $q_t \geq \delta$ [Assumption 5.2(b)]. Hence $\sum_{t \in T^i} \gamma_t^2 < \infty$ and the validity of Assumption 5.2(e) has been established.

5.3.6 Convergence of On-Line Temporal Difference Methods

We now return to on-line temporal difference methods and once more establish convergence. The main idea behind the proof is pretty simple. We show that the update by the on-line method differs from the off-line update by a term which is of second order in the stepsize, thus generalizing the observations made in the context of Example 5.7. We therefore have a stochastic approximation method based on contraction mappings, together with some additional noise terms that are asymptotically insignificant. Convergence follows from our general results in Ch. 4 (Prop. 4.5 in Section 4.3.3).

We start with a precise definition of the on-line algorithm. In the beginning of the tth iteration, we have available a vector J_t^0, we pick an initial state i_0^t, and we simulate a trajectory i_m^t starting from i_0^t. Let $J_{t,m}^0$ be the vector obtained after simulating m transitions of the tth trajectory. The update equations are as follows:

$$
\begin{aligned}
J_{t,0}^0(i) &= J_t^0(i), \qquad \forall\, i,\\
d_{m,t}^0 &= g(i_m^t, i_{m+1}^t) + J_{t,m}^0(i_{m+1}^t) - J_{t,m}^0(i_m^t),\\
J_{t,m+1}^0(i) &= J_{t,m}^0(i) + \gamma_t(i) z_m^t(i) d_{m,t}^0, \qquad \forall\, i,\\
J_{t+1}^0(i) &= J_{t,N_t}^0(i), \qquad \forall\, i,
\end{aligned}
$$

where the superscript 0 is used to indicate that we are dealing with the on-line algorithm, and where N_t is the length of the tth trajectory. Note that the stepsizes $\gamma_t(i)$ are held constant during each trajectory. We then have the following convergence result.

Proposition 5.2: Consider the on-line temporal differences algorithm, as described above and let Assumptions 5.1 and 5.2 hold. Furthermore assume that the eligibility coefficients $z_m^t(i)$ are bounded above by a deterministic constant C. Assume that the policy under consideration is proper. Then, $J_t^0(i)$ converges to $J^\mu(i)$, for all i, with probability 1.

Before proving Prop. 5.2, let us note the additional assumption that the eligibility coefficients $z_m^t(i)$ are bounded. This assumption is satisfied whenever we are dealing with the first-visit or restart variants of TD(λ) because for these variants $z_m^t(i)$ is bounded above by 1. Also, if $\lambda < 1$, it is easily seen that under the every-visit TD(λ) method we have $z_m^t(i) \leq 1/(1-\lambda)$ and our assumption is again satisfied. The only interesting case where the coefficients $z_m^t(i)$ cannot be bounded is the every-visit TD(1) method. We believe that a convergence result is also possible for this case, with a somewhat different line of argument, but the details have not been worked out.

Proof of Prop. 5.2: We introduce the notation $\hat{\gamma}_t = \max_i \gamma_t(i)$. Note that $\sum_t \hat{\gamma}_t^2 < \infty$, as a consequence of Assumptions 5.2(c) and 5.2(e). Throughout this proof, we let $\|\cdot\|$ stand for the maximum norm.

We will compare the evolution of $J^0_{t,m}$ in the course of a single trajectory with a related evolution equation that corresponds to the off-line variant. Recall that the off-line algorithm is given by

$$J_{t+1}(i) = J_t(i) + \gamma_t(i) \sum_{m=0}^{N_t-1} z^t_m(i) d_{m,t}, \tag{5.45}$$

where

$$d_{m,t} = g(i^t_m, i^t_{m+1}) + J_t(i^t_{m+1}) - J_t(i^t_m).$$

We rewrite Eq. (5.45) in incremental form as

$$\begin{aligned} J_{t,0} &= J_t(i), \\ J_{t,m+1}(i) &= J_{t,m}(i) + \gamma_t(i) z^t_m(i) d_{m,t}, \\ J_{t+1}(i) &= J_{t,N_t}(i). \end{aligned}$$

Note that $|d_{m,t}| \leq 2\|J_t\| + G$, where G is a bound on $|g(i^t_m, i^t_{m+1})|$. Using also the assumption that $z^t_m(i) \leq C$, we conclude that

$$\big|J_{t,m}(i) - J_t(i)\big| \leq \gamma_t(i) N_t C \big(2\|J_t\| + G\big), \qquad \forall\, m.$$

Let us consider the two variants, starting with the same vector at the beginning of the tth trajectory, that is, with $J^0_t = J_t$. We will prove, by induction on m, that

$$\big|J^0_{t,m}(i) - J_{t,m}(i)\big| \leq D_m \hat{\gamma}_t \gamma_t(i), \tag{5.46}$$

for some D_m. Note that Eq. (5.46) is true for $m = 0$ with $D_0 = 0$. We assume as an induction hypothesis, that it holds for some m.

Our first observation is that

$$\begin{aligned} \big|J^0_{t,m}(i) - J_t(i)\big| &\leq \big|J^0_{t,m}(i) - J_{t,m}(i)\big| + \big|J_{t,m}(i) - J_t(i)\big| \\ &\leq D_m \hat{\gamma}_t \gamma_t(i) + \gamma_t(i) N_t C \big(2\|J_t\| + G\big). \end{aligned}$$

Hence,

$$\begin{aligned} |d^0_{m,t} - d_{m,t}| &\leq 2\|J^0_{t,m} - J_t\| \\ &\leq 2D_m \hat{\gamma}_t^2 + 2\hat{\gamma}_t N_t C \big(2\|J_t\| + G\big). \end{aligned}$$

We therefore have

$$\begin{aligned} \big|J^0_{t,m+1}(i) - J_{t,m+1}(i)\big| &\leq \big|J^0_{t,m}(i) - J_{t,m}(i)\big| + \big|\gamma_t(i) z^t_m(i) (d^0_{m,t} - d_{m,t})\big| \\ &\leq D_m \hat{\gamma}_t \gamma_t(i) + 2D_m C \hat{\gamma}_t^2 \gamma_t(i) + 2\hat{\gamma}_t \gamma_t(i) N_t C^2 \big(2\|J_t\| + G\big) \\ &= D_m \hat{\gamma}_t \gamma_t(i)(1 + 2\hat{\gamma}_t C) + 2\hat{\gamma}_t \gamma_t(i) N_t C^2 \big(2\|J_t\| + G\big), \end{aligned}$$

which proves Eq. (5.46) for $m+1$, with

$$D_{m+1} = D_m(1 + 2\hat{\gamma}_t C) + 2N_t C^2\big(2\|J_t\| + G\big), \tag{5.47}$$

and the induction is complete.

We now use Eq. (5.46) with $m = N_t$, and obtain

$$\big|J^0_{t+1}(i) - J_{t+1}(i)\big| = \big|J^0_{t,N_t}(i) - J_{t,N_t}(i)\big| \leq D_{N_t}\hat{\gamma}_t\gamma_t(i).$$

Because Eq. (5.47) is a linear recurrence in D_m, we obtain

$$\begin{aligned} D_{N_t} &\leq N_t(1 + 2\hat{\gamma}_t C)^{N_t} 2N_t C^2\big(2\|J_t\| + G\big) \\ &\leq AN_t^2(1 + 2\hat{\gamma}_t C)^{N_t}\big(\|J_t\| + 1\big), \end{aligned}$$

where A is some constant.

Overall, the on-line algorithm is of the form

$$J^0_{t+1}(i) = \big(1 - \gamma_t(i)\big)J^0_t(i) + \gamma_t(i)(H_t J^0_t)(i) + \gamma_t(i)w_t(i) + \gamma_t(i)u_t(i),$$

where H_t and $w_t(i)$ are the same as for the off-line algorithm and $\gamma_t(i)u_t(i) = J^0_{t+1}(i) - J_{t+1}(i)$. Note that $u_t(i)$ is bounded in magnitude by $D_{N_t}\hat{\gamma}_t$. Hence

$$\big|u_t(i)\big| \leq D_{N_t}\hat{\gamma}_t \leq \hat{\gamma}_t AN_t^2(1 + 2\hat{\gamma}_t C)^{N_t}\big(\|J_t\| + 1\big).$$

We can now apply Prop. 4.5 from Section 4.3.3 and establish convergence provided that we can show that $u_t(i)$ satisfies the assumptions required by that proposition. In particular, we need to show that θ_t, defined by

$$\theta_t = \hat{\gamma}_t N_t^2(1 + 2\hat{\gamma}_t C)^{N_t},$$

converges to zero.

Since we are dealing with a proper policy and the state space is finite, the probability that N_t is equal to k is bounded above by an exponentially decaying function of k, such as $B\rho^k$, where $0 \leq \rho < 1$. Let $\gamma^* > 0$ be small enough so that $(1 + 2\gamma^* C)^2\rho < 1$. Then, whenever $\hat{\gamma}_t \leq \gamma^*$, we obtain

$$E\Big[N_t^4(1 + 2\hat{\gamma}_t C)^{2N_t} \mid \mathcal{F}_t\Big] \leq B\sum_{k=1}^{\infty} k^4(1 + 2\gamma^* C)^{2k}\rho^k = K,$$

where K is a deterministic constant. It follows that

$$E[\theta_t^2 \mid \mathcal{F}_t] \leq K\hat{\gamma}_t^2, \qquad \text{if } \hat{\gamma}_t \leq \gamma^*.$$

Let $\chi_t = 1$ if $\hat{\gamma}_t \leq \gamma^*$, and $\chi_t = 0$, otherwise. Let

$$\psi_{t+1} = \sum_{k=0}^{t} \chi_k\theta_k^2.$$

We have

$$E[\psi_{t+1} \mid \mathcal{F}_t] = \psi_t + E[\chi_t\theta_t^2 \mid \mathcal{F}_t] \leq \psi_t + \hat{\gamma}_t^2 K.$$

Since $\hat{\gamma}_t^2$ sums to a finite value, the supermartingale convergence theorem (Prop. 4.2 in Section 4.2.3) applies and shows that ψ_t converges with probability 1. It follows that $\chi_t\theta_t^2$ converges to zero. Since $\hat{\gamma}_t$ converges to zero, χ_t converges to 1. This shows that θ_t converges to zero and the proof is complete. **Q.E.D.**

5.3.7 Convergence for Discounted Problems

We now review convergence results for the case of discounted problems, for the alternatives that were introduced in Section 5.3.4.

Methods Based on Finite-Length Trajectories

The first alternative we have discussed is to convert the problem to a stochastic shortest path problem. In that case, we are back to the situation considered in Sections 5.3.5 and 5.3.6 and we get convergence for both the on-line and the off-line case.

In a second alternative, we only simulate trajectories for a finite number N_t of time steps, which is tantamount to setting the eligibility coefficients $z_m^t(i)$ to zero for $m \geq N_t$. For example, if there are some zero-cost absorbing states, we may let N_t be the number of steps until one of them is reached. Otherwise, we may let N_t be a stopping time, such as the first time that a certain subset of the state space is reached, or a predetermined constant.

The main differences that arise in the discounted case are as follows. First, the discount factor α enters in the definition

$$d_m = g(i_m, i_{m+1}) + \alpha J(i_{m+1}) - J(i_m)$$

of the temporal differences. A second difference is that we replace Assumption 5.1(c) by the condition

$$z_m^t(i) \leq \alpha z_{m-1}^t(i), \quad \text{if } i_m^t \neq i.$$

(Note that this condition is met by all of the special cases we had considered in Section 5.3.4.) By carefully inspecting the proofs in Sections 5.3.5 and 5.3.6, we see that one more condition is required, namely,

$$\mathrm{P}(N_t \geq k \mid \mathcal{F}_t) \leq A\rho^k, \qquad \forall\ k,\ t,$$

where A and ρ are nonnegative constants, with $\rho < 1$. Subject to these additional assumptions, the proofs of Props. 5.1 and 5.2 go through with relatively few modifications and we obtain convergence to J^μ, with probability 1, for both the off-line and the on-line case. The only place where a nontrivial modification is needed is in showing that the mapping A_t in the proof of Prop. 5.1 is a contraction mapping. For this reason, we repeat this part of the proof.

We define

$$(A_t J)(i) = \frac{E\Big[\sum_{m=0}^{N_t-1} z_m^t(i)\big(\alpha J(i_{m+1}^t) - J(i_m^t)\big) \mid \mathcal{F}_t\Big]}{\delta_t(i)} + J(i),$$

Note the presence of α, which comes from the factor α in the definition of the temporal differences for the discounted case. Instead of introducing a weighted maximum norm, we will work with the maximum norm, defined by $\|J\| = \max_i |J(i)|$, and we let $\overline{J} = \|J\|$. We now have the following counterpart of Lemma 5.3.

Lemma 5.4: For every $i \neq 0$ and $t \in T^i$, we have

$$E\left[\sum_{m=0}^{N_t-1} z_m^t(i)\big(\alpha J(i_{m+1}^t) - J(i_m^t)\big) + \sum_{m\in I_t(i)} z_m^t(i)J(i) \;\Big|\; \mathcal{F}_t\right]$$

$$\leq \alpha E\left[\sum_{m\in I_t(i)} z_m^t(i)\overline{J} \;\Big|\; \mathcal{F}_t\right].$$

Proof: Let us fix some state $i \neq 0$ and some $t \in T^i$. As in the proof of Lemma 5.3, we suppress t from our notation, and we do not show explicitly the conditioning on $\mathcal{F}_t$. Furthermore, having fixed i and t, we use the notation z_m to indicate $z_m^t(i)$. Similarly, we use the notation I and N instead of $I_t(i)$ and N_t, respectively.

The left-hand side of the inequality in the statement of the lemma is equal to

$$E\left[\sum_{m=0}^{N-1} z_m\big(\alpha J(i_{m+1}) - J(i_m)\big) + \sum_{m\in I} z_m J(i)\right]$$

$$= E\left[\sum_{m=0}^{N} (\alpha z_{m-1} - z_m)J(i_m) + \sum_{m\in I} z_m J(i)\right], \qquad (5.48)$$

where we have used the properties $z_{-1} = 0$ [cf. Assumption 5.1(b)] and $z_N = 0$. Note that whenever $i_m \neq i$, the coefficient $\alpha z_{m-1} - z_m$ multiplying $J(i_m)$ is nonnegative. Furthermore, whenever $i_m = i$, the coefficient $\alpha z_{m-1} - z_m + z_m$ multiplying $J(i)$ is also nonnegative. Thus, the right-hand side in Eq. (5.48) can be upper bounded by

$$E\left[\sum_{m=0}^{N} (\alpha z_{m-1} - z_m)\overline{J} + \sum_{m\in I} z_m \overline{J}\right] = E\left[\sum_{m=0}^{N-1} z_m(\alpha - 1) + \sum_{m\in I} z_m\right]\overline{J}$$

$$\leq E\left[\sum_{m\in I} z_m(\alpha - 1) + \sum_{m\in I} z_m\right]\overline{J}$$

$$= \alpha E\left[\sum_{m\in I} z_m\right]\bar{J}.$$

Q.E.D.

On-Line Method Based on a Single Trajectory

The last set of alternatives that was introduced in Section 5.3.4 was based on a single infinitely long trajectory and involved the update equation

$$J_{m+1}(i) = J_m(i) + \gamma_m(i) z_m(i) d_m(i), \qquad \forall\ i.$$

We gave there two different formulas for the eligibility coefficients corresponding to the restart and the every-visit variants of TD(λ), respectively [cf. Eqs. (5.33) and (5.34)]. Both of these methods can be shown to converge with probability 1. We omit the arguments for the restart version. The convergence of the every-visit method is proved in Section 6.3.3, in a much more general setting that incorporates a parametric representation of the cost-to-go function.

5.4 OPTIMISTIC POLICY ITERATION

In this section, we discuss a variant of policy iteration whereby we perform policy updates on the basis of only an incomplete evaluation of the current policy. The method is in many respects similar to the asynchronous policy iteration algorithm introduced in Section 2.2.3.

The policy iteration algorithm fixes a policy μ, evaluates the associated cost-to-go function J^μ, possibly using TD(λ), and then performs a policy update. As discussed in Section 2.2.3, methods of this type are often viewed as "actor/critic" systems; cf. Fig. 5.3. The actor uses a policy μ to control the system, while the critic observes the consequences and tries to compute J^μ. For the standard version of policy iteration, the policy μ is fixed for a long time and the critic's computations converge to J^μ; at that point, the limit J^μ is passed to the actor who takes J^μ into account and forms a new policy, by performing the minimization in the right-hand side of Bellman's equation; that is, at each state i, an action u is chosen that minimizes

$$\sum_j p_{ij}(u)\big(g(i,u,j) + J(j)\big),$$

over all $u \in U(i)$. Note that the communication from the critic to the actor is very infrequent.

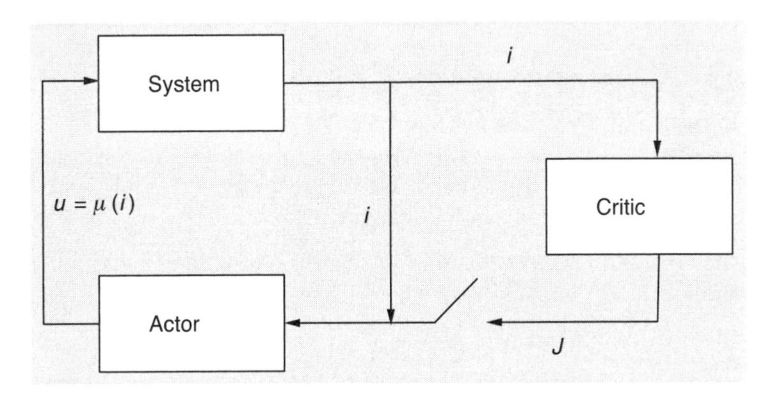

Figure 5.3: An actor/critic system. The critic tries to compute J^μ, e.g., using TD(λ), and occasionally makes a connection with the actor and communicates a J vector. The actor uses the J vectors received from the critic in order to update its policy.

There is an alternative to the standard version of policy iteration, in which we perform policy updates more frequently, without waiting for the policy evaluation algorithm to converge. In terms of the actor/critic framework, this corresponds to more frequent communication from the critic to the actor, with each such communication resulting in a policy update.

An extreme possibility is to perform a policy update subsequent to every update by the policy evaluation algorithm (critic). In the context of simulation-based methods, this means that we may perform a policy update subsequent to the generation of each simulated trajectory, or even subsequent to each simulated transition. Methods of this type, which we will call *optimistic policy iteration*, have been widely used in practice, particularly in conjunction with function approximation, although their convergence properties have never been studied thoroughly. In the remainder of this section, we will describe a variety of methods of this type, and we will present a few results and examples that shed some light on their convergence properties.

Our development will be based on the case where a model of the system is available so that we do not need to work with Q-factors. The generic algorithm to be considered employs TD(λ) in the critic for policy evaluation. In general, there will be a number of TD(λ) iterations under a fixed policy, followed by a policy update based on the vector J currently available at the critic. We observe that there are several degrees of freedom for a method of this type, namely:

(a) the frequency of policy updates (communications from the critic to the actor);

(b) the value of λ;

(c) the choice between alternative versions of TD(λ), e.g., first-visit vs. every-visit or restart, off-line vs. on-line, etc.

It turns out that different choices may result in different convergence behaviors, as will be seen in the remainder of this section. Nevertheless, there are a few simple facts that relate to all methods of this type. More specifically, let us assume the following:

(a) the critic's algorithm is sound in the sense that if μ is held fixed, then the vector J maintained by the critic converges to J^μ;

(b) the critic communicates to the actor an infinite number of times.

Under these assumptions, if the sequence of policies μ generated by the actor converges, then the limit must be an optimal policy. To see this, note that if μ converges, then the critic's vector J must converge to J^μ, by assumption (a) above. Given assumption (b), the vectors J received by the actor also converge to J^μ. Since the policy converges to μ, we conclude that the policy updates based on J^μ and the Bellman equation do not result in any policy changes. Thus, $TJ^\mu = T_\mu J^\mu = J^\mu$, and J^μ is a solution of Bellman's equation, proving that $J^\mu = J^*$.

Thus, we have a general guarantee that if μ converges, we have convergence to the optimal cost-to-go function J^* and to an optimal policy. However, this leaves open the possibility that the algorithm does not converge. Indeed, as will be seen later in this section, there are cases where the sequence of vectors J converges to a limit different than J^* (in which case, the sequence of policies μ must oscillate); there are also cases where J fails to converge.

For the remainder of this section, we assume that we are dealing with a discounted problem, with discount factor $\alpha < 1$, and that the state space is $\{1, \ldots, n\}$. We also occasionally refer to stochastic shortest path problems in which case we assume that the state space is $\{0, 1, \ldots, n\}$, that Assumptions 2.1 and 2.2 are satisfied (there exists a proper policy and every improper policy has infinite cost for some initial state), and we always use the convention $J(0) = 0$.

Visualizing Policy Iteration

In this subsection, we develop a way of visualizing policy iteration, which is particularly useful for the study of its optimistic variants.

Let J be some cost-to-go function. We say that a policy μ is *greedy* with respect to J if μ attains the minimum in the right-hand-side of Bellman's equation, that is, if

$$\sum_j p_{ij}\big(\mu(i)\big)\big(g(i, \mu(i), j) + \alpha J(j)\big) = \min_{u \in U(i)} \sum_j p_{ij}(u)\big(g(i, u, j) + \alpha J(j)\big),$$

for all states i. Using operator notation, a greedy policy μ satisfies

$$T_\mu J = TJ.$$

Note that a greedy policy is not always uniquely defined. In particular, if there is a tie in the minimization with respect to u, there are several greedy policies.

We will now classify vectors J by grouping together all those vectors that lead to the same greedy policy. More specifically, fix some policy μ and let

$$R_\mu = \{J \mid \mu \text{ is greedy with respect to } J\}.$$

Equivalently,

$$J \in R_\mu \quad \text{if and only if} \quad T_\mu J = TJ.$$

In more detail, the above condition becomes

$$\sum_j p_{ij}\big(\mu(i)\big)\big(g\big(i, \mu(i), j\big) + \alpha J(j)\big) \leq \sum_j p_{ij}(u)\big(g(i, u, j) + \alpha J(j)\big),$$

for all i and all $u \in U(i)$. This a system of linear inequalities on the vector J and we conclude that for every policy μ, the set R_μ is a polyhedron. We therefore have a partition of $\Re^n$ (the set of all vectors J) into a collection of polyhedra, and we refer to it as the *greedy partition*. Strictly speaking though, this is not a partition because the different polyhedra share common boundaries. Whenever a vector J leads to more than one greedy policy, that vector belongs to more than one set R_μ.

Let us now recall that a policy μ is optimal if and only if $T_\mu J^\mu = TJ^\mu$; this is equivalent to the requirement $J^\mu \in R_\mu$.

Example 5.11

Consider the discounted, two-state, deterministic problem shown in Fig. 5.4(a). At each one of the two states, there are two possible decisions: stay, at a cost of 1, or move to the other state, at a cost of 0.

Let us identify the greedy policy associated with an arbitrary vector J. At state 1, a greedy policy can choose to stay if and only if

$$1 + \alpha J(1) \leq 0 + \alpha J(2),$$

that is,

$$J(1) \leq J(2) - \frac{1}{\alpha}. \tag{5.49}$$

Similarly, at state 2, a greedy policy can choose to stay if and only if

$$1 + \alpha J(2) \leq 0 + \alpha J(1),$$

that is,

$$J(2) \leq J(1) - \frac{1}{\alpha}. \tag{5.50}$$

We will now identify the sets R_μ associated with the four possible policies.

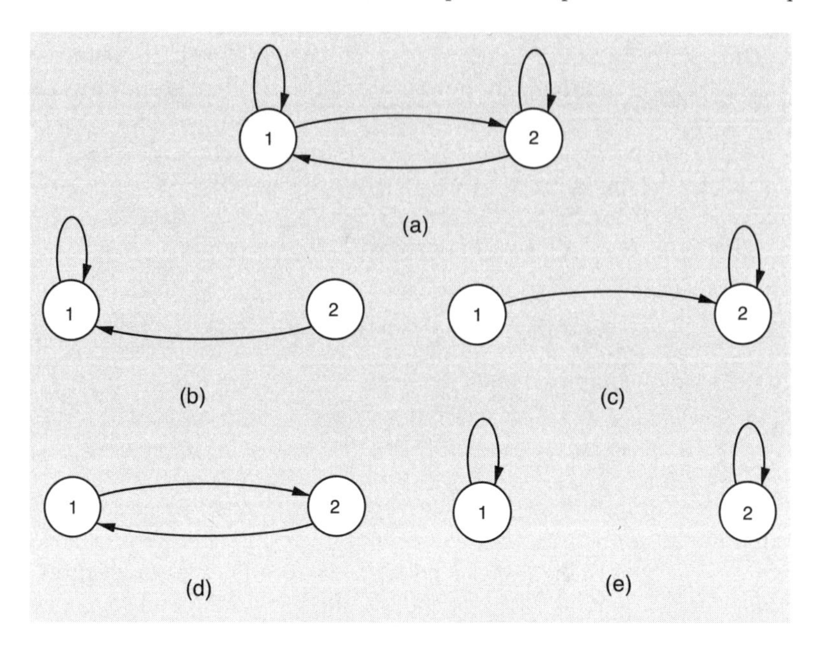

Figure 5.4: A simple deterministic problem is shown in (a). There are four possible policies, μ_l, μ_r, μ_g, μ_w, leading to the transition diagrams shown in parts (b)-(e), respectively.

Consider the worst policy μ_w, which is the one that always decides to stay. Its cost is given by

$$J^{\mu_w}(1) = J^{\mu_w}(2) = \frac{1}{1-\alpha}.$$

For this one to be the greedy policy we need both conditions (5.49) and (5.50) to hold, which is impossible; hence R_{μ_w} is empty.

Consider the "left" policy μ_l that moves from state 2 to state 1 and stays there. Its cost is given by

$$J^{\mu_l}(1) = \frac{1}{1-\alpha}, \qquad J^{\mu_l}(2) = \frac{\alpha}{1-\alpha}.$$

It is easily seen that this is a greedy policy if and only if $J(1) \leq J(2) - 1/\alpha$. By symmetry, the "right" policy μ_r that moves from state 1 to state 2 and stays there has cost

$$J^{\mu_r}(1) = \frac{\alpha}{1-\alpha}, \qquad J^{\mu_r}(2) = \frac{1}{1-\alpha},$$

and is a greedy policy if and only if $J(2) \leq J(1) - 1/\alpha$. Finally, the optimal policy, which is to always move, has zero cost and is a greedy policy if J satisfies $|J(2) - J(1)| \leq 1/\alpha$.

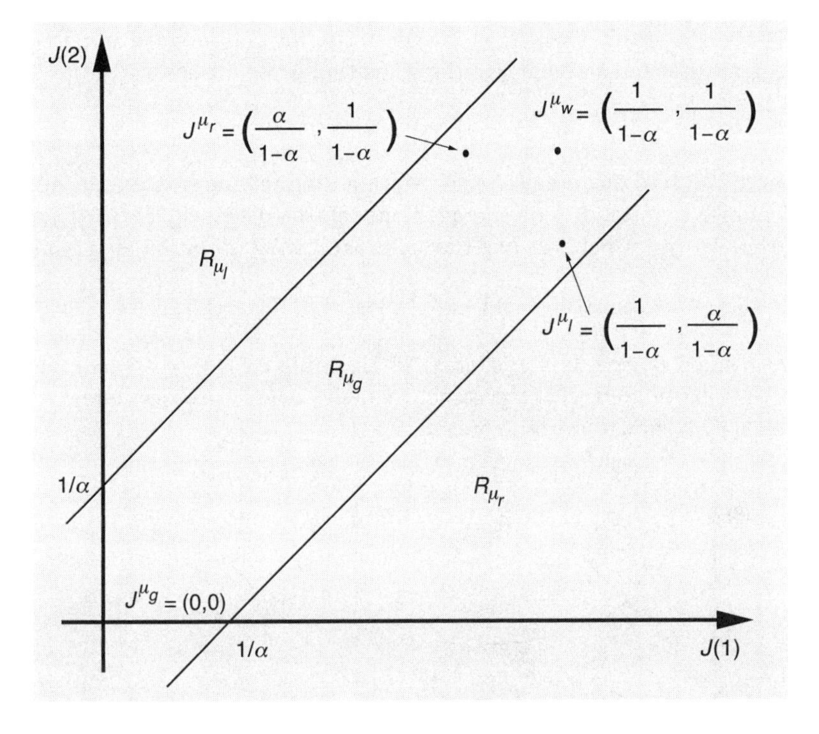

Figure 5.5: Visualization of the greedy partition in Example 5.11.

The vectors J^μ for the different policies and the sets R_μ are shown in Fig. 5.5. Let us now visualize policy iteration.

(a) If we start with a vector J inside R_{μ_g}, the resulting greedy policy is μ_g, and policy evaluation leads to $J^* = (0, 0)$.

(b) If we start with a vector J in R_{μ_l}, the resulting greedy policy is μ_l, and policy evaluation takes us to J^{μ_l}. Since J^{μ_l} belongs to R_{μ_g}, a further policy update leads us to the optimal policy μ_g.

Having developed a way of visualizing the progress of policy iteration, we can now move to the study of its optimistic variants.

Optimistic TD(0)

We consider here a variant of the on-line TD(0) algorithm, for discounted problems. Under a fixed policy μ, the algorithm is of the form

$$\begin{aligned} J(i) &:= J(i) + \gamma\Big(g\big(i, \mu(i), j\big)\big) + \alpha J(j) - J(i)\Big) \\ &= (1-\gamma)J(i) + \gamma\Big(g\big(i, \mu(i), j\big)\big) + \alpha J(j)\Big), \end{aligned}$$

where i is the state at which we choose to perform an update and j is the next state, selected according to the transition probabilities $p_{ij}\big(\mu(i)\big)$. This is of the form

$$J(i) := (1-\gamma)J(i) + \gamma(T_\mu J)(i) + \gamma w, \tag{5.51}$$

where w is a zero mean noise term. With a diminishing stepsize the effects of the noise term tend to be averaged out (cf. the discussion in Section 4.4) and the algorithm behaves like the small-step value iteration algorithm

$$J(i) := (1-\gamma)J(i) + \gamma(T_\mu J)(i),$$

under a fixed policy. In the context of Example 5.11, if γ is small, and if each $J(i)$ is updated with the same frequency, the trajectory followed has the structure shown in Fig. 5.6(a).

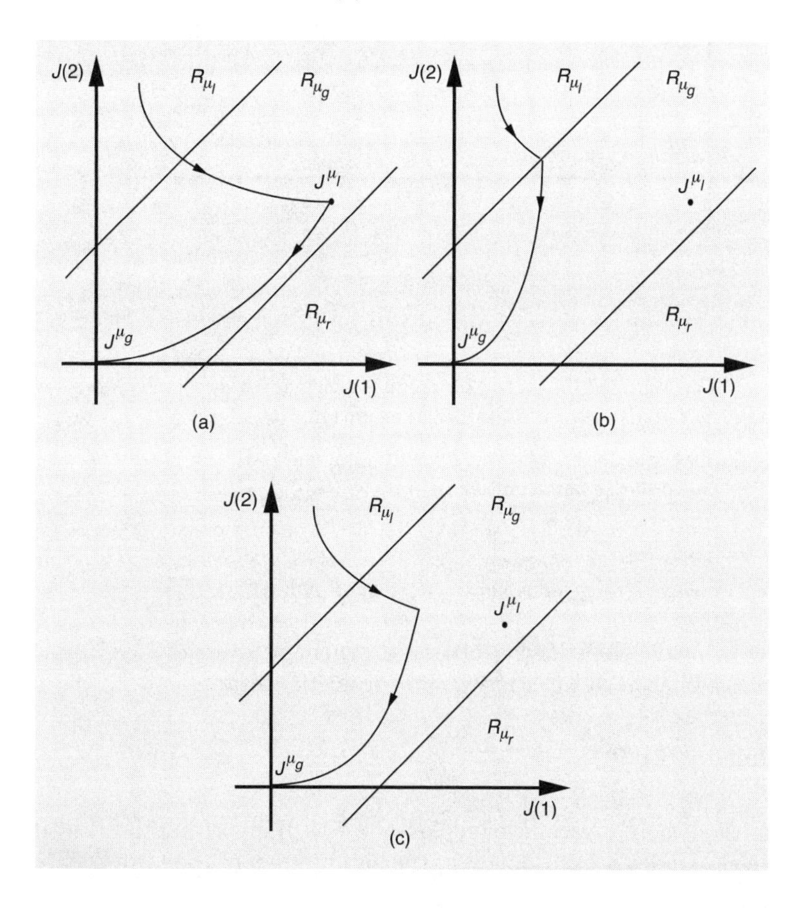

Figure 5.6: The trajectory followed by: (a) policy iteration based on TD(0); (b) optimistic policy iteration based on TD(0); (c) partially optimistic policy iteration based on TD(0). For all three cases, it is assumed that $J(i)$ is updated with the same frequency for all states i.

Let us now consider optimistic TD(0). The update rule is similar to Eq. (5.51), except that the policy μ used at each update is a greedy policy with respect to the current vector J, and satisfies $T_\mu J = TJ$. Thus, the algorithm is of the form

$$J(i) := (1-\gamma)J(i) + \gamma(TJ)(i) + \gamma w, \tag{5.52}$$

where w is again a zero mean noise term. Let us interject here that this version of TD(0) is only of academic interest: if we have a model of the system that allows us to determine a greedy policy μ at each step, this model can also be used to compute $(TJ)(i)$ exactly, instead of using the noisy estimate $g\big(i,\mu(i),j\big) + \alpha J(j)$ of $(TJ)(i)$. In any case, ignoring the noise term, the algorithm tends to behave like a small-step version of value iteration. In the context of Example 5.11, and for small γ, the trajectory followed by this algorithm is of the form shown in Fig. 5.6(b). Note that this trajectory is initially the same as the trajectory followed by (non-optimistic) policy iteration based on TD(0) [cf. Fig. 5.6(a)]. However, as soon as the trajectory crosses from R_{μ_l} into R_{μ_g}, the optimistic method updates the policy from μ_l to μ_g.

The noise term w notwithstanding, convergence results for optimistic TD(0) are immediate consequences of the general results of Section 4.3. This is because the operator T is either a weighted maximum norm contraction (if all policies are proper or if the costs are discounted) or satisfies the monotonicity Assumption 4.4 of Section 4.3.4, and the variance of the noise term w is easily shown to increase only quadratically with J. The only requirement is that an infinite number of updates be carried out at every state and that the stepsizes used satisfy the usual conditions.

Partially Optimistic TD(0)

Let us now consider a partially optimistic variant of TD(0) in which we fix the policy μ, perform a number of updates (5.51) under that policy, and then perform a policy update. In particular, neither do we have a policy update at each step, nor do we wait for convergence to J^μ in order to carry out a policy update. In the context of Example 5.11, the resulting trajectory is of the form shown in Fig. 5.6(c). The breakpoint in that trajectory corresponds to the time at which a policy update is carried out.

For deterministic problems, and if $\gamma = 1$, partially optimistic policy iteration is a special case of the asynchronous policy iteration algorithm introduced in Section 2.2.3. If γ is less than 1, we have a small stepsize variant. Does this algorithm converge? Unfortunately, the picture is rather complicated. The following are known to be true:

(a) For deterministic problems and if the algorithm is initialized with a vector J that satisfies $TJ \le J$, the relation $TJ \le J$ is preserved throughout the algorithm, and we have convergence without any restrictions on the sequence according to which the states are updated (Prop. 2.5 of Section 2.2.3), as long as there is an infinite number of updates at each state. While that proposition was proved for the case $\gamma = 1$, the proof goes through when $0 < \gamma < 1$, or when the costs are discounted.

(b) If the condition $TJ \leq J$ is not initially satisfied, then the algorithm need not converge. A nonconvergent example, involving a deterministic discounted problem, and with stepsize $\gamma = 1$, was given by Williams and Baird [WiB93]. The example is easily modified to yield nonconvergence for smaller stepsizes as well. (A large number of updates at a single state, with a small stepsize, have about the same effect as a single update with $\gamma = 1$.)

(c) If the noise term w is present, as will be the case in problems other than deterministic, the condition $TJ \leq J$ need not be preserved even if it is initially satisfied. In light of item (b) above, there is no reason to expect that convergence is guaranteed.

It should be pointed out that the divergent behavior in the above mentioned examples can be ascribed to a maliciously selected order for choosing states for an update. If states are chosen for updates as they appear along a simulated trajectory, as is the case in TD(0) proper, it is not clear whether nonconvergence is possible.

Optimistic TD(1)

We now consider the case where TD(1) is used for policy evaluation. We focus again on discounted problems, with discount factor $\alpha < 1$.

Suppose that we have a fixed policy μ. We choose an initial state i_0 and generate a trajectory $(i_0, i_1, \ldots)$, simulated according to μ. Having done that, and for every state i_k on the trajectory, the cumulative cost

$$g(i_k, i_{k+1}) + \alpha g(i_{k+1}, i_{k+2}) + \cdots$$

provides us with an unbiased estimate of $J^\mu(i_k)$, i.e., it is equal to $J^\mu(i_k) + w$, where w is a zero mean noise term. Thus, with each cost sample for state i, $J(i)$ is updated with the expected update direction moving it closer to $J^\mu(i)$. However, the nature of the path to be followed by the vector J depends on the frequency with which cost samples are generated for the different states.

The workings of policy iteration based on TD(1) are easier to visualize if we consider the following, somewhat idealized, variant of the method, which we call *synchronous*. We generate one trajectory for each possible initial state, and use a single cost sample ("initial-state-only" variant) to update each $J(i)$; the updates for the different states use the same stepsize. Then, the algorithm is given by

$$J := (1-\gamma)J + \gamma J^{\mu} + \gamma w, \tag{5.53}$$

where w is a zero-mean noise vector.

If $\gamma = 1$ and in the absence of noise, this is simply policy evaluation. For smaller γ, and without noise, the algorithm moves along a straight path from the current J to J^{μ}. In the presence of noise, each step is again targeted at J^{μ}, but noise can cause deviations from the straight path. In the context of Example 5.11, and if γ is very small, the evolution of the algorithm is shown in Fig. 5.7(a). Note that we have a straight trajectory until we reach J^{μ_l}, at which point we have a policy update and we switch to a straight trajectory targeted at J^*.

Let us now consider the optimistic variant of synchronous TD(1). The update equation is the same as (5.53), except that the policy μ being employed is a greedy policy with respect to the current J. In the context of Example 5.11, as soon as the trajectory enters the region R_{μ_g}, the policy is updated to μ_g, and the trajectory starts moving now in the direction of $J^{\mu_g} = J^*$; see Fig. 5.7(b).

Finally, let us consider a partially optimistic variant of TD(1) where a finite number of updates (5.53) are carried out before a policy update. We then obtain a trajectory like the one shown in Fig. 5.7(c). The breakpoint in that trajectory occurs precisely at the time where the policy is updated from μ_l to μ_g.

In turns out that the synchronous variant of optimistic TD(1) is guaranteed to converge to J^* with probability 1, at least for the case of discounted problems [Tsi96]. A similar convergence result is obtained for the initial-state-only variant of optimistic TD(1) if the initial state is chosen randomly, with all states having the same probability of being selected. It is also conjectured that the convergence result extends to cover a suitably defined synchronous optimistic version of TD(λ). We omit the detailed presentation and derivation of these results because they only refer to an idealized version of the algorithm and it is unclear whether they are relevant to more practical variants (as will be discussed next) or to methods involving function approximation.

Synchronous optimistic TD(1) provides some useful insights. However, it is quite restrictive and does not capture the full flavor of the TD(1) methods that are commonly used, for two reasons:

(a) In the synchronous method, every state is chosen to be the initial state with the same frequency whereas, in practice, more freedom is desirable. For example, if there is a natural initial state, we might be inclined to start all or most trajectories from that initial state.

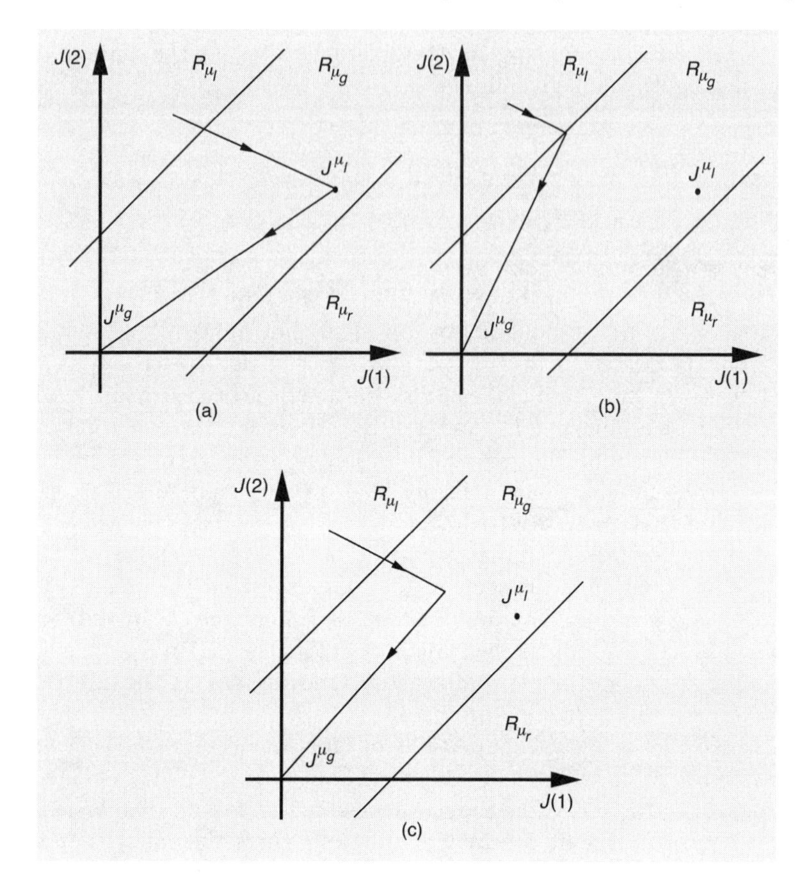

Figure 5.7: The trajectory followed by: (a) policy iteration based on TD(1); (b) optimistic policy iteration based on TD(1); (c) partially optimistic policy iteration based on TD(1). For all three cases, it is assumed that J is updated according to (5.53), but there are differences in the way that the policy μ is selected.

(b) Having generated a trajectory, we may be inclined to use several of the generated cost samples, instead of limiting ourselves to using only the cost sample corresponding to the initial state.

Unfortunately, the more general variants of optimistic TD(1) are much harder to analyze. The example that follows shows that optimistic policy iteration based on (asynchronous) TD(1) can diverge.

Example 5.12

Consider the problem described in Example 5.11. Let us assume that the discount factor α is very close to 1. Then, there exists a vector $\overline{J}$ [see Fig. 5.8(a)] with the following properties:

(a) The vector $\overline{J}$ lies on the boundary of the regions R_{μ_l} and R_{μ_g}, that is, it satisfies $\overline{J}(2) = \overline{J}(1) + (1/\alpha)$.

(b) If we form a 2×1 rectangle centered at $\overline{J}$, any vector J in the interior of that rectangle satisfies

$$0 < J(1) < \frac{1}{1-\alpha},$$

and

$$\frac{1}{\alpha} < J(2) < \frac{\alpha}{1-\alpha}.$$

We partition this 2×1 rectangle into four regions $P_1, \ldots, P_4$, as indicated in Fig. 5.8(a).

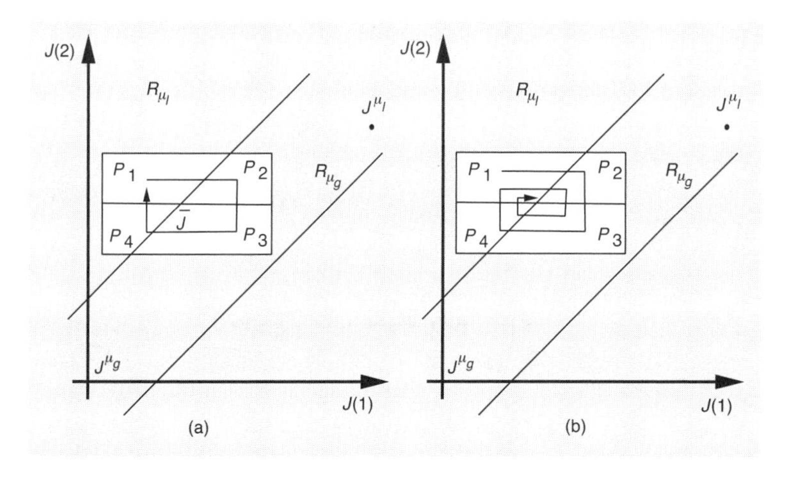

Figure 5.8: (a) Illustration of the failure to converge to J^*. (b) Illustration of convergence to $\overline{J}$ when the stepsize is diminishing.

We consider optimistic policy iteration based on the initial-state-only variant of TD(1). Note that at each iteration, there is a single component of J that gets updated and therefore, an update of $J(i)$ is effected as follows. We generate an infinite trajectory starting at i, using the current policy μ, and evaluate the cost $J^\mu(i)$ of that trajectory. [Since the problem is deterministic, there is no noise in the evaluation of $J^\mu(i)$.] We then update $J(i)$ according to

$$J(i) := (1-\gamma)J(i) + \gamma J^\mu(i).$$

We allow the stepsize to change from one update to the next, but we assume that it is small enough so that $\gamma/(1-\alpha) < 1/2$. This has the following implication. If $0 < J(i) < 1/(1-\alpha)$, we use the fact $0 < J^\mu(i) < 1/(1-\alpha)$ for all μ, to conclude that the step $\gamma\big(J^\mu(i) - J(i)\big)$ is less than $1/2$.

Suppose that we start the algorithm at some J that belongs to P_1 and that we update $J(1)$. For J in P_1, μ_l is a greedy policy and $J^{\mu_l}(1) =$

$1/(1-\alpha) > J(1)$. As a consequence, as long as we are inside P_1, each update increases the value of $J(1)$ and eventually J enters P_2. (Note that J cannot "jump over" P_2 because each step is bounded by $1/2$.)

Inside P_2, μ_g is a greedy policy and $J^{\mu_g}(2) = 0 < J(2)$. Thus, if we perform a number of updates of $J(2)$, then $J(2)$ decreases and J eventually enters P_3. We continue similarly. Once in P_3, we update $J(1)$ until J enters P_4. Once in P_4, we update $J(2)$ until J enters P_1. This process can be continued ad infinitum. The algorithm stays forever in the 2×1 rectangle centered at $\overline{J}$, and does not converge to J^*.

The example we have constructed is compatible with either a fixed stepsize γ or with a diminishing stepsize. With a fixed stepsize, the algorithm could converge to a limit cycle centered at $\overline{J}$, in which case the greedy policy corresponding to the current vector J evolves periodically. However, aperiodic (chaotic) behavior also seems possible.

If a diminishing stepsize is employed, the algorithm can be made to converge to $\overline{J}$, as illustrated in Fig. 5.8(b). If we now recall that there was a lot of freedom in choosing $\overline{J}$, we conclude that there is a large (uncountable) set of possible limit points of the algorithm.

We note a few peculiarities of Example 5.12:

(a) There is no restriction on the relative frequency of updates at states 1 and 2.

(b) Each update is based on the cost J^μ of an infinitely long trajectory.

(c) We use the initial-state-only version of TD(1).

Regarding item (a), we note that the same example goes through if we require each update of $J(1)$ to be followed by an update of $J(2)$ and vice versa, as long as we allow the stepsizes to be arbitrary. If we restrict the updates of $J(1)$ and $J(2)$ to be interleaved, and if we also require the stepsize to be the same for $J(1)$ and $J(2)$, then we obtain an algorithm which is almost the same as synchronous optimistic TD(1) and which is convergent for the case of diminishing stepsize.

Regarding item (b), we could modify our example so that there is an additional termination state, and at each iteration there is a positive probability p of terminating. In that case, each update would involve a trajectory which is finite with probability 1. As long as p is small (especially, if it is much smaller than $1-\alpha$), the character of the problem does not change much and failure to converge should still be possible.

Finally, regarding item (c), we conjecture that other variants of TD(1) (e.g., first-visit or every-visit) are also divergent, especially if there is some latitude in the choice of the stepsizes.

The above example notwithstanding, there is still the possibility that one of the simpler variants of optimistic policy iteration could be guaranteed to converge to J^*. For example, let us suppose that there is a termination state, that we choose the initial state of each trajectory at random, with a fixed probability distribution, and that we use the off-line initial-state-only, or the every-visit TD(1) method. If the fixed distribution is uniform, and if the initial-state-only variant is employed, the method is known to converge as we discussed earlier. If the every-visit method is used or if the initial distribution is non-uniform, the convergence properties of the resulting algorithm are not known.

Summary

A high level summary of the results in this section is as follows. We have a class of methods that are parametrized by λ as well as by the time, call it k, between consecutive policy updates.

When $k = \infty$, we have ordinary policy iteration, with different choices of λ corresponding to different ways of carrying out policy evaluation. We have convergence for all values of λ.

When $k = 1$, we have optimistic policy iteration. For $\lambda = 0$, this is essentially the same as asynchronous value iteration, and we have convergence due to the contraction property. By a continuity argument, it follows that we must also have convergence for small values of λ.

For intermediate values of k and for $\lambda = 0$, we are dealing with noisy versions of modified policy iteration. The method converges for the right initial conditions and in the absence of noise. It can diverge, in general.

For $k = 1$ and $\lambda = 1$, we have optimistic TD(1) which converges synchronously but can diverge asynchronously.

These results seem to favor optimistic TD(0) as well as non-optimistic TD(λ) for any λ. One can never be certain that such theoretical results are indicative of performance on typical practical problems; nevertheless, they provide some indications of what could go wrong with any given method. Of course, all of these results refer to lookup table methods. Once we move into large scale problems and introduce approximations, the picture can only get more complex. Still, results obtained for the case of lookup table representations delineate the range of results that might be possible in a more general context.

5.5 SIMULATION-BASED VALUE ITERATION

We will now investigate simulation-based methods that mimic the value iteration algorithm. We only discuss the undiscounted case, but the methods and the results we present are also applicable to discounted problems. Throughout this section, we assume that an analytical model of the system is available. (The case where such a model is not available is dealt with in the next section.) Thus, the only possible use of simulation is to generate representative states on which the computational effort is to concentrate. This is of interest primarily if the state space is fairly large and we wish to avoid updates at states that are highly unlikely to be ever visited.

The general structure of the algorithm to be considered is as follows. Using random sampling, simulation, or any other method, we obtain a sequence of states $i_0, i_1, \ldots$, and we carry out updates of the form

$$J(i_k) := \min_{u \in U(i_k)} \sum_{j=0}^{n} p_{i_k j}(u)\big(g(i_k, u, j) + J(j)\big). \tag{5.54}$$

An interesting possibility here is to choose the states where we update on the basis of simulated trajectories, but update at the different states in the reverse of the order that they appeared on these trajectories, because this may lead to faster convergence. In any case, the algorithm described here is simply the asynchronous value iteration algorithm introduced in Ch. 2 and is guaranteed to converge to the optimal cost-to-go function, if each state is visited infinitely often (see Prop. 2.3 in Section 2.2.2).

The requirement that each state be visited infinitely often is quite restrictive and may be difficult to satisfy in problems with a large state space. In practice, one may wish to perform updates at only a subset of the state space. Suppose, for example, that we are dealing with a problem in which we always have the same initial state, and that we are only interested in the optimal cost-to-go $J^*(i)$ and in the optimal decisions for those states that may arise while following an optimal policy. If these states are a small subset of the state space, one would hope that an optimal policy can be obtained while performing updates only in that subset. There is a difficulty here because the nature of an optimal policy is not known ahead of time and, therefore, there is no a priori knowledge of the states at which updates should be performed. Clearly, a certain amount of *exploration* of the state space is needed in order to avoid missing a profitable alternative. A nice way out of such dilemmas is provided by the method described in the next subsection.

Simulations Based on a Greedy Policy

We concentrate on the special case where there is a single initial state, say state 1. In our general description of the method (5.54), we had allowed the states $i_0, i_1, \ldots$ to be generated according to some arbitrary mechanism provided that all states are visited infinitely often. We now assume that state trajectories are generated by starting at state $i_0 = 1$ and by following a greedy policy with respect to the currently available cost-to-go function J. A formal statement of the algorithm is as follows. Before a typical update, we are at a state i_k and have available a cost-to-go function J_k. We set

$$J_{k+1}(i_k) = \min_{u \in U(i_k)} \sum_{j=0}^{n} p_{i_k j}(u)\big(g(i_k, u, j) + J_k(j)\big), \tag{5.55}$$

and

$$J_{k+1}(i) = J_k(i), \qquad i \neq i_k.$$

We then choose a value of u that attains the minimum in the right-hand side of Eq. (5.55), and let the next state i_{k+1} be equal to j with probability equal to $p_{i_k j}(u)$. The only exception arises if i_{k+1} is the terminal state, in which case we reset i_{k+1} to be the initial state. The example below shows that this method may suffer from insufficient exploration.

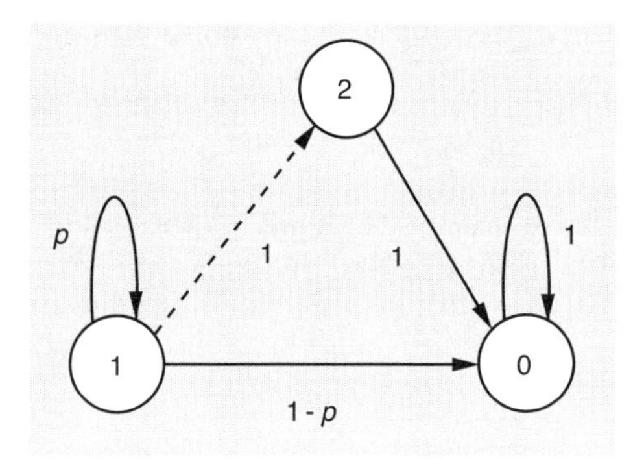

Figure 5.9: A three-state stochastic shortest path problem. State 0 is a cost-free absorbing state. From state 2, we always move to state 0, at zero cost. At state 1, there are two options: under the first, say $u = 1$, we move to state 2 at zero cost; under the second, say $u = 2$, we stay at state 1, with probability p, and move to state 0, with probability $1 - p$; the cost of the second option is equal to 1.

Example 5.13

Consider the three-state stochastic shortest path problem shown in Fig. 5.9, suppose that $p = 1/2$, and let $i_0 = 1$ be the initial state. The optimal policy at state 1 is to move to state 2. Clearly, $J^*(1) = J^*(2) = 0$. Suppose that we initialize the algorithm with $J_0(1) = 2$ and $J_0(2) = 3$; as usual, we also set $J_0(0) = 0$. We have

$$J_1(1) = \min\big\{J_0(2), 1 + J_0(1)/2\big\} = 2,$$

and $J(1)$ remains unchanged. The greedy policy at state 1 is to let $u = 2$ which means that state 2 is not visited. Thus, as long as we keep starting at state 1, the greedy policy does not visit state 2, the values of $J(1)$ and $J(2)$ never change, and we are stuck with a non-optimal policy.

The failure to converge to an optimal policy in Example 5.13 was due to the large initial value of $J(2)$ which prevented us from exploring that state. Suppose now that the algorithm is initialized by setting each $J(i)$ to some value $J_0(i)$ that satisfies

$$J_0(i) \leq J^*(i),$$

for all i. We will prove shortly that the algorithm converges to an optimal policy starting from state 1. The intuition behind this result is as follows. We use monotonicity of the dynamic programming operator to show that the inequality $J(i) \leq J^*(i)$ is maintained throughout the algorithm, and all states appear more profitable than they actually are. Consequently, if a state i is profitably visited under an optimal policy, the cost-to-go $J(i)$ being an underestimate makes that state appear even more profitable, and the greedy policy will eventually take us there.

Example 5.14

Consider the same problem as in Example 5.13 and suppose that $J_0(1)$ and $J_0(2)$ are less than or equal to zero. As long as the greedy policy selects the nonoptimal action at state 1, we update $J(1)$ according to

$$J(1) := 1 + pJ(1).$$

If this were to continue indefinitely, $J(1)$ would eventually become positive, hence larger than $J(2)$. Thus, at some point the greedy policy should switch to the optimal policy.

Suppose now that $p = 1$, $J_0(1) = -100$, and $J_0(2) = 0$. We then see that it takes 100 visits to state 1 before the greedy policy switches to the optimal policy. This indicates that the number of iterations until convergence to an optimal policy can be very large if the initial underestimates J_0 are far below J^*.

We now state a precise convergence result and provide its proof.

Proposition 5.3: Assume that there exists a proper policy and that all improper policies have infinite cost at some state (Assumptions 2.1 and 2.2) Assume that each simulated trajectory starts from the same initial state, say state 1, and that $J_0 \leq J^*$. Then:

(a) The sequence J_k converges to some J_∞.

(b) Let I be the set of states i that are visited infinitely often by the algorithm. For $i \in I$, let $\mu(i)$ be an action that is applied at state i infinitely many times in the course of the algorithm. For $i \notin I$, let $\mu(i)$ be arbitrary. Then,

$$J^\mu(i) = J_\infty(i) = J^*(i), \qquad \forall\, i \in I.$$

Proof: In this proof, we will view the algorithm as an asynchronous value iteration method, limited to the states that are visited infinitely often, and then use a result from Ch. 2 to show that the sequence $J_k(i)$ converges to an underestimate $J_\infty(i)$ of $J^*(i)$. For those states i that are visited infinitely often, we will also show that $J_\infty(i)$ is the cost-to-go of a particular policy, and is therefore no smaller than $J^*(i)$, thus completing the proof.

Our first step is to prove, by induction on k, that $J_k \leq J^*$ for all i. For $k = 0$, this is true by assumption. Suppose this is true for some k. We then have

$$\begin{aligned}
J_{k+1}(i_k) &= \min_{u \in U(i_k)} \sum_{j=0}^{n} p_{i_k j}(u)\big(g(i_k, u, j) + J_k(j)\big) \\
&\leq \min_{u \in U(i_k)} \sum_{j=0}^{n} p_{i_k j}(u)\big(g(i_k, u, j) + J^*(j)\big) \\
&= J^*(i_k),
\end{aligned}$$

and the induction argument has been completed.

Let I be the set of nonterminal states that are visited infinitely often by the algorithm, together with the termination state 0. Let $J_\infty(i)$ be the final values of $J_k(i)$ at the remaining states. Since there exists a time after which we only have updates at states i belonging to I, we will view this as the initial time and assume that $J_k(i) = J_\infty(i)$ for all $i \notin I$ and all k. We are then dealing with a special case of the asynchronous value iteration algorithm for a modified stochastic shortest path problem in which all states $i \notin I$ are treated as terminal states with terminal cost $J_\infty(i)$. Since the original problem satisfied Assumptions 2.1 and 2.2, it is easily shown that Assumptions 2.1 and 2.2 also hold true for the modified problem. (The existence of a proper policy for the original problem implies the existence of a proper policy for the modified problem. Furthermore, an improper policy in the modified problem leads to an improper policy for the original problem and must have infinite cost for at least one initial state in I, hence verifying Assumption 2.2 for the modified problem.) We can now use Prop. 2.3 from Section 2.2.3, on the convergence of asynchronous value iteration, to conclude that J_t converges to some J_∞. Furthermore, for all $i \in I$, $J_\infty(i)$ is equal to the optimal cost-to-go for the modified problem that we have introduced.

Let us fix some $i \in I$, and let $\mu(i)$ be a decision that is applied infinitely many times at state i. At each time k that action $\mu(i)$ is applied at state i, we have

$$\begin{aligned}
J_{k+1}(i) &= \sum_{j=0}^{n} p_{ij}\big(\mu(i)\big)\big(g(i, \mu(i), j) + J_k(j)\big) \\
&= \min_{u \in U(i)} \sum_{j=0}^{n} p_{ij}(u)\big(g(i, u, j) + J_k(j)\big);
\end{aligned}$$

this is because the action that is applied must attain the minimum in Eq. (5.55). Taking the limit as $k \to \infty$, we obtain

$$\begin{aligned} J_\infty(i) &= \sum_{j=0}^{n} p_{ij}\big(\mu(i)\big)\big(g(i,\mu(i),j) + J_\infty(j)\big) \\ &= \min_{u \in U(i)} \sum_{j=0}^{n} p_{ij}(u)\big(g(i,u,j) + J_\infty(j)\big), \qquad \forall\, i \in I. \end{aligned}$$

This implies that $\tilde{\mu} = \{\mu(i) \mid i \in I\}$ is an optimal policy for the modified problem. In particular, $\tilde{\mu}$ is a proper policy for the modified problem.

Consider any $i \in I$ and $j \notin I$. By definition, decision $\mu(i)$ is applied infinitely often at state i; since state j is only visited a finite number of times, we must have $p_{ij}\big(\mu(i)\big) = 0$. Thus, under the policy $\tilde{\mu}$ and starting in I, the states outside I are never reached. Consider now a policy μ for the original problem which coincides with $\tilde{\mu}$ on the set I. Starting from a state $i \in I$, it does not matter whether we are dealing with the original or the modified problem, since states outside I are never reached. Hence, for $i \in I$, the cost of $\tilde{\mu}$ in the modified problem, which is $J_\infty(i)$, is equal to the cost $J^\mu(i)$ of μ in the original problem, and $J_\infty(i) = J^\mu(i) \geq J^*(i)$. On the other hand, we have shown that $J_k \leq J^*$ for all k, implying that $J_\infty = \lim_{k\to\infty} J_k \leq J^*$, and we conclude that $J_\infty(i) = J^*(i)$ for all $i \in I$, and that policy μ is optimal starting from any state $i \in I$. **Q.E.D.**

According to Prop. 5.3, we obtain a policy μ that is optimal for all the states that are visited infinitely often by the algorithm, and $J^\mu(i) = J^*(i)$ for all such states. For other states, it is possible that $J^\mu(i) \neq J^*(i)$, even if these states are visited by some optimal policy other than μ; see the example that follows.

Example 5.15

Consider the stochastic shortest path problem shown in Fig. 5.10, in which all policies are optimal and $J^*(1) = J^*(2) = J^*(3) = 2$. Suppose that the algorithm is initialized with $J_0(1) = 0$, $J_0(2) = 2$, and $J_0(3) = 0$. The greedy policy at state 1 chooses not to go to state 2 and $J(1)$ is updated according to

$$J(1) := 1 + J(1)/2.$$

No matter how many updates of this form are carried out, $J(1)$ is always strictly smaller than 2, which means that the greedy policy never visits state 2 or state 3. We see that $J(1)$ converges to the correct value of 2. However, the value of $J(3)$ is never corrected. Thus, $J(3)$ does not converge to $J^*(3)$ even though state 3 is a state that would be visited under *some* optimal policy.

According to Prop. 5.3, the algorithm provides us, in the limit, with an optimal action and with the value of $J^*(i)$, for all states i that are

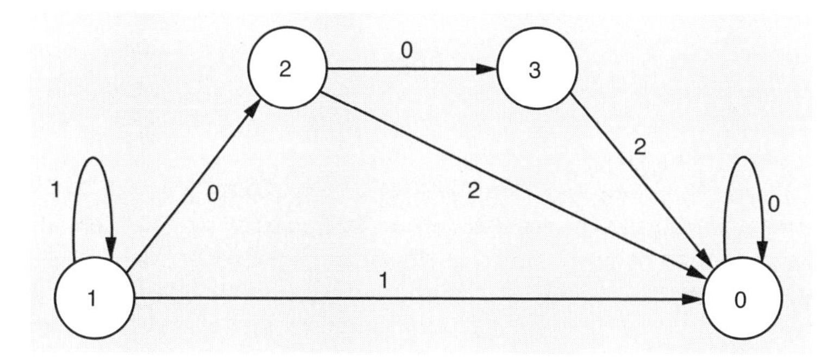

Figure 5.10: A four-state stochastic shortest path problem. The transition costs are indicated next to each arc. At state 2, either arc can be followed. At state 1, there are two options: the first is to move to state 2, at zero cost; under the second option, we stay at state 1 with probability 1/2, or we move to state 0, with probability 1/2.

visited infinitely many times by the algorithm. Is the initial state included in this set of states? The answer is affirmative, but this needs proof. (In principle, the algorithm could consist of a single infinitely long trajectory that never reaches the termination state and, therefore, never revisits the initial state.)

Proposition 5.4: Under the assumptions of Prop. 5.3, the initial state is included in the set of states that are visited infinitely often by the algorithm, and we therefore obtain an optimal policy starting from that state.

Proof: For any state i, let $U^*(i)$ be the set of all actions at state i that satisfy

$$\sum_{j=0}^{n} p_{ij}(u)\big(g(i,u,j) + J^*(j)\big) = \min_{v \in U(i)} \sum_{j=0}^{n} p_{ij}(v)\big(g(i,v,j) + J^*(j)\big).$$

According to Prop. 5.3, if state i is visited an infinite number of times and if action u is selected infinitely often at that state, then u is part of some optimal policy and, therefore, $u \in U^*(i)$. In particular, after some finite time, the algorithm only chooses actions belonging to the sets $U^*(i)$.

Note that any policy that satisfies $\mu(i) \in U^*(i)$ for all i is an optimal policy and must be proper. We apply Prop. 2.2 from Section 2.2.1 to a new problem in which the sets $U(i)$ are replaced by $U^*(i)$. Since all policies are now proper, there exist positive constants $\xi(1), \ldots, \xi(n)$, and a scalar $\beta \in [0, 1)$ such that

$$\sum_{j=0}^{n} p_{ij}(u)\xi(j) \leq \beta\xi(i), \qquad \forall\, i \neq 0,\ u \in U^*(i),$$

where we also use the convention $\xi(0) = 0$. Thus, we see that for all times k greater than some finite time, we have

$$E\big[\xi(i_{k+1}) \mid i_k \neq 0\big] \leq \beta\xi(i_k),$$

where i_k is the state generated by the algorithm at the kth iteration. This means that as long as state 0 is not reached, $\xi(i_k)$ is a positive supermartingale, it must converge with probability 1, and because $\beta < 1$, it can only converge to zero. But this implies that state 0 must be reached with probability 1, and therefore the initial state will be revisited. **Q.E.D.**

The method of this section, specialized to deterministic shortest path problems, amounts to an algorithm for the shortest path problem for a single origin-destination pair (we are looking for a shortest path from state 1 to state 0). As illustrated by Example 5.14 (with $p = 1$), the number of iterations depends heavily on the initial underestimates J_0, and for this reason it is not a polynomial-time algorithm. Nevertheless, it could be an efficient method if tight underestimates are available.

We finally note that the algorithm can be used in the presence of multiple initial states, if we introduce an artificial initial state that leads to any of the true initial states with equal probability.

A Generalization

A careful inspection of the proof of Prop. 5.3 reveals that the particular way in which the system is simulated is not crucial, because we only made use of the following property: if $\mu(i)$ is a decision that is selected infinitely many times at state i and if j is any other state, then either state j is visited infinitely often, or $p_{ij}\big(\mu(i)\big) = 0$. But this property could also be enforced by means other than simulation under the greedy policy. An example is the following: whenever $J(i)$ is updated and a minimizing action u is determined, then every state j for which $p_{ij}(u) > 0$ must be picked for an update of $J(j)$ at some later time. With this observation at hand, we may now define a related algorithm from which simulation has been completely eliminated.

The algorithm maintains a priority queue containing states at which an update must be carried out. During a typical iteration, we pick the state at the top of the queue, perform an update as in Eq. (5.55), choose an action u that attains the minimum in Eq. (5.55), and insert at the bottom (or possibly in the middle) of the queue all states j such that $p_{ij}(u) > 0$.

Proposition 5.3 remains valid for all such variants as long as the algorithm does not concentrate on a subset of the states and leave some other states waiting indefinitely in queue, and as long as we have an infinite number of updates at state 1.

5.6 Q-LEARNING

We now introduce an alternative computational method that can be used whenever there is no explicit model of the system and the cost structure. This method is analogous to value iteration and can be used directly in the case of multiple policies. It updates directly estimates of the Q-factors associated with an optimal policy, thereby avoiding the multiple policy evaluation steps of the policy iteration method.

We assume that we are dealing with a stochastic shortest path problem, with state 0 being a cost-free termination state, and that Assumptions 2.1 and 2.2 of Ch. 2 are satisfied. Let us define the optimal Q-factor $Q^*(i,u)$ corresponding to a pair (i,u), with $u \in U(i)$, by letting $Q^*(0,u) = 0$ and

$$Q^*(i,u) = \sum_{j=0}^{n} p_{ij}(u)\big(g(i,u,j) + J^*(j)\big), \qquad i = 1,\ldots,n. \tag{5.56}$$

Bellman's equation can be written as

$$J^*(i) = \min_{u \in U(i)} Q^*(i,u). \tag{5.57}$$

Combining the above two equations, we obtain

$$Q^*(i,u) = \sum_{j=0}^{n} p_{ij}(u)\left(g(i,u,j) + \min_{v \in U(j)} Q^*(j,v)\right). \tag{5.58}$$

The optimal Q-factors $Q^*(i,u)$ are the unique solution of the above system, as long as we only consider Q-factors that obey the natural condition $Q(0,u) = 0$, which will be assumed throughout. This can be proved as follows. If some $Q(i,u)$ solve the system (5.58), then the vector with components $\min_{u \in U(i)} Q(i,u)$ is seen to satisfy Bellman's equation. Therefore, using the uniqueness of solutions of Bellman's equation, we have $\min_{u \in U(i)} Q(i,u) = \min_{u \in U(i)} Q^*(i,u)$ for all i. Using the assumption that $Q(i,u)$ satisfy Eq. (5.58), we conclude that $Q(i,u) = Q^*(i,u)$.

More insight into the meaning of the Q-factors can be obtained by the following argument. Let us introduce a new system whose states are the original states $0,1,\ldots,n$, together with all pairs (i,u), $u \in U(i)$, $i \neq 0$; see Fig. 5.11. Whenever the state is some (i,u), there are no decisions to

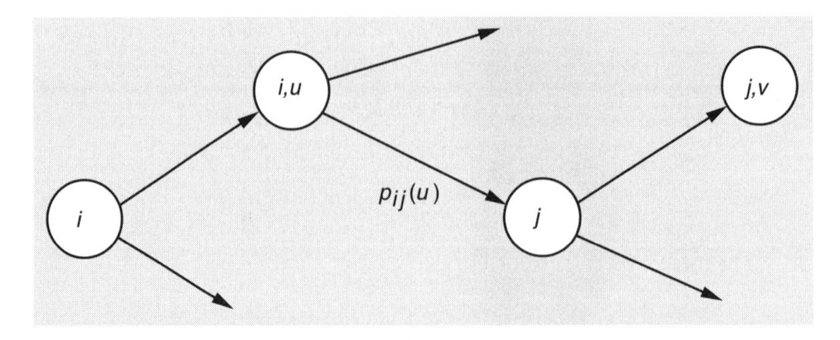

Figure 5.11: An auxiliary problem that provides an interpretation of the Q-factors.

be made, the next state j is chosen according to the probabilities $p_{ij}(u)$, and a cost of $g(i,u,j)$ is incurred. Whenever the state is some i, a decision $u \in U(i)$ is made and the next state is (i,u), deterministically. If we let $J^*(i)$ and $Q^*(i,u)$ be the optimal cost-to-go in this new system, starting from state i or state (i,u), respectively, then Eqs. (5.56)-(5.57) are simply Bellman's equations for the new system and Eq. (5.58) can be viewed as a two-step Bellman equation. Assuming that the original system satisfies Assumptions 2.1 and 2.2, it is easily shown that the same is true for the new system. In particular, it follows from the results of Ch. 2, that Eqs. (5.56)-(5.57) have a unique solution and, therefore, Eq. (5.58) also has a unique solution.

In terms of the Q-factors, the value iteration algorithm can be written as

$$Q(i,u) := \sum_{j=0}^{n} p_{ij}(u)\left(g(i,u,j) + \min_{v \in U(j)} Q(j,v)\right), \qquad \text{for all } (i,u). \quad (5.59)$$

A more general version of this iteration is

$$Q(i,u) := (1-\gamma)Q(i,u) + \gamma \sum_{j=0}^{n} p_{ij}(u)\left(g(i,u,j) + \min_{v \in U(j)} Q(j,v)\right), \quad (5.60)$$

where γ is a stepsize parameter with $\gamma \in (0,1]$, that may change from one iteration to the next. The Q-learning method is an approximate version of this iteration, whereby the expectation with respect to j is replaced by a single sample, i.e.,

$$Q(i,u) := (1-\gamma)Q(i,u) + \gamma\left(g(i,u,j) + \min_{v \in U(j)} Q(j,v)\right). \quad (5.61)$$

Here j and $g(i,u,j)$ are generated from the pair (i,u) by simulation, that is, according to the transition probabilities $p_{ij}(u)$. Thus, Q-learning can be viewed as a combination of value iteration and simulation. Equivalently, Q-learning is the Robbins-Monro stochastic approximation method based on Eq. (5.58).

In the special case where there is a single policy and the minimization with respect to v is omitted, Q-learning uses the same update rule as the TD(0) algorithm of Section 5.3 [cf. Eq. (5.14)]. Once Q-learning is viewed as an extension of TD(0) to the model-free case, one is led to look for a similar extension of TD(λ), for $\lambda > 0$. A key difficulty here is that there is no simple and usable ℓ-step Bellman equation that generalizes Eq. (5.15) to the case of multiple policies. Nevertheless, certain extensions of Q-learning, that incorporate a parameter λ have been discussed in the literature; see, e.g., Watkins [Wat89], and Peng and Williams [PeW94].

The Q-learning algorithm has convergence properties that are fairly similar to those of TD(λ), as long as every state-control pair (i, u) is visited infinitely often and the stepsize γ diminishes to zero at a suitable rate. An example of a stepsize choice that guarantees convergence is the following: if an update corresponds to the mth visit of the pair (i, u), let $\gamma = b/(a + m)$, where a and b are positive constants.

A rigorous convergence proof is provided in the next subsection but its substance is easily summarized. We express the algorithm in the form $Q := (1 - \gamma)Q + \gamma\big(HQ + w\big)$, where HQ is defined to be equal to the right-hand-side of Eq. (5.59), and w is a zero mean noise term. We then show that H satisfies the monotonicity Assumption 4.4 of Section 4.3.4, and appeal to our general results on the convergence of stochastic approximation algorithms. For the special case where all policies are proper, it will be shown that H is also a weighted maximum norm contraction and this leads to somewhat stronger convergence results.

The Convergence of Q-Learning

In this subsection, we prove formally that the Q-learning algorithm converges to the optimal Q-factors $Q^*(i, u)$, with probability 1. We start by writing down a formal description of the algorithm.

We use a discrete variable t to index successive updates and we let $T^{i,u}$ be the set of times t that an update of $Q(i, u)$ is performed. We let $\mathcal{F}_t$ represent the history of the algorithm up to and including the point where the stepsizes $\gamma_t(i, u)$ are chosen, but just before the simulated transitions needed for the updates at time t and their costs are generated. We are dealing with the algorithm:

$$Q_{t+1}(i, u) = \big(1 - \gamma_t(i, u)\big)Q_t(i, u) + \gamma_t(i, u)\left(g(i, u, \bar{i}) + \min_{v \in U(\bar{i})} Q_t(\bar{i}, v)\right),$$

where the successor state $\bar{i}$ is picked at random, with probability $p_{i\bar{i}}(u)$. Furthermore, $\gamma_t(i, u) = 0$ for $t \notin T^{i,u}$. Throughout, we assume that $Q_t(0, u) = 0$ for all t.

Proposition 5.5: Suppose that

$$\sum_{t=0}^{\infty} \gamma_t(i,u) = \infty, \qquad \sum_{t=0}^{\infty} \gamma_t^2(i,u) < \infty, \qquad \forall\ i,\ u \in U(i).$$

Then, $Q_t(i,u)$ converges with probability 1 to $Q^*(i,u)$, for every i and u, in each of the following cases:

(a) If all policies are proper.

(b) If Assumptions 2.1 and 2.2 hold (there exists a proper policy and every improper policy has infinite cost for some initial state), and if $Q_t(i,u)$ is guaranteed to be bounded, with probability 1.

Proof: We define a mapping H that maps a Q-vector to a new Q-vector HQ according to the formula

$$(HQ)(i,u) = \sum_{j=0}^{n} p_{ij}(u)\left(g(i,u,j) + \min_{v \in U(j)} Q(j,v)\right), \qquad i \neq 0,\ u \in U(i). \tag{5.62}$$

Then, the Q-learning algorithm is of the form

$$Q_{t+1}(i,u) = \big(1 - \gamma_t(i,u)\big)Q_t(i,u) + \gamma_t(i,u)\Big((HQ_t)(i,u) + w_t(i,u)\Big),$$

where

$$w_t(i,u) = g(i,u,\bar{\imath}) + \min_{v \in U(\bar{\imath})} Q_t(\bar{\imath},v) - \sum_{j=0}^{n} p_{ij}(u)\Big(g(i,u,j) + \min_{v \in U(j)} Q_t(j,v)\Big).$$

This is precisely the type of algorithm that was considered in Ch. 4.

Note that $E[w_t(i,u) \mid \mathcal{F}_t] = 0$ and that

$$E[w_t^2(i,u) \mid \mathcal{F}_t] \leq K\big(1 + \max_{j,v} Q_t^2(j,v)\big),$$

where K is a constant. Thus, Assumption 4.3 in Section 4.3.1 on the conditional mean and variance of the noise term is satisfied.

It now remains to verify that H has the properties required by the general results of Ch. 4 on the convergence of stochastic approximation algorithms.

We consider first the case where all policies are proper. According to Prop. 2.2 in Section 2.2.1, there exist positive coefficients $\xi(i)$, $i \neq 0$, and a scalar $\beta \in [0,1)$ such that

$$\sum_{j=1}^{n} p_{ij}(u)\xi(j) \leq \beta\xi(i), \qquad \forall\ i,\ u \in U(i).$$

For any vector Q of Q-factors, we define

$$\|Q\|_\xi = \max_{i,u\in U(i)} \frac{|Q(i,u)|}{\xi(i)}.$$

Then, for any two vectors Q and $\overline{Q}$, we have

$$\begin{aligned}\big|(HQ)(i,u) - (H\overline{Q})(i,u)\big| &\le \sum_{j=1}^n p_{ij}(u)\Big|\min_{v\in U(j)} Q(j,v) - \min_{v\in U(j)} \overline{Q}(j,v)\Big| \\ &\le \sum_{j=1}^n p_{ij}(u) \max_{v\in U(j)} \big|Q(j,v) - \overline{Q}(j,v)\big| \\ &\le \sum_{j=1}^n p_{ij}(u)\|Q-\overline{Q}\|_\xi\, \xi(j) \\ &\le \beta\|Q-\overline{Q}\|_\xi\, \xi(i).\end{aligned}$$

We divide both sides by $\xi(i)$ and take the minimum over all i and $u \in U(i)$, to conclude that $\|HQ-H\overline{Q}\|_\xi \le \beta\|Q-\overline{Q}\|_\xi$ and, therefore, H is a weighted maximum norm contraction. Convergence of the Q-learning algorithm now follows from Prop. 4.4 in Section 4.3.2.

Let us now remove the assumption that all policies are proper and impose instead Assumptions 2.1 and 2.2. It is easily seen that H is a monotone mapping, namely, if $Q \le \overline{Q}$, then $HQ \le H\overline{Q}$. Also, if r is a positive scalar and e is the vector with all components equal to 1, then it is easily checked that

$$HQ - re \le H(Q-re) \le H(Q+re) \le HQ + re.$$

Finally, the equation $HQ = Q$, which is the same as Eq. (5.58), has a unique solution, as argued earlier in this section. In conclusion, all of the assumptions of Prop. 4.6 in Section 4.3.4 are satisfied and Q_t converges to Q^* with probability 1, provided that Q_t is bounded with probability 1. **Q.E.D.**

We continue by addressing the boundedness condition in Prop. 5.5. One option is to enforce it artificially using the "projection" method, as discussed in Section 4.3.5. There is also a special case in which boundedness is automatically guaranteed, which is the subject of our next result.

Proposition 5.6: Consider the Q-learning algorithm under the same assumptions as in Prop. 5.5. Furthermore, suppose that all one-stage costs $g(i,u,j)$ are nonnegative, the stepsizes satisfy $\gamma_t(i,u) \le 1$ for all i, u, t, and that the algorithm is initialized so that $Q_0(i,u) \ge 0$ for all (i,u). Then, the sequence Q_t generated by the Q-learning algorithm is bounded with probability 1 and therefore converges to Q^*.

Proof: Given the assumptions $g(i,u,j) \geq 0$ and $Q_0(i,u) \geq 0$, it is evident from Eq. (5.60) and the assumption $\gamma_t(i,u) \leq 1$ that $Q_t(i,u) \geq 0$ for all i, u, and t. This establishes a lower bound on each $Q_t(i,u)$.

Let us now fix a proper policy μ. We define an operator H_μ by letting

$$(H_\mu Q)(i,u) = \sum_{j=0}^{n} p_{ij}(u)\Big(g(i,u,j) + Q\big(j,\mu(j)\big)\Big), \qquad i = 1,\ldots,n.$$

The operator H_μ is the dynamic programming operator for a system with states (i,u) and with the following dynamics: from any state (i,u), we get to state $\big(j,\mu(j)\big)$, with probability $p_{ij}(u)$; in particular, subsequent to the first transition, we are always at a state of the form $\big(i,\mu(i)\big)$ and the first component of the state evolves according to μ. Because μ was assumed proper for the original problem, it follows that the system with states (i,u) also evolves according to a proper policy. Therefore, there exists some $\beta \in [0,1)$ and a weighted maximum norm $\|\cdot\|_\xi$ such that $\|H_\mu Q - Q^\mu\|_\xi \leq \beta\|Q - Q^\mu\|_\xi$ for all vectors Q, where Q^μ is the unique fixed point of H_μ. By comparing the definition (5.62) of H with that of H_μ, we see that for every vector $Q \geq 0$, we have $0 \leq HQ \leq H_\mu Q$. Using this and the triangle inequality, we obtain for every $Q \geq 0$,

$$\begin{aligned}\|HQ\|_\xi \leq \|H_\mu Q\|_\xi &\leq \|H_\mu Q - H_\mu Q^\mu\|_\xi + \|H_\mu Q^\mu\|_\xi \\ &\leq \beta\|Q - Q^\mu\|_\xi + \|Q^\mu\|_\xi \leq \beta\|Q\|_\xi + 2\|Q^\mu\|_\xi.\end{aligned}$$

This establishes that H satisfies the assumptions of Prop. 4.7 in Section 4.3.5 and the result follows. **Q.E.D.**

We close by noting that for the case where the costs $g(i,u,j)$ are allowed to be negative, and without the assumption that all policies are proper, the question of the boundedness of the Q-learning algorithm for stochastic shortest path problems has not yet been settled.

Q-Learning for Discounted Problems

A discounted problem can be handled by converting it to an equivalent stochastic shortest path problem. Alternatively, one can deal directly with the discounted problem, using the algorithm

$$Q(i,u) := (1-\gamma)Q(i,u) + \gamma\Big(g(i,u,j) + \alpha \min_{v\in U(j)} Q(j,v)\Big). \tag{5.63}$$

Here j and $g(i,u,j)$ are generated from the pair (i,u) by simulation, that is, according to the transition probabilities $p_{ij}(u)$, and α is the discount factor. If we assume the same stepsize conditions as in Prop. 5.5, we obtain convergence to the optimal Q-factors $Q^*(i,u)$, with probability 1. The proof is similar to the proof of Prop. 5.5 and is omitted. Suffice to say that the iteration mapping associated to this algorithm is a contraction with respect to the maximum norm, with contraction factor α, and convergence follows by applying Prop. 4.4 in Section 4.3.2.

Exploration

It should be clear that in order for optimistic policy iteration to perform well, all potentially important parts of the state space should be explored. In the case of Q-learning, there is an additional requirement, namely, that all potentially beneficial actions be explored. For example, the convergence theorem requires that all state-action pairs (i, u) are tried infinitely often.

In a common implementation of Q-learning, a sequence of states is generated by simulating the system under the greedy policy provided by the currently available Q-factors, which is similar in spirit to optimistic TD(0). This implies that one only considers state-action pairs of the form $\big(i, \mu(i)\big)$, where μ is the current greedy policy. Even if all states i are adequately explored, it is still possible that certain profitable actions u are never explored. For this reason, several methods have been suggested that occasionally depart from the greedy policy. In one method (see Cybenko, Gray, and Moizumi [CGM95]), one switches between exploration intervals where actions are chosen randomly, and intervals during which greedy actions are employed. In a second method, there is a small positive parameter ϵ: a greedy action is chosen with probability $1 - \epsilon$, and a random action is used with probability ϵ. In a third method, when at state i, an action u is chosen with probability

$$\frac{\exp\{-Q(i,u)/T\}}{\sum_{v \in U(i)} \exp\{-Q(i,v)/T\}},$$

where T is a positive "temperature" parameter; note that when T is chosen very small, almost all of the probability goes to an action u that has the smallest $Q(i, u)$, that is, a greedy action. All three methods involve parameters that control the degree of exploration (the frequency and length of exploration intervals, the probability ϵ, or the temperature T, respectively), and these can be changed in the course of the algorithm. If they are chosen so that enough exploration is guaranteed, convergence to the optimal Q-factors can be established. If at the same time, exploration is gradually eliminated, then we asymptotically converge to a greedy policy, which must be optimal given the convergence of the Q-factors.

5.7 NOTES AND SOURCES

5.2. The comparison of the every-visit and first-visit methods, including Examples 5.4-5.6 is due to Singh and Sutton [SiS96].

5.3. Temporal difference methods have been introduced by Sutton [Sut84], [Sut88], building on related work by Barto, Sutton, and Anderson

[BSA83]. The restart variant was introduced by Singh and Sutton [SiS96], who call it the "replace" method. The different variants are systematically compared by Singh and Dayan [SiD96], on the basis of analytical formulas that they derive. The formalization of general temporal difference methods in Section 5.3.3 is new.

Sutton [Sut88] has shown that under TD(0), the expectation of J_t converges to J^μ. Dayan [Day92] extended this result to the case of general λ. However, this is a very weak notion of convergence. Jaakkola, Jordan, and Singh [JJS94] carried out an argument similar to the one outlined at the end of Section 5.3.2, to show that off-line every-visit TD(λ) is a stochastic approximation algorithm based on a contraction mapping, and then argued (somewhat informally) that this implies convergence. Dayan and Sejnowski [DaS94] sketched another possible approach for proving convergence, based on the ODE approach. Another, rigorous, proof has been provided by Gurvits, Lin, and Hanson [GLH94]. The extension of these proofs to cover cases other than first-visit or every-visit TD(λ), e.g., the restart variant, is not apparent, and this was our motivation for introducing general temporal difference methods and proving their convergence (Prop. 5.1), which is a new result. A (nonprobabilistic) worst-case analysis of the behavior of TD(λ) has been carried out by Schapire and Warmuth [ScW96]. Finally, Barnard [Bar93] shows that, generically, the expected step direction of TD(0) cannot be interpreted as the gradient of any function, in contrast to TD(1).

Proposition 5.2 on the convergence of on-line TD(λ), for $\lambda < 1$, is from Jaakkola et al. [JJS94] who provided an informal proof. The convergence of the on-line method for discounted problems based on a single infinitely long trajectory is a new result.

Bradtke and Barto [BrB96] have introduced a "least squares temporal difference" (LS-TD) method and they prove that under natural assumptions, it converges to the true cost-to-go vector. The update equations in their method are similar to the Kalman filtering recursions, and they are more demanding computationally than TD(0). Recall that the Kalman filter computes at each step the least squares optimal estimates given the available simulation data and therefore converges faster than TD(1), which is an incremental gradient method. For roughly the same reasons, one expects that LS-TD should converge faster than TD(0).

5.4. "Optimistic policy iteration" (the term is nonstandard) has been used explicitly or implicitly in much of the empirical work in reinforcement learning, but not much was available in terms of theoretical analysis, with the exception of the somewhat related work of Williams and Baird [WiB93]. The divergent counterexample (Example 5.12) is new and was based on insights provided by Van Roy. The convergence result for the synchronous case is due to Tsitsiklis [Tsi96].

5.5. Asynchronous value iteration with the greedy policy being used for selecting states to be updated has been called *real-time dynamic programming* by Barto, Bradtke, and Singh [BBS95], who discuss it at some length and also present experimental results for the "racetrack" problem. For deterministic shortest path problems, Prop. 5.3 is due to Korf [Kor90]. The generalization to stochastic shortest path problems, under a positivity assumption on the costs, is due to Barto et al. [BBS95]. The extension to nonpositive costs is a new result.

5.6. The Q-learning algorithm is due to Watkins [Wat89]. Its convergence was established by Watkins and Dayan [WaD92] for discounted problems and for stochastic shortest path problems where all policies are proper. The connection with stochastic approximation was made by Jaakkola et al. [JJS94] and by Tsitsiklis [Tsi94]. The convergence proof without the assumption that all policies are proper is due to Tsitsiklis [Tsi94]. The issue of exploration is discussed by Barto et al. [BBS95] who also provide a number of related references.

That is you can't you know tune in
but it's all right. That is I think it's not too bad.
(The Beatles)

6

Approximate DP with Cost-to-Go Function Approximation

Contents

In this chapter, we discuss several possibilities for the approximate solution of DP problems. Generally, we are interested in approximations of the cost-to-go function J^μ of a given policy μ, of the optimal cost-to-go function J^*, or of the optimal Q-factors $Q^*(i,u)$. This is done using a function which, given a state i, produces an approximation $\tilde{J}(i,r)$ of $J^\mu(i)$, or an approximation $\tilde{J}(i,r)$ of $J^*(i)$, or, given also a control u, an approximation $\tilde{Q}(i,u,r)$ of $Q^*(i,u)$. The approximating function involves a parameter vector r, and may be implemented using a neural network, a feature extraction mapping, or any other suitable architecture; see the discussion in Ch. 3. The parameter vector r is determined by optimization, often using some type of least squares framework.

We observe that the lookup table representation considered in the preceding chapter can be viewed as a limiting form of an approximate representation. In particular, if the dimension of the parameter vector r is the same as the number of non-terminal states and if $\tilde{J}(i,r) = r(i)$ for all i, then maintaining the values of the parameter vector r is the same as keeping the values $J(i)$ in a lookup table. For this limiting case, most of the algorithms to be developed in this chapter degenerate into the methods that were discussed in Ch. 5.

In approximate DP, there are two main choices:

(a) The choice of an approximation architecture to represent various cost-to-go functions or Q-factors.

(b) The choice of a training algorithm, that is, a method for updating the parameter vector r.

These choices are often coupled because some algorithms are guaranteed to work properly only for certain types of architectures. Thus, while much of the discussion in this chapter focuses on algorithmic issues, we will often consider the implications of particular types of architectures, especially linear ones.

In our study of lookup table methods, in Ch. 5, we concentrated on the convergence of different algorithms to the optimal cost-to-go function J^*. Once approximations are introduced, convergence to J^* cannot be expected, for the simple reason that J^* may not be within the set of functions that can be represented exactly by the chosen architecture. Thus, the typical questions to be investigated are:

(a) Does a given algorithm converge?

(b) If an algorithm converges, does the limit possess some desirable properties, e.g., is it close to J^*?

(c) If an algorithm does not converge, does it oscillate within some small neighborhood of J^*?

Our approach to the second and third questions can be roughly described as follows. We introduce a parameter, call it ϵ, that characterizes the *power of the approximation architecture.* For example, we can define ϵ as the minimum of $\|\tilde{J} - J^*\|$, where the minimum is taken over all $\tilde{J}$ that can be represented by the chosen architecture, and where $\|\cdot\|$ is a suitable norm. Given the limitations imposed by an architecture, no algorithm can produce results whose distance from J^* is less than ϵ, and we have an absolute standard against which the performance of different algorithms is to be measured. We say that an algorithm has an *amplification factor* of at most c if the algorithm is guaranteed to converge to a region of radius $c\epsilon$ around J^*.

The power ϵ of an architecture is problem specific; it depends on the number of free parameters (e.g., on the number of "neurons" in a multilayer perceptron), and on the smoothness properties of the function to be approximated. On the other hand, in defining the amplification factor c of an algorithm, we consider the worst case over all possible problems. As a consequence, the amplification factor becomes a problem-independent property of a given algorithm.

We will be deriving error bounds, leading to estimates of the amplification factors, for a variety of algorithms. For discounted problems, we sometimes obtain amplification factors of the form $O\big(1/(1-\alpha)\big)$ or $O\big(1/(1-\alpha)^2\big)$, which may seem disappointing, because α is usually close to 1. Nevertheless, the mere finiteness of the amplification factor can be viewed as a certification that we are dealing with a fundamentally sound algorithm: as the power of the architecture is improved, that is, as $\epsilon \downarrow 0$, we are guaranteed to approach the exact solution. Furthermore, a drastic difference in the amplification factors of two algorithms can be viewed as serious evidence that one is intrinsically better than the other.

In most of this chapter, we stay within our basic framework, which assumes a finite-state controlled Markov chain. However, once an approximation architecture is introduced, finiteness of the state space is no longer required, and most of the algorithms to be presented can be also applied to problems involving infinite, possibly continuous, state spaces.

Chapter Outline

A broad variety of algorithmic ideas and analyses are given in this chapter. We have organized this material in four parts:

(a) Generic issues, applying to all algorithms (Section 6.1).

(b) Methods related to policy iteration (Sections 6.2-6.4).

(c) Methods related to value iteration (Sections 6.5-6.9).

(d) Other methods (Sections 6.10-6.12).

The chapter begins with a discussion, in Section 6.1, of some ways of constructing policies, starting from an approximation of the optimal cost-to-go function. We also provide performance guarantees for such policies.

In Section 6.2, we discuss approximate policy iteration, whereby the performance of any given policy is evaluated approximately, e.g., by means of Monte Carlo simulation followed by a least squares fit. We also derive a bound on the amplification factor of such a method. In Section 6.3, we present temporal difference methods for approximate policy evaluation, and prove convergence for the case of linearly parametrized approximation architectures. Then, in Section 6.4, we discuss optimistic policy iteration. While this method is not completely understood at present, we indicate that it typically exhibits chattering, whereby the policies obtained keep oscillating, even though the parameters of the approximation architecture may converge.

We then turn to algorithms based on value iteration. In Section 6.5, we discuss methods that perform value iterations at some states, followed by a least squares fit, as well as some incremental variants, and we show that divergence is possible. In Section 6.6, we consider the model-free case and present a Q-learning algorithm that is compatible with function approximation, as well as a very brief discussion of advantage updating. Because value iteration can generally diverge, we are motivated to look for special structures under which convergence is guaranteed. In Section 6.7, we establish convergence results and performance guarantees for approximate value iteration, when state aggregation and a piecewise constant approximation of the cost-to-go function is employed. In Section 6.8, similar positive results are obtained under the assumption that the DP operator is a contraction with respect to a weighted Euclidean norm. We show that this assumption is always satisfied for some optimal stopping problems, and we establish convergence of an algorithm of the Q-learning type for such problems. Finally, in Section 6.9, we consider the case where value iterations are carried out only at a set of representative states, as is often done when dealing with discretizations of continuous-state problems. Convergence is again demonstrated under a further assumption on the parametrization of the approximate cost-to-go function at the remaining states.

In Section 6.10, we introduce a method that directly aims at an approximate solution of Bellman's equation. With T being the DP operator, this method tries to minimize the "Bellman error" $\|TJ - J\|$, over the class of functions J that can be represented by the available approximation architecture. Several implementations and the relation with TD(0) are also discussed. In Section 6.11, we note that the "slope" of the cost-to-go function is often very important and we discuss approaches that aim at learning this slope directly. We also discuss some issues that arise from deterministic, continuous-time optimal control problems. The last method of this chapter, which is presented in Section 6.12, is based on linear programming and capitalizes on the linear programming formulation of DP problems. We then conclude in Section 6.13, with an overview of the different methods and of the issues that they raise.

6.1 GENERIC ISSUES – FROM PARAMETERS TO POLICIES

Most of the methods discussed in this chapter eventually lead to an approximate cost-to-go function $\tilde{J}(i,r)$, which is meant to be a good approximation of the optimal cost-to-go function $J^*(i)$. Such a cost-to-go function leads to a corresponding *greedy policy* μ defined by

$$\mu(i) = \arg \min_{u \in U(i)} \sum_j p_{ij}(u)\big(g(i,u,j) + \tilde{J}(j,r)\big), \qquad \forall\ i. \tag{6.1}$$

(For simplicity, we assume here that we are dealing with an undiscounted problem.) In this section, we elaborate on how we can use Eq. (6.1) to obtain an implementable policy. We focus on the situation where the number of states is large and therefore a cost-to-go function cannot be made available as a table with the values of $\tilde{J}(i,r)$ for each i. Instead, we assume that we have chosen a value for the parameter vector r and that we have access to a *subroutine* which on input i outputs $\tilde{J}(i,r)$. Our objective is to use this subroutine as a tool and construct a new one which on input i produces $\mu(i)$.

Note that a subroutine that inputs i and outputs $\mu(i)$, is needed in two different contexts:

(a) Simulation-based methods for the approximate computation of J^* often require us to simulate the system under a policy μ, as in the simulation-based policy iteration method discussed in Ch. 5. In this context, the subroutine must be called after each transition. Given that we may need to simulate a very large number of transitions, the subroutine should be computationally efficient.

(b) Once all training and computation is completed, we will eventually want to implement a policy on a physical system, that runs in real time. Physical considerations often dictate the window available between the time that the system moves into a new state and the time that a decision must be applied. Clearly, the running time of the subroutine for $\mu(i)$ should not be longer than the length of this window.

In the simplest setting, where a model is available, the subroutine for μ works as follows. Once a state i is given, we simply compute the right-hand side of Eq. (6.1); note that doing so requires a number of calls to the subroutine for $\tilde{J}$, with a separate call needed to compute $\tilde{J}(j,r)$ for every j for which $p_{ij}(u)$ is positive. For this to be practically feasible, the number of possible immediate successors j of state i must be reasonably small. If the number of potential successor states j is large, we may have to estimate the sum

$$\sum_j p_{ij}(u)\big(g(i,u,j) + \tilde{J}(j,r)\big)$$

using Monte Carlo simulation, that is, by generating several random samples of j according to the probabilities $p_{ij}(u)$, and by averaging the resulting values of $g(i,u,j) + \tilde{J}(j,r)$.

The issues we have just discussed are greatly simplified if approximate Q-factors $\tilde{Q}(i,u,r)$ are available because for every state i, the corresponding decision $\mu(i)$ can be obtained by evaluating $\tilde{Q}(i,u,r)$ for every $u \in U(i)$ and then taking the minimum.

The preceding comments are predicated on the assumption that the sets $U(i)$ are fairly small in size so that one can afford to compute or estimate

$$\sum_j p_{ij}(u)\big(g(i,u,j) + \tilde{J}(j,r)\big)$$

for every $u \in U(i)$, and then take the minimum. There are also some interesting cases where the set $U(i)$ is large or infinite, but the minimization with respect to u can be carried out analytically (see the example below). This is rarely the case when nonlinear approximation architectures such as multilayer perceptrons are used, but sometimes arises when suitable linear architectures are employed, as shown by the following example.

Example 6.1

Consider the deterministic linear quadratic problem in which we are dealing with the scalar system

$$x_{k+1} = ax_k + bu_k, \qquad x_k, u_k \in \Re,$$

and the cost function

$$g(x,u,y) = Ku^2 + x^2,$$

where K is a positive scalar. Consider an approximation architecture of the form

$$\tilde{J}(x,r) = r(0) + r(1)x + r(2)x^2.$$

Given the current parameters $r(0), r(1), r(2)$, and a state x, the corresponding decision $\mu(x)$ is obtained by minimizing

$$Ku^2 + r(0) + r(1)(ax+bu) + r(2)(ax+bu)^2$$

over all u, which is easy to do analytically.

Generally, in problems where the sets $U(i)$ contain a large number of controls, the implementation of a greedy policy can become quite difficult. For this reason, given such a problem, it may be worthwhile to consider reformulations or approximations that involve a smaller number of controls at each state. For example, in the maintenance problem of Example 2.4 in Ch. 2, it may be worth considering variants where at most one machine can break down at any one time period, so that there are only two possible decisions (repair or not repair). Similarly, the channel allocation problem of Example 2.5 in Ch. 2 greatly simplifies if the arrival, departure, and handoff events that trigger state transitions are assumed to occur one-at-a-time. A more general approach for dealing with problems with a complex control space is considered in Section 6.1.4.

Approximation of Q-Factors

We now mention a variant which requires the solution of an additional approximation problem. In this variant, we form an approximation $\hat{Q}(i,u,s)$, where the parameter vector s is determined by solving the least squares problem

$$\min_s \sum_{(i,u)\in Y} \big(\hat{Q}(i,u,s) - \tilde{Q}(i,u,r)\big)^2, \tag{6.2}$$

where Y is a representative set of state-control pairs (i,u), and $\tilde{Q}(i,u,r)$ is evaluated using the equation

$$\tilde{Q}(i,u,r) = \sum_j p_{ij}(u)\big(g(i,u,j) + \tilde{J}(j,r)\big),$$

and either exact calculation or simulation. Note that the optimization problem (6.2) is solved off-line and the value of s is fixed once and for all. For the resulting policy to be implemented on-line, we need to minimize $\hat{Q}(i,u,s)$ with respect to u whenever the current state is i. This variant could be useful if either a model of the system is unavailable or if for a typical i and u there exist many j with $p_{ij}(u) > 0$.

Policy Approximation

If all of the preceding methods are so computationally demanding that $\mu(i)$ cannot be evaluated on-line, there is another approach that can be followed. Given the cost-to-go function $\tilde{J}$, we use any of the previously discussed methods and the formula

$$\mu(i) = \arg\min_{u\in U(i)} \sum_j p_{ij}(u)\big(g(i,u,j) + \tilde{J}(j,r)\big), \qquad \forall\, i, \tag{6.3}$$

to compute the policy $\mu(i)$ for states i in a representative subset $\hat{S}$. We then "generalize" the decisions $\mu(i)$, $i \in \hat{S}$, to obtain a policy $\tilde{\mu}(i,v)$, which is defined for all states, by introducing a parameter vector v and by solving the least squares problem

$$\min_v \sum_{i\in\hat{S}} \big\|\tilde{\mu}(i,v) - \mu(i)\big\|^2. \tag{6.4}$$

Here, we assume that the controls are elements of some normed space and $\|\cdot\|$ denotes the norm on that space. This approach involves an additional least squares problem but can accelerate the simulation of a policy: once v has been fixed, the decision $\tilde{\mu}(i,v)$ is readily obtained, without having to go through the possibly time consuming calculations in Eq. (6.3). The approximation architecture that provides us with $\tilde{\mu}(i,v)$ is often called an *action network*, to distinguish it from the architecture that provides us with $\tilde{J}(i,r)$, which is often called a *critic* network.

The least squares problem (6.4) can be solved after training is completed and an approximate cost-to-go function $\tilde{J}$ has become available. One reason for doing so could be the need for quick on-line computation of the required decisions when the policy is implemented in real time. However, an action network can also be useful for speeding up the simulation of a policy in the course of a simulation-based method. In that case, the least squares problem (6.4) has to be solved simultaneously with the training of $\tilde{J}$, which means that there will be two approximation architectures (or "neural networks") that are trained in an interleaved or concurrent fashion. Some more details will be provided later on.

6.1.1 Generic Error Bounds

Approximate DP is based on the hypothesis that if J is a good approximation of J^*, then a greedy policy based on J is close to optimal. We provide here some evidence in favor of this hypothesis.

Under our standing assumption that there are finitely many states and that the set of possible decisions at each state is finite, it is easily shown that when J is sufficiently close to J^*, a corresponding greedy policy must be an optimal policy. The result that follows establishes this fact and provides bounds on the performance of a greedy policy when J is not necessarily close to J^*.

Proposition 6.1: Consider a discounted problem, with discount factor $\alpha < 1$, and let $\|\cdot\|$ stand for the maximum norm. Suppose that a vector J satisfies $\|J - J^*\| = \epsilon$ for some $\epsilon \geq 0$. If μ is a greedy policy based on J, then

$$\|J^\mu - J^*\| \leq \frac{2\alpha\epsilon}{1-\alpha}.$$

Furthermore, there exists some $\epsilon_0 > 0$ such that if $\epsilon < \epsilon_0$, then μ is an optimal policy.

Proof: We use the DP mappings T and T_μ defined by

$$(TJ)(i) = \min_{u \in U(i)} \sum_j p_{ij}(u)\big(g(i,u,j) + \alpha J(j)\big), \qquad i = 1, \ldots, n,$$

and

$$(T_\mu J)(i) = \sum_j p_{ij}\big(\mu(i)\big)\big(g(i, \mu(i), j) + \alpha J(j)\big), \qquad i = 1, \ldots, n.$$

We have

$$\begin{aligned}
\|J^\mu - J^*\| &= \|T_\mu J^\mu - J^*\| \\
&\le \|T_\mu J^\mu - T_\mu J\| + \|T_\mu J - J^*\| \\
&\le \alpha\|J^\mu - J\| + \|TJ - J^*\| \\
&\le \alpha\|J^\mu - J^*\| + \alpha\|J^* - J\| + \alpha\|J - J^*\| \\
&= \alpha\|J^\mu - J^*\| + 2\alpha\epsilon,
\end{aligned}$$

and the first result follows. We have used above the facts that both T and T_μ are contractions, and the definition $TJ = T_\mu J$ of a greedy policy.

Let

$$\delta = \min_{\overline{\mu}} \big\|J^{\overline{\mu}} - J^*\big\|,$$

where the minimum is taken over all nonoptimal policies. Since there are finitely many policies, the minimum is attained and $\delta > 0$. If ϵ is small enough so that $2\alpha\epsilon/(1-\alpha) < \delta$, then $\|J^\mu - J^*\| < \delta$, and μ must be optimal. **Q.E.D.**

The bounds provided by Prop. 6.1 are tight, as shown by the example that follows.

Example 6.2

Consider the two-state discounted problem shown in Fig. 6.1, where ϵ is a positive number and $\alpha \in [0, 1)$ is the discount factor. The optimal policy is to move from state 1 to state 2 and the optimal cost-to-go function is $J^*(1) = J^*(2) = 0$. Consider the vector J with $J(1) = -\epsilon$ and $J(2) = \epsilon$. Clearly, $\|J - J^*\| = \epsilon$. The policy μ that decides to stay at state 1 is a greedy policy based on J, because $2\epsilon\alpha + \alpha J(1) = \epsilon\alpha = 0 + \alpha J(2)$. We have $J^\mu(1) = 2\epsilon\alpha/(1-\alpha)$, and the bound given by Prop. 6.1 holds with equality.

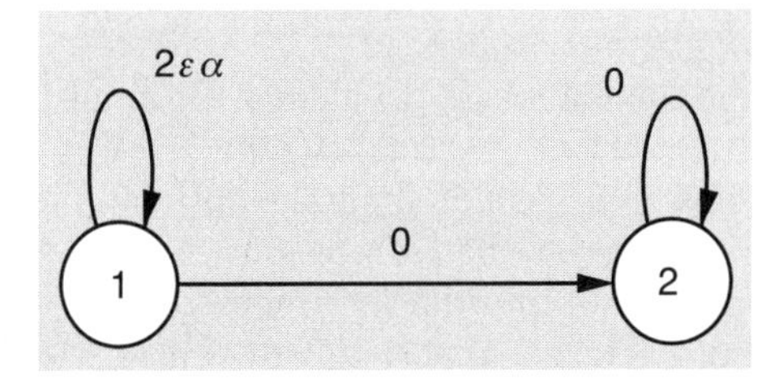

Figure 6.1: A two-state problem. All transitions are deterministic, but at state 1 there are two possible decisions, indicated by the two different outgoing arcs. The cost of each transition is shown next to the corresponding arc.

For undiscounted problems and if all policies are proper, the operators T and T_μ, for any policy μ, are contractions with respect to a (common) weighted maximum norm $\|\cdot\|_\xi$ (Prop. 2.2 in Section 2.2.1). Let ρ be the corresponding contraction factor. If J satisfies $\|J - J^*\|_\xi \leq \epsilon$, and μ is a greedy policy obtained from J, then the proof of Prop. 6.1 applies verbatim and leads to the conclusion

$$\|J^\mu - J^*\|_\xi \leq \frac{2\rho\epsilon}{1-\rho}.$$

For undiscounted problems in which there are improper policies (with every improper policy having infinite cost for some initial states), it can be shown that if J is sufficiently close to J^*, then a greedy policy is optimal. However, if J is not close to J^*, there is no guarantee that a greedy policy will be proper and for this reason there are no performance guarantees (see the example that follows).

Example 6.3

Consider the same problem as in Example 6.2 (cf. Fig. 6.1), except that $\alpha = 1$. Suppose that $J(1) < -2\epsilon$ and $J(2) = 0$. The resulting greedy policy chooses to stay at state 1 and its cost is infinite. On the other hand, if $J(1)$ is sufficiently close to the correct value $J^*(1) = 0$, the resulting greedy policy is optimal.

6.1.2 Multistage Lookahead Variations

To reduce the effect of the approximation error

$$\tilde{J}(i,r) - J^*(i)$$

between true and approximate cost-to-go, one can consider lookahead of several stages in determining a policy μ. The method adopted earlier, namely

$$\mu(i) = \arg \min_{u \in U(i)} \sum_j p_{ij}(u)\big(g(i,u,j) + \tilde{J}(j,r)\big),$$

corresponds to single stage lookahead. At a given state i, it finds the optimal decision for the one-stage problem with immediate cost $g(i,u,j)$ and terminal cost (after the first stage) $\tilde{J}(j,r)$.

An m-stage lookahead version finds the optimal policy for an m-stage problem, whereby we start at the current state i, make the m subsequent decisions with perfect state information, incur the corresponding costs of the m stages, and pay a terminal cost $\tilde{J}(j,r)$, where j is the state after m stages. If $\overline{u}$ is the first decision of the m-stage lookahead optimal policy starting at state i, the policy is defined by

$$\mu(i) = \overline{u}.$$

Note that we need to perform an m-stage lookahead and solve a stochastic shortest path problem each time that a decision $\mu(i)$ needs to be chosen. This stochastic shortest path problem involves a finite horizon (namely, m) and its state space consists of all states that can be reached with positive probability in m or fewer steps, starting from the current state i. Suppose that for any i, there exist at most B states j such that $p_{ij}(u) > 0$ for some u. Then, the stochastic shortest path problem involves $O(B^m)$ states and, depending on the values of B and m, a quick solution may be possible.

If $\tilde{J}(i,r)$ is equal to the exact optimal cost $J^*(i)$ for all states i, that is, if there is no approximation error, then m-stage lookahead gives an optimal decision, no matter what m is; in particular, one-stage lookahead suffices. On the other hand, in the presence of approximation errors, one expects multistage lookahead to yield better performance. For example, in discounted problems, the effects of the approximation error $\|\tilde{J} - J^*\|$ are diminished by a factor of α^k through a k-stage lookahead. Similarly, in a stochastic shortest path problem, as the lookahead becomes larger, so does the probability that the termination state will be reached within the span of the lookahead. Another reason is that the consequences of different actions are often clarified as time progresses, in the sense that the approximation error at the end of the lookahead may tend to be reduced, and this may give multistage lookahead an advantage. In this connection, it is worth mentioning a form of *selective lookahead*, where lookahead is used only at states for which the cost-to-go error is potentially "large." For example, computer chess programs use deeper lookahead in "unclear" positions.

A more detailed study of the effects of lookahead should involve a comparison of the cost-to-go errors

$$\tilde{e}(i,r) = \tilde{J}(i,r) - J^*(i),$$

and the lookahead-induced errors

$$\hat{e}(i,r) = \hat{J}(i,r) - J^*(i),$$

where

$$\hat{J}(i,r) = \min_{u \in U(i)} \sum_j p_{ij}(u)\big(g(i,u,j) + \tilde{J}(j,r)\big).$$

(Note that when two-stage lookahead is used with terminal cost $\tilde{J}$, then $\hat{J}$ plays the role of a one-stage lookahead approximate cost-to-go function.) We may conclude that two-stage lookahead provides some benefit over one-stage lookahead if the error $\hat{e}$ is smaller in some sense than $\tilde{e}$. It can be seen that the maximum norm of $\hat{e}$ is no larger than the maximum norm of $\tilde{e}$. In the absence of additional information, it does not seem possible to draw any further conclusions. However, a statistical comparison of $\tilde{e}$ and $\hat{e}$ could be helpful in specific contexts.

It is well-known that the quality of play of chess programs crucially depends on the size of the lookahead. Multistage lookahead has also been found to be useful to backgammon players, as will be discussed in Section 8.6. This indicates that in many types of problems, multistage lookahead should be much more effective than single stage lookahead. This improvement in performance must of course be weighed against the considerable increase in computation to obtain the decisions $\mu(i)$.

6.1.3 Rollout Policies

Suppose that through some method we have obtained a policy μ that we can simulate in our system. Then it is possible (at least in principle) to implement in real-time, instead of μ, the improved policy $\overline{\mu}$ that is obtained by a single *exact* policy improvement step from μ; that is, for all states i to be encountered in real-time operation of the system, we let

$$\overline{\mu}(i) = \min_{u \in U(i)} Q^{\mu}(i,u),$$

where

$$Q^{\mu}(i,u) = \sum_{j} p_{ij}(u)\big(g(i,u,j) + J^{\mu}(j)\big). \tag{6.5}$$

The key point here is that while our approximate DP method may provide only approximations to $Q^{\mu}(i,u)$, still the exact values of $Q^{\mu}(i,u)$ that are needed in the above policy improvement step can be calculated by Monte-Carlo simulation, as follows. Given any state i that is encountered in real-time operation of the system, and for every possible control $u \in U(i)$, we simulate trajectories that start at i, use decision u at the first stage, and use policy μ for all subsequent stages. The expected cost accumulated during such trajectories is the Q-factor $Q^{\mu}(i,u)$, and by simulating a large number of trajectories, an accurate estimate of $Q^{\mu}(i,u)$ is obtained. We then implement on the actual system a decision u with the smallest Q-factor. We refer to the policy $\overline{\mu}$ as the *rollout policy based on* μ.

Note that it is also possible to define rollout policies based on a policy μ, that make use of multistage (say, m-stage) lookahead. What is different here from the discussion in Section 6.1.2, is that instead of associating a cost-to-go $\tilde{J}(j,r)$ to every state j that can be reached in m steps, we use the exact cost-to-go $J^{\mu}(j)$, as computed by Monte Carlo simulation of several trajectories that start at j and follow policy μ. Clearly, such multistage lookahead involves much more on-line computation, but it may yield better performance than its single-stage counterpart. In what follows, we concentrate on rollout policies with single-stage lookahead.

The viability of a rollout policy depends on how much time is available to make the decision $\overline{\mu}(i)$ following the transition to state i and on how expensive the Monte Carlo evaluation of $Q^{\mu}(i,u)$ is. In particular, it must be possible to perform the Monte Carlo simulations and calculate the rollout control $\overline{\mu}(i)$ within the real-time constraints of the problem. If the problem is deterministic, a single simulation suffices, and the calculations are greatly simplified, but in general, the computational overhead can be substantial. However, it is worth emphasizing that the rollout policy $\overline{\mu}$ will always perform better than the current policy μ, as a consequence of the general results on the policy iteration method (Prop. 2.4).

It is possible to speed up the calculation of the rollout policy if we are willing to accept some potential performance degradation. Here are some possibilities:

(a) Use an approximation $\tilde{J}^\mu(\cdot, r)$ of J^μ to identify a few promising controls through a minimization of the form

$$\min_{u \in U(i)} \sum_j p_{ij}(u)\big(g(i,u,j) + \tilde{J}^\mu(j,r)\big),$$

calculate $Q^\mu(i,u)$ using fairly accurate Monte Carlo simulation for these controls, and approximate $Q^\mu(i,u)$ using relatively few simulation runs for the other controls. Adaptive variants of this approach are also possible, whereby we adjust the accuracy of the Monte Carlo simulation depending on the results of the computation.

(b) Use an approximate representation $\tilde{J}^\mu(\cdot, r)$ of J^μ to approximate $Q^\mu(i,u)$ by using simulation for N stages, and by estimating the cost of the remaining stages as $\tilde{J}^\mu(j_N, r)$, where j_N is the state after N stages.

(c) Identify a subset $\tilde{U}(i)$ of promising controls by using the procedures in (a) above or by using some heuristics, and use the control $\tilde{\mu}(i)$ given by

$$\tilde{\mu}(i) = \arg \min_{u \in \tilde{U}(i) \cup \{\mu(i)\}} Q^\mu(i,u), \tag{6.6}$$

where $Q^\mu(i,u)$ is obtained through accurate Monte-Carlo simulation.

Note that while the policies obtained from possibilities (a) and (b) may perform worse than μ, the policy $\tilde{\mu}$ in possibility (c) can be shown to perform at least as well as μ, i.e., $J^{\tilde{\mu}} \leq J^\mu$. This is because Eq. (6.6) implies that $T_{\tilde{\mu}} J^\mu \leq T_\mu J^\mu = J^\mu$, from which by the Monotonicity Lemma 2.1, we have $J^{\tilde{\mu}} = \lim_{t \to \infty} T_{\tilde{\mu}}^t J^\mu \leq T_{\tilde{\mu}} J^\mu \leq J^\mu$.

It is finally worth mentioning the broad generality of a rollout policy. It can be used to enhance the performance of *any* policy μ, no matter how obtained. In particular, μ does not need to be constructed using NDP techniques, and may be derived using heuristics.

6.1.4 Trading off Control Space Complexity with State Space Complexity

We discussed earlier the difficulty associated with large control constraint sets $U(i)$: the minimization

$$\min_{u \in U(i)} \sum_j p_{ij}(u)\big(g(i,u,j) + \tilde{J}(j,r)\big)$$

required to calculate the greedy policy with respect to a given approximate cost-go function $\tilde{J}(j,r)$ may be very time-consuming. It is thus useful to know that by reformulating the problem, one can in many cases reduce the complexity of the control space by increasing the complexity of the state space. The potential advantage is that the extra state space complexity may still be dealt with by using function approximation.

In particular, suppose that the control u consists of N components,

$$u = (u_1, \ldots, u_N).$$

Then, at a given state i, we can break down the control u into the sequence of the N controls $u_1, u_2, \ldots, u_N$, and introduce artificial intermediate "states" $(i, u_1), (i, u_1, u_2), \ldots, (i, u_1, \ldots, u_{N-1})$, and corresponding transitions to model the effect of these controls. The choice of the last control component u_N at "state" $(i, u_1, \ldots, u_{N-1})$ marks the transition to state j according to the given transition probabilities $p_{ij}(u)$. In this way the control space is simplified at the expense of introducing $N-1$ additional layers of states, and $N-1$ additional cost-to-go functions $J_1(i, u_1)$, $J_2(i, u_1, u_2), \ldots, J_{N-1}(i, u_1, \ldots, u_{N-1})$. The dependence of these cost-to-go functions on the control components is reminiscent of the dependence of the Q-factors on u. The increase in size of the state space can be dealt with by using function approximation, that is, with the introduction of cost-to-go approximations

$$\tilde{J}_1(i, u_1, r_1),\ \tilde{J}_2(i, u_1, u_2, r_2), \ldots, \tilde{J}_{N-1}(i, u_1, \ldots, u_{N-1}, r_{N-1}),$$

in addition to $\tilde{J}(i,r)$.

It is worth noting a few variants of the above procedure. First, instead of selecting the controls in a fixed order, it is possible to leave the order subject to choice. Second, the above procedure can often be simplified by taking advantage of special problem structure. In particular, the intermediate "states" $(i, u_1, \ldots, u_k)$ can be sometimes identified with "regular" problem states, with simplification in the function approximation procedure resulting. As an illustration, consider the maintenance problem of Example 2.4 in Section 2.4. The state here has the form

$$(m_1, \ldots, m_T, y_1, \ldots, y_T, s),$$

where m_t, $t = 1, \ldots, T$, is the number of working machines of type t, y_t, $t = 1, \ldots, T$, is the number of breakdowns of machines of type t, and s is the number of available spare parts. The control is $u = (u_1, \ldots, u_T)$, where u_t is the number of spare parts used to repair breakdowns of machines of type t (each breakdown repair requires one spare part). The control space here is very large if T is large and the number of machines that can break down in one period is large. However, the problem can be reformulated to trade control space complexity with state space complexity as follows.

In the reformulated problem, at any state where $\sum_{t=1}^{T} y_t > 0$, the control choices are to select a particular breakdown type, say t, with $y_t > 0$, and then select between two options:

(1) Leave the breakdown unrepaired, in which case the state evolves to

$$(m_1, \ldots, m_{t-1}, m_t - 1, m_{t+1}, \ldots, m_T, y_1, \ldots, y_{t-1}, y_t - 1, y_{t+1}, \ldots, y_T, s)$$

and the cost C_t of losing the corresponding machine is incurred.

(2) Repair the breakdown, in which case the state evolves to

$$(m_1, \ldots, m_T, y_1, \ldots, y_{t-1}, y_t - 1, y_{t+1}, \ldots, y_T, s - 1),$$

and no cost is incurred. (This option is, of course, available only if $s > 0$.)

Furthermore, at any state where $y_1 = \cdots = y_T = 0$, there is no decision to make, and we simply move to a new state with $\sum_{t=1}^{T} y_t > 0$ according to the given probabilistic structure of the problem.

Note that the complexity of the control space has been greatly reduced, since at each state we have to choose between at most $2T$ options (select a breakdown type, and repair or not repair it). Furthermore, the reformulation did not require an increase in the size of the state space. However, in the reformulated problem, the transitions of the component $(y_1, \ldots, y_T)$ of the state depend on the current value of that component. As a result, the state space reduction that we discussed in Examples 2.2 and 2.4 cannot be used any more.

6.2 APPROXIMATE POLICY ITERATION

We discuss here a generic version of approximate policy iteration, and provide some details on one possible variant, based on Monte Carlo simulation and least squares approximation. (Other variants are considered in Sections 6.3 and 6.4.) We also show that the method is sound in the sense that it has some minimal performance guarantees.

The general structure of approximate policy iteration is the same as for exact policy iteration (see Fig. 6.2). There are two differences, however.

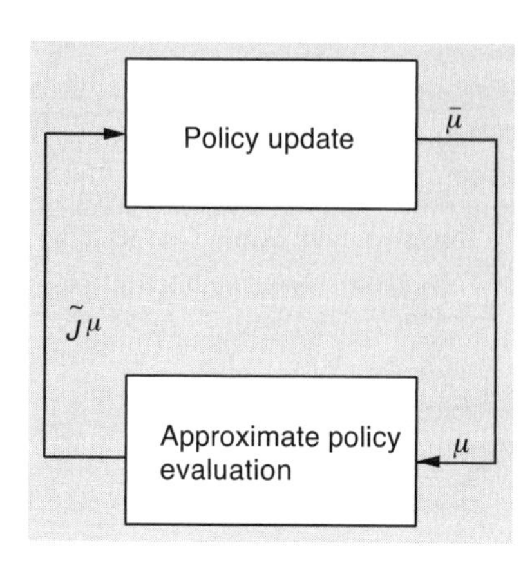

Figure 6.2: Block diagram of approximate policy iteration.

(a) Given the current policy μ, the corresponding cost-to-go function J^μ is not computed exactly. Instead, we compute an approximate cost-to-go function $\tilde{J}^\mu(i, r)$, where r is a vector of tunable parameters. There are two sources of error here. First, the approximation architecture may not be powerful enough to adequately represent J^μ; second, the tuning of the parameter vector r may be based on simulation experiments, and simulation noise becomes a source of error.

(b) Once approximate policy evaluation is completed and $\tilde{J}^\mu(i, r)$ is available, we generate a new policy $\overline{\mu}$ which is greedy with respect to $\tilde{J}^\mu$. As discussed in the preceding section, this can be sometimes done exactly, in which case we do indeed have a greedy policy. On the other hand, there are cases in which we only obtain an approximation of a greedy policy, and this is a new source of error.

In order to run the approximate policy iteration algorithm, an initial policy is required. In practice, it is usually important that this policy be as good as possible. In many problems, one can define a fairly good initial policy through heuristics or other considerations. In the absence of such a policy, we may fix a parameter vector r and then use a corresponding greedy, or approximately greedy, policy.

6.2.1 Approximate Policy Iteration Based on Monte Carlo Simulation

We now focus on a variant of approximate policy iteration that combines Monte Carlo simulation and function approximation for the purpose of policy evaluation. In our discussion we assume that $\alpha = 1$, but the generalization to discounted problems is straightforward. There is also a straightforward generalization to other forms of policy iteration, such as multistage lookahead and λ-policy iteration.

The method uses approximations $\tilde{J}(i,r)$ of the cost-to-go J^μ of stationary policies μ, but approximations $\tilde{Q}(i,u,r)$ of the corresponding Q-factors may also be used. In particular, suppose that we have fixed a proper stationary policy μ, that we have a set of representative states $\tilde{S}$, and that for each $i \in \tilde{S}$, we have $M(i)$ samples of the cost $J^\mu(i)$. The mth such sample is denoted by $c(i,m)$. We consider approximate cost-to-go functions of the form $\tilde{J}(i,r)$, where r is a parameter vector to be determined by solving the least squares optimization problem

$$\min_r \sum_{i\in\tilde{S}} \sum_{m=1}^{M(i)} \big(\tilde{J}(i,r) - c(i,m)\big)^2. \tag{6.7}$$

Once an optimal value of r has been determined, we can approximate the cost-to-go $J^\mu(i)$ of the policy μ by $\tilde{J}(i,r)$. Then, approximate Q-factors can be evaluated using the formula

$$\tilde{Q}(i,u,r) = \sum_j p_{ij}(u)\big(g(i,u,j) + \tilde{J}(j,r)\big), \tag{6.8}$$

and one can obtain an improved policy $\overline{\mu}$ using the formula

$$\begin{aligned} \overline{\mu}(i) &= \arg \min_{u\in U(i)} \tilde{Q}(i,u,r) \\ &= \arg \min_{u\in U(i)} \sum_j p_{ij}(u)\big(g(i,u,j) + \tilde{J}(j,r)\big), \qquad \text{for all } i. \end{aligned} \tag{6.9}$$

As discussed in the preceding section, the policy $\overline{\mu}$ is "obtained" in the sense that we have a subroutine which, given a state i, produces the value of $\overline{\mu}(i)$.

We thus have an algorithm that alternates between approximate policy evaluation steps and policy improvement steps, as illustrated in Fig. 6.2. The algorithm requires a single least-squares optimization of the form (6.7) per policy iteration, to generate the approximation $\tilde{J}(i,r)$ associated with the current policy μ. The parameter vector r determines the Q-factors via Eq. (6.8) and the next policy $\overline{\mu}$ via Eq. (6.9).

For another view of the approximate policy iteration algorithm, note that it consists of four modules (see Fig. 6.3):

(a) The *simulator*, which given a state-decision pair (i,u), generates the next state j according to the correct transition probabilities.

(b) The *decision generator*, which generates the decision $\overline{\mu}(i)$ of the improved policy at the current state i [cf. Eq. (6.9)] for use in the simulator.

(c) The *cost-to-go approximator*, which is the function $\tilde{J}(j,r)$ that is consulted by the decision generator for approximate cost-to-go values to use in the minimization of Eq. (6.9).

(d) The *least squares solver*, which accepts as input the sample trajectories produced by the simulator and solves a least squares problem of the form (6.7) to obtain the approximation $\tilde{J}(i, \overline{r})$ of the cost function of the improved policy $\overline{\mu}$, as implemented through the decision generator.

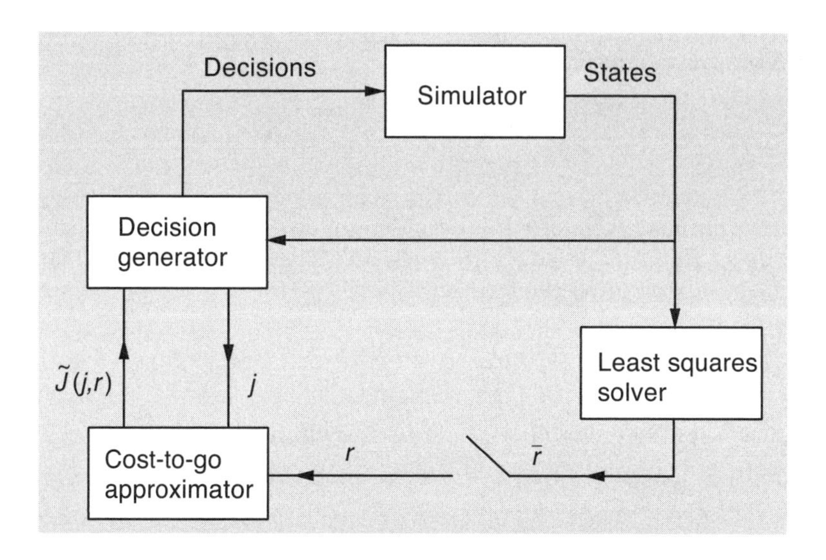

Figure 6.3: Structure of the approximate policy iteration algorithm. The communication from the least squares solver to the cost-to-go approximator only takes place after a policy has been fully evaluated and an optimal value of $\overline{r}$ has been determined, at which point the cost-to-go approximation $\tilde{J}(i, r)$ is replaced by $\tilde{J}(i, \overline{r})$.

An issue critical to the success of this method is that the least squares solution r constructed by the cost-to-go approximator will usually depend strongly on the sampling method, that is, on the set of representative states for which we collect cost samples. For this reason, we may wish to pay special attention to the sampling process, e.g., by employing the iterative resampling method discussed at the end of Section 5.2.2.

We comment here on a minor issue that arises in the context of discounted problems. The cost-to-go $J^\mu(i)$ is the suitably discounted expected cost of an infinite trajectory and the generation of the cost samples requires an infinitely long simulation. A common device here is to use finite horizon approximations of the infinite horizon cost samples. In particular, we may simulate a trajectory for a finite number of steps, say $\overline{N}$, and accumulate the discounted costs of only the first $\overline{N} - N$ visited states, using the transition sequences starting from these states up to the end of the $\overline{N}$-step horizon. Since each cost sample obtained in this way involves at least N transitions, the expected value of the resulting sample cost-to-go is within

$G\alpha^N/(1-\alpha)$ of J^μ, where G is an upper bound on $|g(i,u,j)|$. By taking N sufficiently large, the effects of using a finite trajectory can be made arbitrarily small. An alternative approach is to use a transformation to a stochastic shortest path problem (cf. Section 2.3), and to simulate trajectories until termination occurs. While this leads to unbiased estimates of $J^\mu(i)$, the randomness of the termination time adds to the variance of the cost samples, and it is unclear whether this approach offers any advantages.

Using an Action Network

In our discussion of policy approximation in the preceding section, we have pointed out that the computation of an improved policy $\overline{\mu}$ based on Eq. (6.9) may be too time consuming to be useful in a simulation context. In that case, we could compute $\overline{\mu}(i)$ only at a set $\hat{S}$ of sample states and use an approximation architecture $\tilde{\mu}(i,v)$ to represent $\overline{\mu}$, where v is a vector of tunable parameters. Assuming, for simplicity, that the possible decisions u are scalars, this can be done by posing the least squares problem

$$\min_v \sum_{i\in\hat{S}} \big(\tilde{\mu}(i,v)-\overline{\mu}(i)\big)^2,$$

which can be solved by means of the incremental gradient iteration

$$v := v + \gamma\nabla_v\tilde{\mu}(i,v)\big(\overline{\mu}(i)-\tilde{\mu}(i,v)\big),$$

or by any of the other methods discussed in Ch. 3. In this case, approximate policy iteration alternates between solutions of two approximation problems. Given a policy μ, we simulate it and perform an approximate policy evaluation to compute an approximation $\tilde{J}$ of J^μ. Given $\tilde{J}$, we compute the improved policy $\overline{\mu}$ for some states and then solve a policy approximation problem to obtain the new policy $\tilde{\mu}(\cdot,v)$.

Suppose now that computing $\overline{\mu}(i)$, according to Eq. (6.9), is difficult even if we were to consider only a limited set of representative states i. In that case, samples of the improved policy $\overline{\mu}$ are unavailable. Even so, we may sometimes be able to evaluate the derivative

$$\frac{\partial}{\partial u}\sum_j p_{ij}(u)\big(g(i,u,j)+\tilde{J}(j,r)\big)$$

at $u=\tilde{\mu}(i,v)$. (Here, we are departing from our standing assumption of a finite control set, and assume that u is free to take any scalar value.) The partial derivative with respect to u suggests a direction along which $\mu(i)$ should be changed to obtain a performance improvement. We can then update the parameters v of the action network so that $\tilde{\mu}(i,v)$ moves in the desired direction. The resulting update rule is

$$v := v - \gamma\nabla_v\tilde{\mu}(i,v)\frac{\partial}{\partial u}\sum_j p_{ij}(u)\big(g(i,u,j)+\tilde{J}(j,r)\big), \tag{6.10}$$

where the partial derivative is evaluated at $u = \tilde{\mu}(i, v)$. Note that Eq. (6.10) is basically a gradient step that tries to minimize

$$\sum_j p_{ij}\big(\tilde{\mu}(i,v)\big)\Big(g\big(i,\tilde{\mu}(i,v),j\big) + \tilde{J}(j,r)\Big),$$

with respect to v. Once several updates of the form (6.10) have been carried out, we can assume that we have constructed a new policy, which we can then proceed to evaluate, thus initiating the next stage of approximate policy iteration.

Note that the methods that we have described here involve two approximation architectures that are trained in an interleaved fashion. Policy evaluation involves the training of the parameter vector r of the approximation $\tilde{J}(i, r)$, and a policy update involves the training of the parameter vector v used to represent the new policy. An alternative approach in which the two approximation architectures are trained simultaneously will be discussed in Section 6.4, in the context of optimistic policy iteration.

Solving the Least-Squares Problem

The least squares problem (6.7) that is used to evaluate a policy μ can be solved by an incremental gradient method, of the type discussed in Ch. 3. Given a sample state trajectory $(i_0, i_1, \ldots, i_N)$ generated using the policy μ, the parameter vector r is updated by

$$r := r - \gamma \sum_{k=0}^{N-1} \nabla \tilde{J}(i_k, r)\left(\tilde{J}(i_k, r) - \sum_{m=k}^{N-1} g\big(i_m, \mu(i_m), i_{m+1}\big)\right), \qquad (6.11)$$

where γ is a stepsize. (Here, as well as throughout this chapter, the gradient is the vector of partial derivatives with respect to the components of the parameter vector r, unless a subscript is used to indicate otherwise.) The term multiplied by γ in the right-hand side above is a sample gradient corresponding to only some of the terms in the sum of squares in problem (6.7), namely, the cost samples $c(i_k, \cdot)$ associated with the states i_k visited by the trajectory under consideration.

Note that the update (6.11) assumes that for each trajectory, we obtain and use a cost sample that starts from every state visited by the trajectory (even if the same state is visited twice). In a "first-visit" variant of the method, the summation from $k = 0$ to $N - 1$ could be replaced by a sum over those k such that the state i_k is visited at time k for the first time. For problems involving a large number of states, multiple visits to the same state by a single trajectory are unlikely and this issue becomes less of a concern. Finally, for the case where $\tilde{J}(i, r) = r(i)$ (lookup table representation), $\nabla\tilde{J}(i, r)$ is the ith unit vector and the method (6.11) reduces to the method discussed in Section 5.2.

The algorithm (6.11) is the incremental gradient method, with each iteration corresponding to a new simulated trajectory. Another incremental method that can be used in this context (cf. Ch. 3) is the extended Kalman filter, which could lead to faster convergence, at the expense of more computation per iteration. For the case where the dependence on the parameter vector r is linear, one can use exact Kalman filtering, which leads to an optimal solution with a single pass through the training data. A key advantage of incremental algorithms is that they allow us to decide whether or not to simulate more trajectories, depending on the quality of the results obtained so far.

An alternative set of computational methods is obtained if we simulate a number of trajectories, collect the results, formulate the problem (6.7), and then solve it in batch mode. For example, if we are dealing with a linear least squares problem, the optimality conditions lead to a linear system of equations in the unknown vector r that can be solved using the methods discussed in Ch. 3, or simply by applying routines available in numerical linear algebra packages.

6.2.2 Error Bounds for Approximate Policy Iteration

In this section, we investigate the theoretical guarantees of approximate policy iteration. We assume that all policy evaluations and all policy updates are performed with a certain error tolerance of ϵ and δ, respectively. We then show that approximate policy iteration will produce policies whose performance differs from the optimal by a factor that diminishes to zero as ϵ and δ are decreased, as long as improper policies are avoided. The error bounds to be obtained refer to the worst possible case and may be too pessimistic to be practically relevant. Nevertheless, they remain of conceptual value in that they establish the fundamental soundness of approximate policy iteration.

Discounted Problems

Let us assume a discounted problem, with discount factor α. We consider an approximate policy iteration algorithm that generates a sequence of stationary policies μ_k and a corresponding sequence of approximate cost-to-go functions J_k satisfying

$$\max_i \left|J_k(i) - J^{\mu_k}(i)\right| \le \epsilon, \qquad k = 0, 1, \ldots \tag{6.12}$$

and

$$\max_i \left|(T_{\mu_{k+1}} J_k)(i) - (TJ_k)(i)\right| \le \delta, \qquad k = 0, 1, \ldots \tag{6.13}$$

where ϵ and δ are some positive scalars, and μ_0 is some policy. The scalar ϵ is an assumed worst-case bound on the error incurred during policy evaluation (the critic's error). It includes the random errors due to simulation plus any additional errors due to function approximation. In practice, the value of ϵ is rarely known; in fact, if there is a small probability of large simulation errors, ϵ will have to be set to a rather large value. The scalar δ is a bound on the error incurred in the course of the computations required for a policy update. If a model of the system is available, these calculations can be often done exactly, and δ can be taken to be very small or zero.

The following proposition shows that the algorithm eventually produces policies whose performance differs from the optimal by at most a constant multiple of ϵ and δ.

Proposition 6.2: The sequence of policies μ_k generated by the approximate policy iteration algorithm satisfies

$$\limsup_{k\to\infty} \max_{i=1,\ldots,n} \left|J^{\mu_k}(i) - J^*(i)\right| \leq \frac{\delta + 2\alpha\epsilon}{(1-\alpha)^2}. \tag{6.14}$$

Proof: Our first step is to establish the following lemma, which shows that a policy generated by a policy update cannot be much worse than the previous policy.

Lemma 6.1: Let μ be some policy and suppose that a vector J satisfies $\|J - J^\mu\| \leq \epsilon$ for some $\epsilon \geq 0$, where $\|\cdot\|$ is the maximum norm. Let $\overline{\mu}$ be a policy that satisfies

$$(T_{\overline{\mu}}J)(i) \leq (TJ)(i) + \delta, \qquad \forall\, i,$$

for some $\delta \geq 0$. Then,

$$J^{\overline{\mu}}(i) \leq J^\mu(i) + \frac{\delta + 2\alpha\epsilon}{1-\alpha}, \qquad \forall\, i.$$

Proof: Let

$$\xi = \max_{i=1,\ldots,n} \left(J^{\overline{\mu}}(i) - J^\mu(i)\right)$$

and let e be the vector with all entries equal to 1. Note that

$$J^{\overline{\mu}} \leq J^\mu + \xi e$$

and

$$J^{\overline{\mu}} = T_{\overline{\mu}}J^{\overline{\mu}} \leq T_{\overline{\mu}}(J^\mu + \xi e) = T_{\overline{\mu}}J^\mu + \alpha\xi e. \tag{6.15}$$

Using our assumption on $\overline{\mu}$, we have

$$0 \leq -T_{\overline{\mu}}J + TJ + \delta e \leq -T_{\overline{\mu}}J + T_{\mu}J + \delta e.$$

Using this inequality, together with Eq. (6.15), we obtain

$$\begin{aligned} J^{\overline{\mu}} - J^{\mu} &\leq T_{\overline{\mu}}J^{\mu} + \alpha\xi e - J^{\mu} \\ &\leq T_{\overline{\mu}}J^{\mu} - T_{\overline{\mu}}J + T_{\mu}J - J^{\mu} + \alpha\xi e + \delta e \\ &\leq \alpha\|J^{\mu} - J\|e + \alpha\|J - J^{\mu}\|e + \alpha\xi e + \delta e \\ &\leq 2\alpha\epsilon e + \alpha\xi e + \delta e. \end{aligned}$$

Thus,

$$\xi \leq \alpha\xi + 2\alpha\epsilon + \delta,$$

or

$$\xi \leq \frac{\delta + 2\alpha\epsilon}{1 - \alpha},$$

and the desired result follows. **Q.E.D.**

Let

$$\xi_k = \max_{i=1,\ldots,n} \big(J^{\mu_{k+1}}(i) - J^{\mu_k}(i)\big).$$

We apply Lemma 6.1 with $\mu = \mu_k$ and $\overline{\mu} = \mu_{k+1}$, and obtain

$$\xi_k \leq \frac{\delta + 2\alpha\epsilon}{1 - \alpha}. \tag{6.16}$$

We now let

$$\zeta_k = \max_{i=1,\ldots,n} \big(J^{\mu_k}(i) - J^*(i)\big).$$

The lemma that follows provides us with some bounds for ζ_k.

Lemma 6.2: We have

$$\zeta_{k+1} \leq \alpha\zeta_k + \alpha\xi_k + \delta + 2\alpha\epsilon, \qquad \forall\ k.$$

Proof: We first note that

$$J^{\mu_k} \leq J^* + \zeta_k e,$$

which leads to

$$TJ^{\mu_k} \leq T(J^* + \zeta_k e) = TJ^* + \alpha\zeta_k e = J^* + \alpha\zeta_k e.$$

We then have

$$\begin{aligned}
T_{\mu_{k+1}} J^{\mu_k} &\leq T_{\mu_{k+1}}(J_k + \epsilon e) \\
&= T_{\mu_{k+1}} J_k + \alpha\epsilon e \\
&\leq T J_k + \delta e + \alpha\epsilon e \\
&\leq T(J^{\mu_k} + \epsilon e) + \delta e + \alpha\epsilon e \\
&= T J^{\mu_k} + \delta e + 2\alpha\epsilon e \\
&\leq T(J^* + \zeta_k e) + \delta e + 2\alpha\epsilon e \\
&= J^* + \alpha\zeta_k e + \delta e + 2\alpha\epsilon e.
\end{aligned}$$

Thus,

$$\begin{aligned}
J^{\mu_{k+1}} &= T_{\mu_{k+1}} J^{\mu_{k+1}} \\
&\leq T_{\mu_{k+1}}(J^{\mu_k} + \xi_k e) \\
&= T_{\mu_{k+1}} J^{\mu_k} + \alpha\xi_k e \\
&\leq J^* + \alpha\zeta_k e + (\delta + 2\alpha\epsilon)e + \alpha\xi_k e,
\end{aligned}$$

which leads to the desired result. **Q.E.D.**

We now start with the inequality provided by Lemma 6.2, take the limit superior as $k \to \infty$, and also use Eq. (6.16), to obtain

$$(1-\alpha) \limsup_{k\to\infty} \zeta_k \leq \alpha \frac{\delta + 2\alpha\epsilon}{1-\alpha} + \delta + 2\alpha\epsilon.$$

This relation simplifies to

$$\limsup_{k\to\infty} \zeta_k \leq \frac{\delta + 2\alpha\epsilon}{(1-\alpha)^2},$$

which proves the proposition. **Q.E.D.**

We note that if for some reason we can guarantee that ξ_k is of the order of $\delta + 2\alpha\epsilon$, then the result of Lemma 6.2 easily leads to the conclusion that ζ_k eventually becomes of the order of $(\delta + 2\alpha\epsilon)/(1-\alpha)$. In particular, if the method converges, we have $\xi_k = 0$ for sufficiently large k, and we obtain

$$\limsup_{k\to\infty} \zeta_k \leq \frac{\delta + 2\alpha\epsilon}{1-\alpha}.$$

However, such convergence is quite uncommon unless ϵ and δ are very close to zero.

Stochastic Shortest Path Problems

Suppose now that we are dealing with a stochastic shortest path problem, under the usual assumptions (there exists a proper policy and improper policies have infinite cost for some starting state). We assume that the state space is $\{0, 1, \ldots, n\}$, with state 0 being a cost-free termination state.

We consider the same approximate policy iteration algorithm as in the preceding subsection [cf. Eqs. (6.12)-(6.13)]. One difficulty with this algorithm is that, even if the current policy μ_k is proper, the next policy μ_{k+1} need not be proper. In this case, we have $J^{\mu_{k+1}}(i) = \infty$ for some i, and the method breaks down. A possible way of dealing with this difficulty (in theory) is to reduce the tolerances ϵ and δ in Eqs. (6.12) and (6.13). It can be shown, that for sufficiently small ϵ and δ, the next policy will be proper. In any case, we will analyze the method under the assumption that all policies generated by the algorithm are proper. This assumption is of course satisfied if every policy is proper.

The following proposition provides an estimate of the difference $J^{\mu_k} - J^*$ in terms of the scalar

$$\rho = \max_{\substack{i=1,\ldots,n \\ \mu:\ \text{proper}}} \mathrm{P}(i_n \neq 0 \mid i_0 = i, \mu).$$

Note that for every proper policy μ and state i, we have $\mathrm{P}(i_n \neq 0 \mid i_0 = i, \mu) < 1$ by the definition of a proper policy, and since the number of proper policies is finite, we have $\rho < 1$.

Proposition 6.3: Assume that the stationary policies μ_k generated by the approximate policy iteration algorithm are all proper. Then

$$\limsup_{k\to\infty} \max_{i=1,\ldots,n} \left| J^{\mu_k}(i) - J^*(i) \right| \leq \frac{n(1-\rho+n)(\delta+2\epsilon)}{(1-\rho)^2}. \tag{6.17}$$

Proof: From Eqs. (6.12) and (6.13), we have for all k,

$$T_{\mu_{k+1}} J^{\mu_k} - \epsilon e \leq T_{\mu_{k+1}} J_k \leq T J_k + \delta e,$$

where $e = (1, 1, \ldots, 1)$, while from Eq. (6.12), we have for all k,

$$T J_k \leq T J^{\mu_k} + \epsilon e.$$

By combining these two relations, we obtain for all k,

$$T_{\mu_{k+1}} J^{\mu_k} \leq T J^{\mu_k} + (\delta + 2\epsilon) e \leq T_{\mu_k} J^{\mu_k} + (\delta + 2\epsilon) e. \tag{6.18}$$

From Eq. (6.18) and the equation $T_{\mu_k} J^{\mu_k} = J^{\mu_k}$, we have

$$T_{\mu_{k+1}} J^{\mu_k} \leq J^{\mu_k} + (\delta + 2\epsilon)e.$$

By subtracting from this relation the equation $T_{\mu_{k+1}} J^{\mu_{k+1}} = J^{\mu_{k+1}}$, we obtain

$$T_{\mu_{k+1}} J^{\mu_k} - T_{\mu_{k+1}} J^{\mu_{k+1}} \leq J^{\mu_k} - J^{\mu_{k+1}} + (\delta + 2\epsilon)e.$$

This relation can be written as

$$J^{\mu_{k+1}} - J^{\mu_k} \leq P_{\mu_{k+1}}(J^{\mu_{k+1}} - J^{\mu_k}) + (\delta + 2\epsilon)e, \tag{6.19}$$

where $P_{\mu_{k+1}}$ is the $n \times n$ transition probability matrix corresponding to μ_{k+1}.

Lemma 6.3: Let P be the $n \times n$ transition probability matrix corresponding to some proper policy μ and let c be a nonnegative scalar.

(a) If a vector x satisfies

$$x \leq Px + ce,$$

then

$$x(i) \leq \frac{nc}{1 - \rho}, \qquad \forall\ i.$$

(b) If a sequence of vectors x_k satisfies

$$x_{k+1} \leq Px_k + ce, \qquad \forall\ k,$$

then

$$\limsup_{k \to \infty} x_k(i) \leq \frac{nc}{1 - \rho}, \qquad \forall\ i.$$

Proof: (a) Let $y(i) = \max\{0, x(i)\}$, $i = 1, \ldots, n$. Then, $x \leq Px + ce \leq Py + ce$, which together with the relation $0 \leq Py + ce$, implies $y \leq Py + ce$. We then have

$$y \leq P(Py + ce) + ce \leq P^2 y + 2ce.$$

By repeating this process for a total of $n - 1$ times, we have

$$y \leq P^n y + nce.$$

By the definition of ρ, we have

$$P^n y \leq \rho\big(\max_i y(i)\big)e,$$

and it follows that

$$\max_i y(i) \leq \rho \max_i y(i) + nc.$$

Hence, $x(i) \leq \max_i y(i) \leq nc/(1-\rho)$, as desired.

(b) Proceeding as in part (a), we obtain

$$\max_i y_{k+n}(i) \leq \rho \max_i y_k(i) + nc, \qquad \forall\ k,$$

where $y_k(i) = \max\{0, x_k(i)\}$. Hence,

$$\limsup_{k\to\infty} \big(\max_i y_{k+n}(i)\big) \leq \rho \limsup_{k\to\infty} \big(\max_i y_k(i)\big) + nc,$$

and the result follows. **Q.E.D.**

Let

$$\xi_k = \max_{i=1,\ldots,n} \big(J^{\mu_{k+1}}(i) - J^{\mu_k}(i)\big).$$

Then Eq. (6.19) and Lemma 6.3(a) yield

$$\xi_k \leq \frac{n(\delta + 2\epsilon)}{1-\rho}. \tag{6.20}$$

Let μ^* be an optimal stationary policy and note that μ^* must be proper. From Eq. (6.18), we have

$$\begin{aligned} T_{\mu_{k+1}} J^{\mu_k} &\leq T_{\mu^*} J^{\mu_k} + (\delta + 2\epsilon)e \\ &= T_{\mu^*} J^{\mu_k} - T_{\mu^*} J^{\mu^*} + J^* + (\delta + 2\epsilon)e \\ &= P_{\mu^*}(J^{\mu_k} - J^*) + J^* + (\delta + 2\epsilon)e. \end{aligned}$$

We also have

$$T_{\mu_{k+1}} J^{\mu_k} = J^{\mu_{k+1}} + T_{\mu_{k+1}} J^{\mu_k} - T_{\mu_{k+1}} J^{\mu_{k+1}} = J^{\mu_{k+1}} + P_{\mu_{k+1}}(J^{\mu_k} - J^{\mu_{k+1}}).$$

By subtracting the last two relations and by using Eq. (6.20), we obtain

$$\begin{aligned} J^{\mu_{k+1}} - J^* &\leq P_{\mu^*}(J^{\mu_k} - J^*) + P_{\mu_{k+1}}(J^{\mu_{k+1}} - J^{\mu_k}) + (\delta + 2\epsilon)e \\ &\leq P_{\mu^*}(J^{\mu_k} - J^*) + \xi_k P_{\mu_{k+1}} e + (\delta + 2\epsilon)e \\ &\leq P_{\mu^*}(J^{\mu_k} - J^*) + \xi_k e + (\delta + 2\epsilon)e \\ &\leq P_{\mu^*}(J^{\mu_k} - J^*) + \frac{(1-\rho+n)(\delta + 2\epsilon)}{1-\rho} e. \end{aligned} \tag{6.21}$$

We now use Lemma 6.3(b) with $x_k = J^{\mu_k} - J^*$, and we obtain

$$\limsup_{k\to\infty} \big(J^{\mu_k}(i) - J^*(i)\big) \leq \frac{n(1-\rho+n)(\delta + 2\epsilon)}{(1-\rho)^2}, \qquad \forall\ i,$$

which is the desired result. **Q.E.D.**

The error bound (6.17) uses the default estimate of the number of stages required to reach the termination state 0 with positive probability, which is n. We can strengthen the error bound if we have a better estimate. In particular, for all $m \geq 1$, let

$$\rho_m = \max_{\substack{i=1,\ldots,n \\ \mu:\ \text{proper}}} \mathrm{P}(i_m \neq 0 \mid i_0 = i, \mu),$$

and let τ be any m for which $\rho_m < 1$. Then the proof of Prop. 6.3 can be modified to show that

$$\limsup_{k\to\infty} \max_{i=1,\ldots,n} \left| J^{\mu_k}(i) - J^*(i) \right| \leq \frac{\tau(1-\rho_\tau+\tau)(\delta+2\epsilon)}{(1-\rho_\tau)^2}. \tag{6.22}$$

Note that this bound only depends on τ and not on n. In addition, if there exists some $N > 0$ such that termination occurs within at most N stages for all proper policies, then $\rho_N = 0$ and we obtain

$$\limsup_{k\to\infty} \max_{i=1,\ldots,n} \left| J^{\mu_k}(i) - J^*(i) \right| \leq N(1+N)(\delta+2\epsilon).$$

If we have a guarantee that the sequence of policies μ_k eventually settles on some limit policy μ, our bounds can be improved with the denominator term $(1-\rho)^2$ being replaced by $1-\rho$. This can be seen from Eq. (6.21), as follows. If policies converge, then eventually ξ_k becomes zero, and we obtain

$$J^{\mu_{k+1}} - J^* \leq P_{\mu^*}(J^{\mu_k} - J^*) + (\delta + 2\epsilon)e.$$

We then use Lemma 6.3(b) and obtain

$$\limsup_{k\to\infty} \max_{i=1,\ldots,n} \big(J^{\mu_k}(i) - J^*(i) \big) \leq \frac{n(\delta+2\epsilon)}{1-\rho}, \qquad \forall\ i.$$

However, experience with the algorithm indicates that the sequence of policies rarely converges, when ϵ and δ are substantially larger than zero.

6.2.3 Tightness of the Error Bounds and Empirical Behavior

We provide here an example of a discounted problem in which the performance loss due to approximations in policy iteration comes arbitrarily close to $(\delta + 2\alpha\epsilon)/(1-\alpha)^2$, proving that the bound in Prop. 6.2 is tight. The same example can be also viewed as a stochastic shortest path problem, by interpreting $1-\alpha$ as a termination probability. In reference to the notation used in the bound (6.22), we have $\tau = 1$ and $\rho_\tau = \alpha$. We thus conclude that the bound (6.22) is also tight, within a small constant factor.

Example 6.4

Consider a discounted problem involving states $1, \ldots, n$. All transitions are deterministic. At state 1, the only option is to stay there at zero cost. At each state $i \geq 2$ there is a choice of staying, at cost g_i, or of moving to state $i-1$, at zero cost. The one-stage costs g_i of staying are defined recursively by

$$g_2 = \delta + 2\alpha\epsilon, \qquad g_{i+1} = \alpha g_i + \delta + 2\alpha\epsilon, \quad i \geq 2,$$

where α is the discount factor and ϵ, δ are positive constants. Clearly, the optimal cost-to-go is zero for all states.

Consider an approximate policy iteration method in which each policy evaluation is allowed to be incorrect by ϵ, as in Eq. (6.12), and the policy update may involve an error of δ, as in Eq. (6.13). We will show, using induction, that after $k-1$ iterations $(k \geq 2)$, we can obtain a policy that decides to stay at state k, and moves to state $i-1$ for every $i \neq k$, $i \geq 2$.

Let us start with the optimal policy, and suppose that $J(1)$ is incorrectly evaluated to be ϵ, and $J(2)$ is incorrectly evaluated to be $-\epsilon$. For a policy update at state 2, we compare $0+\alpha J(1)$, which is $\alpha\epsilon$, with $g_2+\alpha J(2)$, which is $\delta+2\alpha\epsilon+\alpha(-\epsilon) = \delta+\alpha\epsilon$. Since we allow an error of δ in the policy update, we can choose the policy that stays at state 2. For every state $i > 2$, suppose that $J(i)$ has been correctly evaluated to be 0. Then, the policy update compares $g_i + \alpha J(i)$, which is positive, with $\alpha J(i-1)$, which is nonpositive. Thus, if the error in the policy update is zero, the new policy will choose to move to state $i-1$. This establishes our claim for $k = 2$.

Let now $k \geq 2$ be arbitrary and suppose that we have reached a policy μ that decides to stay at state k but for all other states $i \geq 2$, it decides to move to state $i-1$. Under that policy, we have $J^\mu(k) = g_k/(1-\alpha)$ and $J^\mu(k+1) = \alpha g_k/(1-\alpha)$. Suppose that policy evaluation overestimates $J^\mu(k)$ to be $J(k) = g_k/(1-\alpha) + \epsilon$, underestimates $J^\mu(k+1)$ to be $J(k+1) = \alpha g_k/(1-\alpha) - \epsilon$, and that $J^\mu(i)$ is correctly evaluated for all other states i. It is easily seen that for all states i different than 1 or $k+1$, the resulting greedy policy chooses to move to state $i-1$. Let us now consider state $k+1$. To determine a greedy policy, we need to compare $0 + \alpha J(k)$, which is $\alpha g_k/(1-\alpha) + \alpha\epsilon$, with $g_{k+1} + \alpha J(k+1)$, which is $g_{k+1} + \alpha^2 g_k/(1-\alpha) - \alpha\epsilon$. Using the definition of g_{k+1}, it is seen that the choice of moving to state k dominates by

$$\left(g_{k+1} + \frac{\alpha^2 g_k}{1-\alpha} - \alpha\epsilon\right) - \left(\frac{\alpha g_k}{1-\alpha} + \alpha\epsilon\right) = \delta.$$

However, since we are allowed an error of δ in the policy update, the algorithm is in fact free to choose the policy that stays at state $k+1$.

Once, we reach the policy that stays at n, and for $i \neq 1, n$, moves to $i-1$, a subsequent policy update takes us back to the optimal policy, and the process can be repeated. We thus obtain a periodic sequence of policies and once every n steps, we have a policy whose cost is $g_n/(1-\alpha)$, followed by the optimal policy whose cost is zero. Using the recursion for the one-stage costs g_k, we see that for n large, g_n approaches $(\delta+2\alpha\epsilon)/(1-\alpha)$. This means that by taking the number of states to be large enough, we have an oscillation whose amplitude approaches $(\delta + 2\alpha\epsilon)/(1-\alpha)^2$, and therefore the bound in Prop. 6.2 is tight.

The behavior of approximate policy iteration in practice is in agreement with the insights provided by our error bounds. In the beginning, the method tends to make rapid and fairly monotonic progress, but eventually it gets into an oscillatory pattern. For discounted problems, presumably this happens after $\tilde{J}$ gets within $O\big((\delta+2\epsilon)/(1-\alpha)^2\big)$ of J^*; thereafter, $\tilde{J}$ oscillates fairly randomly within that zone. This behavior will be illustrated by our case studies in Ch. 8.

Our bounds, referring to the worst case, are rather pessimistic, especially as far as the $(1-\alpha)^2$ term in the denominator is concerned. In practice, the deviation from J^* is smaller than what this theory seems to predict. Let us also keep in mind that out of the sequence of policies generated by the algorithm (whose performance will generally oscillate), one should choose to implement the policy that had the best performance, and this may lead to further reduction of the distance from optimality.

As indicated by Example 6.4, a policy update in the course of the approximate policy iteration algorithm, can lead to a policy which is substantially worse than the previous one. This suggests that one should exercise caution and move from one policy to the next one in a gradual fashion. This argument leads to the optimistic policy iteration method, to be discussed in Section 6.4.

6.3 APPROXIMATE POLICY EVALUATION USING TD(λ)

In the preceding section, we discussed one possible method for approximate policy evaluation, namely Monte Carlo simulation, followed by least squares approximation. In this section, we develop an alternative implementation, in terms of temporal differences, that generalizes the TD(1) method introduced in Ch. 5. We will see in Section 6.3.1 that TD(1) can be interpreted as a type of incremental gradient method for solving the least squares problem associated with the cost-to-go approximation.

As in Ch. 5, we modify TD(1) using a parameter λ, and we obtain a family of methods similar to the TD(λ) methods of that chapter. The motivation for TD(λ) is not as clear as in Ch. 5, and in fact we will argue that as λ becomes smaller, there is a tendency for the quality of the cost-to-go approximation to deteriorate. Nonetheless, there are reasons for our interest in TD(λ), for $\lambda < 1$: there is empirical evidence that for some problems, particularly those where the costs-to-go have large variance, it converges faster and leads to better performance than that obtained from TD(1).

For the case of lookup table representations, we have seen that TD(λ) converges to the correct vector J^μ, for all values of λ. However, TD(λ) combined with function approximation, may converge to a different limit for different values of λ. In fact, even the issue of convergence is much more complex. The only available convergence results refer to the case of linearly parametrized approximation architectures and, unlike the results of Section 5.3, they do not allow for much freedom in the choice of the eligibility coefficients.

Throughout this section, we assume that a policy μ has been fixed. For this reason, we do not need to show the dependence of various quantities on μ, and we employ the notation $g(i,j)$ and p_{ij} instead of $g\big(i,\mu(i),j\big)$ and $p_{ij}\big(\mu(i)\big)$, respectively.

6.3.1 Approximate Policy Evaluation Using TD(1)

We will now develop TD(1) and interpret it as a gradient-like method for least squares approximation. We consider a stochastic shortest path problem, with 0 being a cost-free absorbing state, and we assume that μ is a proper policy. We use an approximate cost-to-go function $\tilde{J}(i,r)$ to approximate $J^\mu(i)$, where r is a vector of tunable parameters. Consistently with our convention of fixing $J(0)$ to zero, we assume that $\tilde{J}(0,r)=0$ for all r.

In one version of the Monte Carlo method that was discussed in Section 6.2, we pick an initial state i_0 and generate a sample trajectory $(i_0,i_1,\ldots,i_N)$, where i_N is the first time that state 0 is reached. Once this trajectory is available, we carry out an incremental gradient update according to the formula [cf. Eq. (6.11)]

$$r := r - \gamma \sum_{k=0}^{N-1} \nabla\tilde{J}(i_k,r)\left(\tilde{J}(i_k,r) - \sum_{m=k}^{N-1} g(i_m,i_{m+1})\right). \tag{6.23}$$

Just as there is a temporal difference implementation of Monte-Carlo simulation for the lookup table case, there is also a temporal difference implementation of the incremental gradient iteration (6.23). We define temporal differences d_k by

$$d_k = g(i_k,i_{k+1}) + \tilde{J}(i_{k+1},r) - \tilde{J}(i_k,r), \qquad k=0,\ldots,N-1, \tag{6.24}$$

and, using the fact $\tilde{J}(i_N,r)=\tilde{J}(0,r)=0$, the iteration (6.23) becomes

$$r := r + \gamma \sum_{k=0}^{N-1} \nabla\tilde{J}(i_k,r)(d_k + d_{k+1} + \cdots + d_{N-1}).$$

Instead of waiting for the end of the trajectory to update r as above, we may perform the part of the update involving d_k as soon as d_k becomes available. This leads to the update equation

$$r := r + \gamma d_k \sum_{m=0}^{k} \nabla \tilde{J}(i_m, r), \qquad k = 0, 1, \ldots, N-1.$$

For example, following the state transition (i_0, i_1), we set

$$r := r + \gamma d_0 \nabla \tilde{J}(i_0, r). \tag{6.25}$$

Following the state transition (i_1, i_2), we set

$$r := r + \gamma d_1 \big(\nabla \tilde{J}(i_0, r) + \nabla \tilde{J}(i_1, r)\big), \tag{6.26}$$

and similarly for subsequent transitions. Finally, following the state transition (i_{N-1}, i_N), we set

$$r := r + \gamma d_{N-1} \big(\nabla \tilde{J}(i_0, r) + \nabla \tilde{J}(i_1, r) + \cdots + \nabla \tilde{J}(i_{N-1}, r)\big). \tag{6.27}$$

For the special case where r has dimension equal to the number of nonterminal states, and if $\tilde{J}(i, r) = r(i)$ (which is a lookup table representation), we recover the TD(1) method of Ch. 5, with $r(i)$ playing the role of $J(i)$.

If all updates of the vector r are performed at the end of a trajectory, we have a generalization of the *off-line* method described in Section 5.3. If on the other hand, an update is performed subsequent to each transition, according to Eqs. (6.25)-(6.27), we have a generalization of the *on-line* method of Section 5.3. There is a small complication having to do with the evaluation of the gradient terms $\nabla \tilde{J}(i_k, r)$. While the same gradient term may appear in several update formulas [e.g., $\nabla \tilde{J}(i_0, r)$ appears in all of them], it would be very cumbersome to reevaluate these gradients at each step, at the current value of r. A more practical alternative is to evaluate the gradient $\nabla \tilde{J}(i_k, r)$ as soon as the state i_k is generated, at the prevailing value of r, and use that value of the gradient in subsequent updates, even though r may have changed in between. For the special case of linear architectures, i.e., if $\tilde{J}(i, r)$ is a linear function of r, the value of the gradient does not depend on r and this issue does not arise. In any case, the effect of such minor changes as well as the difference between the off-line and the on-line algorithm are of second order in the stepsize and have negligible effect on convergence properties, for reasons similar to those discussed in Section 5.3.

Note that in order to expect any kind of convergent behavior, the stepsize γ should diminish over time. A popular choice is to let $\gamma = c/(m + d)$ in the course of the mth trajectory, where c and d are some positive constants.

We should stress once more the importance of proper sampling. In particular, the method used for picking the initial states can be crucial to the quality of the cost-to-go approximation constructed by TD(1). One generally expects the approximation to be better for those states that are more frequently visited, and the identity of these states depends strongly on the initialization of the simulated trajectories.

6.3.2 TD(λ) for General λ

As in Ch. 5, we now generalize from TD(1) to TD(λ), where λ is a parameter in $[0, 1]$. In an off-line version of TD(λ), a trajectory $i_0, \ldots, i_N$ is simulated and afterwards the parameter vector r is updated by letting

$$r := r + \gamma \sum_{m=0}^{N-1} \nabla \tilde{J}(i_m, r) \sum_{k=m}^{N-1} d_k \lambda^{k-m}.$$

In the on-line version, the terms involving d_k are used for a partial update of r, as soon as d_k becomes available. More specifically, for $k = 0, \ldots, N-1$, following the state transition (i_k, i_{k+1}), we set

$$r := r + \gamma d_k \sum_{m=0}^{k} \lambda^{k-m} \nabla \tilde{J}(i_m, r).$$

For the case of a lookup table representation, it is easy to check that this method reduces to the TD(λ) method of Section 5.3.

While this method has received wide attention, its convergence behavior is unclear, unless r contains enough parameters to make possible an exact representation of $J^\mu(i)$ by $\tilde{J}(i, r)$ for all states i. In general, when $\lambda < 1$, TD(λ) resembles an incremental gradient method for the objective function $\sum_i \big(\tilde{J}(i, r) - J^\mu(i)\big)^2$, except that some error terms are added to the gradient. [Only TD(1) uses the correct gradient (modulo a zero mean simulation error).] These error terms depend on r as well as λ, and their effect cannot be neglected unless $\lambda = 1$ or a lookup table representation is used. For this reason, the limit obtained by TD(λ), depends in general on λ.

To amplify this point and to enhance our understanding, let us consider in more detail the algorithm TD(0) obtained by setting $\lambda = 0$. In the on-line version of the algorithm, a transition from i to j leads to an update of the form

$$r := r + \gamma d(i, j, r) \nabla \tilde{J}(i, r),$$

where $d(i, j, r)$ is the temporal difference $g(i, j) + \tilde{J}(j, r) - \tilde{J}(i, r)$. Recall that in the lookup table case, TD(0) can be thought of as a stochastic approximation method for solving the system of equations

$$J(i) = \sum_{j=0}^{n} p_{ij} \big(g(i, j) + J(j)\big), \qquad i = 1, \ldots, n.$$

With a parametric representation $\tilde{J}(i, r)$, we could similarly try to minimize

$$\sum_{i=1}^{n} \left(\tilde{J}(i, r) - \sum_{j=0}^{n} p_{ij} \big(g(i, j) + \tilde{J}(j, r)\big) \right)^2,$$

with respect to r. An incremental gradient update based on the ith term of this least squares criterion is of the form

$$r := r + \gamma \sum_{j=0}^{n} p_{ij} d(i,j,r) \left(\nabla \tilde{J}(i,r) - \sum_{k=0}^{n} p_{ik} \nabla \tilde{J}(k,r) \right).$$

The term $\sum_{j=0}^{n} p_{ij} d(i,j,r)$ is the expectation of $d(i,j,r)$ and can be replaced by a single sample $d(i,j,r)$, where j is the outcome of a transition simulated according to the transition probabilities p_{ij}, leading to the update

$$r := r + \gamma d(i,j,r) \left(\nabla \tilde{J}(i,r) - \sum_{k=0}^{n} p_{ik} \nabla \tilde{J}(k,r) \right).$$

Thus, TD(0) could be explained as an incremental gradient algorithm, except that the term $d(i,j,r) \sum_{k=0}^{n} p_{ik} \nabla \tilde{J}(k,r)$ is omitted. The consequences of omitting this term are not easy to predict.

The example that follows shows that TD(λ) not only gives in the limit a vector $r(\lambda)$ that depends on λ, but also that the quality of $\tilde{J}\big(i, r(\lambda)\big)$ as an approximation to $J^\mu(i)$ can get worse as λ becomes smaller than 1. In particular, the approximation provided by TD(0) can be very poor. This casts some doubt on the suitability of TD(0) for obtaining good approximations of the cost-to-go function.

Example 6.5

In this example there is only one policy, and the state transitions and associated costs are deterministic. In particular, from state $i \geq 1$ we move to state $i-1$ with a given cost g_i. We use a linear approximation of the form

$$\tilde{J}(i,r) = ir$$

for the cost-to-go function. Let all simulation runs start at state n and end at 0 after visiting all the states $n-1, n-2, \ldots, 1$ in succession. The temporal difference associated with the transition from i to $i-1$ is

$$g_i + \tilde{J}(i-1,r) - \tilde{J}(i,r) = g_i - r,$$

and the corresponding gradient is

$$\nabla \tilde{J}(i,r) = i.$$

The iteration of TD(λ) corresponding to a complete trajectory is given by

$$r := r + \gamma \sum_{k=1}^{n} (g_k - r)\big(\lambda^{n-k} n + \lambda^{n-k-1}(n-1) + \cdots + k\big), \tag{6.28}$$

and is linear in r.

Suppose that the stepsize γ is either constant and satisfies

$$0 < \gamma < 2\left(\sum_{k=1}^{n}\left(\lambda^{n-k}n + \lambda^{n-k-1}(n-1) + \cdots + k\right)\right)^{-1},$$

[in which case the coefficient of r in iteration (6.28) is in the range $(-1, 1)$ and the iteration is contracting], or else γ is diminishing at a rate that is inversely proportional to the number of simulation runs performed thus far. Then, the TD(λ) iteration (6.28) converges to the scalar $\hat{r}(\lambda)$ for which the increment in the right-hand side of Eq. (6.28) is zero, that is,

$$\sum_{k=1}^{n}\big(g_k - \hat{r}(\lambda)\big)\left(\lambda^{n-k}n + \lambda^{n-k-1}(n-1) + \cdots + k\right) = 0.$$

In particular, we have

$$\hat{r}(1) = \frac{n(g_1 + \cdots + g_n) + (n-1)(g_1 + \cdots + g_{n-1}) + \cdots + g_1}{n^2 + (n-1)^2 + \cdots + 1}, \tag{6.29}$$

$$\hat{r}(0) = \frac{ng_n + (n-1)g_{n-1} + \cdots + g_1}{n + (n-1) + \cdots + 1}. \tag{6.30}$$

It can be seen that $\hat{r}(1)$ minimizes over r the sum of squared errors

$$\sum_{i=1}^{n}\big(J(i) - \tilde{J}(i,r)\big)^2, \tag{6.31}$$

where

$$J(i) = g_1 + \cdots + g_i, \qquad \tilde{J}(i,r) = ir, \qquad i = 1, \ldots, n.$$

Indeed the optimality condition for minimization of the function (6.31) over r is

$$\sum_{i=1}^{n} i(g_1 + \cdots + g_i - ir) = 0,$$

which when solved for r gives a solution equal to $r(1)$ as given by Eq. (6.29).

Figures 6.4 and 6.5 show the form of the cost-to-go function $J(i)$, and the approximate representations $\tilde{J}\big(i, \hat{r}(1)\big)$ and $\tilde{J}\big(i, \hat{r}(0)\big)$ provided by TD(1) and TD(0), respectively, for $n = 50$ and for the following two cases:

(1) $g_1 = 1$ and $g_i = 0$ for all $i \neq 1$.

(2) $g_n = -(n-1)$ and $g_i = 1$ for all $i \neq n$.

It can be seen that TD(0) can yield a very poor approximation to the cost function.

The above example can be generalized with similar results. For instance, the cost of a transition from i to $i-1$ may be random, in which case the costs g_i must be replaced by their expected values in Eqs. (6.29) and (6.30). Also the results are qualitatively similar if the successor state of state i is randomly chosen among the states j with $j < i$. Finally, the trajectories need not all start at state n.

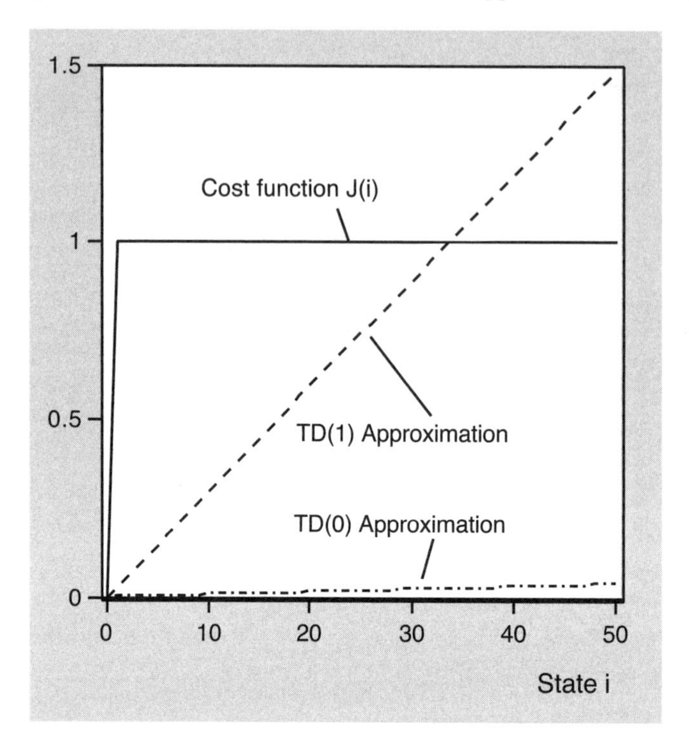

Figure 6.4: Form of the cost-to-go function $J(i)$, and the linear representations $\tilde{J}\big(i,\hat{r}(1)\big)$, and $\tilde{J}\big(i,\hat{r}(0)\big)$ provided by TD(1) and TD(0), respectively, for the case

$$g_1 = 1, \qquad g_i = 0, \quad \forall\, i \neq 1.$$

Discounted Problems

For discounted problems, the TD(λ) update rule is almost the same as for the undiscounted case. As in Ch. 5, we need a small modification in the definition of the temporal differences which are now given by

$$d_k = g(i_k, i_{k+1}) + \alpha \tilde{J}(i_{k+1}, r) - \tilde{J}(i_k, r),$$

and the (on-line) update rule following the transition from i_k to i_{k+1} becomes

$$r := r + \gamma d_k \sum_{m=0}^{k} (\alpha\lambda)^{k-m} \nabla \tilde{J}(i_m, r).$$

In the absence of an absorbing termination state, the trajectory never terminates and the entire algorithm involves a single infinitely long trajectory. In this case, and in order to have any hope for convergence, it is necessary to gradually reduce γ towards zero as the algorithm progresses.

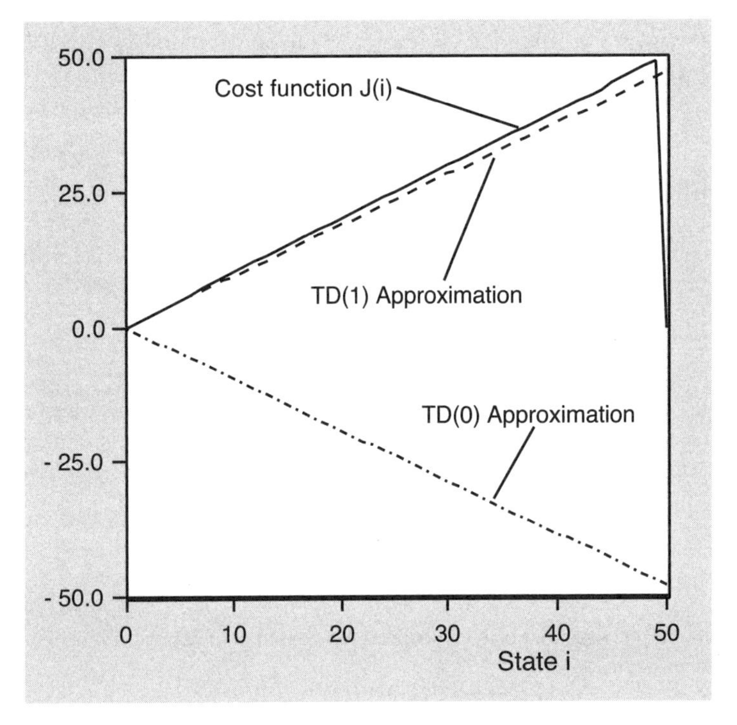

Figure 6.5: Form of the cost-to-go function $J(i)$, and the linear representations $\tilde{J}\big(i, \hat{r}(1)\big)$, and $\tilde{J}\big(i, \hat{r}(0)\big)$ provided by TD(1) and TD(0), respectively, for the case

$$g_n = -(n-1), \qquad g_i = 1, \quad \forall\, i \neq n.$$

The use of a single infinite trajectory may be inappropriate in some cases. For example, if some of the important states of the Markov chain are transient (e.g., if there is a fixed initial state which is never revisited), there will be very little training at these states, with a poor approximation resulting. If this is the case, the trajectory should be frequently reinitialized.

As a last comment, if α is set to a value substantially smaller than 1, the choice of λ has a diminished effect on the update equation, and there is less of a motivation for using values of λ smaller than 1.

TD(λ) Can Diverge for Nonlinear Architectures

We will now show that when $\lambda = 0$, the TD(λ) method described in this section can fail to converge, even with an appropriately diminishing stepsize, when the parameter r affects the approximate cost-to-go function $\tilde{J}(i, r)$ nonlinearly. This is to be contrasted with the case of linear parametrizations for which convergence guarantees will be provided in the next subsection. By a continuity argument, divergence is also possible for TD(λ), for small values of λ. At this point it is not known whether divergence is possible for all values of λ smaller than 1.

Example 6.6

We consider a Markov chain with three states, and a transition probability matrix given by

$$P = \begin{bmatrix} 1/2 & 0 & 1/2 \\ 1/2 & 1/2 & 0 \\ 0 & 1/2 & 1/2 \end{bmatrix}.$$

We let the costs per stage be identically equal to zero; we also let $\alpha < 1$ be the discount factor.

Let Q be the matrix given by

$$Q = \begin{bmatrix} 1 & 1/2 & 3/2 \\ 3/2 & 1 & 1/2 \\ 1/2 & 3/2 & 1 \end{bmatrix},$$

and let ϵ be a small positive scalar. We consider an approximate cost-to-go vector

$$\tilde{J}(r) = \big(\tilde{J}(1,r), \tilde{J}(2,r), \tilde{J}(3,r)\big),$$

parametrized by a single scalar r. Let $e = (1,1,1)$. The form of $\tilde{J}$ is defined for $r \geq 0$ by letting $\tilde{J}(0)$ be some nonzero vector that satisfies $e'\tilde{J}(0) = 0$, and by requiring that $\tilde{J}(r)$ be the (unique) solution of the linear differential equation

$$\frac{d\tilde{J}}{dr}(r) = (Q + \epsilon I)\tilde{J}(r), \tag{6.32}$$

where I is the 3×3 identity matrix.

We make a few observations on the form of $\tilde{J}(r)$.

(a) From Eq. (6.32) we obtain

$$e'\frac{d\tilde{J}}{dr}(r) = e'(Q + \epsilon I)\tilde{J}(r) = (3 + \epsilon)e'\tilde{J}(r).$$

Since the initial condition $\tilde{J}(0)$ satisfies $e'\tilde{J}(0) = 0$, it follows that $e'\tilde{J}(r) = 0$ for all $r > 0$.

(b) It is seen that $Q + Q' = 2ee'$. Since $e'\tilde{J}(r) = 0$ for all r, we obtain $\tilde{J}(r)'(Q + Q')\tilde{J}(r) = 2\tilde{J}(r)'ee'\tilde{J}(r) = 0$, for all r, which also yields

$$\tilde{J}(r)'Q\tilde{J}(r) = 0, \qquad \forall\, r. \tag{6.33}$$

Using Eq. (6.32), we obtain

$$\frac{d}{dr}\|\tilde{J}(r)\|^2 = \tilde{J}(r)'(Q + Q')\tilde{J}(r) + 2\epsilon\|\tilde{J}(r)\|^2 = 2\epsilon\|\tilde{J}(r)\|^2, \tag{6.34}$$

where $\|\cdot\|$ stands for the Euclidean norm. In particular, $\|\tilde{J}(r)\|$ diverges to infinity as r increases.

The TD(0) algorithm, based on a single infinitely long trajectory leads to the update equation

$$r_{t+1} = r_t + \gamma_t \frac{d\tilde{J}}{dr}(i_t, r_t)\big(\alpha\tilde{J}(i_{t+1}, r_t) - \tilde{J}(i_t, r)\big),$$

where i_t is the state visited by the trajectory at time t. We argue somewhat heuristically, along the lines of the discussion of the ODE approach in Section 4.4, although the argument can be made fully rigorous. Since the stepsize is decreasing to zero, we can think of r_t as being temporarily fixed at some value r while the Markov chain i_t reaches steady-state. We can then compute the steady-state expectation of the update and this is the average direction of motion for r.

We note that the steady-state distribution resulting from P is uniform. Hence, within a factor of 3, the steady-state expectation of the update direction is given by

$$\sum_{i=1}^{3} \frac{d\tilde{J}}{dr}(i, r)\left(\alpha \sum_{j=1}^{3} p_{ij}\tilde{J}(j, r) - \tilde{J}(i, r)\right).$$

This is the inner product of the vector $d\tilde{J}/dr$, which is $(Q + \epsilon I)\tilde{J}(r)$, with the vector with components $\alpha\sum_{j=1}^{3} p_{ij}\tilde{J}(j, r) - \tilde{J}(i, r)$, which is the vector $\alpha P\tilde{J}(r) - \tilde{J}(r)$. We conclude that the average direction of motion of the parameter r is captured by the ordinary differential equation

$$\frac{dr}{dt} = \big((Q + \epsilon I)\tilde{J}(r)\big)'(\alpha P - I)\tilde{J}(r) = \tilde{J}(r)'(Q' + \epsilon I)(\alpha P - I)\tilde{J}(r).$$

For $\epsilon = 0$, we have, using Eq. (6.33),

$$\frac{dr}{dt} = \tilde{J}(r)'Q'(\alpha P - I)\tilde{J}(r) = \alpha\tilde{J}(r)'Q'P\tilde{J}(r) = \frac{\alpha}{2}\tilde{J}(r)'(Q'P + P'Q)\tilde{J}(r).$$

Note that

$$Q'P + P'Q = \begin{bmatrix} 2.5 & 1.75 & 1.75 \\ 1.75 & 2.5 & 1.75 \\ 1.75 & 1.75 & 2.5 \end{bmatrix},$$

which is easily verified to be positive definite. Hence, there exists a positive constant c such that

$$\frac{dr}{dt} \geq c\big\|\tilde{J}(r)\big\|^2. \tag{6.35}$$

By a continuity argument, this inequality remains true (possibly with a smaller positive constant c) if ϵ is positive but sufficiently small. By combining Eqs. (6.34) and (6.35), we conclude that both r and $\|\tilde{J}(r)\|$ diverge to infinity.

6.3.3 TD(λ) with Linear Architectures – Discounted Problems

We now focus on the case where the approximate cost-to-go is a linear function of the parameter vector. In this subsection, we consider discounted problems and we focus on an on-line method involving the simulation of a single infinitely long trajectory. We will show that TD(λ) is guaranteed to converge but, in contrast with the results of Ch. 5, convergence rests on a specific method for simulating the system and for setting the eligibility coefficients.

We consider a linear parametrization of the form

$$\tilde{J}(i,r) = \sum_{k=1}^{K} r(k)\phi_k(i).$$

Here, $r = \big(r(1),\ldots,r(K)\big)$ is a vector of tunable parameters and $\phi_k(\cdot)$ are fixed scalar functions defined on the state space. The functions ϕ_k are the basis functions in a linearly parametrized approximation architecture; alternatively, they can be viewed as features of the state that are combined linearly to form an estimate of the cost-to-go.

It is convenient to define for each state i, a vector $\phi(i)$ by letting

$$\phi(i) = \big(\phi_1(i),\ldots,\phi_K(i)\big).$$

We also define a matrix Φ, of dimensions $n \times K$, whose ith row is the row vector $\phi(i)'$ and whose kth column is the vector $\phi_k = \big(\phi_k(1),\ldots,\phi_k(n)\big)$. That is,

$$\Phi = \begin{bmatrix} | & & | \\ \phi_1 & \cdots & \phi_K \\ | & & | \end{bmatrix} = \begin{bmatrix} - & \phi(1)' & - \\ \cdots & \cdots & \cdots \\ - & \phi(n)' & - \end{bmatrix}.$$

Our approximation architecture can be described concisely in the form

$$\tilde{J}(r) = \big(\tilde{J}(1,r),\ldots,\tilde{J}(n,r)\big) = \Phi r,$$

and it is seen that

$$\nabla \tilde{J}(i,r) = \phi(i).$$

We now develop a mathematically precise description of the algorithm. Starting from some initial state i_0, we generate a single infinitely long trajectory $(i_0, i_1, \ldots)$. Suppose that at some time t, the current value of the parameter vector r is r_t. As soon as the transition from i_t to i_{t+1} is simulated, we evaluate the temporal difference

$$d_t = g(i_t, i_{t+1}) + \alpha\tilde{J}(i_{t+1}, r_t) - \tilde{J}(i_t, r_t),$$

and update r_t according to the formula

$$\begin{aligned} r_{t+1} &= r_t + \gamma_t d_t \sum_{k=0}^{t} (\alpha\lambda)^{t-k} \nabla \tilde{J}(i_k, r_t) \\ &= r_t + \gamma_t d_t \sum_{k=0}^{t} (\alpha\lambda)^{t-k} \phi(i_k), \end{aligned}$$

where γ_t is a scalar stepsize whose properties will be specified later.

A more convenient representation of the method is obtained if we define an *eligibility vector* z_t by

$$z_t = \sum_{k=0}^{t} (\alpha\lambda)^{t-k} \phi(i_k).$$

Note that z_t is of dimension K, with its ℓth component $z_t(\ell)$ being associated with the ℓth basis function $\phi_\ell(\cdot)$. With this new notation, the TD(λ) updates are given by

$$r_{t+1} = r_t + \gamma_t d_t z_t, \tag{6.36}$$

and the eligibility vectors are updated according to

$$z_{t+1} = \alpha\lambda z_t + \phi(i_{t+1}). \tag{6.37}$$

We introduce the following assumptions.

Assumption 6.1:

(a) The stepsizes γ_t are positive and deterministic (predetermined). Furthermore, we have $\sum_{t=0}^{\infty} \gamma_t = \infty$ and $\sum_{t=0}^{\infty} \gamma_t^2 < \infty$.

(b) There exist positive numbers $\pi(j)$ such that

$$\lim_{t\to\infty} \mathrm{P}(i_t = j \mid i_0 = i) = \pi(j), \qquad \forall\, i, j.$$

(c) We have $K \leq n$ (no more parameters than states) and the matrix Φ has full rank.

Under Assumption 6.1(b), we are dealing with an aperiodic Markov chain with a single recurrent class and no transient states. Furthermore, all states are visited an infinite number of times during an infinitely long trajectory. Besides the aperiodicity requirement, this assumption results in no loss of generality, as far as the validity of the convergence result is concerned. This is because the states that are only visited a finite number of times can be simply eliminated. From a pragmatic point of view, however, the cost-to-go approximation provided by the limiting vector r will usually be very poor at the transient states.

In connection with Assumption 6.1(b), we note that the vector $\pi = \big(\pi(1), \ldots, \pi(n)\big)$ of steady-state probabilities satisfies

$$\pi' P = \pi'.$$

Furthermore, it is well known from Markov chain theory that the transition probabilities $\mathrm{P}(i_t = j \mid i_0 = i)$ converge to $\pi(j)$ at the rate of a geometric progression. These facts will prove useful for our convergence analysis.

Assumption 6.1(c) requires that $\phi_1(\cdot), \ldots, \phi_K(\cdot)$ be linearly independent functions on the state space $\{1, \ldots, n\}$. This results in no loss of generality because if some function $\phi_k(\cdot)$ is a linear combination of the others, it can be eliminated without changing the power of the approximation architecture.

We now move towards a convergence result. Its proof is based on a technique that is very different from those used in Ch. 5, and leads to some new insights. Recall that the transition probability matrix P satisfies $\|PJ\|_\infty \leq \|J\|_\infty$ for all J, where $\|\cdot\|_\infty$ is the maximum norm. This property has been used repeatedly in our earlier development and underlies the contraction property of the DP operator T, for discounted problems. In contrast, the proof that follows relies on a different and less known property of P, namely, the relation $\|PJ\|_D \leq \|J\|_D$, where $\|\cdot\|_D$ is the *weighted quadratic norm* defined by

$$\|J\|_D^2 = J' D J = \sum_{i=1}^{n} \pi(i) J(i)^2,$$

where D is a diagonal matrix with diagonal entries $\pi(1), \pi(2), \ldots, \pi(n)$, and where $\pi(i)$ is the steady-state probability of state i [cf. Assumption 6.1(b)].

For an outline of the proof, we first show that TD(λ) can be expressed in the form

$$r_{t+1} = r_t + \gamma_t\big(A(X_t) r_t + b(X_t)\big),$$

studied in Section 4.4.1, where X_t is a suitably defined auxiliary Markov process. We then derive formulas for the steady-state expectation of $A(X_t)$ and $b(X_t)$ (Lemma 6.5). Finally, we use an argument involving the norm $\|\cdot\|_D$, to show that the steady-state average of $A(X_t)$ is negative definite (Lemma 6.6). Convergence then follows once we verify an additional key assumption of Prop. 4.8 from Section 4.4.1, which is done in Lemma 6.7.

Proposition 6.4: Under Assumption 6.1, the sequence r_t generated by Eqs. (6.36) and (6.37) converges, with probability 1.

Proof: As mentioned earlier, the proof rests heavily on the lemma that follows.

Lemma 6.4: For any $J \in \Re^n$, we have $\|PJ\|_D \leq \|J\|_D$.

Proof: Let p_{ij} be the ijth entry of P. Then,

$$\begin{aligned}
\|PJ\|_D^2 &= J'P'DPJ \\
&= \sum_{i=1}^{n} \pi(i) \left(\sum_{j=1}^{n} p_{ij} J(j) \right)^2 \\
&\leq \sum_{i=1}^{n} \pi(i) \sum_{j=1}^{n} p_{ij} J(j)^2 \\
&= \sum_{j=1}^{n} \sum_{i=1}^{n} \pi(i) p_{ij} J(j)^2 \\
&= \sum_{j=1}^{n} \pi(j) J(j)^2 \\
&= J'DJ \\
&= \|J\|_D^2.
\end{aligned}$$

We have used here the inequality $E\big[J(i_{t+1}) \mid i_t = i\big]^2 \leq E\big[J(i_{t+1})^2 \mid i_t = i\big]$, as well as the property $\pi'P = \pi'$ of the vector of invariant probabilities. **Q.E.D.**

Let $X_t = (i_t, i_{t+1}, z_t)$. It is easily seen that X_t is a Markov process, with an infinite state space (since z_t takes real values). Indeed, z_{t+1} and i_{t+1} are deterministic functions of X_t, and the distribution of i_{t+2} only depends on i_{t+1}. This Markov process has a steady-state version, which can be constructed as follows. We first construct a stationary Markov chain i_t, $-\infty < t < \infty$, whose transition probabilities are given by P, and we then let

$$z_t = \sum_{\tau=-\infty}^{t} (\alpha\lambda)^{t-\tau} \phi(i_\tau). \tag{6.38}$$

It is easily seen that the resulting process X_t is stationary, i.e., it is a Markov chain in steady-state. We will use $E_0[\cdot]$ to denote expectation with respect to the steady-state distribution of X_t.

Note that

$$\begin{aligned}
d_t &= g(i_t, i_{t+1}) + \alpha \tilde{J}(i_{t+1}, r_t) - \tilde{J}(i_t, r_t) \\
&= g(i_t, i_{t+1}) + \alpha \phi(i_{t+1})' r_t - \phi(i_t)' r_t.
\end{aligned}$$

Hence, the step $d_t z_t$ involved in the update of r_t is of the form

$$d_t z_t = A(X_t) r_t + b(X_t),$$

where $b(X_t)$ is a K-dimensional vector defined by

$$b(X_t) = z_t g(i_t, i_{t+1}),$$

and $A(X_t)$ is a $K \times K$ matrix defined by

$$A(X_t) = z_t \big(\alpha \phi(i_{t+1})' - \phi(i_t)'\big).$$

Let

$$b = E_0\big[b(X_t)\big],$$

and

$$A = E_0\big[A(X_t)\big].$$

Lemma 6.5: We have

$$A = \Phi' D (M - I) \Phi,$$

where

$$M = (1 - \lambda) \sum_{m=0}^{\infty} \lambda^m (\alpha P)^{m+1}.$$

Furthermore,

$$b = \Phi' D q,$$

where

$$q = \sum_{m=0}^{\infty} (\alpha \lambda P)^m \overline{g},$$

and where $\overline{g}$ is a vector in $\Re^n$ whose ith component is given by $\overline{g}(i) = \sum_{j=1}^{n} p_{ij} g(i, j)$.

Proof: We observe that for any two vectors J and $\overline{J}$, we have

$$\begin{aligned} E_0\big[J(i_0)\overline{J}(i_m)\big] &= \sum_{i=1}^{n} \pi(i) \sum_{j=1}^{n} \mathrm{P}(i_m = j \mid i_0 = i) J(i) \overline{J}(j) \\ &= \sum_{i=1}^{n} \pi(i) J(i) [P^m \overline{J}](i) \\ &= J' D P^m \overline{J}, \end{aligned} \tag{6.39}$$

where the notation $[P^m\overline{J}](i)$ stands for the ith component of the vector $P^m\overline{J}$. By specializing to the case where $J = \Phi r$, $\overline{J} = \Phi\overline{r}$, we have $J(i_0) = \phi(i_0)'r$ and $\overline{J}(i_m) = \phi(i_m)'\overline{r}$, and we obtain

$$E_0\big[r'\phi(i_0)\phi(i_m)'\overline{r}\big] = r'\Phi' DP^m\Phi\overline{r}.$$

Since the vectors r and $\overline{r}$ are arbitrary, it follows that

$$E_0\big[\phi(i_0)\phi(i_m)'\big] = \Phi' DP^m\Phi.$$

Using Eq. (6.38), with $t = 0$ and $\tau = -m$, we obtain

$$\begin{aligned} E_0\big[z_0\phi(i_0)'\big] &= E_0\left[\sum_{m=0}^{\infty}\phi(i_{-m})(\alpha\lambda)^m\phi(i_0)'\right] \\ &= \Phi' D\sum_{m=0}^{\infty}(\alpha\lambda)^m P^m\Phi. \end{aligned}$$

Using a similar argument, we also obtain

$$E_0[z_0\phi(i_1)'] = \Phi' D\sum_{m=0}^{\infty}(\alpha\lambda)^m P^{m+1}\Phi.$$

Hence,

$$\begin{aligned} A &= E_0\Big[z_0\big(\alpha\phi(i_1)' - \phi(i_0)'\big)\Big] \\ &= \Phi' D\sum_{m=0}^{\infty}(\alpha\lambda)^m(\alpha P^{m+1} - P^m)\Phi \\ &= \Phi' D\left((1-\lambda)\sum_{m=0}^{\infty}\lambda^m(\alpha P)^{m+1} - I\right)\Phi \\ &= \Phi' D(M - I)\Phi. \end{aligned}$$

We now use Eq. (6.39), with $J = \Phi r$ and $\overline{J} = \overline{g}$, to obtain

$$\begin{aligned} r'b &= E_0\big[r'z_0 g(i_0, i_1)\big] \\ &= E_0\big[r'z_0\overline{g}(i_0)\big] \\ &= E_0\left[r'\sum_{m=0}^{\infty}(\alpha\lambda)^m\phi(i_{-m})\overline{g}(i_0)\right] \\ &= r'\Phi' D\sum_{m=0}^{\infty}(\alpha\lambda P)^m\overline{g}. \end{aligned}$$

Since the vector r is arbitrary, the desired formula for b follows. **Q.E.D.**

Lemma 6.6:

(a) For every vector $J \in \Re^n$, we have

$$\|MJ\|_D \leq \frac{\alpha(1-\lambda)}{1-\alpha\lambda}\|J\|_D.$$

(b) The matrix A is negative definite.

Proof:

(a) Recall that

$$\|PJ\|_D \leq \|J\|_D, \qquad \forall\ J.$$

From this, it easily follows that

$$\|P^m J\|_D \leq \|J\|_D, \qquad \forall\ J,\ m \geq 0.$$

Using the definition of M and the triangle inequality, we obtain

$$\|MJ\|_D \leq (1-\lambda)\sum_{m=0}^{\infty}\lambda^m\alpha^{m+1}\|J\|_D = \frac{\alpha(1-\lambda)}{1-\alpha\lambda}\|J\|_D.$$

(b) Let $\|\cdot\|$ stand for the Euclidean norm. Using the Schwartz inequality, we have

$$\begin{aligned} J'DMJ &= J'D^{1/2}D^{1/2}MJ \\ &\leq \|D^{1/2}J\| \cdot \|D^{1/2}MJ\| \\ &= \|J\|_D \cdot \|MJ\|_D \\ &\leq \alpha\|J\|_D \cdot \|J\|_D \\ &= \alpha J'DJ. \end{aligned}$$

Thus,

$$J'D(M-I)J \leq -(1-\alpha)J'DJ < 0, \qquad \forall\ J \neq 0,$$

which proves that $D(M-I)$ is negative definite. Using the assumption that the matrix Φ has full rank, it follows that $A = \Phi'D(M-I)\Phi$ is also negative definite. **Q.E.D.**

We have established so far that the algorithm is of the form

$$r_{t+1} = r_t + \gamma_t\big(A(X_t)r_t + b(X_t)\big),$$

and that the steady-state average of $A(X_t)$ is negative definite. If $A(\cdot)$ and $b(\cdot)$ did not depend on X_t, convergence would follow at once by considering the potential function $\|r - r^*\|^2$, where $\|\cdot\|$ is the Euclidean norm and r^* is the solution of $Ar + b = 0$. However, because of the dependence on the Markov process X_t, we need to resort to the convergence results of Section 4.4. These results require that the dependence of $A(X_t)$ and $b(X_t)$ on X_τ, for $\tau \leq t$, be exponentially decreasing. We will show that this is indeed the case, using the fact that the dependence of i_t on i_0 is exponentially decreasing in ergodic finite state Markov chains. In addition, the formula for z_t involves an exponentially decaying dependence on the past. These considerations are translated into a formal argument in the lemma that follows.

Lemma 6.7: There exists some $\rho < 1$ and some $C > 0$ such that for every $t \geq 0$ and every X,

$$\left\|E\left[A(X_t) \mid X_0 = X\right] - A\right\| \leq C\rho^t,$$

and

$$\left\|E\left[b(X_t) \mid X_0 = X\right] - b\right\| \leq C\rho^t.$$

Proof: Let B be a constant such that $\|\phi_k(i)\| \leq B$ for all k and i. Note that every component of z_t is bounded in magnitude by $B/(1 - \alpha\lambda)$.

We have

$$E\left[A(X_t) \mid X_0\right] = E\Big[z_t\big(\alpha\phi(i_{t+1}) - \phi(i_t)\big)' \mid i_0, i_1, z_0\Big].$$

Let us focus on the term $E\left[z_t\phi(i_t)' \mid i_0, i_1, z_0\right]$. We have

$$\begin{aligned}
E\left[z_t\phi(i_t)' \mid i_0, i_1, z_0\right] &= E\left[\sum_{m=0}^{t} \phi(i_m)(\alpha\lambda)^{t-m}\phi(i_t)' \mid i_0, i_1, z_0\right] \\
&= E_0\left[\sum_{m=-\infty}^{t} \phi(i_m)(\alpha\lambda)^{t-m}\phi(i_t)'\right] + f_1(t) + f_2(t) \\
&= E_0\left[z_t\phi(i_t)'\right] + f_1(t) + f_2(t),
\end{aligned}$$

where

$$\begin{aligned}
f_1(t) = E&\left[\sum_{m=0}^{t} \phi(i_m)(\alpha\lambda)^{t-m}\phi(i_t)' \mid i_0, i_1, z_0\right] \\
&- E_0\left[\sum_{m=0}^{t} \phi(i_m)(\alpha\lambda)^{t-m}\phi(i_t)'\right],
\end{aligned}$$

and

$$f_2(t) = -E_0\left[\sum_{m=-\infty}^{-1} \phi(i_m)(\alpha\lambda)^{t-m}\phi(i_t)'\right].$$

We will now show that $f_1(t)$ and $f_2(t)$ decay exponentially as t tends to infinity. Regarding $f_2(t)$, this is fairly obvious because all entries of $\phi(i_m)\phi(i_t)'$ are bounded by a deterministic constant and because

$$\sum_{m=-\infty}^{-1} (\alpha\lambda)^{t-m} = (\alpha\lambda)^t \sum_{m=1}^{\infty} (\alpha\lambda)^m,$$

which is an exponentially decaying function of t.

Regarding $f_1(t)$, we first consider the difference

$$E\big[\phi(i_m)\phi(i_t)' \mid i_0, i_1, z_0\big] - E_0\big[\phi(i_m)\phi(i_t)'\big].$$

For $1 \leq m \leq t$, this is equal to

$$\sum_{j=1}^{n} \mathrm{P}(i_m = j \mid i_1)\phi(j)E\big[\phi(i_t)' \mid i_m = j\big] - \sum_{j=1}^{n} \pi(j)\phi(j)E\big[\phi(i_t)' \mid i_m = j\big].$$

Because the m-step transition probabilities $\mathrm{P}(i_m = j \mid i_1)$ converge to the steady-state probabilities $\pi(j)$ exponentially fast in m, we conclude that this difference is bounded in magnitude by $C_1\rho_1^t$ for some $\rho_1 < 1$ and some C_1. For $m = 0$, the difference under consideration is clearly bounded by a constant C_1. We then see that the magnitude of $f_1(t)$ is bounded above by

$$C_1 \sum_{m=0}^{t} (\alpha\lambda)^{t-m}\rho_1^m,$$

which decays exponentially to zero as t converges to infinity.

By a similar argument, we also obtain

$$E\big[z_t\alpha\phi(i_{t+1})' \mid i_0, i_1, z_0\big] = E_0\big[z_t\alpha\phi(i_{t+1})'\big] + f_3(t),$$

where $f_3(t)$ decays exponentially with t. It then follows immediately that the difference between $E\big[A(X_t) \mid X_0\big]$ and A decays exponentially with t.

The proof for $b(X_t)$ is similar and is omitted. **Q.E.D.**

We can now invoke Prop. 4.8 from Section 4.4.1 and conclude that r_t indeed converges to the unique solution of the system $Ar+b=0$. **Q.E.D.**

Interpretation of the Algorithm and Error Bounds

As shown in the proof of Prop. 6.4, the sequence of parameter vectors r_t converges to some limit, call it r^*, that satisfies $Ar^* + b = 0$. We now set out to develop an interpretation of this limiting vector r^*, and to derive a bound on the magnitude of the difference between the approximate cost-to-go Φr^* and the true cost-to-go J^μ of the policy under consideration.

For $0 \leq \lambda \leq 1$, we define an operator $T^{(\lambda)}$ by letting

$$T^{(\lambda)}J = (1-\lambda)\sum_{m=0}^{\infty} \lambda^m (\alpha P)^{m+1} J + \sum_{m=0}^{\infty} (\alpha\lambda P)^m \overline{g} = MJ + q, \tag{6.40}$$

where M and q have been defined in Lemma 6.5. Using simple algebra, it can be verified that for $\lambda < 1$ we have

$$(T^{(\lambda)}J)(i) = (1-\lambda)\sum_{m=0}^{\infty} \lambda^m E\left[\sum_{t=0}^{m} \alpha^t g(i_t, i_{t+1}) + \alpha^{m+1} J(i_{m+1}) \;\Big|\; i_0 = i\right],$$

and for $\lambda = 1$,

$$(T^{(1)}J)(i) = \sum_{m=0}^{\infty} \alpha^m E\big[g(i_m, i_{m+1}) \mid i_0 = i\big] = J^\mu(i).$$

It is seen that $T^{(\lambda)}J^\mu = J^\mu$. Furthermore, because of the result in Lemma 6.6(a), the mapping $T^{(\lambda)}$ is a contraction mapping with respect to the norm $\|\cdot\|_D$.

We have seen in Section 5.3 that the TD(λ) method, for the case of a lookup table representation, is a Robbins-Monro stochastic approximation method for solving the system of equations $T^{(\lambda)}J = J$, and J^μ is the unique solution [cf. Eq. (5.29)]. In particular, given a cost-to-go estimate J, $T^{(\lambda)}J$ is a new, and hopefully better, estimate based on a weighted average of m-step Bellman equations, for all nonnegative m.

We now return to the case where function approximation is used, and where we have convergence to the solution of the system $Ar + b = 0$. Note that

$$\begin{aligned} Ar + b &= \Phi' D(M - I)\Phi r + \Phi' D q \\ &= \Phi' D(M\Phi r + q) - \Phi' D\Phi r \\ &= \Phi' D T^{(\lambda)}(\Phi r) - \Phi' D \Phi r. \end{aligned}$$

The matrix $\Phi' D \Phi$ is invertible because D is a diagonal matrix with positive diagonal entries and the matrix Φ is assumed to have full rank. Hence, the system $Ar + b = 0$ can be rewritten as

$$\Phi(\Phi' D\Phi)^{-1}\Phi' D T^{(\lambda)}(\Phi r) = \Phi r,$$

or

$$\Pi T^{(\lambda)}(\Phi r) = \Phi r,$$

where Π is the matrix defined by

$$\Pi = \Phi(\Phi' D \Phi)^{-1}\Phi' D.$$

The lemma that follows provides an interpretation of the matrix Π; it shows that it corresponds to a projection on the space of all vectors of the form Φr, with respect to the norm $\|\cdot\|_D$.

Lemma 6.8: Assume that the matrix Φ has linearly independent columns and let $\Pi = \Phi(\Phi' D \Phi)^{-1}\Phi' D$. For every vector J, we have

$$\|\Pi J - J\|_D = \min_r \|\Phi r - J\|_D.$$

Proof: We minimize the objective function $\|\Phi r - J\|_D^2$, with respect to r, by writing it in the form $r'\Phi' D \Phi r - 2r'\Phi' DJ + J'DJ$. We take the gradient, and set it to zero, to obtain that the optimal solution Φr is given by ΠJ. **Q.E.D.**

Having come to the conclusion that r^* satisfies $\Pi T^{(\lambda)}(\Phi r^*) = \Phi r^*$, we can imagine the following algorithm. Start with a vector r and the corresponding approximate cost-to-go function Φr, and then compute $T^{(\lambda)}(\Phi r)$. Because the latter vector need not be within the set of vectors that can be represented by the approximation architecture, we project it on the space of possible approximate cost-to-go vectors, to obtain $\Pi T^{(\lambda)}(\Phi r)$. This algorithm, namely,

$$J := \Pi T^{(\lambda)}(\Phi r),$$

is guaranteed to converge because $T^{(\lambda)}$ is a contraction, and Π is a nonexpansion in the sense that $\|\Pi J\|_D \leq \|J\|_D$ (this is a generic property of projections). The TD(λ) algorithm can be now understood as a method that leads to the same limit point, by taking small steps based on simulation, as opposed to using exact calculations and exact projections.

Our next step is to develop error bounds. Because a similar situation will be encountered in later sections, we first develop a general result, and then specialize it to the current context.

Lemma 6.9: Suppose that $T : \Re^n \mapsto \Re^n$ is a contraction mapping, with respect to the norm $\|\cdot\|_D$, with contraction factor $\beta < 1$, and with a fixed point J^*. Let Π be the projection on some subspace of $\Re^n$, with respect to the same norm. Then, the system of equations $\Pi T J = J$ has a unique solution $\overline{J}$ that satisfies

$$\|\overline{J} - J^*\|_D \leq \frac{\|\Pi J^* - J^*\|_D}{1 - \beta}.$$

Proof: Existence and uniqueness of a solution $\overline{J}$ follows from the contraction mapping theorem (see e.g., [BeT89], p. 182, or [Lue69], p. 272), because the mapping $J \mapsto \Pi T J$ is a contraction. We now have

$$\begin{aligned}\|\overline{J} - J^*\|_D &\leq \|\overline{J} - \Pi J^*\|_D + \|\Pi J^* - J^*\|_D \\ &= \|\Pi T\overline{J} - \Pi T J^*\|_D + \|\Pi J^* - J^*\|_D \\ &\leq \beta\|\overline{J} - J^*\|_D + \|\Pi J^* - J^*\|_D,\end{aligned}$$

and the result follows. **Q.E.D.**

We now use Lemma 6.9 to obtain error bounds for TD(λ). Recall that the operator $T^{(\lambda)}$ is a contraction (with respect to $\|\cdot\|_D$), with contraction factor $\beta = \alpha(1-\lambda)/(1-\alpha\lambda)$ (cf. Lemma 6.6), and its fixed point is J^μ. Furthermore, the temporal difference algorithm converges to a vector Φr^* such that $\Pi T^{(\lambda)}(\Phi r^*) = \Phi r^*$. We observe that $1-\beta = 1-\alpha(1-\lambda)/(1-\alpha\lambda) = (1-\alpha)/(1-\alpha\lambda)$, and we have the following result.

Proposition 6.5: Let r^* be the limit of the sequence r_t generated by the TD(λ) algorithm described by Eqs. (6.36) and (6.37). We then have

$$\|\Phi r^* - J^\mu\|_D \leq \frac{1-\alpha\lambda}{1-\alpha}\|\Pi J^\mu - J^\mu\|_D.$$

Note the term $\|\Pi J^\mu - J^\mu\|_D$. When the error is measured according to the norm $\|\cdot\|_D$, ΠJ^μ is the best possible approximation of J^μ, given the available approximation architecture, and this sets a lower bound for the approximation error. For the case $\lambda = 1$, this lower bound matches the upper bound, and we have $\Phi r^* = \Pi J^\mu$. As λ decreases, the numerator term increases and the bound deteriorates. The worst bound, namely, $\|\Pi J^\mu - J^\mu\|_D/(1-\alpha)$ is obtained when $\lambda = 0$. Although this is only a bound, it strongly suggests that higher values of λ can be expected to produce more accurate approximations of J^*. This is consistent with the conclusions that were reached in Example 6.5.

The Case of an Infinite State Space

The case of an infinite state space (either discrete or continuous) has not been addressed before in this book. However, TD(λ) is often used in problems involving infinite state spaces and for this reason, it is important to point out that Props. 6.4-6.5 extend to that case, under mild assumptions (see Tsitsiklis and Van Roy [TsV96b]). Some, but not all, of the assumptions required are of the following type (see [TsV96b] for exact statements).

(a) We need $E\big[A(X_t) \mid X_0\big]$ to converge to its steady-state expectation A sufficiently fast, which is the case in many applications of practical interest (compare with Lemma 6.7).

(b) The "basis functions" $\phi_k(\cdot)$ need not be bounded, as long as their rate of increase is smaller than the decay rate of the steady-state probabilities. For example, if the state space is the set of all $x \in \Re^n$ and if the steady-state probability distribution decays exponentially with x, the basis functions are allowed to grow polynomially.

Remarks on TD(0) and the Possibility of Divergence

According to our results, for the case of linear architectures, TD(0) is guaranteed to converge as long as our updates are based on complete trajectories, generated according to the policy under consideration. Let us consider now a version of TD(0) that rests on the simulation of individual, possibly unrelated transitions.

At stage t of the algorithm, we have available a parameter vector r_t. We choose some state i_t and simulate a transition to a next state $\bar{i}_t$. (Unlike our previous development, i_{t+1} need not be the same as $\bar{i}_t$.) We then update r_t according to the TD(0) update rule, which is

$$\begin{aligned} r_{t+1} &= r_t + \gamma_t \phi(i_t) d_t \\ &= r_t + \gamma_t \phi(i_t)\big(g(i_t, \bar{i}_t) + \alpha\phi(\bar{i}_t)'r - \phi(i_t)'r\big). \end{aligned} \tag{6.41}$$

If the states i_t are independent identically distributed samples, drawn according to the steady-state probabilities $\pi(i)$, the convergence result of Prop. 6.4 remains valid. What happens is that the random matrices $A(X_t)$ are replaced by independent identically distributed random matrices A_t satisfying $E[A_t] = A$, and similarly for $b(X_t)$. Thus, the algorithm is of the form

$$r_{t+1} = r_t + \gamma_t(A_t r_t + b_t),$$

with $E[A_t]$ negative definite, and convergence follows easily from a potential function argument. This suggests that there is no imperative to select sample states by simulating trajectories. Any other method that leads to sample states drawn according to the steady-state probabilities would do. Realistically, however, simulation of long trajectories is the only viable method for generating samples of the steady-state distribution.

If we sample states with frequencies other than the steady-state probabilities, the expectation $E[A_t]$ changes, and there is no guarantee that it is negative definite. In particular, the algorithm can diverge, as shown by the example that follows.

Example 6.7

Consider the two-state Markov chain shown in Fig. 6.6. Assume that all transitions carry zero cost and let $\alpha < 1$ be the discount factor. Consider an approximate cost-to-go function $\tilde{J}(i, r)$ involving a single scalar parameter r, of the form $\tilde{J}(1, r) = r$ and $\tilde{J}(2, r) = 2r$. Note that $\phi(1) = 1$ and $\phi(2) = 2$.

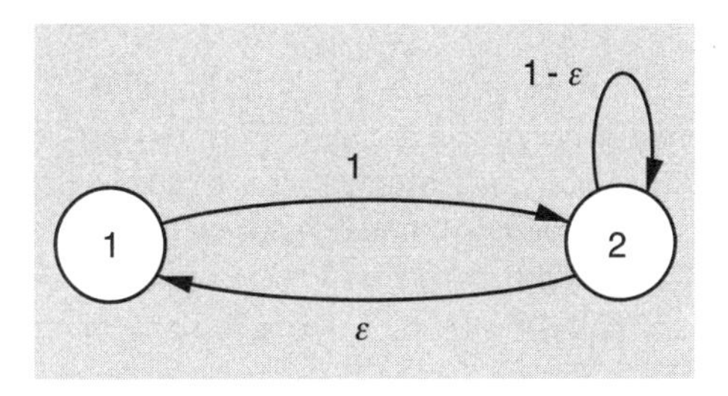

Figure 6.6: A two-state Markov chain.

At a typical iteration of the algorithm, we let r_t be the current value of the parameter r and choose a state i_t which is either 1 or 2 with equal probability. We assume that the choices of i_t for different times t are independent. We then simulate a transition to a next state $\bar{i}_t$, form the temporal difference d_t, and update r according to the TD(0) update rule (6.41). By taking expectations, and using the fact that $i_t = 1$ with probability 1/2, we obtain

$$\begin{aligned} E[r_{t+1}] &= E[r_t] + \gamma_t \frac{1}{2}(2\alpha - 1)E[r_t] + \gamma_t \frac{1}{2} 2\big((1-\epsilon)(2\alpha - 2) + \epsilon(\alpha - 2)\big)E[r_t] \\ &= E[r_t] + \frac{\gamma_t}{2}\big(6\alpha - 5 + O(\epsilon)\big)E[r_t], \end{aligned}$$

where $O(\epsilon)$ is a term which is linear in ϵ. We observe that if $\alpha > 5/6$ and ϵ is small enough, $E[r_t]$ is multiplied by a constant larger than 1 and therefore diverges.

This is to be contrasted with the trajectory-based TD(0) method. Assuming that ϵ is very small, the vast majority of the transitions are from state 2 to itself and we have

$$E[r_{t+1}] = E[r_t] + \gamma_t 2(2\alpha - 2)E[r_t] + O(\epsilon)E[r_t],$$

which converges because $\alpha < 1$, in agreement with the results in this section.

A similar potential for divergence exists for TD(λ), with $\lambda > 0$. Our convergence results assume that we generate temporal differences along a complete trajectory (for discounted problems, an infinitely long trajectory, and for stochastic shortest path problems, a trajectory that reaches the termination state). With other mechanisms, however, divergence is possible. Suppose for example that we pick a random initial state (according to a fixed distribution), that we run the temporal differences algorithm for N steps (where N is a fixed constant), and then restart. This algorithm is more general than the one considered in Example 6.7 (where we had $\lambda = 0$ and $N = 1$), and can therefore diverge.

6.3.4 TD(λ) with Linear Architectures – Stochastic Shortest Path Problems

We will now derive convergence results for the TD(λ) algorithm applied to stochastic shortest path problems. We assume that state 0 is absorbing and cost-free, and that we are dealing with a fixed proper policy. As in the preceding section, we consider a linear approximation architecture of the form

$$\tilde{J}(i,r) = \phi(i)'r, \qquad i = 0, 1, \ldots, n,$$

together with the convention $\phi(0) = 0$.

We focus on an off-line method in which an update of the parameter vector r is carried out whenever a trajectory is completed. A general iteration of the algorithm works as follows. We choose an initial state i_0 randomly and independently from the past of the algorithm, according to some fixed probability distribution. We then generate a trajectory $(i_0, i_1, \ldots, i_N)$, where N is the first time that state 0 is reached. Given the current parameter vector r, the temporal differences d_t are defined by

$$d_t = g(i_t, i_{t+1}) + \phi(i_{t+1})'r - \phi(i_t)'r, \qquad t = 0, 1, \ldots, N-1.$$

We also define eligibility coefficients z_t, which are initialized with

$$z_0 = \phi(i_0),$$

and are updated according to

$$z_{t+1} = \lambda z_t + \phi(i_{t+1}), \qquad t = 0, 1, \ldots, N-2.$$

At the end of a trajectory, we update the parameter vector r according to

$$\begin{aligned} r &:= r + \gamma \sum_{t=0}^{N-1} z_t d_t \\ &= r + \gamma \sum_{t=0}^{N-1} z_t \big(g(i_t, i_{t+1}) + \phi(i_{t+1})'r - \phi(i_t)'r\big). \end{aligned} \tag{6.42}$$

We assume that the stepsize remains constant during each trajectory and we let γ_k be the stepsize used in the course of the kth trajectory.

The convergence result that follows will be proved by an argument similar to the proof of Prop. 6.4, but there are some differences that make this proof a little easier. In Prop. 6.4, we used t to index individual transitions of the Markov chain, whereas here t is used to index trajectories. Because different trajectories are independently generated, the algorithm will be expressed in the form

$$r := r + \gamma(Ar + b + w),$$

where w is a zero mean noise term, which is independent from one iteration to the next. We argue as in the proof of Prop. 6.4 to obtain a formula for A and show that A is negative definite. Convergence is then established using the results of Ch. 4.

Proposition 6.6: Suppose that the stepsizes γ_k are nonnegative and satisfy $\sum_{k=0}^{\infty} \gamma_k = \infty$, and $\sum_{k=0}^{\infty} \gamma_k^2 < \infty$. Suppose also that all states have positive probability of being visited by the algorithm. Finally, suppose that the columns of the matrix Φ are linearly independent. Then, the sequence of vectors r generated by Eq. (6.42) converges with probability 1.

Proof: We define a matrix A and a vector b by letting

$$A = E\left[\sum_{t=0}^{N-1} z_t\big(\phi(i_{t+1})' - \phi(i_t)'\big)\right],$$

and

$$b = E\left[\sum_{t=0}^{N-1} z_t g(i_t, i_{t+1})\right].$$

Then, the algorithm is of the form

$$r := r + \gamma(Ar + b + w),$$

where w is zero-mean noise, which is independent from one iteration to the next. Furthermore, w is affine in r and therefore its variance increases linearly with $\|r\|^2$. In particular, the assumptions on w that were introduced in Section 4.2 are satisfied.

We will prove that the matrix A is negative definite. This will imply that the expected direction of update $Ar + b$ is a descent direction, with respect to the potential function $\|r - r^*\|^2$, where r^* satisfies $Ar^* + b = 0$, and where $\|\cdot\|$ is the Euclidean norm. Convergence will then follow from Prop. 4.1 in Section 4.2.

We have

$$\begin{aligned}
A &= E\left[\sum_{t=0}^{N-1} z_t\big(\phi(i_{t+1})' - \phi(i_t)'\big)\right] \\
&= E\left[\sum_{t=0}^{N-1}\sum_{m=0}^{t} \lambda^{t-m}\phi(i_m)\big(\phi(i_{t+1})' - \phi(i_t)'\big)\right] \\
&= E\left[\sum_{t=0}^{\infty}\sum_{m=0}^{t} \lambda^{t-m}\phi(i_m)\big(\phi(i_{t+1})' - \phi(i_t)'\big)\right],
\end{aligned}$$

where the last step uses the convention $\phi(i_k) = \phi(0) = 0$ for $k \geq N$.

Let P be the $n \times n$ matrix with the transition probabilities p_{ij}, $i, j = 1, \ldots, n$. Since the policy is proper, P^m converges geometrically to zero. Let $q_0(i) = \mathrm{P}(i_0 = i)$; these are the probabilities according to which the initial state i_0 is chosen. Similarly, let $q_m(i) = \mathrm{P}(i_m = i)$. For $m = 0, 1, \ldots$, let Q_m be a diagonal matrix whose diagonal entries are $q_m(i)$.

Using the same calculation as in the proof of Lemma 6.5, we obtain

$$E\big[\phi(i_m)\phi(i_t)'\big] = \Phi' Q_m P^{t-m}\Phi, \qquad \forall\, t \geq m.$$

Using this relation, we conclude that

$$A = \Phi' \sum_{t=0}^{\infty}\sum_{m=0}^{t} Q_m(\lambda P)^{t-m}(P - I)\Phi.$$

Let

$$B = \sum_{t=0}^{\infty}\sum_{m=0}^{t} Q_m(\lambda P)^{t-m}(P - I).$$

Lemma 6.10: The matrix B is negative definite.

Proof: Rearranging the order of the summation, and letting $k = t - m$, we obtain

$$\begin{aligned}
B &= \sum_{m=0}^{\infty} Q_m \sum_{k=0}^{\infty}(\lambda P)^k(P - I) \\
&= Q\sum_{k=0}^{\infty}(\lambda P)^k(P - I) \\
&= Q\left((1-\lambda)\sum_{k=0}^{\infty}\lambda^k P^{k+1} - I\right) \\
&= Q(M - I),
\end{aligned}$$

where

$$Q = \sum_{m=0}^{\infty} Q_m,$$

and

$$M = (1-\lambda) \sum_{k=0}^{\infty} \lambda^k P^{k+1}.$$

(Note that Q is finite because the diagonal entries of Q_m decay exponentially with m.) By our assumption that all states have positive probability of being visited, Q has positive diagonal entries and is positive definite. Note that if $\lambda = 1$, then $B = -Q$ and we are done. We therefore assume that $\lambda < 1$.

Let q (respectively, q_m) be a vector whose components are the diagonal entries of Q (respectively, Q_m). It is easily verified that $q_m' P = q_{m+1}'$ and therefore $q'P = q' - q_0' \leq q'$; that is,

$$\sum_{i=1}^{n} q(i) p_{ij} \leq q(j), \qquad \forall\ j.$$

We define a norm $\|\cdot\|_Q$ by letting $\|J\|_Q^2 = J'QJ$. Using an argument similar to the proof of Lemma 6.4, we obtain

$$\|PJ\|_Q \leq \|J\|_Q, \qquad \forall\ J,$$

which leads to

$$\|P^{k+1}J\|_Q \leq \|J\|_Q, \qquad \forall\ J,\ k \geq 0.$$

Let us focus on the case $0 < \lambda < 1$. Since M is a weighted average of the matrices P^{k+1}, we obtain $\|MJ\|_Q \leq \|J\|_Q$. For any $J \neq 0$, $P^k J$ converges to zero, which means that the inequality $\|P^{k+1}J\|_Q \leq \|J\|_Q$ is strict for some k. It then follows that $\|MJ\|_Q < \|J\|_Q$ for all $J \neq 0$, which implies that there exists some $\rho < 1$ such that $\|MJ\|_Q \leq \rho \|J\|_Q$ for all J. From here on, we follow the exact same steps as in the proof of Lemma 6.6(b), and conclude that B is negative definite.

We now consider the case $\lambda = 0$. (This case needs separate treatment because we have $M = P$ and the inequality $\|PJ\|_Q < \|J\|_Q$ is not necessarily true for all $J \neq 0$.) Using the Schwartz inequality and the fact $\|PJ\|_Q \leq \|J\|_Q$, we obtain

$$J'QPJ \leq \|J\|_Q \cdot \|PJ\|_Q \leq \|J\|_Q^2 = J'QJ. \tag{6.43}$$

The inequality (6.43) is strict unless J and PJ are colinear and $\|PJ\|_Q = \|J\|_Q$. This implies that the inequality is strict unless $PJ = J$ or $PJ = -J$. In either case, we obtain $P^{2m}J = J$ for all $m \geq 0$. Since we are dealing with a proper policy, P^{2m} converges to zero, and we must have $J = 0$.

We conclude that the inequality (6.43) is strict for all $J \neq 0$. Therefore, the matrix $Q(P-I)$ is negative definite, and the desired result has been established for the case $\lambda = 0$ as well. **Q.E.D.**

Due to the full column rank assumption for Φ, negative definiteness of $\Phi' B \Phi = A$ follows, and the proof of the proposition is complete. **Q.E.D.**

Example 6.7 can be reformulated as a stochastic shortest path problem, by interpreting $1-\alpha$ as a termination probability. Once more, we see that divergence is possible if the states are sampled according to a mechanism different than the one assumed in Prop. 6.6.

Error Bounds

We can define an operator $T^{(\lambda)}$ using the same formula as in the discounted case [cf. Eq. (6.40)], except that we set $\alpha = 1$. We can then follow the same line of argument and show that the algorithm converges to a vector r^* that satisfies $\Pi T^{(\lambda)}(\Phi r^*) = \Phi r^*$, where Π is now the projection with respect to the norm $\|\cdot\|_Q$. We can also duplicate the argument that led to Prop. 6.5 and arrive at the error bound

$$\|\Phi r^* - J^\mu\|_Q \leq \frac{\|\Pi J^\mu - J^\mu\|_Q}{1-\beta},$$

where β is the contraction factor of the operator $T^{(\lambda)}$. Similar to the discounted case, this is the same as the contraction factor of the matrix

$$M = (1-\lambda)\sum_{k=0}^{\infty} \lambda^k P^{k+1}.$$

It is possible to develop estimates for the contraction factor of M, in terms of the rate of convergence of P^k to zero, but we do not pursue this subject any further. We observe, however, that when $\lambda = 1$, we have $M = 0$ and $\beta = 0$, and we obtain the most favorable bound.

6.4 OPTIMISTIC POLICY ITERATION

In the approximate policy iteration approach discussed in Section 6.2, we fix a policy, and construct an approximation $\tilde{J}(\cdot, r)$ of J^μ, by solving a least squares problem or by running TD(λ). We then perform a policy update by switching to a new policy $\overline{\mu}$ which is greedy with respect to $\tilde{J}(\cdot, r)$, that is, a policy $\overline{\mu}$ that satisfies $T\tilde{J} = T_{\overline{\mu}}\tilde{J}$. An alternative is to perform a policy update earlier, before the approximate evaluation of J^μ converges. An extreme possibility, in the spirit of the optimistic policy iteration method of

Section 5.4, is to use a simulation-based method for approximating J^μ and to replace μ with $\overline{\mu}$ subsequent to the simulation of every state trajectory or even subsequent to every state transition. The resulting type of method, which we will again call optimistic policy iteration, is often used in practice, sometimes with success, even though there is little understanding of its convergence behavior.

There are many possible variations of optimistic policy iteration. We mention a few below.

Optimistic TD(λ) for Discounted Problems

Consider a discounted problem and let i_k stand for the simulated sequence of states. We will describe here the TD(λ)-based variant of optimistic policy iteration that updates the current policy after every state transition.

At a typical iteration, we are at some state i_k and we have a current parameter vector r_k. We select $u_k \in U(i_k)$ such that

$$u_k = \arg \min_{u \in U(i_k)} \sum_{j=1}^{n} p_{i_k j}(u)\big(g(i_k, u, j) + \alpha \tilde{J}(j, r_k)\big), \tag{6.44}$$

we generate the next state i_{k+1} by simulating a transition according to the transition probabilities $p_{i_k j}(u_k)$, and finally we perform the TD(λ) update

$$r_{k+1} = r_k + \gamma_k d_k \sum_{m=0}^{k} (\alpha\lambda)^{k-m} \nabla \tilde{J}(i_m, r_k),$$

where d_k is the temporal difference

$$d_k = g(i_k, u_k, i_{k+1}) + \alpha \tilde{J}(i_{k+1}, r_k) - \tilde{J}(i_k, r_k).$$

In fact, this method is not easily implemented because we need to evaluate a different gradient $\nabla \tilde{J}(i_m, r_k)$ for each k and m. For this reason, one uses instead the update rule

$$r_{k+1} = r_k + \gamma_k d_k \sum_{m=0}^{k} (\alpha\lambda)^{k-m} \nabla \tilde{J}(i_m, r_m). \tag{6.45}$$

The gradient $\nabla \tilde{J}(i_m, r_m)$ is evaluated just once, when state i_m is reached. If we define the eligibility vector

$$z_k = \sum_{m=0}^{k} (\alpha\lambda)^{k-m} \nabla \tilde{J}(i_m, r_m),$$

the algorithm takes the simpler form

$$r_{k+1} = r_k + \gamma_k d_k z_k,$$

where z_k is updated by

$$z_{k+1} = \alpha\lambda z_k + \nabla\tilde{J}(i_{k+1}, r_{k+1}).$$

Note that if we set $\lambda = 0$, we get an algorithm of the form

$$r_{k+1} = r_k + \gamma_k d_k \nabla\tilde{J}(i_k, r_k).$$

We observe that the expectation of d_k given the state of the algorithm up to time k, is given by

$$\begin{aligned} E[d_k \mid i_k, r_k] &= \sum_{j=1}^{n} p_{i_k j}(u_k)\big(g(i_k, u_k, j) + \alpha\tilde{J}(j, r_k)\big) - \tilde{J}(i_k, r_k) \\ &= \big(T\tilde{J}(\cdot, r_k)\big)(i_k) - \tilde{J}(i_k, r_k), \end{aligned}$$

where u_k is the action chosen by the greedy policy and T is the DP operator. Since $\big(T\tilde{J}(\cdot, r_k)\big)(i_k)$ has to be computed anyway in order to determine the greedy action u_k, it can be used to replace d_k by its expectation, thus removing some of the noise in the algorithm, and leading to the deterministic update rule

$$r_{k+1} = r_k + \gamma_k \nabla\tilde{J}(i_k, r_k)\Big(\big(T\tilde{J}(\cdot, r_k)\big)(i_k) - \tilde{J}(i_k, r_k)\Big).$$

In this case, the only role played by simulation is to generate the trajectory and select the states i_k, but otherwise the algorithm is deterministic, similar to the simulation-based value iteration algorithm discussed in Section 5.5.

Optimistic TD(λ) for Stochastic Shortest Path Problems

For a different variant, suppose that we are dealing with a stochastic shortest path problem, with all policies proper, that we wish to use TD(λ), and that we update the policy at the end of each trajectory. Let r_t be the value of the parameter vector at the beginning of the tth trajectory. We pick an initial state i_0 and simulate a trajectory $i_0, i_1, \ldots, i_N$, using the policy μ_t defined by

$$\mu_t(i) = \arg\min_{u \in U(i)} \sum_j p_{ij}(u)\big(g(i, u, j) + \tilde{J}(i, r_t)\big), \qquad \forall\ i.$$

Of course, we do not need to generate the policy for all states i; it is only when the trajectory reaches a state i that we compute $\mu_t(i)$ as above. Then, at the end of the trajectory, we update the parameter vector r by letting

$$\begin{aligned} r_{t+1} &= r_t + \gamma_t \sum_{k=0}^{N-1} d_k \sum_{m=0}^{k} \lambda^{k-m} \nabla \tilde{J}(i_m, r_t) \\ &= r_t + \gamma_t \sum_{k=0}^{N-1} \nabla \tilde{J}(i_k, r_t)(d_k + \lambda d_{k+1} + \cdots + \lambda^{N-1-k} d_{N-1}). \end{aligned} \tag{6.46}$$

As discussed earlier (cf. Sections 6.2.1 and 6.3.1), if we were dealing with a fixed policy and $\lambda = 1$, this would be an incremental gradient method for solving a least squares approximation problem.

Optimistic λ-Policy Iteration

Let us discuss another possibility for optimistic policy iteration, which is attractive primarily in the case of linear architectures. Consider the stochastic shortest path problem, let r_t be the parameter vector after t policy updates, and let μ_t be the greedy policy with respect to r_t. We then simulate a trajectory $i_0, i_1, \ldots, i_N$ using policy μ_t, and we form the temporal differences

$$d_k = g\big(i_k, \mu_t(i_k), i_{k+1}\big) + \tilde{J}(i_{k+1}, r_t) - \tilde{J}(i_k, r_t).$$

We then evaluate μ_t using a least squares minimization that approximates the policy evaluation step of the λ-policy iteration method of Section 2.3.1, where λ is a scalar satisfying $0 \leq \lambda < 1$. In particular, we calculate

$$\hat{r}_t = \arg\min_r \sum_{k=0}^{N-1} \left(\tilde{J}(i_k, r) - \sum_{m=k}^{N-1} \lambda^{m-k} d_m - \tilde{J}(i_k, r_t) \right)^2 . \tag{6.47}$$

The new parameter vector r_{t+1} is then obtained by interpolation between r_t and $\hat{r}_t$, i.e.,

$$r_{t+1} = r_t + \gamma_t(\hat{r}_t - r_t), \tag{6.48}$$

where γ_t is a stepsize that satisfies $0 < \gamma_t < 1$ and diminishes as t increases.

A gradient iteration for performing the least squares minimization (6.47) has the form

$$\tilde{r}_t = r_t + \gamma_t \sum_{k=0}^{N-1} \nabla \tilde{J}(i_k, r_t) \sum_{m=k}^{N-1} \lambda^{m-k} d_m, \tag{6.49}$$

which is identical to the optimistic TD(λ) iteration (6.46). Thus, we can view the optimistic λ-policy iteration (6.48) as a version of optimistic

TD(λ) where the gradient increment in Eq. (6.49) is replaced by the least squares minimization increment $\hat{r}_t - r_t$ in Eq. (6.48).

Note that there are variants of the optimistic λ-policy iteration method that use multiple simulated trajectories at each iteration, and there are also discounted cost variants. In the extreme case of a very large number of trajectories per iteration and a stepsize $\gamma_t = 1$, we obtain an approximate (nonoptimistic) version of the λ-policy iteration method of Section 2.3.1.

Optimistic Policy Iteration Using Kalman Filtering

Recall that the TD(1) update rule has been motivated as an incremental gradient method for solving the least squares problem associated with approximate policy evaluation. As discussed in Section 6.2.1, the extended Kalman filter can be used for the same purpose (or the ordinary Kalman filter, if the approximation architecture is linear). Kalman filters can also be used to perform parameter updates in optimistic policy iteration; the algorithm is essentially the same as the one used when simulated trajectory costs are generated under a fixed policy. Here, one may wish to employ a weighted least squares criterion (e.g., the exponentially weighted cost discussed in Section 3.2), that places more emphasis on recently simulated trajectory costs. As policies change in the course of optimistic policy iteration, the Kalman filtering algorithm will then attempt to provide a good fit to the more recent simulated trajectory costs, and will therefore reflect the costs-to-go of recently used policies. Still, if there is to be a hope of parameter convergence, weighting should be gradually eliminated as the algorithm progresses.

Optimistic Policy Iteration with an Action Network

As discussed in Section 6.2, approximate policy iteration may sometimes require the use of an action network to represent the current policy. We then noted that (nonoptimistic) approximate policy iteration consists of a succession of phases: the cost-to-go approximator is trained during one phase and the action network during the next. With optimistic policy iteration, however, the two approximation architectures can be trained concurrently, as we proceed to illustrate.

Suppose that u is allowed to take arbitrary scalar values and let us use an approximation architecture $\tilde{\mu}(i, v)$ to represent policies. One method, based on optimistic TD(λ) is as follows. At time k, we have a policy $\tilde{\mu}(\cdot, v_k)$, an approximate cost-to-go function $\tilde{J}(\cdot, r_k)$, and the current state is i_k. We apply the decision $u_k = \tilde{\mu}(i_k, v_k)$, simulate a transition, and update r_k on the basis of the observed temporal difference d_k, as in Eq. (6.45). In addition, we compute

$$\frac{\partial}{\partial u} \sum_j p_{i_k j}(u)\big(g(i_k, u, j) + \tilde{J}(j, r_k)\big), \tag{6.50}$$

evaluated at $\overline{u} = \tilde{\mu}(i_k, v_k)$, and update the action network according to

$$v_{k+1} = v_k - \gamma \nabla_v \tilde{\mu}(i_k, v_k) \frac{\partial}{\partial u} \sum_j p_{i_k j}(u)\big(g(i_k, u, j) + \tilde{J}(j, r_k)\big). \tag{6.51}$$

Similar to the update equation (6.10) in Section 6.2.1, the form of the update equation (6.51) can be motivated as a gradient descent step for minimizing

$$\sum_j p_{i_k j}\big(\tilde{\mu}(i_k, v)\big)\Big(g\big(i_k, \tilde{\mu}(i_k, v), j\big) + \tilde{J}(j, r_k)\Big),$$

with respect to v. This method is of interest if the expression (6.50) is easier to compute than the greedy decision given by Eq. (6.44).

Partially Optimistic Policy Iteration

Consistently with the discussion in Section 5.4, there is an intermediate class of algorithms whereby we update the policy frequently (before the approximate policy evaluation converges) but not as frequently as every transition or every trajectory. In particular, for the case of a stochastic shortest path problem, each iteration may consist of the generation of a fixed number of trajectories using the current policy, and a corresponding parameter vector update using one of the algorithms discussed so far.

Exploration

Optimistic policy iteration relies on a greedy policy for simulating the system, which raises the possibility of inadequate exploration of the state space, the issues being the same as those discussed in Section 5.5. In particular, it is possible that certain profitable alternatives are never explored and remain undiscovered. In Section 5.5, this difficulty was circumvented by starting with an underestimate of the optimal cost-to-go function J^*, and then ensuring that we have an underestimate throughout the duration of the algorithm. However, once function approximation is introduced, the initial condition $\tilde{J} \leq J^*$ cannot guarantee that this relation will remain true, and inadequate exploration is possible. One possibility for guaranteeing adequate exploration of the state space is to frequently restart the trajectory and to ensure that the initial states employed form a rich and representative subset of the state space, or to introduce some artificial noise in the decisions. Exploration is somewhat less of an issue in problems with a lot of randomness; it is a major concern, however, for deterministic or close-to-deterministic problems.

6.4.1 Analysis of Optimistic Policy Iteration

We have seen in Section 6.2 that (nonoptimistic) approximate policy iteration offers some performance guarantees but on the other hand, it has the potential for oscillation. For example, starting with a moderately good policy, and with $\tilde{J}$ fairly close to J^*, the next policy can be much worse, and the evaluation of the new policy can lead to a vector $\tilde{J}$ which is further away from J^*. In contrast, with optimistic policy iteration, once a bad policy is adopted, $\tilde{J}$ starts moving gradually away from J^* and there is the hope that while this happens, the policy will be updated again to avoid a move far away from J^*. It is unclear whether such a self-correcting mechanism will always be operative, but the example that follows gives some insight into such a phenomenon.

Example 6.8

Consider a discounted problem with the structure shown in Fig. 6.7(a). Suppose that we start with the (optimal) policy μ^*, which decides to go to state 2. With approximate policy iteration, we simulate this policy a few times and average the simulated costs to form an estimate of the cost-to-go $J^*(1)$. While the mean value of the estimate is 0, suppose that the randomness in the simulation leads us to $J(1) < -\epsilon/\alpha$. We then adopt a new policy μ that chooses to stay at state 1. The cost-to-go $J(1)$ under policy μ is $\epsilon/(1-\alpha)$ and for α close to 1, the error in the evaluation of the previous policy is greatly amplified [cf. Fig. 6.7(b)]. A subsequent policy update based on $J(1) = \epsilon/(1-\alpha)$, $J(2) = 0$, takes us back to the optimal policy μ^*.

Let us now consider optimistic policy iteration and suppose that at some point, $J(1)$ becomes less than $-\epsilon/\alpha$, so that the policy μ of staying at state 1 is preferred. As long as $J(1) < -\epsilon/\alpha$, this policy remains in effect and the TD(λ) updates gradually increase $J(1)$. As soon as $J(1)$ becomes larger than $-\epsilon/\alpha$, we switch back to the optimal policy μ^* [cf. Fig. 6.7(c)]. The key difference between the two methods is that under optimistic policy iteration, the selection of an erroneous policy was corrected much faster and the size of the oscillation of $J(1)$ became much smaller.

We have argued that the behavior of optimistic policy iteration may be less oscillatory relative to its nonoptimistic variant and we can ask whether this behavior can be mathematically substantiated. We thus consider the following two questions:

(a) Assuming that the generated policies converge, how close to optimal is the limit policy?

(b) How common is policy oscillation, and what are the properties of the policies involved in the oscillation when policy oscillation occurs?

For the lookup table case, we saw in Section 5.4 that if the policies generated by optimistic policy iteration converge, and if the policy evaluation method employed is sound, then the method converges to an optimal

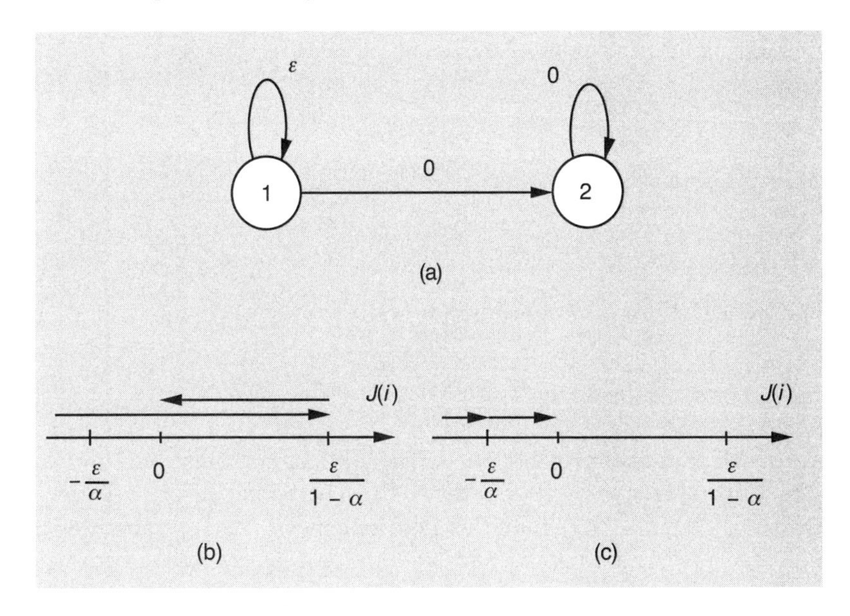

Figure 6.7: (a) There are two decisions at state 1: stay at a cost of $\epsilon > 0$, or move to state 2 and incur a random cost whose mean is equal to 0. Once at state 2, we must stay there, at zero cost. (b) The behavior of approximate policy iteration. (c) The behavior of optimistic policy iteration.

policy. We now develop an analogous result that takes function approximation into account. We only discuss the case of discounted problems.

Suppose that the approximation architecture is rich enough and that the policy evaluation algorithm is sound, in the following sense: if the policy μ is held fixed, the algorithm converges to some $\tilde{J}$ that satisfies

$$\max_i \left| \tilde{J}(i) - J^\mu(i) \right| \leq \epsilon, \tag{6.52}$$

where ϵ is some constant [compare with Eq. (6.12)]. Furthermore, suppose that the sequence of policies generated by optimistic policy iteration converges to some limit policy μ. Using our assumption above, this implies that the sequence of vectors $\tilde{J}$ converges. In addition, the limit policy μ is greedy with respect to the limit vector $\tilde{J}$, because otherwise it would have been revised; hence,

$$T\tilde{J} = T_\mu \tilde{J}. \tag{6.53}$$

We rewrite Eq. (6.52) in the form

$$J^\mu - \epsilon e \leq \tilde{J} \leq J^\mu + \epsilon e,$$

where e is the vector with all components equal to 1. Using also Eq. (6.53), we obtain

$$TJ^\mu \geq T(\tilde{J} - \epsilon e) = T\tilde{J} - \alpha\epsilon e = T_\mu \tilde{J} - \alpha\epsilon e \geq T_\mu J^\mu - 2\alpha\epsilon e = J^\mu - 2\alpha\epsilon e.$$

By operating on both sides of this inequality with T, we get

$$T^2 J^\mu \geq T J^\mu - 2\epsilon\alpha^2 e \geq J^\mu - 2\epsilon(\alpha + \alpha^2)e.$$

Continuing similarly, we obtain

$$T^k J^\mu \geq J^\mu - 2\epsilon(\alpha + \alpha^2 + \cdots + \alpha^k)e, \qquad \forall\, k.$$

Using the fact that $T^k J^\mu$ converges to J^*, we obtain

$$J^\mu \leq J^* + \frac{2\epsilon\alpha}{1-\alpha} e. \tag{6.54}$$

Recall that the error bounds for nonoptimistic policy iteration were of the form $O\big(1/(1-\alpha)^2\big)$ (cf. Prop. 6.2). The bound (6.54) is better by a factor of $O\big(1/(1-\alpha)\big)$, but this does not mean that optimistic policy iteration is preferable, because our assumption that we have a convergent sequence of policies typically fails to hold. In fact, under the comparable assumption that the policies generated by nonoptimistic approximate policy iteration converge, the same argument as above leads to the same bound (6.54). Thus, under the unrealistic assumption of policy convergence the bounds for the optimistic and nonoptimistic variants are the same.

On the other hand, we will explain in the next subsection that failure of policies to converge is prevalent. This leaves us with a major theoretical question. Is there some variant of optimistic policy iteration that is guaranteed to lead to policies whose performance is within $O\big(\epsilon/(1-\alpha)\big)$, or even $O\big(\epsilon/(1-\alpha)^2\big)$ of the optimal, where ϵ is some constant that characterizes the power of the approximation architecture? As of this writing, the answer is not known.

6.4.2 Oscillation of Policies in Optimistic Policy Iteration

We mentioned earlier that a basic characteristic of nonoptimistic policy iteration is that the parameter vector as well as the corresponding policy may change substantially from one iteration to the next, and that a sustained policy oscillation will often result in the limit. One may hope that in optimistic policy iteration, where parameter changes are more gradual and tend to zero because of a typically diminishing stepsize, some form of convergence of policies may be obtained.

Unfortunately, however, it turns out that there is a fundamental structure, common to all approximate policy iteration methods, that causes oscillations. In fact we will see that in optimistic policy iteration, oscillations are more complex and less well understood than in its nonoptimistic version. In particular, we will encounter a phenomenon, called *chattering*, whereby there is *simultaneously oscillation in policy space and convergence*

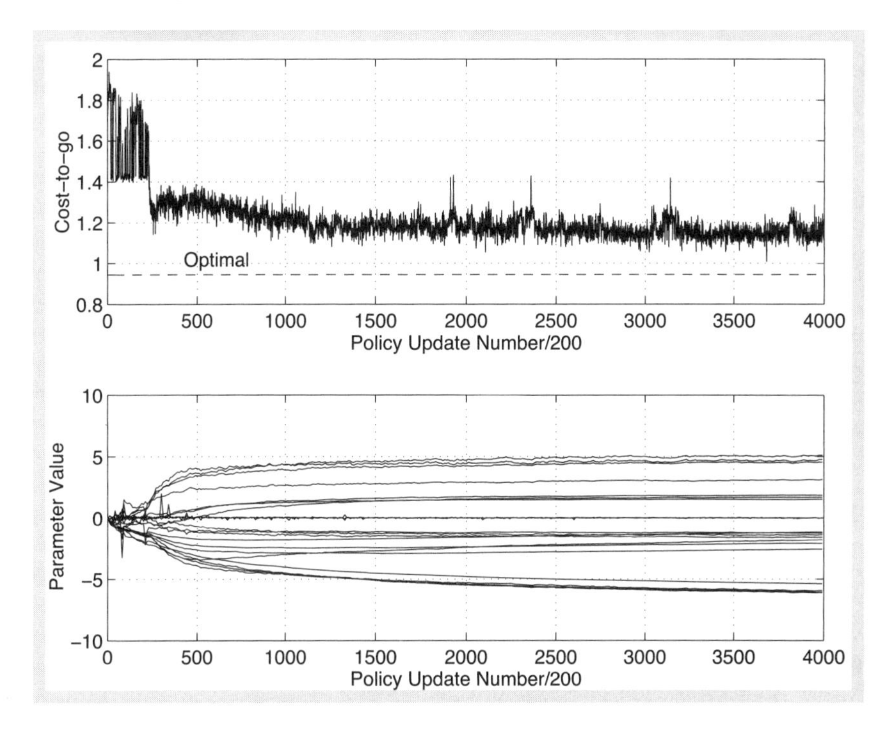

Figure 6.8: Chattering phenomenon in optimistic policy iteration. The plot at the top shows the oscillatory performance of the sequence of policies generated by optimistic policy iteration. The plot relates to the football problem discussed in detail in Section 8.2. Each data point gives the cost-to-go from a fixed initial state, calculated accurately by Monte Carlo simulation using many thousands of trajectories. The plot at the bottom shows the generated sequence of parameters of a linear architecture based on quadratic polynomials, which is used to approximate the cost-to-go of the different policies. While there is apparent parameter convergence, the limit parameters do not correspond to any of the generated policies, and cannot be used to approximate the optimal cost-to-go function.

in parameter space. Furthermore, the limit to which the parameter sequence converges need not correspond to any of the policies of the problem. It may instead be a boundary point on some surface that separates policies in parameter space. Figure 6.8 illustrates the chattering phenomenon with an example from the football case study (cf. Section 8.2).

To understand the chattering phenomenon, let us consider the stochastic shortest path problem and let us fix the approximation architecture $\tilde{J}(i,r)$. We note that $\tilde{J}$ defines for each stationary policy μ, a set R_μ of parameter vectors r for which μ is greedy with respect to $\tilde{J}(\cdot,r)$, that is,

$$R_\mu = \left\{ r \;\middle|\; \mu(i) = \arg\min_{u\in U(i)} \sum_{j=0}^{n} p_{ij}(u)\big(g(i,u,j) + \tilde{J}(j,r)\big),\ i = 1,\ldots,n \right\}.$$

These sets form a partition of the parameter space into the union $\cup_\mu R_\mu$, which we refer to as the *greedy partition*, and which is a generalization of the greedy partition introduced in Section 5.4.1.

Let us also consider an iterative policy evaluation method within an approximate policy iteration scheme, such as for example TD(λ) or Monte-Carlo simulation, and for simplicity, let us assume that if we fix μ, then the policy evaluation method converges to a unique parameter vector r_μ. Then the trajectory of nonoptimistic policy iteration is specified by the choice of the initial policy μ_0 [or alternatively the initial parameter vector r_0, which specifies μ_0 as a greedy policy with respect to $\tilde{J}(\cdot, r_0)$]. The method first generates r_{μ_0} by using the given iterative policy evaluation method, and then finds a policy μ_1 that is greedy with respect to $\tilde{J}(\cdot, r_{\mu_0})$, i.e., a μ_1 such that

$$r_{\mu_0} \in R_{\mu_1}.$$

The process is then repeated with μ_1 replacing μ_0. If some policy μ_k satisfying

$$r_{\mu_k} \in R_{\mu_k} \tag{6.55}$$

is encountered, the method keeps generating that policy. In fact the above condition is necessary and sufficient for policy convergence in the nonoptimistic policy iteration method. In the case of a lookup table representation where the parameter vectors r_μ are equal to the cost-to-go vector J^μ, the condition $r_{\mu_k} \in R_{\mu_k}$ is equivalent to $T_{\mu_k} J^{\mu_k} = T J^{\mu_k}$ and is satisfied if and only if μ_k is optimal. When there is function approximation, however, this condition need not be satisfied for any policy, and it can be seen that in general the algorithm ends up repeating some cycle of policies $\mu_k, \mu_{k+1}, \ldots, \mu_{k+m}$ with

$$r_{\mu_k} \in R_{\mu_{k+1}}, r_{\mu_{k+1}} \in R_{\mu_{k+2}}, \ldots, r_{\mu_{k+m-1}} \in R_{\mu_{k+m}}, r_{\mu_{k+m}} \in R_{\mu_k}; \tag{6.56}$$

(see Fig. 6.9). Furthermore, there may be several cycles of this type that the method may end up converging to. The actual cycle obtained may depend on the initial policy μ_0.

In the case of optimistic policy iteration, the trajectory of the method is less predictable and depends on the fine details of the iterative policy evaluation method, such as the stepsize used and the frequency of the policy updates. Generally, given the current policy μ, optimistic policy iteration will move towards the corresponding "target" parameter r_μ, for as long as μ continues to be greedy with respect to the current cost-to-go approximation $\tilde{J}(\cdot, r)$, that is, for as long as the current parameter vector r belongs to the set R_μ. Once, however, the parameter r crosses into another set, say $R_{\overline{\mu}}$, the policy $\overline{\mu}$ becomes greedy, and r changes course and starts moving towards the new "target" $r_{\overline{\mu}}$. Thus, the "targets" r_μ of the method, and the corresponding policies μ and sets R_μ may keep changing, similar to nonoptimistic policy iteration. Simultaneously, the parameter

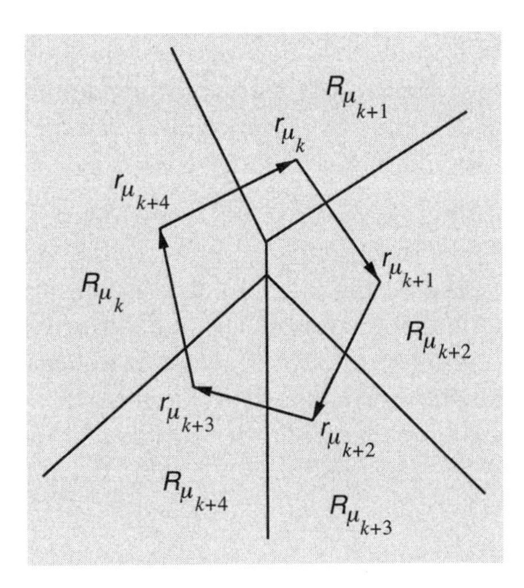

Figure 6.9: Greedy partition and cycle of policies generated by nonoptimistic policy iteration. The algorithm ends up repeating a cycle $\mu_k, \mu_{k+1}, \ldots, \mu_{k+m}$ with

$$r_{\mu_k} \in R_{\mu_{k+1}},\ r_{\mu_{k+1}} \in R_{\mu_{k+2}}, \ldots, r_{\mu_{k+m-1}} \in R_{\mu_{k+m}},\ r_{\mu_{k+m}} \in R_{\mu_k}.$$

vector r will "chatter" along the boundaries that separate the regions R_μ that the method visits (see Fig. 6.10). Furthermore, as Fig. 6.10 shows, if diminishing parameter changes are made between policy updates (such as for example when a diminishing stepsize is used by the policy evaluation method) and the method eventually cycles between several policies, the parameter vectors will tend to converge to the common boundary of the regions R_μ corresponding to these policies.

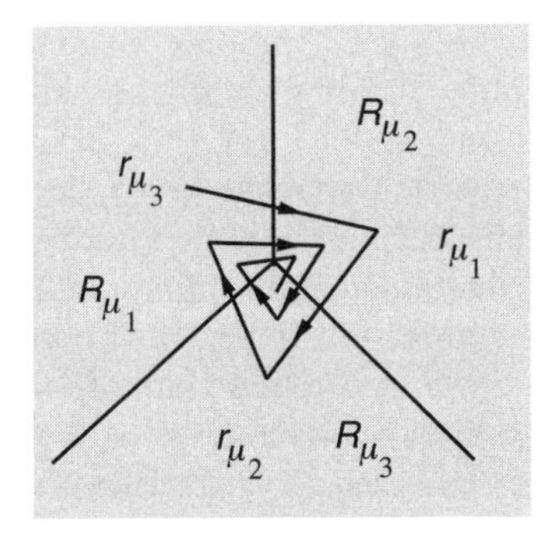

Figure 6.10: Illustration of the trajectory of optimistic policy iteration. The algorithm settles into an oscillation between policies μ_1, μ_2, μ_3 with $r_{\mu_1} \in R_{\mu_2}$, $r_{\mu_2} \in R_{\mu_3}$, $r_{\mu_3} \in R_{\mu_1}$. The parameter vectors converge to the common boundary of these policies.

The insight provided by the preceding interpretations leads to a number of conjectures and practical conclusions:

(a) Even though in optimistic policy iteration the parameter vectors tend to converge, their limit tends to be on the common boundary of several subsets of the greedy partition and cannot always be used to construct an approximation of the cost-to-go of any policy or the optimal cost-to-go.

(b) The error bounds on $\limsup_{k\to\infty} \|J^{\mu_k} - J^*\|$ are often comparable for optimistic, nonoptimistic, and partially optimistic methods, since all of these methods operate on the same greedy partition and often settle into an oscillation between cycles of policies with the property (6.56). Experimental results seem to confirm this conjecture, but for optimistic policy iteration, the derivation of error bounds that are comparable to the ones of Props. 6.2 and 6.3 for the nonoptimistic version is a major open research question.

(c) At the end of optimistic policy iteration and in contrast with the nonoptimistic version, one must go back and perform a screening process; that is, evaluate by simulation the many policies generated by the method starting from the initial conditions of interest and select the most promising one. This is a disadvantage of optimistic policy iteration that may nullify whatever practical rate of convergence advantages the method may have over its nonoptimistic counterpart.

(d) The choice of the iterative policy evaluation method [e.g., TD(0) versus TD(1) versus Monte-Carlo] does not seem crucial for the quality of the final policy obtained. Using a different policy evaluation method changes the targets r_μ somewhat, but leaves the greedy partition unchanged. As a result, different methods "fish in the same waters" and tend to yield similar ultimate cycles of policies. This is in contrast with the case where we want to evaluate a single policy, where different policy evaluation methods provide different error bound guarantees, as we saw in Section 6.3.

(e) Experiments with simple examples involving relatively few states show that optimistic policy iteration can exhibit very complex behavior. This is true for any type of plausible policy evaluation method such as TD(0), TD(1), Monte-Carlo simulation, λ-policy iteration, or gradient methods based on the Bellman equation error approach to be discussed in Section 6.10. In particular, optimistic policy iteration methods may have multiple stable and unstable equilibrium points in parameter space. Some of the stable equilibrium points may correspond to a limit cycle of policies (lie on the common boundary of some of the sets R_μ of the greedy partition), and others may correspond to single policies (lie in the interior of some set R_μ). Furthermore, the limit cycle of policies or the stable equilibrium point to which the method converges may depend on the initial choice of parameter vector. The following examples illustrate the various possibilities:

Example 6.9

Consider a stochastic shortest path problem, where in addition to the termination state 0, there are two states 1 and 2. At state 2 there is one control that moves us to state 1 with cost -1. At state 1 there are two controls: the first moves us to 0 with cost 1, and the second keeps us at 1 with probability p and cost c, and moves us to 0 with probability $1-p$ and cost 1. There are two policies: μ that tries to keep us in state 1, and μ^* that moves us from 1 to 0 (policy μ^* is optimal if and only if $c \geq 0$). We consider the linear approximation architecture

$$\tilde{J}(i,r) = ir,$$

involving the scalar parameter r. The policy μ is greedy if and only if the parameter r is such that at state 1, the expected immediate cost plus the estimated cost-to-go of μ, which is $pc + 1 - p + pr$, is no more than the expected immediate cost plus estimated cost-to-go of μ^*, which is 1. Thus, μ is greedy if and only if $r \leq 1 - c$, and the greedy partition consists of the sets

$$R = (-\infty, 1-c], \qquad R^* = [1-c, \infty);$$

μ is greedy in R while μ^* is greedy in R^*. The "targets" r_μ and r_{μ^*}, to be denoted by r and r^*, respectively, depend on the method used for policy evaluation. Let us consider TD(λ) with states generated by simulating full trajectories starting from state 2. It can be verified that for the case of the first visit TD(1) method, we have

$$r = \frac{1}{5} + \frac{3pc}{5(1-p)}, \qquad r^* = \frac{1}{5}.$$

[To verify this, note that under μ^* the sample costs from states 1 and 2 are 1 and 0, respectively, and in first visit TD(1), r^* minimizes the least squares cost $(r-1)^2 + (2r)^2$, while under μ the expected sample costs from states 1 and 2 are $1 + pc/(1-p)$ and $pc/(1-p)$, respectively, and r minimizes the expected least squares cost $E[(r - \text{cost}(1))^2 + (2r - \text{cost}(2))^2]$.] For the case of every visit TD(1), it can be verified that

$$r = \frac{1 - p + pc(3-2p)}{(1-p)(5-4p)}, \qquad r^* = \frac{1}{5}.$$

For the case of every visit TD(0), it can be verified that

$$r = -\frac{1}{3} + \frac{pc}{3(1-p)}, \qquad r^* = -\frac{1}{3}.$$

There are four cases, only the first three of which are possible for first visit TD(1), every visit TD(1), and every visit TD(0), with appropriate choice of the parameters p and $c > 0$ (see Fig. 6.11):

(1) r and r^* belong to R^*. Here nonoptimistic policy iteration converges to r^* and generates μ^* in at most one iteration. Optimistic policy

iteration generates a sequence r_k that converges to r^* and generates μ^* at all except finitely many iterations.

(2) r and r^* belong to R. This is case (1) above with the roles of r and r^* reversed.

(3) r belongs to R^* and r^* belongs to R. Here nonoptimistic policy iteration oscillates between μ and μ^*. Optimistic policy iteration also oscillates between μ and μ^*, and generates a sequence r_k that converges to the boundary point $1 - c$ between R and R^*. What happens here is that when r_k gets into R it is pulled back towards r and into R^*, and reversely, when r_k gets into R^* it is pulled back towards r^*.

(4) r belongs to R and r^* belongs to R^*. Here nonoptimistic policy iteration converges to r and μ if it is started in R, and converges to r^* and μ^* if it is started in R^*. Optimistic policy iteration generates a sequence r_k that may converge to either r or r^*.

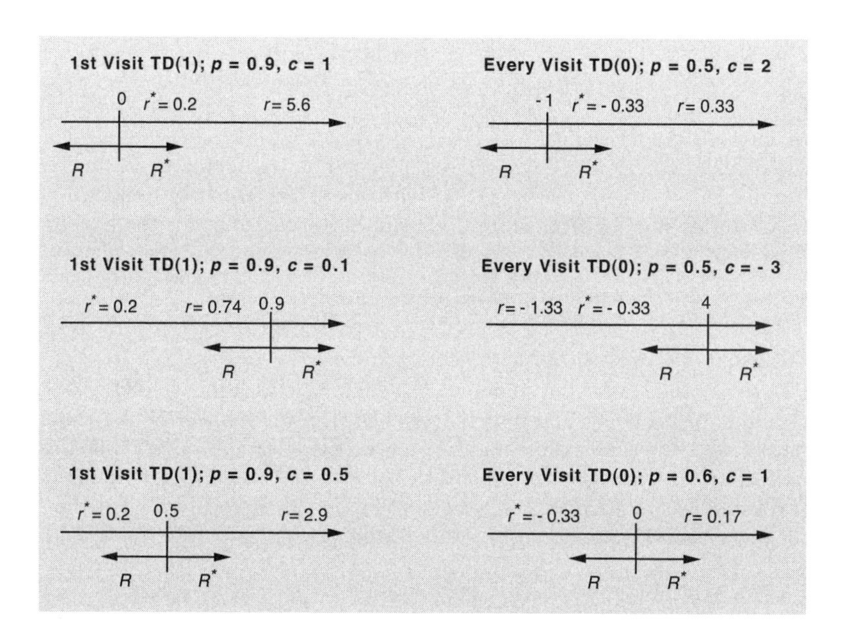

Figure 6.11: Illustration of the greedy partition and the "targets" of first visit TD(1) and every visit TD(0) in the first three cases in Example 6.9. The fourth case ($r \in R$, $r^* \in R^*$) is not possible for first visit and every visit TD(1), or for every visit TD(0), in this example. Oscillation of policies occurs in the two bottom cases where $r \in R^*$ and $r^* \in R$.

Example 6.10

Consider a stochastic shortest path problem, where in addition to the termi-

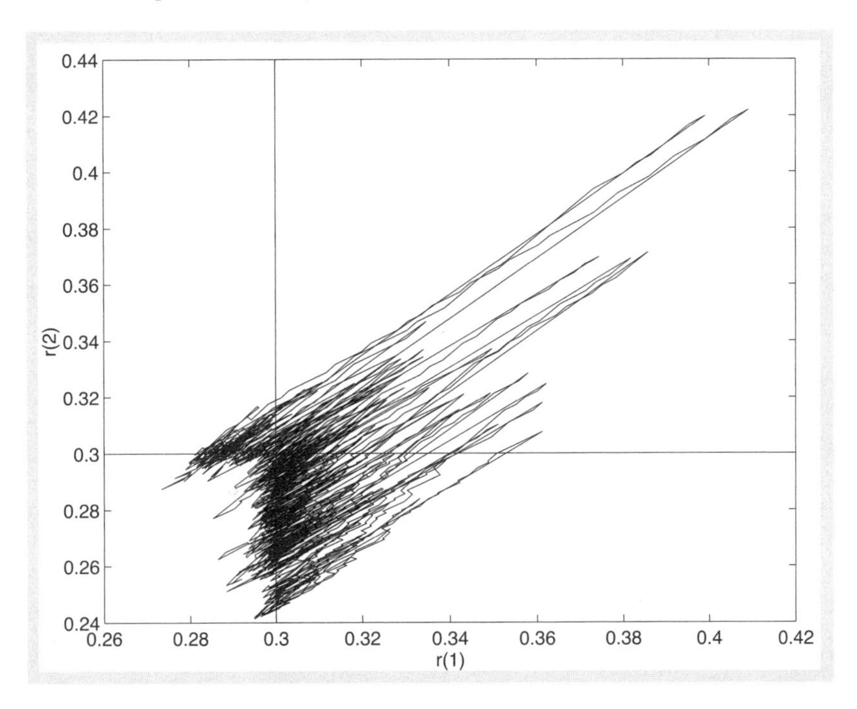

Figure 6.12: Illustration of the greedy partition and optimistic policy iteration for Example 6.10. There are four policies: (1) μ^* which moves to 0 at both states 1 and 2, (2) μ which at state 1 tries to move to state 2, and at state 2 tries to move to state 1, (3) μ_1 which at state 1 tries to move to state 2, and at state 2 moves to state 0, (4) μ_2 which at state 2 tries to move to state 1, and at state 1 moves to state 0. The problem data are $p = 0.7$ and $c = 0.7$. The partition is given by

$$R_{\mu^*} = \{r \mid r(1) \leq 0.3,\ r(2) \leq 0.3\}, \qquad R_{\mu} = \{r \mid r(1) \geq 0.3,\ r(2) \geq 0.3\},$$

$$R_{\mu_1} = \{r \mid r(1) \geq 0.3,\ r(2) \leq 0.3\}, \qquad R_{\mu_2} = \{r \mid r(1) \leq 0.3,\ r(2) \geq 0.3\}.$$

We show the iterates of iterations 17,001 to 20,000 generated by optimistic policy iteration using every visit TD(1) with states generated by simulating full trajectories starting from state 3. Each iteration involves a single trajectory; that is, there is both a parameter and a policy update at the end of each trajectory. The chattering at the parameter values $\overline{r}(1) = \overline{r}(2) = 0.3$ involves all four policies.

nation state 0, there are three states 1, 2, and 3. This example is obtained by adaptation of the preceding example (state 1 is essentially split in two states). At state 3 there is one control that moves us to states 1 and 2 with equal probability 1/2 and with cost -1. At state 2 there are two controls: the first moves us to 0 with cost 1, and the second moves us to 1 with probability p and cost $c > 0$, and moves us to 0 with probability $1-p$ and cost 1. Similarly, at state 1 there are two controls: the first moves us to 0 with cost 1, and the second moves us to 2 with probability p and cost $c > 0$, and moves us to

0 with probability $1-p$ and cost 1. We consider the linear approximation architecture

$$\tilde{J}\big(i, r(1), r(2)\big) = \begin{cases} r(1) & \text{if } i = 1, \\ r(2) & \text{if } i = 2, \\ 2\big(r(1) + r(2)\big) & \text{if } i = 3, \end{cases}$$

involving the parameter vector $r = (r(1), r(2))$. If the problem data are $p = 0.7$ and $c = 0.7$, it can be verified that nonoptimistic policy iteration oscillates. Figure 6.12 shows the trajectory of optimistic policy iteration for this choice of p and c.

We end this section with some speculation on the nature of the oscillation when chattering occurs. Suppose that we have convergence to a parameter vector $\overline{r}$ and that there is a steady-state policy oscillation involving a collection of policies $\mathcal{M}$. Then, all the policies in $\mathcal{M}$ are greedy with respect to $\tilde{J}(i, \overline{r})$, which implies that there is a subset S of states such that for each $i \in S$, there are at least two different controls $\mu_1(i)$ and $\mu_2(i)$ satisfying

$$\begin{aligned} \min_{u \in U(i)} \sum_j p_{ij}(u)\big(g(i,u,j) + \tilde{J}(j,\overline{r})\big) \\ &= \sum_j p_{ij}\big(\mu_1(i)\big)\Big(g\big(i, \mu_1(i), j\big) + \tilde{J}(j, \overline{r})\Big) \qquad (6.57) \\ &= \sum_j p_{ij}\big(\mu_2(i)\big)\Big(g\big(i, \mu_2(i), j\big) + \tilde{J}(j, \overline{r})\Big). \end{aligned}$$

Each equation of the type above can be viewed as a constraining relation on the parameter vector $\overline{r}$. Thus, if m is the dimension of $\overline{r}$, then excluding singular situations, there will be at most m relations of the form (6.57) holding. This implies that there will be at most m "ambiguous" states where more than one control is greedy with respect to $\tilde{J}(\cdot, \overline{r})$.

Now assume that we have a problem where the total number of states is much larger than m and, furthermore, there are no "critical" states; that is, the cost consequences of changing a policy in just a small number of states (say, of the order of m) is relatively small. It then follows that all policies in the set $\mathcal{M}$ involved in chattering have roughly the same cost. Furthermore, if TD(1) is used, the cost approximation $\tilde{J}(\cdot, \overline{r})$ is close to the cost approximation $\tilde{J}(\cdot, r_\mu)$ that would be generated by TD(1) for any of the policies $\mu \in \mathcal{M}$. From this, we can construct an argument that, for most states i, and all policies $\mu \in \mathcal{M}$,

$$J^\mu(i) - J^*(i) = O\left(\frac{\epsilon}{1-\alpha}\right),$$

where ϵ is the power of the architecture [compare with the discussion preceding Eq. (6.54)]. Note, however, that the assumption of "no critical states," aside from the fact that it may not be easily quantifiable, will not be true for many problems.

In a practical implementation where the stepsize does not quite approach zero, the parameter vector may end up oscillating within a small neighborhood instead of converging to a limit $\bar{r}$. In that case, the number of policies involved in chattering may be much larger than the number m assumed above. This raises the question whether all of these policies have roughly similar costs-to-go; if yes, the argument given in the preceding paragraph may still have some validity. An interesting case seems to arise when the control variable u is continuous rather than discrete. We can expect that for many problems, small changes in r can only cause small changes in the corresponding greedy policy, which would then imply that the costs-to-go of the policies involved in any chattering phenomenon are close to each other.

6.5 APPROXIMATE VALUE ITERATION

The value iteration algorithm is of the form

$$J(i) := \min_{u \in U(i)} \sum_j p_{ij}(u)\big(g(i,u,j) + J(j)\big),$$

or, in more abstract notation, $J(i) := (TJ)(i)$, where T is the DP operator. In this section, we discuss a number of different ways that the value iteration algorithm can be adapted to a setting involving parametric representations of the optimal cost function. Our discussion revolves primarily around the case where a model of the system is available, so that the transition probabilities $p_{ij}(u)$ and the one-step costs $g(i,u,j)$ are known. We will discuss first the finite horizon case and then move on to the case of infinite horizon problems.

6.5.1 Sequential Backward Approximation for Finite Horizon Problems

Finite horizon problems, where there is a fixed number N of stages, can be viewed as a special case of stochastic shortest path problems, by introducing a cost-free termination state to which the system moves immediately after the Nth state transition. For such problems the finite horizon DP algorithm can be used. It takes the form

$$J_1^*(i) = \min_{u \in U(i)} \sum_j p_{ij}(u)\big(g(i,u,j) + G(j)\big),$$

where $G(j)$ is a given terminal cost for arriving at state j following the last stage, and

$$J_k^*(i) = \min_{u \in U(i)} \sum_j p_{ij}(u)\big(g(i,u,j) + J_{k-1}^*(j)\big), \qquad k = 2, \ldots, N.$$

Here $J_k^*(i)$ is the optimal cost-to-go associated with a k-stage problem, when the starting state is i.

We can introduce approximations to the cost-to-go functions J_k^*, by proceeding recursively. In particular, we approximate the function $J_1^*(i)$ by a function

$$\tilde{J}_1(i, r_1),$$

where r_1 is a parameter vector, which can be obtained by solving the problem

$$\min_r \sum_{i \in S_1} v(i)\big(J_1^*(i) - \tilde{J}_1(i, r)\big)^2, \tag{6.58}$$

where S_1 is a representative set of states and $v(i)$ are some predefined positive weights indicating the importance of each state i. The least-squares problem (6.58) can be solved using any of the methods discussed in Ch. 3. Furthermore, if a linear architecture is used, this is simply a linear least squares problem.

Once an approximating function $\tilde{J}_1(i, r_1)$ has been obtained, it can be used to similarly obtain an approximating function $\tilde{J}_2(i, r_2)$. In particular, (approximate) cost-to-go function values $\hat{J}_2(i)$ are obtained for a representative subset of states S_2 through the (approximate) DP formula

$$\hat{J}_2(i) = \min_{u \in U(i)} \sum_j p_{ij}(u)\big(g(i, u, j) + \tilde{J}_1(j, r_1)\big), \qquad i \in S_2.$$

These values are then used to compute an approximation of the cost-to-go function $J_2^*(i)$ by a function

$$\tilde{J}_2(i, r_2),$$

where r_2 is a parameter vector, which is obtained by solving the problem

$$\min_r \sum_{i \in S_2} v(i)\big(\hat{J}_2(i) - \tilde{J}_2(i, r)\big)^2.$$

The process can be similarly continued to obtain $\tilde{J}_k(i, r_k)$ up to $k = N$, by solving for each k the problem

$$\min_r \sum_{i \in S_k} v(i)\big(\hat{J}_k(i) - \tilde{J}_k(i, r)\big)^2.$$

Given approximate cost-to-go functions $\tilde{J}_1(i, r_1), \ldots, \tilde{J}_N(i, r_N)$, we obtain a suboptimal policy by making the decision

$$\tilde{\mu}_k(i) = \arg \min_{u \in U(i)} \sum_j p_{ij}(u)\big(g(i, u, j) + \tilde{J}_{k-1}(j, r_{k-1})\big), \tag{6.59}$$

if we are at state i and there are k stages remaining. [For $k = 1$, $\tilde{J}_{k-1}(j, r_{k-1})$ is replaced by the terminal cost $G(j)$.] This control can be calculated on-line, once the state becomes known, as discussed in Section 6.1.

For the method that we have described, one can prove formally that if the approximation architecture is rich enough, and if the training method is successful in accurately solving the least squares problem, then the function $\tilde{J}_k$ generated at each stage is a close approximation of the optimal cost-to-go function J_k^*. Based on that, one can also show that the performance of the policy defined by Eq. (6.59) is close to optimal. We do not provide any details because such results can be obtained as special cases of the results derived in Section 6.5.3 for infinite horizon stochastic shortest path problems.

6.5.2 Sequential Approximation in State Space

We present another approximation method that is similar to (and in fact generalizes) the one just presented for finite horizon problems. In this method, we gradually approximate the cost of a policy or the optimal cost over the state space rather than over time. For this, we need a somewhat restrictive assumption on the structure of the problem. In particular, as in Section 2.2.2, we assume that there is a finite sequence $S_1, S_2, \ldots, S_M$ of subsets of the state space such that each state $1, \ldots, n$ belongs to one and only one of these subsets, and the following property holds:

> For all $m = 1, \ldots, M$, all states $i \in S_m$, and all choices of the control $u \in U(i)$, the successor state j is either the termination state 0 or else belongs to one of the subsets $S_{m-1}, \ldots, S_1$.

Under the preceding assumption, the problem decomposes over the state space. In particular, for $i \in S_1$, we have $J^*(i) = \min_u g(i, u, 0)$. Then given the optimal cost-to-go $J^*(i)$ for $i \in S_1$, we can determine the optimal cost-to-go $J^*(i)$ for states i in S_2 by stipulating that following a transition to a state j in S_1, there immediately follows a transition to the terminal state with the (already known) cost-to-go $J^*(j)$. Similarly, the optimal cost-to-go can be determined for all states in the subsets $S_3, \ldots, S_M$.

The preceding procedure can be approximated similar to the finite horizon DP algorithm. First an approximation of the optimal cost-to-go function over the states in S_1 will be computed, then an approximation over S_2, etc.

6.5.3 Sequential Backward Approximation for Infinite Horizon Problems

We now turn to infinite horizon problems. The algorithms that follow incorporate a discount factor α, but we can always let $\alpha = 1$ and apply them to stochastic shortest path problems. The main algorithm is essentially the same as the finite horizon algorithm of Section 6.5.1, except that we keep iterating until the algorithm converges. A formal description is as follows. The algorithm is initialized with a parameter vector r_0 and a corresponding cost-to-go function $\tilde{J}(i, r_0)$. At a typical iteration, we have a parameter vector r_k, we select a set S_k of representative states, and we compute estimates of the cost-to-go from the states in S_k by letting

$$\hat{J}_{k+1}(i) = \min_{u \in U(i)} \sum_j p_{ij}(u)\big(g(i,u,j) + \alpha \tilde{J}(j, r_k)\big), \qquad i \in S_k.$$

We then determine a new set of parameters r_{k+1} by minimizing with respect to r the quadratic cost criterion

$$\sum_{i \in S_k} v(i)\big(\hat{J}_{k+1}(i) - \tilde{J}(i,r)\big)^2, \tag{6.60}$$

where $v(i)$ are some predefined positive weights. In the special case where the approximation architecture $\tilde{J}$ is linear, we are dealing with a linear least squares problem that can be solved efficiently, using the algorithms of Ch. 3.

We now discuss the soundness of this method. We first argue that under some circumstances the algorithm performs well. We then continue with a counterexample demonstrating that the algorithm can diverge.

Performance Guarantees

We provide here some error bounds on the cost-to-go functions generated by approximate value iteration. We first discuss the case of stochastic shortest path problems, and then consider discounted problems. In order to simplify the argument, and for the purposes of this subsection only, we use the notation J_k to indicate the cost-to-go function $\tilde{J}(\cdot, r_k)$ obtained after k iterations of the algorithm.

Let N be such that $\|T^N J_0 - J^*\|_\infty \leq \delta$, where $\|\cdot\|_\infty$ is the maximum norm; such an N exists since $T^k J$ converges to J^* for every function J. Suppose in addition, that the approximation architecture is flexible enough and that the states sampled are rich enough so that at each stage of the algorithm we have

$$\|J_{k+1} - TJ_k\|_\infty \leq \epsilon.$$

Intuitively, this requires that carrying out a value iteration at the sample states and then fitting the results to obtain J_{k+1} is not very different from the result TJ_k of exact value iteration.

We first prove, by induction on k, that $\|T^k J_0 - J_k\|_\infty \leq k\epsilon$ for all k. This is certainly true for $k = 0$. Assuming it is true for some k, we obtain

$$\begin{aligned}
\|T^{k+1} J_0 - J_{k+1}\|_\infty &\leq \|T^{k+1} J_0 - TJ_k\|_\infty + \|TJ_k - J_{k+1}\|_\infty \\
&\leq \|T^k J_0 - J_k\|_\infty + \epsilon \\
&\leq k\epsilon + \epsilon,
\end{aligned}$$

and the induction argument is complete. We conclude that

$$\|J_N - J^*\|_\infty \le \|J_N - T^N J_0\|_\infty + \|T^N J_0 - J^*\|_\infty \le N\epsilon + \delta.$$

This shows that when δ and ϵ are small, and N is not very large, a close approximation of J^* will be obtained.

For the case of discounted problems, more refined error bounds are possible, as we now proceed to show. Let $\alpha \in (0,1)$ be the discount factor and let us again assume that

$$\|J_{k+1} - TJ_k\|_\infty \le \epsilon, \tag{6.61}$$

for all k. We will show that the sequence of cost-to-go functions J_k, approaches J^*, within an error of $\epsilon/(1-\alpha)$. To see this, note that Eq. (6.61) yields

$$TJ_0 - \epsilon e \le J_1 \le TJ_0 + \epsilon e,$$

where $e = (1,1,\ldots,1)$. By applying T to this relation, we obtain

$$T^2 J_0 - \alpha\epsilon e \le TJ_1 \le T^2 J_0 + \alpha\epsilon e.$$

Using Eq. (6.61), we have

$$TJ_1 - \epsilon e \le J_2 \le TJ_1 + \epsilon e,$$

and we obtain

$$T^2 J_0 - \epsilon(1+\alpha)e \le J_2 \le T^2 J_0 + \epsilon(1+\alpha)e.$$

Proceeding similarly, we obtain for all $k \ge 1$,

$$T^k J_0 - \epsilon(1+\alpha+\cdots+\alpha^{k-1})e \le J_k \le T^k J_0 + \epsilon(1+\alpha+\cdots+\alpha^{k-1})e.$$

By taking the limit superior and the limit inferior as $k \to \infty$, and by using the fact $\lim_{k\to\infty} T^k J_0 = J^*$, we see that

$$J^* - \frac{\epsilon}{1-\alpha}e \le \liminf_{k\to\infty} J_k \le \limsup_{k\to\infty} J_k \le J^* + \frac{\epsilon}{1-\alpha}e.$$

It is worth noting that the assumption $\|J_{k+1} - TJ_k\|_\infty \le \epsilon$ [cf. Eq. (6.61)] that we have used here is qualitatively different from the one used in our analysis of approximate policy iteration [Eqs. (6.12) and (6.13)]. In Section 6.2, we had assumed that the cost-to-go J^μ of any generated policy can be approximated within ϵ. In contrast, here we assume that we can closely approximate functions of the form TJ_k, for any J_k generated by the method, which could be a larger class.

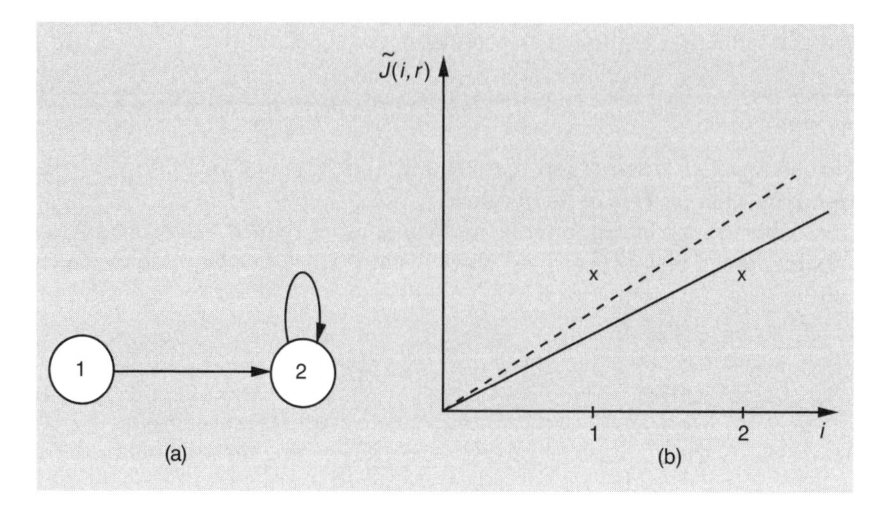

Figure 6.13: (a) A deterministic system in which there is a single policy. The cost per stage is zero. (b) Illustration of divergence.

The Possibility of Divergence

Unfortunately, the example that follows shows that approximate value iteration suffers from potential divergence.

Example 6.11

Consider the deterministic two-state system shown in Fig. 6.13(a). State 2 is absorbing and all transitions are cost-free. Let $\alpha < 1$ be the discount factor. Obviously, we have $J^*(1) = J^*(2) = 0$. Consider a linear function approximator, with a single scalar parameter r, of the form $\tilde{J}(1,r) = r$ and $\tilde{J}(2,r) = 2r$. Given a value r_k of r, we obtain $\hat{J}(1) = \hat{J}(2) = 2\alpha r_k$. We form the least squares problem

$$\min_r \left((r - 2\alpha r_k)^2 + (2r - 2\alpha r_k)^2 \right),$$

and by setting the derivative to zero, we obtain $r_{k+1} = 6\alpha r_k/5$. Hence, if $\alpha > 5/6$, the algorithm diverges.

A graphical illustration of divergence is provided in Fig. 6.13(b). The solid line assumes a given value of r and is a plot of $\tilde{J}(i,r)$ as a function of i. The values of $\hat{J}(1)$ and $\hat{J}(2)$, indicated by crosses in the figure, "pull" r in opposite directions. For $\alpha > 5/6$, the upward pull by $\hat{J}(1)$ is "stronger" and r moves further away from zero, as indicated by the dashed line. Note that in this example there is no $\epsilon > 0$ such that the assumption $\|J_{k+1} - TJ_k\|_\infty \leq \epsilon$ [cf. Eq. (6.61)] holds, and so the performance guarantees that were derived earlier do not apply.

The same example can be used to show divergence in stochastic shortest path problems, by letting the discount factor be unity and introducing a termination probability of $1 - \alpha$ at each state.

We note that the function approximation architecture used in Example 6.11 was rich enough to allow an exact representation of the optimal cost-to-go function J^*. As the example demonstrates, this property is not enough to guarantee the soundness of the algorithm. Apparently, a stronger condition is needed: the function approximator must be able to closely represent all of the intermediate cost-to-go functions obtained in the course of the value iteration algorithm. Given that such a condition is in general very difficult to verify, one must either accept the risk of a divergent algorithm or else restrict to particular types of function approximators under which divergent behavior becomes impossible. The latter possibility will be pursued in Sections 6.7 and 6.9.

6.5.4 Incremental Value Iteration

The approximate value iteration algorithm that we have described above proceeds in phases: we obtain estimates $\hat{J}(i)$ at a number of states i and then compute a new parameter vector r by solving the least squares problem (6.60). This could be done, for example, by means of a gradient algorithm like

$$\begin{aligned} r &:= r + \gamma \sum_{i \in S_k} \nabla \tilde{J}(i,r)\big(\hat{J}(i) - \tilde{J}(i,r)\big) \\ &= r + \gamma \sum_{i \in S_k} \nabla \tilde{J}(i,r) \left(\min_{u \in U(i)} \sum_j p_{ij}(u)\big(g(i,u,j) + \alpha \tilde{J}(j,r_k)\big) - \tilde{J}(i,r) \right). \end{aligned} \tag{6.62}$$

If the incremental gradient algorithm is used, this update equation is replaced by updates of the form

$$r := r + \gamma \nabla \tilde{J}(i,r) \left(\min_{u \in U(i)} \sum_j p_{ij}(u)\big(g(i,u,j) + \alpha \tilde{J}(j,r_k)\big) - \tilde{J}(i,r) \right), \tag{6.63}$$

where we are to cycle over the states i in the set S_k. Note that throughout this incremental gradient method, the vector r_k used to evaluate $\tilde{J}(j, r_k)$ remains constant. Once the algorithm converges, the limiting value of r is denoted by r_{k+1}, and we are ready to start the next phase of the algorithm.

In an alternative, fully incremental, version of the algorithm, we stop distinguishing between different phases and instead of freezing r_k throughout an entire phase, we always use the most recently computed value of r. In particular the iteration (6.63) is replaced by

$$r := r + \gamma \nabla \tilde{J}(i,r) \left(\min_{u \in U(i)} \sum_j p_{ij}(u)\big(g(i,u,j) + \alpha \tilde{J}(j,r)\big) - \tilde{J}(i,r) \right). \tag{6.64}$$

The iteration (6.64) is to be carried out at a sequence of states i that can be generated in a number of different ways. For example, we may choose states randomly according to some distribution, or we may simulate the system under some policy and perform updates at the states visited by the simulation.

In the special case where there is a single policy, the sum in Eq. (6.64) can be replaced by a single sample estimate, leading to the update equation

$$r := r + \gamma \nabla \tilde{J}(i,r)\big(g(i,u,j) + \alpha \tilde{J}(j,r) - \tilde{J}(i,r)\big),$$

where j is sampled according to the probabilities $p_{ij}(u)$. We note that this has the same form as the update equation used in the TD(0) algorithm. The only difference is that in TD(0), it is assumed that one generates an entire state trajectory and performs updates along all states on the trajectory, whereas the sequence of states used for the updates in incremental value iteration are arbitrary.

In the special case where the dimension of r is the same as the dimension of the state space and $\tilde{J}(i,r) = r(i)$ for each i, the algorithm (6.64) is a small stepsize version of the asynchronous value iteration algorithm discussed in Ch. 2, and again in Section 5.5. It is known to converge as long as every state is visited an infinite number of times and the stepsize satisfies $0 < \gamma \leq 1$. In the presence of function approximation, a diminishing stepsize is necessary in order to have any hope of convergence. Unfortunately, this incremental approximate value iteration algorithm suffers from the same drawbacks as the original approximate value iteration algorithm, as shown next.

Example 6.12

Consider the incremental approximate value iteration algorithm applied to the same discounted problem as in Example 6.11. Note that $\nabla \tilde{J}(1,r) = 1$ and $\nabla \tilde{J}(2,r) = 2$. Suppose that we alternate between updates at states 1 and 2. Then, the evolution of r can be described by

$$\begin{aligned} r_{k+1} &= r_k + \gamma_k(2\alpha r_k - r_k), \\ r_{k+2} &= r_{k+1} + \gamma_{k+1} 2(2\alpha r_{k+1} - 2r_{k+1}), \end{aligned}$$

for every odd k. Assuming that $\gamma_k = \gamma_{k+1}$ for every odd k, we have

$$r_{k+2} = r_k\big(1 + \gamma_k(2\alpha - 1)\big)\big(1 + 4\gamma_k(\alpha - 1)\big) = r_k\big(1 + \gamma_k(6\alpha - 5)\big) + O(\gamma_k^2).$$

It is seen that if γ_k is small enough so as to neglect the $O(\gamma_k^2)$ term, and if $\alpha > 5/6$, we obtain

$$r_{k+2} \geq r_k(1 + c\gamma_k),$$

where c is some positive constant. Under the usual assumption that $\sum_k \gamma_k = \infty$, the algorithm diverges.

6.6 Q-LEARNING AND ADVANTAGE UPDATING

Recall the form of the Q-learning algorithm under a lookup table representation. We choose a state-action pair (i, u) and simulate a transition to a next state j according to the transition probabilities $p_{ij}(u)$. We then update $Q(i, u)$ according to

$$Q(i, u) := Q(i, u) + \gamma\Big(g(i, u, j) + \alpha \min_{v \in U(j)} Q(j, v) - Q(i, u)\Big). \qquad (6.65)$$

For discounted problems (as well as for certain undiscounted problems), we have seen in Section 5.6 that this algorithm converges to the optimal Q-factors, with probability 1, provided that all state-action pairs are considered infinitely many times and the stepsizes satisfy the usual conditions.

We now generalize to the case where we have approximate Q-factors of the form $\tilde{Q}(i, u, r)$, where r is a parameter vector. Equation (6.65) provides us with a desired update direction for $\tilde{Q}(i, u, r)$. Since we can only control the parameter r, we compute $\nabla\tilde{Q}(i, u, r)$, which indicates in which way r should be updated so that $\tilde{Q}(i, u, r)$ moves in the desired direction. Based on this heuristic argument, the following iteration has been suggested:

$$r := r + \gamma\nabla\tilde{Q}(i, u, r)\Big(g(i, u, j) + \alpha \min_{v \in U(j)} \tilde{Q}(j, v, r) - \tilde{Q}(i, u, r)\Big). \qquad (6.66)$$

This algorithm is frequently used, especially in the model-free case where the transition probabilities $p_{ij}(u)$ are unknown. Nevertheless, theoretical justification for this algorithm and convergence results are limited to the following cases:

(a) The lookup table case (Section 5.6).

(b) For the case of a single policy, it is essentially the same as TD(0) and it converges as long as we use a linear parametric representation and we generate sample states by simulating complete trajectories according to the given policy (Section 6.3).

(c) The case where the approximation architecture corresponds to state aggregation (Section 6.7.7).

(d) Some specially structured problems such as optimal stopping problems, to be discussed in Section 6.8, where a variant of Q-learning will be shown to converge.

Other than these convergence results, little can be said, mainly because there is no interpretation of the method as a stochastic gradient algorithm.

Whenever Q-learning is used in practice, its performance depends crucially on the sampling mechanism that is being used. One wants to concentrate on the most interesting states, and one possibility is to generate states by simulating a reasonably good policy; sometimes a greedy policy is used. On the other hand, one needs to explore the entire state space as well as all possible decisions (rather than restricting to the decisions given by a greedy policy). For this reason, the algorithm is often a combination of steps simulated according to a greedy policy, and of random steps involving arbitrary decisions u, whose purpose is to enhance exploration.

Of course, Q-learning can also be applied even if a model of a system is available and this raises the question whether this would be a better method than, say, optimistic policy iteration, which tries to learn only the cost-to-go function J^*. The key difference between these two alternatives is in the way that the simulated system is controlled and the potential for exploration. With optimistic policy iteration based on TD(λ), the degree of exploration depends largely on the nature of the greedy policies arrived at by the algorithm, which is outside our control. It is only when the problem contains a lot of randomness, guaranteeing adequate exploration independent of policy, that this issue can be bypassed. Thus, when there is a need for explicitly forcing exploration, Q-learning may be more suitable.

6.6.1 Q-Learning and Policy Iteration

We now discuss some variations of Q-learning that are closer in spirit to the approximate policy iteration methods discussed in Sections 6.2-6.4. We start by considering the case of a fixed policy.

Suppose that we generate a state trajectory $i_0, i_1, \ldots$, according to a fixed policy μ, that is, by employing the actions $u_k = \mu(i_k)$. Then, the Q-learning algorithm (6.66) becomes

$$r := r + \gamma \nabla \tilde{Q}(i_k, u_k, r)\big(g(i_k, u_k, i_{k+1}) + \alpha \tilde{Q}(i_{k+1}, u_{k+1}, r) - \tilde{Q}(i_k, u_k, r)\big). \tag{6.67}$$

By identifying $\tilde{Q}\big(i, \mu(i), r\big)$ with $\tilde{J}(i, r)$, we see that it is identical to the TD(0) policy evaluation algorithm studied in Section 6.3. In particular, it is guaranteed to converge for the case of a linear architecture. Extensions involving a λ factor, as in TD(λ), are possible, but we do not discuss them here in order to keep the presentation simple. The algorithm (6.67) can be used for policy evaluation, followed by a policy update to a new policy $\overline{\mu}$ defined by

$$\overline{\mu}(i) = \arg \min_{u \in U(i)} \tilde{Q}(i, u, r),$$

and the same process can be repeated. There is a difficulty here because the algorithm (6.67) is trained only with the actions chosen by the policy μ, and the approximation provided for other actions will be poor. For this reason, the policy μ is usually replaced by a *randomized* policy, call it $\hat{\mu}$, which deviates from μ and applies a random action, with some positive probability ϵ. One then hopes that the parameter vector r will converge to a value such that $\tilde{Q}(i, u, r)$ is a good approximation of

$$\sum_j p_{ij}(u)\big(g(i, u, j) + \alpha J^{\hat{\mu}}(i)\big),$$

for all actions u.

If this algorithm is accepted as a legitimate policy evaluation method to be used in approximate policy iteration, one may venture into the next step and consider the following optimistic variant. Let ϵ be a small positive constant. At a typical step, we have a current parameter vector r_k, we are at some state i_k, and we have chosen an action u_k. We then simulate a transition to the next state i_{k+1}. Finally, we generate the decision u_{k+1} to be used at the next state i_{k+1} as follows. With probability ϵ, we choose an action u_{k+1} randomly and uniformly from the set $U(i_{k+1})$ (other probabilistic mechanisms are also possible), and with probability $1-\epsilon$, we choose an action u_{k+1} that minimizes $\tilde{Q}(i_{k+1}, u, r_k)$ over all $u \in U(i_{k+1})$. At that point, we update r_k according to the formula

$$\begin{aligned} r_{k+1} = r_k + \gamma \nabla \tilde{Q}(i_k, u_k, r_k)\big(g(i_k, u_k, i_{k+1}) &+ \alpha \tilde{Q}(i_{k+1}, u_{k+1}, r_k) \\ &- \tilde{Q}(i_k, u_k, r_k)\big), \end{aligned} \tag{6.68}$$

and we are ready to continue with the next iteration.

The algorithm that we have just described is often called SARSA, which stands for State-Action-Reward-State-Action, because a typical update involves the state i_k, the action u_k, the cost $g(i_k, u_k, i_{k+1})$, the next state i_{k+1}, and the next action u_{k+1}. The update equation (6.68) is almost the same as for ordinary Q-learning, except that the term $\min_{v \in U(i_{k+1})} \tilde{Q}(i_{k+1}, v, r_k)$ in the Q-learning algorithm (6.66) is replaced by $\tilde{Q}(i_{k+1}, u_{k+1}, r_k)$, and the two are different with probability ϵ.

This form of SARSA is essentially a variant of optimistic policy iteration that can be applied without the need for an explicit model of the system. As such, it can be expected to have more or less the same convergence behavior as the optimistic policy iteration methods of Section 6.4. In particular, there are no convergence guarantees.

6.6.2 Advantage Updating

A Q-factor $Q(i, u)$ can always be expressed in the form

$$Q(i, u) = A(i, u) + J(i),$$

where

$$J(i) = \min_{v \in U(i)} Q(i, v)$$

is an estimate of the optimal cost-to-go at state i and

$$A(i,u) = Q(i,u) - \min_{v \in U(i)} Q(i,v),$$

can be viewed as the *disadvantage* of action u, when compared to the best possible action, that is, the action v that minimizes $Q(i,v)$.

Let $J^*(i)$ be the optimal cost-to-go function, let $Q^*(i,u)$ be the corresponding optimal Q-factors, and let $A^*(i,u)$ be the resulting disadvantages. It is clear that in order to construct an optimal policy, we only need to know the disadvantages $A^*(i,u)$. While learning the disadvantages without simultaneously estimating the optimal cost-to-go function or the Q-factors does not seem possible, there have been arguments in favor of using a separate approximation architecture for A^* and J^* [Bai93]. The main idea is that in some problems, especially continuous time problems and their discretizations, the range of $J^*(\cdot)$ and $Q^*(\cdot,\cdot)$ is very large compared to the range of the disadvantage function $A^*(\cdot,\cdot)$. Learning algorithms, based on least squares criteria attempt to capture the large scale behavior of the function being estimated and would therefore make no attempt to capture the finer scale variations of $Q^*(\cdot,\cdot)$. In particular, the dependence of Q^* on u may be lost.

Motivated by the preceding arguments, "advantage updating" algorithms have been proposed by Baird [Bai93], that separately update the cost-to-go function and the (dis)advantages. With a lookup table representation, advantage updating has been shown to be mathematically equivalent to Q-learning [Bai93] and therefore inherits the convergence properties of Q-learning. However, once function approximation is introduced, the properties of advantage updating methods are largely unknown.

We close with an example which is meant to provide a more concrete feeling for the issues being discussed here.

Example 6.13

Consider a scalar continuous-state discrete-time system described by the equation

$$x_{t+1} = x_t + \delta u_t, \qquad t \geq 0,$$

where $x_t \in \Re$ is the state, $u_t \in \Re$ is the control, and δ is a small positive scalar. We let the cost per stage be of the form

$$g(x,u) = \delta(x^2 + u^2).$$

When δ is small, it is interpreted as a small time increment, and the problem can be viewed as a discrete time approximation of the problem of minimizing

$$\int_0^\infty (x_t^2 + u_t^2)\, dt,$$

for the continuous-time system described by

$$\frac{dx_t}{dt} = u_t.$$

Let us consider for simplicity an approach based on policy iteration. We start with the policy μ given by $\mu(x) = -2x$. Under this policy, we have

$$x_t = (1-2\delta)^t x_0, \qquad u_t = -2(1-2\delta)^t x_0,$$

and the cost-to-go is given by

$$J^\mu(x) = 5\delta x^2 \sum_{t=0}^{\infty} (1-2\delta)^{2t} = \frac{5\delta x^2}{1-(1-2\delta)^2} \approx \frac{5}{4}(1+\delta)x^2,$$

where we have ignored terms that are of the order of δ^2 or smaller. Ignoring second order terms in δ, the Q-factors Q^μ corresponding to this policy are given by

$$\begin{aligned} Q^\mu(x,u) &\approx \delta(x^2+u^2) + \frac{5}{4}(1+\delta)(x+\delta u)^2 \\ &\approx \frac{5}{4}x^2 + \delta\left(\frac{9}{4}x^2 + u^2 + \frac{5}{2}xu\right). \end{aligned} \tag{6.69}$$

Suppose now that we were to train a function approximator so as to learn the function Q^μ. Assuming that δ is very small, the parameters of the architecture will be tuned so as to obtain the best possible fit to the function $5x^2/4$. Note that all of the information relevant to a policy update is contained in the term $\delta\big((9x^2/4) + u^2 + (5xu/2)\big)$. However, this latter term would be treated as insignificant by the training algorithm and we would therefore lack any meaningful basis for carrying out a policy update.

The right-hand side of Eq. (6.69) can be rewritten in the form

$$Q^\mu(x,u) \approx \left(\frac{5}{4} + \frac{11\delta}{16}\right)x^2 + \delta\left(u + \frac{5}{4}x\right)^2.$$

A policy improvement step would lead to $u = -5x/4$ and we recognize $\delta\big(u + (5x/4)\big)^2$ as the disadvantage $A(x,u)$ of any decision u. A function approximator which is to learn the term $\delta\big(u + (5x/4)\big)^2$ will not be affected by the presence of δ, as long as the inputs and outputs of the approximation architecture are suitably scaled. As a consequence, the quality of the policy obtained after a policy update will not be affected by the size of δ.

6.7 VALUE ITERATION WITH STATE AGGREGATION

In this section, we consider the fully incremental approximate value iteration method together with a function approximator derived from state aggregation. We will show that convergence is guaranteed no matter how states are sampled, and we derive some error bounds. We concentrate on the case of stochastic shortest path problems, but almost everything generalizes to the case of discounted problems.

We consider a partition of the set $\{0, 1, \ldots, n\}$ of states into disjoint subsets $S_0, S_1, \ldots, S_K$, with $S_0 = \{0\}$. We introduce a K-dimensional parameter vector r whose kth component is meant to approximate the value function for all states $i \in S_k$, $k \neq 0$. In other words, we are dealing with the piecewise constant approximation

$$\tilde{J}(i, r) = r(k), \qquad \text{if } i \in S_k,\ k \neq 0,$$

together with our usual convention $\tilde{J}(0, r) = 0$. As discussed in Ch. 3, this is a feature-based architecture whereby we assign a common value $r(k)$ to all states i that share a common feature vector. Note that such a representation is able to closely approximate J^* as long as J^* does not vary too much within each subset. In particular, let us define

$$\epsilon(k) = \max_{i,j \in S_k} \left| J^*(i) - J^*(j) \right|,$$

which is a measure of the variation of J^* on the set S_k. If we choose r so that

$$r(k) = \min_{i \in S_k} J^*(i) + \frac{\epsilon(k)}{2}, \qquad \forall\ k,$$

it is easily seen that

$$\max_i \left| \tilde{J}(i, r) - J^*(i) \right| = \max_k \frac{\epsilon(k)}{2}.$$

6.7.1 A Method Based on Value Iteration

We now consider the fully incremental approximate value iteration method considered in the previous section [cf. Eq. (6.64)]. We start by noting that $\partial \tilde{J}(i, r)/\partial r(k)$ is equal to 1 if $i \in S_k$, and is equal to zero, otherwise. Thus, Eq. (6.64) becomes

$$r(k) := (1 - \gamma) r(k) + \gamma \min_{u \in U(i)} \sum_{j=0}^{n} p_{ij}(u)\big(g(i, u, j) + \tilde{J}(j, r)\big), \qquad \text{if } i \in S_k. \tag{6.70}$$

Let us use $k(j)$ to denote the index k of the subset S_k to which j belongs; that is, $j \in S_{k(j)}$. In order to have a more precise description, let us assume that at every iteration t the algorithm considers a state i_t and performs the corresponding update. We then have

$$\begin{aligned} r_{t+1}(k) &= \big(1 - \gamma_t(k)\big) r_t(k) + \gamma_t(k) \min_{u \in U(i_t)} \sum_{j=0}^{n} p_{i_t j}(u)\big(g(i_t, u, j) + \tilde{J}(j, r_t)\big) \\ &= \big(1 - \gamma_t(k)\big) r_t(k) + \gamma_t(k) \min_{u \in U(i_t)} \sum_{j=0}^{n} p_{i_t j}(u)\Big(g(i_t, u, j) + r_t\big(k(j)\big)\Big), \end{aligned} \tag{6.71}$$

if $i_t \in S_k$, and

$$r_{t+1}(k) = r_t(k), \qquad \text{if } i_t \notin S_k.$$

Here, $\gamma_t(k)$ is a positive stepsize parameter and we make the usual assumptions that

$$\sum_{t \in T^k} \gamma_t(k) = \infty, \qquad \sum_{t \in T^k} \gamma_t^2(k) < \infty, \qquad \forall\, k, \tag{6.72}$$

where T^k is the set of times such that $i_t \in S_k$. The state i_t used for the update at time t can be chosen by a variety of mechanisms. In order to simplify the analysis, we will assume that it is chosen by random sampling of the type described below. We first choose, in some arbitrary manner, a subset S_k to be sampled, and then select a random element i of S_k with probability $q(i \mid k)$, independently of the past of the algorithm. In what follows, we show that the resulting algorithm can be interpreted in terms of an auxiliary problem, and that it possesses some desirable theoretical guarantees.

6.7.2 Relation to an Auxiliary Problem

The algorithm described in this section was developed as an approximate version of value iteration, based on state aggregation, for solving the original problem. We now provide additional insight by demonstrating that it can also be viewed as an exact value iteration algorithm for an auxiliary aggregated problem.

We construct an auxiliary stochastic shortest path problem as follows. We introduce two types of states:

(1) the states $i = 0, 1, \ldots, n$ of the original problem, and

(2) an additional K states, denoted by $s_1, \ldots, s_K$; each s_k is viewed as an aggregate state representing the subset S_k of the state space.

The dynamics of the auxiliary system are as follows.

(a) Whenever we are at a state s_k, there are no decisions to be made, and a zero-cost transition to a state $i \in S_k$ takes place, with probability $q(i \mid k)$.

(b) Whenever the state is some $i \neq 0$ and a decision $u \in U(i)$ is made, the next state is s_k with probability $\sum_{j \in S_k} p_{ij}(u)$, in which case a cost equal to $\sum_{j \in S_k} p_{ij}(u) g(i, u, j)$, is incurred. Alternatively, the next state could be state 0, with probability $p_{i0}(u)$, and the associated transition cost is $g(i, u, 0)$.

If we use $R^*(s_k)$ and $J^*(i)$ to denote the cost-to-go for the two types of states in the auxiliary problem, Bellman's equation takes the form

$$R^*(s_k) = \sum_{i \in S_k} q(i \mid k) J^*(i),$$

$$J^*(i) = \min_{u \in U(i)} \sum_{k=0}^{K} \sum_{j \in S_k} p_{ij}(u)\big(g(i,u,j) + R^*(s_k)\big)$$
$$= \min_{u \in U(i)} \sum_{j=0}^{n} p_{ij}(u)\big(g(i,u,j) + R^*(s_{k(j)})\big),$$

where we make use of the convention $R^*(s_0) = 0$. Using the second equation to eliminate $J^*(i)$ from the first, we obtain

$$R^*(s_k) = \sum_{i \in S_k} q(i \mid k) \min_{u \in U(i)} \sum_{j=0}^{n} p_{ij}(u)\big(g(i,u,j) + R^*(s_{k(j)})\big). \tag{6.73}$$

This equation can be written in the form $R^* = HR^*$, where $R^* = \big(R^*(s_1), \ldots, R^*(s_K)\big)$ and where H is a nonlinear mapping such that the kth component of HR^* is equal to the right-hand side in Eq. (6.73). Note that the summation over all $i \in S_k$ in Eq. (6.73) amounts to taking expectation, with respect to the probabilities $q(i \mid k)$, $i \in S_k$. The Robbins-Monro stochastic approximation algorithm based on Eq. (6.73), in which this expectation is replaced by a single sample i drawn according to the probabilities $q(i \mid k)$, is given by

$$R(s_k) := (1-\gamma)R(s_k) + \gamma \min_{u \in U(i)} \sum_{j=0}^{n} p_{ij}(u)\big(g(i,u,j) + R(s_{k(j)})\big).$$

We recognize this as being the same as the algorithm (6.71) introduced in the beginning of this section, provided that we identify $R(s_k)$ with $r(k)$.

We have therefore succeeded in describing the method of this section as a stochastic approximation algorithm that attempts to solve *exactly* the auxiliary problem. This provides us with two advantages. First, it allows us to obtain an intuitive feeling for the workings of the algorithm. If we believe, based on some understanding of the original problem, that the auxiliary problem is a reasonable approximation of the original, then we can expect the method to work well. Second, by realizing that we are dealing with the Robbins-Monro stochastic approximation method based on Eq. (6.73), the general results of Ch. 4 can be brought to bear in order to establish convergence.

6.7.3 Convergence Results

Suppose that the auxiliary problem is such that all policies are proper. It follows that the DP operator for the auxiliary system is a contraction mapping with respect to a suitably weighted maximum norm (Prop. 2.2 in Section 2.2). The operator H is essentially a two-step DP operator for the auxiliary system and it follows easily that H has the same contraction property. (The formal proof of this statement is identical to the argument carried out during the proof of Prop. 5.5, in the context of the Q-learning algorithm, when all policies are proper.) We can now invoke Prop. 4.4 of Section 4.3 and conclude that the algorithm (6.71) is guaranteed to converge to the unique solution of Eq. (6.73).

Let us now make the weaker assumption that there exists at least one proper policy for the auxiliary system and that every improper policy has infinite cost. It then follows from Prop. 4.6 of Section 4.3 that if the sequence r_t is guaranteed to be bounded with probability 1, then the algorithm again converges. In particular, if all of the costs $g(i,u,j)$ are nonnegative and if the algorithm is initialized with a nonnegative vector r_0, convergence is guaranteed. (The proof is the same as the one given for Prop. 5.6 in Section 5.6.)

Note that the convergence results stated in the preceding two paragraphs require assumptions on the policies associated to the auxiliary system. It is important to realize that these assumptions for the auxiliary system do not necessarily follow from corresponding assumptions for the original system. This is illustrated by the following example.

Example 6.14

Consider a stochastic shortest path problem in which we have a single (proper) policy and all transitions are deterministic, as shown in Fig. 6.14(a).

All transition costs starting from states 1,2,3, are unity and state 0 is a zero-cost absorbing state. We let $S_1 = \{1,3\}$ and $S_2 = \{2\}$. Suppose that whenever we sample the set S_1, we always choose state 1; that is, $q(1 \mid 1) = 1$. Then, the update equations become:

$$r(1) := (1-\gamma)r(1) + \gamma\bigl(1 + r(2)\bigr),$$
$$r(2) := (1-\gamma)r(2) + \gamma\bigl(1 + r(1)\bigr).$$

This algorithm cannot converge, since by adding the above two equations we obtain $r(1)+r(2) := r(1)+r(2)+2\gamma$. To understand the reason for divergence, we form the auxiliary problem; it has the structure shown in Fig. 6.14(b) and we see that we are dealing with an improper policy. In particular, the equation $R = HR$ does not have a solution.

Let us change the sampling distribution so that $q(1 \mid 1) = q > 0$ and $q(3 \mid 1) = 1 - q > 0$. The resulting auxiliary system has the structure shown in Fig. 6.14(c). We are now dealing with a proper policy and it can be seen that the algorithm is guaranteed to converge to the unique solution of the system

$$r(1) = 1 + qr(2),$$
$$r(2) = 1 + r(1),$$

which is given by $r(1) = (1+q)/(1-q)$ and $r(2) = 2/(1-q)$. Clearly, as q approaches 1, the quality of the limit, as an approximation of J^*, deteriorates.

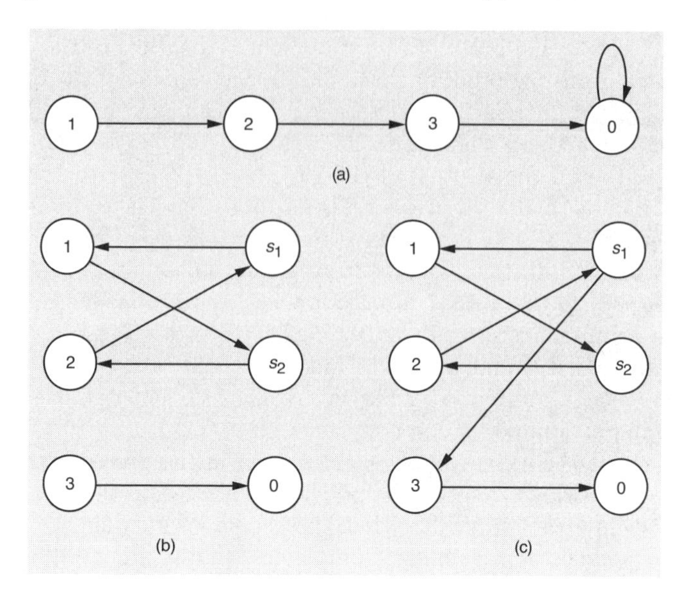

Figure 6.14: (a) A simple deterministic problem with a single policy. (b) The transition graph of the auxiliary problem for $S_1 = \{1,3\}$, $S_2 = \{2\}$, and $q(1 \mid 1) = 1$; only nonzero probability transitions are shown. (c) The auxiliary problem when $q > 0$.

The preceding example amplifies the importance of the sampling distribution and shows that it can significantly affect the limiting value of $\tilde{J}(i, r_t)$. Nevertheless, as long as all states are sampled with positive probability, the algorithm does not diverge, as will be shown in the proposition that follows.

Proposition 6.7: Consider the algorithm described by Eq. (6.71). Suppose that all states are sampled with positive probability, and that the stepsize condition (6.72) is satisfied.

(a) If all policies in the original problem are proper, the same is true for the auxiliary problem and r_t converges with probability 1 to the unique solution of the system (6.73).

(b) Suppose that (for the original problem) there exists a proper policy and all improper policies have infinite cost for some initial state. Furthermore, suppose that $g(i, u, j) \geq 0$ for all i, u, j, and that the algorithm is initialized with a nonnegative vector r_0. Then, the sequence r_t converges with probability 1 to the unique solution of the system (6.73).

Proof: *(Abbreviated)* Note that there is a one-to-one correspondence between the policies of the original and of the auxiliary problem. This is because, in the auxiliary problem, we only get to make decisions at states $1, \ldots, n$, but not at the states s_k. So, let us fix a policy for the two problems. Let p_{ij} be the transition probabilities in the original problem and let $\tilde{p}_{ij}$ be the probability that starting from $i = 0, 1, \ldots, n$, we end up at some $j = 0, 1, \ldots, n$, after two transitions in the auxiliary problem. Note that if $p_{ij} > 0$, then there is positive probability of moving from i to $s_{k(j)}$ in the auxiliary problem and, given that we are at $s_{k(j)}$, there is positive probability that state j is sampled. We conclude that whenever p_{ij} is positive, $\tilde{p}_{ij}$ is also positive. This implies that any proper policy in the original problem is also proper in the auxiliary problem.

Suppose that all policies in the original problem are proper. Then, the same is true for the auxiliary problem and as argued earlier in this subsection, the operator H is a weighted maximum norm contraction. Since we are dealing with a Robbins-Monro stochastic approximation algorithm based on H, Prop. 4.4 in Section 4.3 applies and proves convergence. For a complete proof, we also need to verify that the noise variance does not grow faster that $\|r\|^2$, but this is straightforward and similar to the argument used in the convergence proof for Q-learning (Prop. 5.5 in Section 5.6).

We now move to the case where improper policies are possible. Consider an improper policy for the auxiliary problem and let us focus on a recurrent class of states that does not include the termination state. Let S be the set of states $i \in \{1, \ldots, n\}$ in that recurrent class. Then, we must have $\tilde{p}_{ij} = 0$ for all $i \in S$ and $j \notin S$. This implies that we also have $p_{ij} = 0$ for all $i \in S$ and $j \notin S$. In particular, we have an improper policy μ for the original problem under which the state stays forever in the set S. Since all improper policies have infinite cost for some initial state (in the original problem), $\sum_j p_{ij} g\big(i, \mu(i), j\big)$ must be positive for at least one of the states in S. Since the two problems have the same expected costs per stage, the expected cost per stage is positive for at least one nontransient state in the auxiliary problem. This argument shows that the auxiliary problem satisfies the usual assumptions (there exists a proper policy and for every improper policy some initial state has infinite cost).

Having verified the above, it is easily shown that the iteration mapping H has a unique fixed point (because it is closely related to the DP operator for the auxiliary problem). We can then apply our convergence results for stochastic approximation algorithms based on monotone mappings. Proposition 4.6 in Section 4.3 establishes convergence as long as the iterates stay bounded. Boundedness follows from the existence of a proper policy and the nonnegativity of the costs; the details of the proof are the same as in the proof of convergence of Q-learning (Prop. 5.6 in Section 5.6). **Q.E.D.**

The nonnegativity assumption in part (b) of the preceding proposition cannot be relaxed. The example that follows shows that divergence is possible in the absence of that assumption. The reason is that even if all improper policies in the original problem have infinite cost, the auxiliary problem may possess improper policies with finite cost.

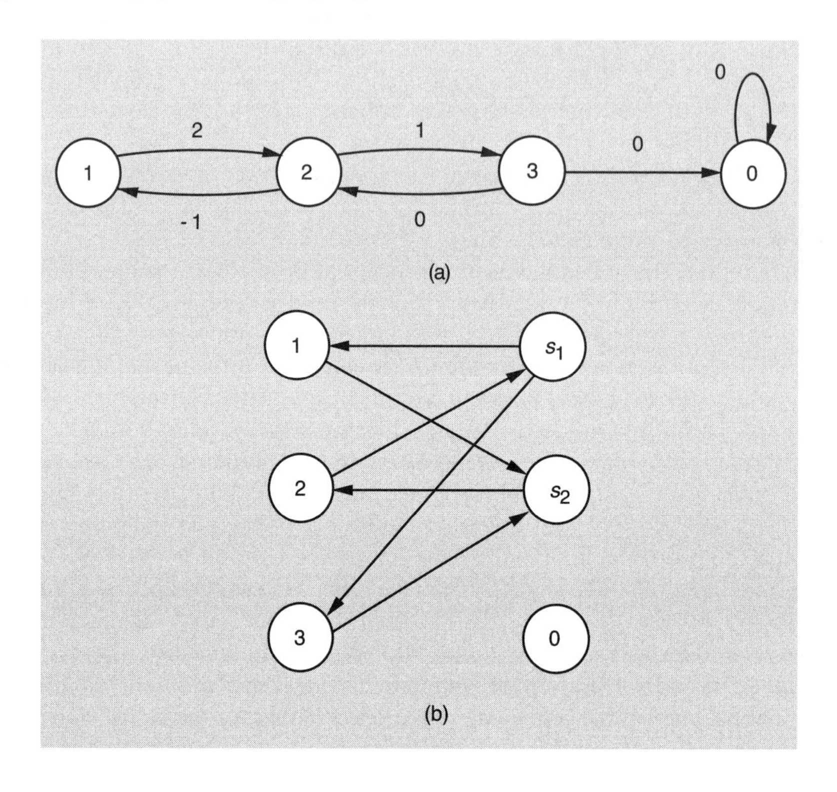

Figure 6.15: (a) The system in Example 6.15. (b) The transition graph of the auxiliary system under the policy that chooses to go from state 3 to state 2 and from state 2 to state 1.

Example 6.15

Consider the system shown in Fig. 6.15(a). All transitions are deterministic but at states 2 and 3, there are two possible decisions, as indicated in the figure. The labels next to each arc indicate the transition costs. Note that every improper policy has infinite cost because the sum of the costs along every cycle are positive. Let $S_1 = \{1, 3\}$ and $S_2 = \{2\}$. Let $q(1 \mid 1) = q > 0$ and $q(3 \mid 1) = 1 - q > 0$. The equation $r = Hr$ takes the form

$$r(1) = q\big(2 + r(2)\big) + (1 - q) \min \big\{0, r(2)\big\},$$

$$r(2) = \min \big\{ -1 + r(1), 1 + r(1)\big\} = -1 + r(1).$$

Using simple algebra, it can be verified that if $q < 1/2$, the equation $r = Hr$ has no solutions. Since the algorithm could only converge to a solution of this equation, it follows that the algorithm does not converge.

A better understanding of the divergent behavior can be obtained by considering the policy that chooses to go from state 3 to state 2 and from state 2 to state 1. Under that policy, the auxiliary system has the structure shown in Fig. 6.15(b). The expected cost incurred between two consecutive visits to state 2 in the auxiliary system, is equal to $-1 + 2q$. (From state 2, we move to s_1, at a cost of -1; once at s_1, we choose state 1 with probability q and then incur a cost of 2, or choose state 3, with probability $1 - q$, and incur zero cost.) Thus, if $q < 1/2$, the total expected cost of this policy is $-\infty$, which violates the usual assumption that every improper policy must have cost $+\infty$.

Discounted Problems

The algorithm of this section is easily adapted to discounted problems; all that is required is to insert a factor of α at the usual place in the update equation, and H becomes a maximum norm contraction. The auxiliary problem is also a discounted problem. Hence, the equation $r = Hr$ has a unique solution and the algorithm converges to it with probability 1, as long as each subset S_k is sampled infinitely often, and under the usual assumptions on the stepsize.

6.7.4 Error Bounds

We continue by introducing some stronger assumptions on the original problem that will allow us to establish bounds on the quality of the approximate cost-to-go function obtained in the limit. We assume that there exists a positive constant α such that for every state i and every decision $u \in U(i)$, we have

$$p_{i0}(u) \geq 1 - \alpha. \tag{6.74}$$

We note that such an assumption is satisfied whenever we start with a discounted problem (with discount factor α) and convert it to a stochastic shortest path problem, as discussed in Ch. 2. The bounds to be derived are also valid for discounted problems that are treated as such (without converting them to undiscounted ones); the arguments are entirely similar and will be omitted.

Under condition (6.74), it is easily seen that every policy is proper. Using our earlier discussion, the mapping H described by Eq. (6.73) is a contraction and the algorithm converges, with probability 1, to the unique solution of the system $r = Hr$, to be denoted by r^*. Let

$$\epsilon(k) = \max_{i,j \in S_k} \big|J^*(i) - J^*(j)\big|,$$

and $\epsilon = \max_k \epsilon(k)$. We then have the following result.

Proposition 6.8: Under condition (6.74), the unique solution r^* of the system $r = Hr$ [cf. Eq. (6.73)] satisfies

$$r^*(k) - \frac{\epsilon}{1-\alpha} \leq J^*(i) \leq r^*(k) + \frac{\epsilon}{1-\alpha}, \qquad \forall\, k,\ i \in S_k.$$

Proof: In order to eliminate some details, we make the minor assumption that $g(i,u,0) = 0$ for all i and $u \in U(i)$. The proof can be easily extended to the case where this assumption does not hold. Let us define

$$\overline{r}(k) = \min_{i \in S_k} J^*(i) + \frac{\epsilon}{1-\alpha}, \qquad k = 1, \ldots, K.$$

We then have

$$\begin{aligned}
(H\overline{r})(k) &= \sum_{i \in S_k} q(i \mid k) \min_{u \in U(i)} \sum_{j=1}^n p_{ij}(u)\Big(g(i,u,j) + \overline{r}\big(k(j)\big)\Big) \\
&\leq \sum_{i \in S_k} q(i \mid k) \min_{u \in U(i)} \sum_{j=1}^n p_{ij}(u)\Big(g(i,u,j) + J^*(j) + \frac{\epsilon}{1-\alpha}\Big) \\
&= \sum_{i \in S_k} q(i \mid k) \left(J^*(i) + \min_{u \in U(i)} \sum_{j=1}^n p_{ij}(u) \frac{\epsilon}{1-\alpha}\right) \\
&\leq \min_{i \in S_k} \big(J^*(i) + \epsilon\big) + \frac{\alpha\epsilon}{1-\alpha} \\
&= \min_{i \in S_k} J^*(i) + \frac{\epsilon}{1-\alpha} \\
&= \overline{r}(k).
\end{aligned}$$

We have therefore established that $H\overline{r} \leq \overline{r}$. Since $r^* = \lim_{t\to\infty} H^t \overline{r}$, the monotonicity of H implies that $r^* \leq \overline{r}$. This proves the left-hand side inequality in the proposition's statement. The right-hand side inequality is proved similarly by defining

$$\underline{r}(k) = \max_{i \in S_k} J^*(i) - \frac{\epsilon}{1-\alpha},$$

and establishing that $H\underline{r} \geq \underline{r}$. **Q.E.D.**

It is known that the bounds provided by Prop. 6.8 are tight. In particular, an example is given by Tsitsiklis and Van Roy [TsV96a] where the approximation error is as large as $\epsilon/(1-\alpha)$.

Using the bounds in Prop. 6.8, we can also obtain bounds on the quality of the resulting policy. Let μ be a greedy policy, defined by

$$\mu(i) = \arg \min_{u \in U(i)} \sum_{j=0}^{n} p_{ij}(u)\Big(g(i,u,j) + r^*\big(k(j)\big)\Big), \qquad \forall\ i.$$

Then, Prop. 6.1 in Section 6.1 shows that

$$J^\mu(i) \leq J^*(i) + \frac{2\alpha\epsilon}{(1-\alpha)^2}, \qquad \forall\ i.$$

We note that this is similar to the performance guarantees of approximate policy iteration (cf. Prop. 6.2 in Section 6.2). However, the two results are not directly comparable. In approximate policy iteration, we assumed the ability to approximate any J^μ within ϵ, where ϵ is the power of the approximation architecture. In contrast, Prop. 6.8 only assumes that J^* can be approximated with accuracy $\epsilon/2$, but with a special type of architecture, the class of functions that are piecewise constant over the sets S_k.

6.7.5 Comparison with TD(0)

Let us now assume that there is a single policy μ, which would be the case, for example, if we were trying to evaluate a policy within a policy iteration scheme. The algorithm and the convergence results remain applicable, with J^* replaced by J^μ. For the single policy case, the algorithm can be viewed as an adaptation of TD(0) to a particular type of approximation architecture.

In order to make a closer connection with the results of Section 6.3, we define the "projection" ΠJ^μ of J^μ by

$$\|\Pi J^\mu - J^\mu\|_\infty = \min_r \|\tilde{J}(\cdot, r) - J^\mu\|_\infty,$$

and it is easily seen that $\|\Pi J^\mu - J^\mu\|_\infty = \epsilon/2$. Thus, the error bound of Prop. 6.8 is of the form

$$\|J^\mu - \tilde{J}(\cdot, r^*)\|_\infty \leq \frac{2\|\Pi J^\mu - J^\mu\|_\infty}{1-\alpha},$$

which is very similar to the error bound $\|\Pi J^\mu - J^\mu\|_D/(1-\alpha)$ derived for TD(0) (cf. Prop. 6.5 in Section 6.3). Despite the apparent similarity, the two results are not comparable, neither one implying the other. Leaving aside the fact that different norms are involved, the convergence of TD(0) is based on a particular sampling mechanism (simulate the policy under consideration), but it allows for a much more general approximation architecture than the one considered here.

6.7.6 Discussion of Sampling Mechanisms

The algorithm considered in this section and our convergence results are a bit restrictive because of the requirement that the different samples from the same set S_k are independent identically distributed. In a simulation-based implementation one might wish to generate random trajectories (under some policy) and perform updates at the states visited by the trajectory; in that case, the independence assumption would be violated. Fortunately, our convergence results remain valid under much more general sampling mechanisms.

To be more specific, suppose that there is an underlying finite-state, stationary, ergodic Markov process X_t and that the state to be sampled at time t is a function of X_t. For example, X_t could be the state of the system under consideration, when controlled under a fixed policy. We then define $q(i \mid k)$ as the steady-state probability that state i is sampled, given that the partition S_k is selected. In this context, the convergence behavior of the algorithm is the same as in the case of independent sampling, except that the proof requires somewhat more sophisticated machinery.

One could also consider a nonstationary sampling mechanism, that evolves based on the progress of the algorithm. For example, if our sampling is based on simulations of the system using a greedy policy, we get a method resembling optimistic policy iteration using TD(0). The convergence of the resulting method is questionable.

In practice, one would want to combine some aspects of optimistic policy iteration, so as to ensure that the most important states are visited, together with some random sampling to ensure that no regions of the state space remain unexplored. An alternative is to heuristically design a reasonably good policy and let the sample states be the states visited by that policy.

6.7.7 The Model-Free Case

If a model of the system is unavailable, but the set $U(i)$ is the same for all states i in the same subset S_k, we can introduce approximate Q-factors that remain constant within each subset S_k, of the form

$$\tilde{Q}(i,u) = Q(S_k,u), \qquad i \in S_k,\ u \in U(i).$$

We then obtain an algorithm that updates the Q-factors $Q(S_k,u)$, using the equation

$$Q(S_k,u) := (1-\gamma)Q(S_k,u) + \gamma\Big(g(i,u,j) + \min_{v \in U(j)} Q(S_{k(j)},v)\Big),$$

where i is an element of S_k chosen with probability $q(i \mid k)$, and j is the outcome of a transition simulated according to the transition probabilities $p_{ij}(u)$. It is not difficult to see that this algorithm coincides with

Q-learning, with a lookup table representation, applied to the auxiliary problem. With a suitably decreasing stepsize and if all pairs (S_k, u) are considered an infinite number of times, the convergence results for Q-learning apply (cf. Section 5.6).

We note that if the set $U(i)$ is different for different states $i \in S_k$, or if the same u has drastically different effects at different states i in S_k, a method of this type cannot be useful.

6.8 EUCLIDEAN CONTRACTIONS AND OPTIMAL STOPPING

In this section, we focus on the case of a linearly parametrized approximation architecture. We impose an additional assumption on the DP operator T and then proceed to show that approximate value iteration becomes a convergent algorithm. We close by demonstrating the convergence of Q-learning when applied to a broad class of optimal stopping problems.

In order to motivate the nature of our assumptions on the DP operator T, recall that approximate value iteration can be viewed as a combination of two elements: updates of cost-to-go values, according to the value iteration formula, which amounts to applying the operator T; "projections" on the set of cost-to-go functions that can be represented by the approximation architecture. For discounted problems, the DP operator is a contraction (with respect to the maximum norm) and projection operators are nonexpansive (with respect to the Euclidean norm). However, in order to assert that the composition of the two is also a contraction, we need their properties to refer to the same norm and this is essentially what will be assumed in this section. We will be working in terms of weighted quadratic norms; recall that such norms played a key role in establishing the convergence of TD(λ) for the case of a single policy and linear parametrizations (Section 6.3).

6.8.1 Assumptions and Main Convergence Result

Consider a stochastic shortest path problem with state space $\{0, 1, \ldots, n\}$, where 0 is a cost-free termination state. We assume a linearly parametrized approximation of the cost-to-go, of the form

$$\tilde{J}(i, r) = \phi(i)'r, \qquad i = 1, \ldots, n,$$

where r and each $\phi(i)$ are K-dimensional vectors, and where $\phi(0) = 0$. Let Φ be the $n \times K$ matrix whose ith row is the row vector $\phi(i)'$. We may then write, in vector notation, $\tilde{J}(r) = \Phi r$, where $\tilde{J}(r)$ is the vector $\big(\tilde{J}(1, r), \ldots, \tilde{J}(n, r)\big)$.

Let $\pi = \big(\pi(1), \ldots, \pi(n)\big)$ be a vector with positive components that sum to 1, and let D be an $n \times n$ diagonal matrix with diagonal entries

$\pi(1), \ldots, \pi(n)$. As in Section 6.3, we define the weighted quadratic norm $\|\cdot\|_D$ by $\|J\|_D^2 = J'DJ$, for all $J \in \Re^n$.

Let us consider the fully incremental approximate value iteration algorithm of Section 6.5.4 [cf. Eq. (6.64)], specialized to the case of a linear approximation architecture. Since $\nabla \tilde{J}(i,r) = \phi(i)$, Eq. (6.64) becomes

$$\begin{aligned} r &:= r + \gamma\phi(i)\left(\min_{u\in U(i)} \sum_{j=0}^{n} p_{ij}(u)\big(g(i,u,j) + \phi(j)'r\big) - \phi(i)'r\right) \\ &= r + \gamma\phi(i)\Big(\big(T(\Phi r)\big)(i) - \phi(i)'r\Big). \end{aligned}$$

A more precise description of the algorithm is

$$r_{t+1} = r_t + \gamma_t\phi(i_t)\Big(\big(T(\Phi r_t)\big)(i_t) - \phi(i_t)'r_t\Big), \tag{6.75}$$

where i_t is a state chosen for an update at iteration t.

We now introduce the main assumption of this section.

Assumption 6.3: The operator T is a contraction with respect to the norm $\|\cdot\|_D$, namely, there exists some $\beta < 1$ such that

$$\|TJ - T\bar{J}\|_D \le \beta\|J - \bar{J}\|_D, \qquad \forall\ J,\ \bar{J}.$$

The result that follows shows that if the states i_t are independently sampled according to the probability distribution specified by the vector π, then the algorithm converges, with probability 1. The proof is in some respects similar to the convergence proofs for TD(λ) (Props. 6.4 and 6.6), specialized to the case $\lambda = 0$. One difference here is that the update formula (6.75) depends nonlinearly on r. Nevertheless, the contraction property that we assume can be used to show that the expected update direction has a descent property with respect to a certain quadratic potential function.

Proposition 6.9: Consider the algorithm described by Eq. (6.75) and assume the following.

(a) The stepsizes γ_t are deterministic, nonnegative, and satisfy $\sum_{t=0}^{\infty} \gamma_t = \infty$ and $\sum_{t=0}^{\infty} \gamma_t^2 < \infty$.

(b) The states i_t are independently chosen and $\mathrm{P}(i_t = i) = \pi(i)$ for all i and all t.

(c) The columns of the matrix Φ are linearly independent.

(d) Assumption 6.3 holds.

Then, with probability 1, the sequence r_t converges to r^*, the unique solution of the system of equations

$$r^* = (\Phi' D \Phi)^{-1} \Phi' D T(\Phi r^*). \tag{6.76}$$

Proof: Let us define a matrix Π by

$$\Pi = \Phi(\Phi' D \Phi)^{-1} \Phi' D. \tag{6.77}$$

As shown in Lemma 6.8 of Section 6.3.3, for any given vector J, ΠJ is a vector of the form Φr, where r is chosen so as to minimize $\|\Phi r - J\|_D$, and in that sense, Π is a projection matrix. It then follows that Π is nonexpansive, in the sense that $\|\Pi J\|_D \leq \|J\|_D$, for all $J \in \Re^n$. While this is evident from the geometric interpretation of projections, it can also be proved algebraically as follows. We have

$$\begin{aligned} \|J\|_D^2 &= \|\Pi J + (I - \Pi)J\|_D^2 \\ &= \|\Pi J\|_D^2 + \|(I - \Pi)J\|_D^2 + 2J'\Pi' D(I - \Pi)J \\ &= \|\Pi J\|_D^2 + \|(I - \Pi)J\|_D^2 \\ &\geq \|\Pi J\|_D^2, \end{aligned}$$

where the last equality follows from the relation $\Pi' D = \Pi' D \Pi$, which is an easy consequence of Eq. (6.77). We conclude that $\|\Pi J\|_D \leq \|J\|_D$ for all J.

Consider now the mapping that maps a vector J to the vector $\Pi T J$. Since T is a contraction and Π is nonexpansive (with respect to the norm $\|\cdot\|_D$), it follows that this mapping is a contraction mapping. We now invoke the contraction mapping theorem (see e.g., [BeT89], p. 182, or [Lue69], p. 272), which asserts that a contraction mapping from $\Re^n$ into itself has a unique fixed point; we conclude that there exists a unique vector $\overline{J}$ that satisfies $\overline{J} = \Pi T \overline{J}$. Since $\overline{J}$ is in the range of Π, it follows that there exists some r^* such that $\overline{J} = \Phi r^*$. Using Eq. (6.77), we have

$$\Phi r^* = \Pi T(\Phi r^*) = \Phi(\Phi' D \Phi)^{-1} \Phi' D T(\Phi r^*). \tag{6.78}$$

Since Φ has linearly independent columns, we obtain

$$r^* = (\Phi' D \Phi)^{-1} \Phi' D T(\Phi r^*),$$

and therefore Eq. (6.76) has a solution. [This can also be proved by multiplying both sides of Eq. (6.78) by $(\Phi' D \Phi)^{-1} \Phi' D$.]

We now prove the uniqueness of r^*. If r is another vector satisfying $r = (\Phi' D\Phi)^{-1}\Phi' DT(\Phi r)$, then both Φr and Φr^* are solutions (for J) of the equation $J = \Pi TJ$ which, as shown earlier, has a unique solution. Thus, $\Phi r = \Phi r^*$ and since Φ has full column rank, we conclude that $r = r^*$.

We now continue with the proof of convergence. Let s_t be defined by

$$s_t = \phi(i_t)\Big(\big(T(\Phi r_t)\big)(i_t) - \phi(i_t)' r_t\Big),$$

and note that the algorithm we are studying is of the form

$$r_{t+1} = r_t + \gamma_t s_t.$$

Some simple algebra yields

$$E[s_t \mid r_t] = \sum_{i=1}^{n} \pi(i)\phi(i)\Big(\big(T(\Phi r_t)\big)(i) - \phi(i)' r_t\Big) = \Phi' DT(\Phi r_t) - \Phi' D\Phi r_t.$$

Our convergence proof is based on a stochastic descent property with respect to the potential function

$$f(r) = \frac{1}{2}\|r - r^*\|_2^2,$$

where $\|\cdot\|_2$ is the (unweighted) Euclidean norm. As a first step, we establish that $\nabla f(r_t)' E[s_t \mid r_t]$ is negative whenever $r_t \neq r^*$. Omitting, for simplicity, the subscript t from our notation, we have

$$\begin{aligned}
\nabla f(r)' E[s \mid r] &= (r - r^*)'\Phi' D\big(T(\Phi r) - \Phi r\big) \\
&= (r - r^*)'\Phi' D\big(\Pi T(\Phi r) + (I - \Pi)T(\Phi r) - \Phi r\big) \\
&= (\Phi r - \Phi r^*)' D\big(\Pi T(\Phi r) - \Phi r\big),
\end{aligned}$$

where the last equality follows because $\Phi' D\Pi = \Phi' D$ [cf. Eq. (6.77)]. Using the definition of r^* [cf. Eq. (6.76)], the contraction assumption on T, and the nonexpansion property of Π, we have

$$\begin{aligned}
\|\Pi T(\Phi r) - \Phi r^*\|_D &= \|\Pi T(\Phi r) - \Pi T(\Phi r^*)\|_D \\
&\leq \|T(\Phi r) - T(\Phi r^*)\|_D \\
&\leq \beta\|\Phi r - \Phi r^*\|_D.
\end{aligned} \tag{6.79}$$

We have, using Eq. (6.79) to obtain the first inequality,

$$\begin{aligned}
\beta^2\|\Phi r - \Phi r^*\|_D^2 &\geq \|\Pi T(\Phi r) - \Phi r^*\|_D^2 \\
&= \|\Pi T(\Phi r) - \Phi r + (\Phi r - \Phi r^*)\|_D^2 \\
&= \|\Pi T(\Phi r) - \Phi r\|_D^2 + \|\Phi r - \Phi r^*\|_D^2 \\
&\quad + 2\big(\Pi T(\Phi r) - \Phi r\big)' D(\Phi r - \Phi r^*) \\
&\geq \|\Phi r - \Phi r^*\|_D^2 + 2\big(\Pi T(\Phi r) - \Phi r\big)' D(\Phi r - \Phi r^*),
\end{aligned}$$

from which it follows that

$$\nabla f(r)'E[s \mid r] = \big(\Pi T(\Phi r) - \Phi r\big)' D(\Phi r - \Phi r^*) \leq -\frac{1-\beta^2}{2}\|\Phi r - \Phi r^*\|_D^2 \leq 0. \tag{6.80}$$

Since Φ has full column rank, $\Phi' D\Phi$ is positive definite and, therefore, there exists some $c > 0$ such that $r'\Phi' D\Phi r \geq c r' r$, for all r. Using this inequality with $r - r^*$ in place of r, we obtain $\|\Phi r - \Phi r^*\|_D^2 \geq c\|r - r^*\|_2^2$. We conclude that

$$\nabla f(r)'E[s \mid r] \leq -\frac{c(1-\beta^2)}{2}\|r - r^*\|_2^2 = -\frac{c(1-\beta^2)}{2}\|\nabla f(r)\|_2^2.$$

In order to apply the convergence results from Ch. 4, we also need a bound on $E\big[\|s_t\|_2^2 \mid r_t\big]$. We have

$$E\big[\|s_t\|_2^2 \mid r_t\big] = \sum_{i=1}^{n} \pi(i)\Big\|\phi(i)\Big(\big(T(\Phi r_t)\big)(i) - \phi(i)' r_t\Big)\Big\|_2^2.$$

Using the contraction property of T, it is not hard to show that

$$\big\|T(\Phi r_t)\big\|_2 \leq A\big(1 + \|\Phi r_t\|_2\big) \leq B\big(1 + \|r_t\|_2\big),$$

for some constants A and B. Using this relation, it follows easily that

$$\begin{aligned} E\big[\|s_t\|_2^2 \mid r_t\big] &\leq C\big(1 + \|r_t\|_2\big)^2 \\ &\leq C\big(1 + \|r_t - r^*\|_2 + \|r^*\|_2\big)^2 \\ &\leq F\big(1 + \|r_t - r^*\|_2^2\big) \\ &= F\big(1 + \|\nabla f(r_t)\|_2^2\big), \end{aligned}$$

for some constants C and F.

We can now apply Prop. 4.1 from Section 4.2. We obtain that $\nabla f(r_t) = r_t - r^*$ converges to zero, and the proof is complete. **Q.E.D.**

6.8.2 Error Bounds

We continue with a result that shows that if the approximation architecture $\tilde{J} = \Phi r$ is capable of closely approximating J^*, then the limit Φr^* provided by the algorithm provides a close approximation.

Proposition 6.10: Under the assumptions of Prop. 6.9, the limit r^* to which the algorithm converges satisfies

$$\|\Phi r^* - J^*\|_D \leq \frac{1}{1-\beta}\|\Pi J^* - J^*\|_D.$$

Proof: Recall that Φr^* has been characterized as the solution of the equation $\Phi r^* = \Pi T(\Phi r^*)$. Using the contraction assumption on T, the result follows immediately from Lemma 6.9 in Section 6.3.3. **Q.E.D.**

6.8.3 Applicability of the Result

The main difficulty in applying the convergence result of this section is in defining a suitable norm $\|\cdot\|_D$, verifying that T is a contraction with respect to this norm, and finally generating the sample states according to the distribution $\pi(\cdot)$.

One case arises when we are dealing with a single policy and we have a discounted problem. Then, we can let $\pi(\cdot)$ be the steady-state distribution. As shown in Section 6.3, the transition probability matrix P satisfies $\|PJ\|_D \leq \|J\|_D$ for every vector J and it follows that T is a contraction, with a contraction factor equal to the discount factor. In that case, we recover the convergence result for TD(0), already shown in Section 6.3. A second case where Prop. 6.9 applies is discussed in Van Roy and Tsitsiklis [VaT96], which deals with systems where the states are embedded in a Euclidean space and all transitions are local. Yet another case is the subject of the next subsection.

Regarding the sampling mechanism, the discussion in Section 6.7.6 applies. In particular, a convergence result can be proved without the assumption that the different samples i_t are independent, as long as they are generated according to some Markov process with steady-state distribution $\pi(\cdot)$; for example, by generating independent trajectories according to some fixed policy. The proof, however, requires more sophisticated machinery.

6.8.4 Q-Learning for Optimal Stopping Problems

Optimal stopping problems are a special case of DP problems in which we only get to choose the time at which the process terminates. They arise in many contexts (see e.g. [Ber95a]), such as search problems, sequential hypothesis testing, and pricing of derivative financial instruments.

We are given a Markov chain with state space $\{1, \ldots, n\}$, described by a transition probability matrix P, and which has a steady-state distribution $\pi(\cdot)$. We assume that $\pi(i) > 0$ for all states i. Given the current state i, we assume that we have two options: to stop and incur a termination cost $h(i)$, or to continue and incur a cost $g(i, j)$, where j is the next state. We adopt a discounted formulation, and let $\alpha \in (0, 1)$ be the discount factor.

We digress to note that a similar development is possible without the positivity assumption on the steady-state probabilities, but the algorithm has to be different (we need to keep reinitializing the trajectory so as to ensure that all transient states are visited infinitely often), and the weights used to define the norm $\|\cdot\|_D$ must be defined differently. (The same choice of norm as the one that was used in Prop. 6.6 of Section 6.3.4 will do.) Furthermore, the case of $\alpha = 1$ can also be handled under suitable assumptions.

The DP operator T for the problem that we have defined is given by

$$(TJ)(i) = \min\left\{ h(i),\ \sum_{j=1}^{n} p_{ij}\big(g(i,j) + \alpha J(j)\big) \right\},$$

or, in vector notation,

$$TJ = \min\{h, g + \alpha PJ\},$$

where $h = \big(h(1), \ldots, h(n)\big)$, and g is a vector whose ith component is equal to $\sum_{j=1}^{n} p_{ij} g(i,j)$.

In the spirit of the Q-learning algorithm, we associate a Q-factor with each of the two possible decisions. The Q-factor in the case where the decision is to stop is equal to the termination cost $h(i)$. The Q-factor in the case where the decision is to continue is denoted by $Q(i)$. This latter Q-factor satisfies

$$Q(i) = \sum_{j=1}^{n} p_{ij}\Big(g(i,j) + \alpha \min\big\{h(j), Q(j)\big\}\Big),$$

and the Q-learning algorithm is

$$Q(i) := Q(i) + \gamma\Big(g(i,j) + \alpha \min\big\{h(j), Q(j)\big\} - Q(i)\Big),$$

where i is the state at which we choose to update and j is a successor state, generated randomly according to the transition probabilities p_{ij}.

Let us now adopt an approximation architecture and introduce an approximation $\tilde{Q}(i,r)$ of the Q-factors $Q(i)$, where r is a K-dimensional parameter vector. We assume that the approximation architecture is linear and that

$$\tilde{Q}(i,r) = \phi(i)'r,$$

where $\phi(i)$ is a K-dimensional feature vector associated with state i. In the presence of such function approximation, we can construct the following Q-learning algorithm [cf. Eq. (6.66) in Section 6.6]. At the beginning of the tth iteration we have access to a parameter vector r_t. We select the stepsize γ_t and a state i_t, simulate a transition to a state j_t, according to the transition probabilities $p_{i_t j}$, and perform the update

$$r_{t+1} = r_t + \gamma_t \phi(i_t)\Big(g(i_t, j_t) + \alpha \min\big\{h(j_t), \tilde{Q}(j_t, r_t)\big\} - \tilde{Q}(i_t, r_t)\Big). \tag{6.81}$$

We rewrite the algorithm in the form

$$\begin{aligned} r_{t+1} &= r_t + \gamma_t \phi(i_t)\Big(g(i_t, j_t) + \alpha \min\big\{h(j_t), \phi(j_t)'r_t\big\} - \phi(i_t)'r_t\Big) \\ &= r_t + \gamma_t \phi(i_t)\Big(\big(T(\Phi r_t)\big)(i_t) - \phi(i_t)'r_t\Big) + \gamma_t \phi(i_t) w_t, \end{aligned}$$

where

$$(TQ)(i) = \sum_{j=1}^{n} p_{ij}\Big(g(i,j) + \alpha \min\big\{h(j), Q(j)\big\}\Big),$$

and

$$\begin{aligned} w_t = g(i_t, j_t) &+ \alpha \min\big\{h(j_t), \phi(j_t)'r_t\big\} \\ &- \sum_{j=1}^{n} p_{i_t j}\Big(g(i_t, j) + \alpha \min\big\{h(j), \phi(j)'r_t\big\}\Big), \end{aligned}$$

is a zero mean noise term.

Let D be the diagonal matrix associated with the steady-state probabilities $\pi(i)$ and consider the norm $\|\cdot\|_D$. We claim that the operator T is a contraction with respect to this norm. Indeed, for any two vectors Q and $\overline{Q}$, we have

$$\begin{aligned} \big|(TQ)(i) - (T\overline{Q})(i)\big| &\le \alpha \sum_{j=1}^{n} p_{ij}\big|\min\{h(j), Q(j)\} - \min\{h(j), \overline{Q}(j)\}\big| \\ &\le \alpha \sum_{j=1}^{n} p_{ij}\big|Q(j) - \overline{Q}(j)\big|, \end{aligned}$$

or

$$|TQ - T\overline{Q}| \le \alpha P|Q - \overline{Q}|,$$

where we use the notation $|x|$ to denote a vector whose components are the absolute values of the components of x. Hence,

$$\|TQ - T\overline{Q}\|_D \le \alpha \big\|P|Q - \overline{Q}|\,\big\|_D \le \alpha\|Q - \overline{Q}\|_D,$$

where the last step used the fact that $\|PJ\|_D \le \|J\|_D$ for every vector J (Lemma 6.4 in Section 6.3.3). We conclude that T has the claimed contraction property.

If we now assume that the states i_t are independently sampled according to the steady-state distribution, we see that the assumptions of Prop. 6.9 are satisfied and the Q-learning algorithm converges with probability 1, except for the presence of the additional term w_t. Fortunately, this additional noise term does not pose any difficulties. Going back to the proof of Prop. 6.9, we see that the noise term w_t being zero-mean does not affect the expected direction of motion $E[s_t \mid r_t]$ and we still have a descent direction on the average. The variance of w_t is easily seen to be proportional to $1 + \|r_t\|^2$. Hence, Prop. 6.9 does extend and covers the present algorithm as well. In addition, the error bounds of Prop. 6.10 also apply. We summarize this discussion in the following result.

Proposition 6.11: Suppose that the stepsizes γ_t are nonnegative, deterministic, and satisfy $\sum_{t=0}^{\infty} \gamma_t = \infty$ and $\sum_{t=0}^{\infty} \gamma_t^2 = \infty$. Suppose also that the states i_t are independently chosen according to the steady-state probabilities $\pi(i)$ and that the columns of the matrix Φ [with rows $\phi(i)'$] are linearly independent. Then, the sequence r_t generated by the Q-learning algorithm (6.81) converges with probability 1 and the limit r^* satisfies

$$\|\Phi r^* - Q^*\|_D \leq \frac{1}{1-\alpha} \|\Pi Q^* - Q^*\|_D,$$

where Π is the projection with respect to the norm $\|\cdot\|_D$ and Q^* is the vector of optimal Q-factors.

In practice, it is difficult to construct independent samples from the steady-state distribution of the (unstopped) Markov chain, unless one simulates the Markov chain sufficiently long. Carrying out an independent long simulation to generate each random sample would be too demanding computationally. Instead, the most natural version of the algorithm is the following. Carry out a single, infinitely long simulation of the (unstopped) Markov chain, let i_t be the sequence of visited states, and let $j_t = i_{t+1}$. Then, at each time t, update r_t according to the update rule in Eq. (6.81). For this version, the sample states i_t are obviously dependent and a key assumption of Prop. 6.9 is violated. Nevertheless, this variant is guaranteed to converge, with probability 1, to the same limit as the variant based on independent identically distributed samples. One approach for obtaining such a result rests on the averaging ideas discussed in Section 4.4 and used in the convergence proof of on-line TD(λ) in Section 6.3.3. In contrast to these earlier results, the expected update direction involves the nonlinear DP operator T and is a nonlinear function of the parameter vector r. Thus, Prop. 4.8 of Section 4.4 is not enough to establish convergence, but more powerful results from [BMP90] will do. In an alternative, more elementary approach to proving convergence, we may add ("lump") all the updates that take place between successive visits to a fixed reference state. Conditional on the current time t, the direction of a lumped update that starts at time t is statistically independent from previous lumped updates. Using calculations similar to those in the proof of Prop. 6.9, this is a descent direction with respect to the potential function $\|r - r^*\|_2^2$ and convergence follows.

6.9 VALUE ITERATION WITH REPRESENTATIVE STATES

Suppose that a feature vector $\phi(i)$ has been associated with each state i. If the set of all possible feature vectors $\phi(i)$ is rather small, we can lump together all states sharing the same feature vector and use the aggregation method described in Section 6.7. However, the set of possible feature vectors is often huge making it impossible to associate a separate value with each one of them. This leads us to approximation architectures involving fewer parameters such as the linear architecture

$$\tilde{J}(i,r) = \phi(i)'r = \sum_{k=1}^{K} \phi_k(i)r(k),$$

where $\phi(i) = \big(\phi_1(i), \ldots, \phi_K(i)\big)$. In this section, we discuss such a linear approximation architecture, with some additional special structure, which leads to demonstrably convergent algorithms.

Let there be a total of $n+1$ states with state 0 being, as usual, a zero-cost absorbing state. We single out K states ($K < n$) considered to be sufficiently representative of the entire state space. By possibly renumbering the states, we can and will assume that the representative states are the states $1, \ldots, K$. For these states, we use a lookup table representation, involving the variables $\tilde{J}(k)$, $k = 1, \ldots, K$. For the remaining states, we introduce a linear parametrization of the form

$$\tilde{J}(i) = \sum_{k=1}^{K} \theta_k(i)\tilde{J}(k), \qquad i = K+1, \ldots, n, \tag{6.82}$$

where the $\theta_k(i)$ are some fixed nonnegative coefficients that satisfy

$$\sum_{k=1}^{K} \theta_k(i) = 1, \qquad i = K+1, \ldots, n.$$

Thus, the cost-to-go $\tilde{J}(i)$ of any state $i > K$ is expressed as a weighted average of the costs-to-go $\tilde{J}(k)$ of the representative states. The coefficient $\theta_k(i)$ may be viewed as a measure of similarity or proximity of state i to the representative state k. This approximation architecture is very similar to what has been traditionally used in the discretization or finite-element approximation of continuous-state stochastic control problems [KuD92], but can also be applied to large scale discrete problems, as long as a meaningful set of representative states can be generated. The example that follows is meant to amplify the connection with continuous-state problems.

Example 6.16

Suppose that each state i can be identified with a point x_i in two-dimensional Euclidean space. (The discussion below extends to higher dimensions in an obvious manner.) Let us form a triangulation of the state space. That is, we form a set of triangles with disjoint interiors such that every state lies in one of these triangles. Furthermore, we require that every vertex of each triangle is one of the states, and we let the vertices be our representative states. Then, every nonrepresentative state x_i lies in some triangle with vertices x_{i_1}, x_{i_2}, x_{i_3}, and there exist nonnegative coefficients $\theta_{i_1}(i)$, $\theta_{i_2}(i)$, and $\theta_{i_3}(i)$, that sum to 1, and such that

$$x_i = \theta_{i_1}(i)x_{i_1} + \theta_{i_2}(i)x_{i_2} + \theta_{i_3}(i)x_{i_3}.$$

The parametric representation we have adopted is of the form

$$\tilde{J}(i) = \theta_{i_1}(i)\tilde{J}(i_1) + \theta_{i_2}(i)\tilde{J}(i_2) + \theta_{i_3}(i)\tilde{J}(i_3).$$

In effect, we are considering cost-to-go functions that are linear in each triangle and, therefore, completely determined by the values at the triangle vertices.

The algorithm we have in mind updates the coefficients $\tilde{J}(k)$, $k = 1,\ldots,K$, by carrying out value iterations at the representative states, and uses the parametric representation in Eq. (6.82) to obtain the necessary costs-to-go at the nonrepresentative states. Mathematically, the algorithm is of the form

$$\tilde{J}_{t+1}(i) = \min_{u\in U(i)} \sum_{j=0}^{n} p_{ij}(u)\left(g(i,u,j) + \sum_{k=1}^{K} \theta_k(j)\tilde{J}_t(k)\right), \qquad i = 1,\ldots,K. \tag{6.83}$$

If this algorithm converges, then the values $\tilde{J}(i)$, $i = 1,\ldots,K$, obtained in the limit satisfy

$$\begin{aligned} \tilde{J}(i) &= \min_{u\in U(i)} \sum_{j=0}^{n} p_{ij}(u)\big(g(i,u,j) + \tilde{J}(j)\big) \\ &= \min_{u\in U(i)} \sum_{j=0}^{n} p_{ij}(u)\left(g(i,u,j) + \sum_{k=1}^{K} \theta_k(j)\tilde{J}(k)\right), \qquad i = 1,\ldots,K. \end{aligned} \tag{6.84}$$

Regarding the implementation of the algorithm, if a representative state has a large number of possible successor states [states j for which $p_{ij}(u)$ is positive], then the outer summation over j in the right-hand side of Eq. (6.83) (which is really an expectation over j) may have to be replaced by an estimate obtained by Monte Carlo simulation.

Relation to an Auxiliary Problem

Equation (6.84) can be interpreted as the Bellman equation for an auxiliary stochastic shortest path problem with state space $\{0, 1, \ldots, K\}$, defined as follows. Given a current state $i \leq K$, we make a decision u and choose some j according to the probabilities $p_{ij}(u)$. If $j \leq K$, then j is our next state. If $j > K$, the next state k is chosen at random according to the probabilities $\theta_k(j)$. It is seen that the algorithm defined by Eq. (6.83) is simply the ordinary value iteration algorithm for the auxiliary problem. Consequently, as long as the auxiliary problem satisfies the usual assumptions (there exists a proper policy and every improper policy has infinite cost for some initial state), the algorithm is guaranteed to converge to the optimal cost-to-go function $\big(\tilde{J}^*(1), \ldots, \tilde{J}^*(K)\big)$ for the auxiliary problem, which is the unique solution of the system of equations (6.84).

As in Section 6.7, stronger results are possible under some more assumptions on the problem structure. For example, for discounted problems, convergence does not require any assumptions on the nature of the different policies. Furthermore, error bounds similar to the ones in Prop. 6.8 are obtained by following more or less the same lines of argument (see Tsitsiklis and Van Roy [TsV96a]).

6.10 BELLMAN ERROR METHODS

One possibility for approximation of the optimal cost by a function $\tilde{J}(i, r)$, where r is a parameter vector, is based on minimizing the error in Bellman's equation; for example, by solving the problem

$$\min_r \sum_{i \in \tilde{S}} \left(\tilde{J}(i, r) - \min_{u \in U(i)} \sum_j p_{ij}(u)\big(g(i, u, j) + \tilde{J}(j, r)\big) \right)^2, \tag{6.85}$$

where $\tilde{S}$ is a suitably chosen subset of "representative" states. Equivalently, if we define the error $D(i, r)$ in Bellman's equation by

$$D(i, r) = \min_{u \in U(i)} \sum_j p_{ij}(u)\big(g(i, u, j) + \tilde{J}(j, r)\big) - \tilde{J}(i, r),$$

we are dealing with the problem

$$\min_r \sum_{i \in \tilde{S}} D^2(i, r).$$

This minimization may be attempted by using some type of gradient or Gauss-Newton method; some possibilities will be discussed shortly. We observe that if $\tilde{S}$ is the entire state space and if the cost in the problem (6.85) can be brought down to zero, then $\tilde{J}(i,r)$ solves Bellman's equation and we have $\tilde{J}(i,r) = J^*(i)$ for all i. The set $\tilde{S}$ of representative states may be selected by means of regular or random sampling of the state space, or by using simulation to help us focus on the more significant parts of the state space.

The cost function in the problem (6.85) is the sum of several cost terms. We may therefore use the incremental gradient method which, instead of computing the gradient of the full cost function, it bases each update on the gradient of a single cost term. The resulting update equation is given by

$$\begin{aligned} r :&= r - \gamma D(i,r)\nabla D(i,r) \\ &= r - \gamma D(i,r)\Big(\sum_j p_{ij}(\overline{u})\nabla\tilde{J}(j,r) - \nabla\tilde{J}(i,r)\Big), \end{aligned} \tag{6.86}$$

where $\overline{u}$ is given by

$$\overline{u} = \arg \min_{u\in U(i)} \sum_j p_{ij}(u)\big(g(i,u,j) + \tilde{J}(j,r)\big),$$

and γ is a stepsize parameter. The method should perform many such iterations at each of the representative states. One possibility is to cycle through the set of representative states in some order, which may change from one cycle to the next, as discussed in Section 3.2. In a second alternative, we can let $\tilde{S}$ be the entire state space and choose the state at which to carry out an update at random according to the uniform distribution; in that case, the incremental gradient method is a stochastic gradient method, as discussed in Example 4.4 of Section 4.2. Finally, random sampling under distributions other than the uniform is possible. For example if state i is chosen for the next update with probability $v(i)$, then the expected update direction is equal to

$$-\sum_i v(i)D(i,r)\nabla D(i,r),$$

which indicates that we are dealing with a stochastic gradient algorithm for the weighted Bellman error criterion

$$\sum_i v(i)D^2(i,r).$$

Note that in iteration (6.86) we approximate the gradient of the term

$$\min_{u\in U(i)} \sum_j p_{ij}(u)\big(g(i,u,j) + \tilde{J}(j,r)\big) \tag{6.87}$$

by

$$\sum_j p_{ij}(\overline{u})\nabla \tilde{J}(j,r),$$

which can be shown to be correct when the above minimum is attained at a unique $\overline{u} \in U(i)$. The set of values of r for which $D(i,r)$ is nondifferentiable is typically a zero volume set and, in the presence of noise, one can often safely assume that the algorithm stays clear of points of nondifferentiability. On the other hand, some of the key assumptions used in the convergence analysis of Ch. 3 are violated and the convergence properties of the iteration (6.86) should be analyzed using the theory of nondifferentiable optimization methods. One possibility to avoid this complication is to replace the nondifferentiable term (6.87) by a smooth approximation (see [Ber82b], Ch. 3, or [Ber95b], Section 1.10).

6.10.1 The Case of a Single Policy

The special case where we want to approximate the cost function of a given policy μ is particularly interesting. For any given state i that has been chosen for an update, the iteration (6.86) takes the form

$$\begin{aligned} r :&= r - \gamma E\big[d(i,j,r) \mid i,\mu\big] E\big[\nabla d(i,j,r) \mid i,\mu\big] \\ &= r - \gamma E\big[d(i,j,r) \mid i,\mu\big] \Big(E\big[\nabla \tilde{J}(j,r) \mid i,\mu\big] - \nabla \tilde{J}(i,r)\Big), \end{aligned} \tag{6.88}$$

where

$$d(i,j,r) = g\big(i,\mu(i),j\big) + \tilde{J}(j,r) - \tilde{J}(i,r),$$

and $E[\cdot \mid i,\mu]$ denotes expected value over j using the transition probabilities $p_{ij}\big(\mu(i)\big)$. There is a simpler version of iteration (6.88) that does not require averaging over the successor states j. In this version, the two expected values in iteration (6.88) are replaced by two independent single sample values. In particular, r is updated by

$$r := r - \gamma d(i,j,r)\big(\nabla \tilde{J}(\overline{j},r) - \nabla \tilde{J}(i,r)\big), \tag{6.89}$$

where j and $\overline{j}$ correspond to two independent transitions starting from i. It is necessary to use two independently generated states j and $\overline{j}$ in order for the expected value (over j and $\overline{j}$, given i) of the product

$$d(i,j,r)\big(\nabla \tilde{J}(\overline{j},r) - \nabla \tilde{J}(i,r)\big),$$

to be equal to the term

$$E\big[d(i,j,r) \mid i,\mu\big]\Big(E\big[\nabla \tilde{J}(j,r) \mid i,\mu\big] - \nabla \tilde{J}(i,r)\Big)$$

appearing in the right-hand side of Eq. (6.88).

If the state i at which an update is carried out is chosen randomly at each iteration, the incremental gradient method (6.88) can be viewed as a stochastic gradient method, as discussed in the preceding subsection. By replacing the expectations in Eq. (6.88) by random samples, an additional noise source is introduced, but the expected direction of update is not changed. For that reason, the method (6.89) is also a stochastic gradient method.

6.10.2 Approximation of the Q-Factors

The algorithm (6.86) requires a model of the system and the algorithm in the last subsection deals with the case of a fixed policy. If we wish to construct an optimal (or close to optimal) policy but a model of the system is unavailable or is too complex to be useful, we need to work in terms of approximate Q-factors. We are thus led to consider versions of the above iterations that update Q-factor approximations rather than cost-to-go approximations.

Let us introduce an approximation $\tilde{Q}(i,u,r)$ to the optimal Q-factor $Q^*(i,u)$, where r is a parameter vector. Bellman's equation for the Q-factors is given by [cf. Eq. (5.58) in Section 5.6]

$$Q^*(i,u) = \sum_j p_{ij}(u)\Big(g(i,u,j) + \min_{v\in U(j)} Q^*(j,v)\Big).$$

In analogy with problem (6.85), we determine the parameter vector r by solving the least squares problem

$$\min_r \sum_{(i,u)\in\tilde{V}} \left(\tilde{Q}(i,u,r) - \sum_j p_{ij}(u)\Big(g(i,u,j) + \min_{v\in U(j)} \tilde{Q}(j,v,r)\Big)\right)^2, \tag{6.90}$$

where $\tilde{V}$ is a suitably chosen subset of "representative" state-control pairs. The analog of the incremental gradient methods (6.86) and (6.88) is given by

$$\begin{aligned} r &:= r - \gamma E\big[d^u(i,j,r) \mid i,u\big] E\big[\nabla d^u(i,j,r) \mid i,u\big] \\ &= r - \gamma E\big[d^u(i,j,r) \mid i,u\big] \left(\sum_j p_{ij}(u)\nabla\tilde{Q}(j,\overline{u},r) - \nabla\tilde{Q}(i,u,r)\right), \end{aligned}$$

where $d^u(i,j,r)$ is given by

$$d^u(i,j,r) = g(i,u,j) + \min_{v\in U(j)} \tilde{Q}(j,v,r) - \tilde{Q}(i,u,r),$$

$\overline{u}$ is obtained by

$$\overline{u} = \arg\min_{v\in U(j)} \tilde{Q}(j,v,r),$$

and γ is a stepsize parameter. In analogy with Eq. (6.89), the two-sample version of this iteration is given by

$$r := r - \gamma d^u(i,j,r)\big(\nabla\tilde{Q}(\overline{j},\overline{u},r) - \nabla\tilde{Q}(i,u,r)\big),$$

where j and $\overline{j}$ are two states independently generated from i according to the transition probabilities corresponding to u, and

$$\overline{u} = \arg\min_{v\in U(j)} \tilde{Q}(\overline{j},v,r).$$

6.10.3 Another Variant

Note that the iteration (6.86), which was based on approximation of the optimal cost-to-go, does not lead to a two-sample algorithm, and this is one advantage of using Q-factor approximation. The point is that it is possible to use single-sample or two-sample approximations in gradient-like methods for terms of the form $E\big[\min\{\cdot\}\big]$, such as the one appearing in Eq. (6.90), but not for terms of the form $\min\big\{E[\cdot]\big\}$, such as the one appearing in Eq. (6.85).

For another example to which the two-sample approximation idea applies, suppose that we are dealing with a Bellman equation of the form

$$J(i) = \sum_y p(y) \min_{u\in U(i)} \Big(g(i,y,u) + J\big(f(i,y,u)\big)\Big),$$

where $p(\cdot)$ is a probability mass function over a finite set of possible values for y, and $f(i,y,u)$ is a successor state obtained deterministically from i, y, and u. As seen in Section 2.4, such an equation arises in some problems involving systems with uncontrollable state components (see Examples 2.3-2.5 in Section 2.4). This leads us to the optimization problem

$$\min_r \sum_{i\in\tilde{S}} \left(\tilde{J}(i,r) - \sum_y p(y) \min_{u\in U(i)} \Big(g(i,y,u) + \tilde{J}\big(f(i,y,u),r\big)\Big)\right)^2,$$

where $\tilde{S}$ is a suitably chosen set of "representative" states. Because this problem involves a term of the form $E\big[\min\{\cdot\}\big]$, a two-sample gradient-like method is possible. It has the form

$$r := r - \gamma d(i,y,r)\Big(\nabla\tilde{J}\big(f(i,\overline{y},\overline{u}),r\big) - \nabla\tilde{J}(i,r)\Big),$$

where y and $\overline{y}$ are two independent random samples drawn according to the distribution $p(\cdot)$,

$$d(i,y,r) = \min_{u\in U(i)} \Big(g(i,y,u) + \tilde{J}\big(f(i,y,u),r\big)\Big) - \tilde{J}(i,r),$$

and

$$\overline{u} = \arg\min_{v\in U(i)} \Big(g(i,\overline{y},v) + \tilde{J}\big(f(i,\overline{y},v),r\big)\Big).$$

6.10.4 Discussion and Related Methods

The approximation methods discussed so far are easily generalized to the case of discounted problems; the only change required is to replace $\tilde{J}(j,r)$ throughout by $\alpha\tilde{J}(j,r)$. An extension is also possible to the case of average cost problems and will be discussed in Ch. 7.

How good are the policies that result from Bellman error minimization? If we have a discounted problem, with discount factor α, and if the method converges to some J that satisfies $\|J - TJ\|_\infty \leq \epsilon$, it is easily shown that $\|J - J^*\|_\infty \leq \epsilon/(1-\alpha)$. However, the method of this section is geared towards minimizing the Euclidean norm of the error vector $J - TJ$. Whether and under what circumstances this will also lead to a small value of $\|J - TJ\|_\infty$ is unclear.

The cost function (6.85) minimized by Bellman error methods can have multiple local minima (with the exception of problems involving a single policy and a linear parametrization). Thus, even if a very good fit is possible, there is no guarantee that an incremental gradient method will converge to it. In fact, it is even possible that the approximation architecture can represent J^* exactly, and the optimal cost in the problem (6.85) is zero, but the method converges to some nonoptimal local minimum with nonzero cost.

A form of the chattering phenomenon discussed in Section 6.4.2 may also occur when the Bellman equation error method is used for the case of multiple policies. For example, there may exist a local minimum $\bar{r}$ of the cost function

$$\sum_{i\in\tilde{S}}\left(\tilde{J}(i,r) - \min_{u\in U(i)}\sum_{j} p_{ij}(u)\big(g(i,u,j) + \tilde{J}(j,r)\big)\right)^2,$$

that lies on a boundary of the greedy partition, so that there is more than one policy that is greedy with respect to $\tilde{J}(\cdot,\bar{r})$. If a gradient method, used to minimize the Bellman equation error, happens to converge to $\bar{r}$, then chattering between these multiple greedy policies is likely to occur.

There is a close relationship between Bellman error methods and TD(0) (cf. the discussion in Section 6.3.2). For example, in the single policy case, TD(0) involves the same update equation as Eq. (6.89), except that the term $\nabla\tilde{J}(\bar{j},r)$ is omitted. When that term is included, convergence is much more robust because we are dealing with a gradient-based method. When that term is omitted, as in TD(0), the method can diverge but whenever it does converge, convergence is often faster. For this reason, Baird has proposed [Bai95] combining the best aspects of the two methods, e.g., by using an update equation of the form

$$r := r - \gamma d(i,j,r)\big(\delta\nabla\tilde{J}(\bar{j},r) - \nabla\tilde{J}(i,r)\big),$$

where δ is a constant between zero and one, to be chosen experimentally. One would like δ to be sufficiently large so as to preserve the descent property of the update direction, but also small enough so that some of the accelerated convergence of TD(0) is present. Alternatively, δ could be changed adaptively as the algorithm progresses. For example, it could be set initially to a small value, so as to obtain rapid convergence, and then increased towards 1 in order to enhance the accuracy of the final approximation.

6.11 CONTINUOUS STATES AND THE SLOPE OF THE COST-TO-GO

In many problems of practical importance, the states can be associated with grid points in some Euclidean space, and state transitions are only possible between neighboring states. This is the case, for example, when dealing with a discretization of a continuous-time problem. In such problems, the slope of the cost-to-go function becomes important, and we are led to some new methods that are briefly explored in this section.

As a motivating example, if the state is an n-dimensional vector indicating the number of customers in each of n queues, transitions usually correspond to a customer arriving, departing, or moving to a different queue, which means that at most two components of the state will be incremented or decremented by one. For a more concrete illustration, suppose that we have n_1 customers in queue 1, n_2 customers in queue 2, and that the optimal cost-to-go function J^* is known. If a new customer arrives who can be routed to either queue, a decision can be made by comparing $J^*(n_1+1, n_2)$ to $J^*(n_1, n_2+1)$. If only an estimate $\tilde{J}$ of J^* is available, a good decision can be made provided that $\tilde{J}(n_1+1, n_2) - \tilde{J}(n_1, n_2+1)$ is close to $J^*(n_1+1, n_2) - J^*(n_1, n_2+1)$. It is therefore desirable that $\tilde{J}$ captures correctly the slope of J^* in the vicinity of (n_1, n_2), but there is no reason to require that $\tilde{J}$ be close to J^*. Thus, the slope of J^* can be more important than J^* itself.

In order to amplify the importance of the slope of J^*, let us consider a continuous-time, continuous-state, deterministic optimal control problem of the form

$$
\begin{aligned}
\text{minimize} \quad & \int_0^\infty g(x_t, u_t)\, dt \\
\text{subject to} \quad & \frac{dx_t}{dt} = f(x_t, u_t), \\
& u_t \in U,
\end{aligned}
$$

where x_t and u_t belong to Euclidean spaces $\Re^n$ and $\Re^m$, respectively. If we let δ be a small time increment and discretize time, we are led to the problem of minimizing

$$\delta \sum_{t=0}^{\infty} g(x_t, u_t),$$

subject to

$$x_{t+1} = x_t + \delta f(x_t, u_t).$$

The Bellman equation for the latter problem is

$$J^*(x) = \min_{u \in U} \Big(\delta g(x, u) + J^*\big(x + \delta f(x, u)\big) \Big).$$

We approximate $J^*\big(x + \delta f(x, u)\big)$ by $J^*(x) + \delta \nabla_x J^*(x)' f(x, u)$, and divide by δ to arrive at the *Hamilton-Jacobi equation*

$$0 = \min_{u \in U} \big(g(x, u) + \nabla_x J^*(x)' f(x, u) \big),$$

which is a continuous-time analog of the Bellman equation. Evidently, u should be chosen so as to minimize $g(x, u) + \nabla_x J^*(x)' f(x, u)$, and it is clear that only the gradient of J^* matters. (Note that the argument we just gave is rather informal, bypassing several technical issues, such as the differentiability of J^*, for which we have to refer the reader to the technical literature; see e.g., Fleming and Rishel [FlR75].)

To gain some more insight, let us follow the steps involved in policy iteration. Given a policy μ, the resulting cost-to-go function J^μ satisfies

$$0 = g\big(x, \mu(x)\big) + \nabla_x J^\mu(x)' f\big(x, \mu(x)\big), \tag{6.91}$$

(this is the Hamilton-Jacobi equation, specialized to the single policy case), and a policy update leads to the new policy

$$\overline{\mu}(x) = \arg \min_{u \in U} \big(g(x, u) + \nabla_x J^\mu(x)' f(x, u) \big).$$

It is then clear that approximate policy evaluation should strive to learn the function $\nabla_x J^\mu$ rather than J^μ.

Learning the Slope Under a Fixed Policy

Suppose that we are using some form of approximate policy iteration and that we have fixed the current policy μ. Furthermore, suppose that due to reasons such as the ones just discussed, we wish to modify our training method so that it tries to approximate the function $\nabla_x J^\mu$. We will discuss here a number of possibilities.

The first question is related to the nature of the approximation architecture. With the state x assumed n-dimensional, $\nabla_x J^\mu(x)$ is also n-dimensional which might suggest a general approximation architecture with n outputs instead of one (e.g., a multilayer perceptron with n units at the output layer). However, the fact that $\nabla_x J^\mu$ is the gradient of a scalar function contains useful information that should be exploited. The most natural way of doing that is to start with a parametric representation $\tilde{J}(x,r)$ of $\tilde{J}^\mu(x)$ and then view $\nabla_x \tilde{J}(x,r)$ as the parametric representation of $\nabla_x J^\mu(x)$.

A second question is related to the availability of training data based on which the parameter vector r is to be tuned. In this respect, we observe from Eq. (6.91), that $-g\big(x,\mu(x)\big)$ provides us with training data for learning the directional derivative $\nabla_x J^\mu(x)' f\big(x,\mu(x)\big)$. Suppose that $g\big(x,\mu(x)\big)$ and $f\big(x,\mu(x)\big)$ have become available, possibly through simulation, at a collection X of representative elements x of the state space. We are then led to the least squares problem

$$\min_r \sum_{x\in X} \Big(g\big(x,\mu(x)\big) + \nabla_x \tilde{J}(x,r)' f\big(x,\mu(x)\big)\Big)^2,$$

which can be solved using the incremental gradient method

$$r := r - \gamma \nabla_r \Big(\nabla_x \tilde{J}(x,r)' f\big(x,\mu(x)\big)\Big) \Big(g\big(x,\mu(x)\big) + \nabla_x \tilde{J}(x,r)' f\big(x,\mu(x)\big)\Big). \tag{6.92}$$

It turns out that this method is essentially the same as the Bellman error method for the case of a single policy (cf. Section 6.10.1). We compare with the two-sample gradient method (6.89) and note that the term $g\big(x,\mu(x)\big) + \nabla_x \tilde{J}(x,r)' f\big(x,\mu(x)\big)$ plays the role of the temporal difference $d(i,j,r)$ whereas $\nabla_r\big(\nabla_x \tilde{J}(x,r)' f\big(x,\mu(x)\big)\big)$ can be viewed as the limit of $\nabla_r \tilde{J}(\bar{j},r) - \nabla_r \tilde{J}(i,r)$. In conclusion, the objective of Bellman error methods is very similar to the goal of approximating $\nabla_x J^\mu$. Similar arguments also apply to the multiple policy case.

The method we have just described is trained on the basis of the directional derivatives of J^μ along the direction of motion, which is $f\big(x,\mu(x)\big)$. It is plausible that a better approximation will result if we use training data for other directions as well. One possibility for generating such training data is to simulate two trajectories x_t and y_t, starting from two different nearby states x_0 and y_0, and use the difference of the simulated costs to form an estimate of the directional derivative $\nabla_x J^\mu(x)'(y_0 - x_0)$. In fact, by taking y_0 very close to x_0, it is sufficient to simulate one trajectory x_t and estimate y_t using the formula

$$y_t \approx x_t + \frac{\partial x_t}{\partial x_0}(y_0 - x_0).$$

Here, $\partial x_t / \partial x_0$ is an $n \times n$ matrix of partial derivatives that can be computed based on the formula

$$x_{t+1} = x_t + \delta f\big(x_t, \mu(x_t)\big), \tag{6.93}$$

and the chain rule.

To be more specific, suppose that we have generated K trajectories, indexed by i, and let x_0^i be the initial state of the ith trajectory. Furthermore, suppose that by proper application of the chain rule, we have calculated $\nabla_x J^\mu(x_0^i)$. We may then pose the least squares problem

$$\min_r \sum_{i=1}^{K} \left\| \nabla_x \tilde{J}(x_0^i, r) - \nabla_x J^\mu(x_0^i) \right\|^2,$$

which can be solved using the methods of Ch. 3.

Once this methodology is adopted for policy evaluation, it can be used in conjunction with optimistic policy iteration whereby each trajectory is simulated using a greedy policy, that is, a policy that at every state x minimizes over all u the expression

$$g(x, u) + \nabla_x \tilde{J}(x, r)' f(x, u),$$

where r is the current parameter vector. However, the convergence behavior of such a method has not been analyzed.

Let us interject here that estimates of the gradient $J^\mu(x_0)$ can also be obtained for certain types of stochastic systems, on the basis of a single simulation, using a technique known as Infinitesimal Perturbation Analysis; see, e.g., Ho and Cao [HoC91] or Glasserman [Gla91].

We have discussed a method for fitting $\nabla_x J^\mu(x)$ based on sample values of this function obtained by simulating complete trajectories, followed by the solution of a least squares problem. In that sense, the method is like TD(1) applied to the estimation of $\nabla_x J^\mu$. Methods analogous to TD(0) are also possible, but we do not pursue this subject any further.

If a model of the system (e.g., the form of the functions f and g) is not available, knowing the optimal cost-to-go function, or its gradient, is not enough for making good decisions. As should be clear from the Hamilton-Jacobi equation, we also need to learn the quantity $g(x, u) + \nabla_x J^*(x)' f(x, u)$. The latter quantity, when multiplied by δ, is identical to the disadvantages introduced in Section 6.6.2, hence the motivation for advantage updating methods.

We have argued in this section that estimating the gradient of J^* is sometimes more important than estimating J^* itself. One then asks whether there are any special circumstances under which an algorithm that learns J^* well is also guaranteed to learn $\nabla_x J^*$ well. In some cases this is true. For example, suppose that the optimal cost-to-go function is quadratic, say $J^*(x) = x'Ax + b'x + c$ for some A, b, and c. Suppose also that we use a quadratic approximation architecture of the form $\tilde{J}(x) = x'\tilde{A}x + \tilde{b}'x + \tilde{c}$, where $\tilde{A}$, $\tilde{b}$, and $\tilde{c}$ are the tunable parameters. If the approximation $\tilde{J}$ provides a good fit of J^*, the parameters in the approximation must be close to the true parameters. It follows that $\nabla_x \tilde{J}$ is a good approximation of

$\nabla_x J^*$, and we need not make an additional effort to learn $\nabla_x J^*$. However, this example is somewhat misleading, because the entire argument is based on the possibility of a perfect fit of J^* by the approximation architecture. In problems where a perfect fit is impossible, there is no a priori reason why $\nabla_x J^*$ will be well approximated; it is such problems that should be the litmus test of new algorithms.

Using an Action Network

When the current policy is represented by an approximation architecture $\tilde{\mu}(\cdot, v)$, where v is a parameter vector, we sometimes need to update the policy parameter v according to the formula

$$v := v - \gamma \nabla_v \tilde{\mu}(i, v) \frac{\partial}{\partial u} \sum_j p_{ij}(u)\big(g(i, u, j) + \tilde{J}(j, r)\big),$$

where i is the current state and where the partial derivative with respect to u is evaluated at $u = \tilde{\mu}(i, v)$ [cf. Eq. (6.51) in Section 6.4]. For simplicity, we are assuming here that u is scalar.

We wish to specialize this update to the continuous-time deterministic problem discussed in this section. The state i is replaced by x, the cost per stage $g(i, u, j)$ becomes $\delta g(x, u)$, the next state j becomes $x + \delta f(x, u)$, and $\tilde{J}(j, r)$ is approximated by $\tilde{J}(x, r) + \delta \nabla_x \tilde{J}(x, r)' f(x, u)$. With these substitutions, we end up with the update rule

$$v := v - \gamma \nabla_v \tilde{\mu}(x, v) \Big(\frac{\partial g}{\partial u}(x, u) + \nabla_x \tilde{J}(x, r)' \frac{\partial f}{\partial u}(x, u) \Big),$$

which is just a gradient iteration aimed at minimizing

$$g\big(x, \tilde{\mu}(x, v)\big) + \nabla_x \tilde{J}(x, r)' f\big(x, \tilde{\mu}(x, v)\big).$$

This update rule is sometimes described as a "backpropagation" through a sequence of three structures. Given a small change in the parameter v in the action network, $\nabla_v \tilde{\mu}(x, v)$ gives us the direction of change of the control variable; multiplication by $\partial f / \partial u$ provides us with the direction of change of the next state; finally, multiplication by $\nabla_x \tilde{J}$ gives us the change of the cost-to-go of the next state. To summarize, we are determining a direction along which v should be updated so as to improve the quality of the next state, as evaluated by its approximate cost-to-go. The derivatives $\nabla_v \tilde{\mu}(i, v)$ and $\nabla_x \tilde{J}(x, r)$ are easily computed for most approximation architectures; for example, if a multilayer perceptron is employed, these derivatives are obtained with an easy application of the chain rule. The derivatives $\partial g / \partial u$ and $\partial f / \partial u$ can be obtained from an analytical model, if available. Otherwise, one can simulate the system and train separate neural networks to construct models of these functions.

The above update rule for v is based on a limited lookahead, as we use $\tilde{J}$ to evaluate the state that results δ time units later. For deterministic systems, multistage or "deeper" lookahead is also possible without much additional effort. The deeper the lookahead, the smaller the importance of the "critic" $\tilde{J}$. In fact, if one uses infinite lookahead, that evaluates each action in terms of the exact cost to be accumulated in the long run, the critic $\tilde{J}$ can be eliminated altogether. One then recovers more traditional methods such as "backpropagation through time" [Hay94], that can be viewed as incremental gradient methods in policy space.

In more detail, the parameter v determines a policy $\tilde{\mu}(\cdot, v)$, which in turn determines a cost-to-go function $\hat{J}(x, v)$, defined by

$$\hat{J}(x, v) = J^{\tilde{\mu}(\cdot, v)}(x).$$

Note that $\hat{J}(x, v)$ is available through simulation and $\nabla_v \hat{J}(x, v)$ can be computed using the chain rule along the simulated trajectory. We can then use the update equation

$$v := v - \gamma \nabla_v \hat{J}(x, v),$$

which is an incremental gradient method for minimizing over v a cost function of the form $\sum_{x \in X} \hat{J}(x, v)$, where X is a set of representative states.

6.12 APPROXIMATE LINEAR PROGRAMMING

As discussed in Ch. 2, the optimal cost-to-go function J^* in stochastic shortest path problems can be obtained by formulating and solving the linear programming problem

$$\begin{aligned} &\text{maximize} \quad \sum_{i=1}^{n} v(i) J(i) \\ &\text{subject to} \quad J(i) \leq \sum_{j=0}^{n} p_{ij}(u)\big(g(i, u, j) + J(j)\big), \qquad \forall\ i,\ u \in U(i), \end{aligned} \tag{6.94}$$

where the $v(i)$ are arbitrary positive weights. When the state space is large, such a linear programming problem is difficult to handle because there is a large number of variables and a large number of constraints. We can develop, however, an approximate linear programming problem, based on a parametric representation of the function J. We restrict attention to linear parametrizations of the form

$$\tilde{J}(i, r) = \sum_{k=1}^{K} r(k) \phi_k(i) = \phi(i)' r, \tag{6.95}$$

where $r = \big(r(1), \ldots, r(K)\big)$ is a vector of free parameters, $\phi_k(i)$ is the kth feature of state i, and $\phi(i) = \big(\phi_1(i), \ldots, \phi_K(i)\big)$, is an associated feature vector. By substituting the functional form (6.95) in the problem (6.94), we obtain a linear programming problem in the variables $r(1), \ldots, r(K)$:

$$\begin{aligned} \text{maximize} \quad & \sum_{i=1}^{n} v(i) \sum_{k=1}^{K} r(k)\phi_k(i) \\ \text{subject to} \quad & \sum_{k=1}^{K} r(k)\phi_k(i) \leq \sum_{j=0}^{n} p_{ij}(u) \left(g(i,u,j) + \sum_{k=1}^{K} r(k)\phi_k(j) \right), \\ & \qquad\qquad \forall\ i,\ u \in U(i). \end{aligned} \tag{6.96}$$

Intuitively, the original problem (6.94) involves an optimization over a polyhedron P defined by the constraints in that problem. Equation (6.95) can be viewed as an additional constraint that restricts us to a K-dimensional subspace S. Then, problem (6.96) is essentially the same as problem (6.94) except that we are now optimizing over the smaller set $S \cap P$. If the true optimal cost-to-go function J^* can be exactly represented in the form (6.95), and therefore lies in S, the additional constraint is inconsequential and $\sum_{k=1}^{K} r^*(k)\phi_k(i) = J^*(i)$, where $r^*(1), \ldots, r^*(K)$ is an optimal solution of (6.96). If J^* lies close to the subspace S, it is easily shown that the optimal costs of the two linear programming problems are also close; however, the optimal solutions can be quite far and the quality of the resulting policies cannot be easily predicted.

The linear programming problem (6.96) is easier to solve than the original one because the number of variables to be optimized has been reduced to K. Unfortunately, the number of constraints is still large. This difficulty can be handled in two different, though related, ways that we discuss next.

Restriction to a Few Representative States

Suppose that we have selected a relatively small subset I of the state space consisting of states that are considered sufficiently representative. We may then approximate the problem (6.96) by the following:

$$\begin{aligned} \text{maximize} \quad & \sum_{i \in I} v(i) \sum_{k=1}^{K} r(k)\phi_k(i) \\ \text{subject to} \quad & \sum_{k=1}^{K} r(k)\phi_k(i) \leq \sum_{j=0}^{n} p_{ij}(u) \left(g(i,u,j) + \sum_{k=1}^{K} r(k)\phi_k(j) \right), \\ & \qquad\qquad \forall\ i \in I,\ u \in U(i). \end{aligned} \tag{6.97}$$

This problem can be solved using any general purpose method for linear programming.

Cutting Plane Method

We first discuss some issues related to the cost function in the problem (6.96). We note that the cost coefficients $\sum_{i=1}^{n} v(i)\phi_k(i)$ associated with each variable $r(k)$ are difficult to compute because a summation over the entire state space is needed. This difficulty can be easily bypassed by resorting to a Monte Carlo approach. Suppose that we are able to generate a sequence $i_1, \ldots, i_M$ of independent sample states that are uniformly distributed over the set of nonterminal states. We may then form the expression $(n/M)\sum_{\ell=1}^{M} v(i_\ell)\phi_k(i_\ell)$ and use it to approximate $\sum_{i=1}^{n} v(i)\phi_k(i)$. As M increases, this estimate is guaranteed to converge to the correct value of $\sum_{i=1}^{n} v(i)\phi_k(i)$.

Alternatively, if the weights $v(i)$ sum to 1, we may generate sample states $i_1, \ldots, i_M$, according to the probability distribution determined by these weights, and form the sum $(1/M)\sum_{\ell=1}^{M} \phi_k(i_\ell)$. With the probability distribution we are using, we have

$$E\big[\phi_k(i_\ell)\big] = \sum_{i=1}^{n} v(i)\phi_k(i),$$

which means that we have again an unbiased estimate of the desired cost coefficient; this estimate also converges to its mean value as M increases.

The effect of different choices for the weights $v(i)$ are far from understood. To appreciate the nature of this problem, we observe that by suitably choosing $v(i)$, we can get the cost coefficients $\sum_{i=1}^{n} v(i)\phi_k(i)$ in the reduced linear program (6.96) to be pretty arbitrary. Thus, a lot of experimentation may be needed to determine a "good" choice for the weights.

Let us now assume that, one way or another, we have determined the cost coefficients to be employed. We still have a problem with a large number of constraints, but problems of this type can be often handled efficiently using the cutting plane method, which is a general purpose algorithm for dealing with linear programming problems involving a large number of constraints. At present, there is very little experience with the linear programming approach outlined in this section, and further study and experimentation is needed.

6.13 OVERVIEW

In this chapter, we have introduced a large variety of methods, and most of them present us with a large number of further choices. It would be therefore proper to pause and take inventory of the available options. It would also be desirable to make an assessment of the relative merits of the various methods. In this respect, theory can be useful in providing insights and in indicating what kind of convergence behavior can be expected from any given method. However, a complete answer can only rely on practical experience and may often depend on the specifics of the problem at hand.

Unfortunately, there is no work in the literature that has attempted a systematic and exhaustive comparison of all solution approaches. Most of the available experience relies on variants of approximate policy iteration, both optimistic and nonoptimistic, as well as on Q-learning. Temporal difference methods figure prominently here. One of the reasons is that gradient-based methods such as TD(1) have some convergence guarantees but can be very slow. On the other hand, TD(λ) with $\lambda < 1$, is often surprisingly effective, despite the fact that it has less of a mathematical basis.

Some insights into the behavior of the different methods will be provided by the case studies to be presented in Ch. 8. Nevertheless, much more experimentation is needed before a complete picture starts to emerge, including, for example, more applications of approximate value iteration or linear programming.

We conclude by listing of a few different issues and choices, together with some brief comments.

(a) *Approximation architecture*
An approximation architecture must be chosen with due regard to whatever is known about the problem structure. Whenever possible, linear architectures are preferable, because they can be trained with faster and more reliable methods.

(b) *Representation of policies*
Depending on the nature of the action space, an action network may be needed, leading to a parametric class of policies. An action network tends to have some advantages whenever the action space is continuous, or whenever the minimization in the right-hand side of Bellman's equation is computationally prohibitive.

(c) *Model-based versus model-free methods*
Whenever a model of the system is available or can be estimated, it seems plausible that model-based methods will make better use of available information. In the absence of a model, however, a method based on Q-factors becomes necessary.

(d) *The value of λ in temporal difference methods*
Practical experience indicates that intermediate values of λ, between 0 and 1, often work best. This is particularly so when the cost-to-go samples obtained through simulation have a large variance. For any given problem, however, some experimentation with different values of λ is always called for.

(e) *The degree of optimism in approximate policy iteration*
In a sense, most approximate policy iteration methods are partially optimistic, since we cannot run an infinite number of iterations during policy evaluation. So, the choice of when to switch to a new policy is a matter of degree. Once more, this is something that calls for experimentation.

(f) *Exploration*
There are two somewhat separate issues here. First, one must explore all "important" regions of the state space, which requires either a lot of randomness in the problem itself or, else, avoiding strict adherence to a rigid policy. Alternatively, the freedom in initializing the different trajectories can be exploited. Second, in the absence of a model, one must also ensure that a rich enough set of actions is tried.

(g) *Deterministic problems*
Deterministic problems are special in that many quantities of interest can be computed exactly on the basis of a single trajectory, as opposed to Monte Carlo simulation. On the other hand, in the absence of randomness, the need for adequate exploration may become more pressing.

(h) *Learning the slope of the cost-to-go function*
If the states are elements of a Euclidean space and if transitions are mostly local, the slope of the cost-to-go function can be of great importance, and in such cases, one has the choice of using methods that explicitly attempt to provide a good fit of the slope.

Besides the above catalog, there are many other important issues, such as the choice of training algorithms (e.g., incremental gradient methods versus Kalman filters), the choice of stepsizes, the option of using rollout policies with or without multistage lookahead, etc. Any list of this type is bound to raise more questions than there are answers. Nevertheless, it is useful to be aware of the different possibilities and possess as much insight as theory can provide.

6.14 NOTES AND SOURCES

The methods discussed in this chapter involve several elements whose seeds can be found in early work in the fields of reinforcement learning, artificial intelligence, and control. However, different communities placed emphasis on different aspects of the problem, and due to the absence of interaction, it took until the late eighties for a coherent picture to start emerging. In what follows, we provide a very rough schematic description of the different intellectual threads involved.

Some of the central ideas in the field have their origins in the work of Samuel [Sam59], [Sam67], who built a checkers-playing program that made decisions based on multistage lookahead and an approximate cost-to-go function, represented as a linear combination of a number of features. (An element of "feature iteration" was also present, whereby some features would be eliminated and some would be added, out of a fixed pool, depending on their contribution.) Samuel's key innovation was that he trained the tunable parameters of the approximation architecture, using temporal differences to guide the direction of update of each parameter. (For some more recent works using methods related to those of Samuel, see Christensen and Korf [ChK86], and Holland [Hol86].)

In Samuel's work, the cost-to-go approximator was trained using suitable "reinforcement signals"; once this training was complete, obtaining good decisions was straightforward. This is to be contrasted to another trend in reinforcement learning in which a policy or an "action network" is the central element. The key question then is finding a promising way for training the parameters of the action network. It was gradually realized that a separate "critic network" (see e.g., Widrow, Gupta, and Maitra [WGM73]) could be useful for tuning the action network. The work of Barto, Sutton, and Anderson [BSA83] combined an action and a critic network in a meaningful fashion and used TD(λ) for training the critic network. As the field developed, it was eventually understood that the critic has to take the central role, whereas an action network is an auxiliary device that may or may not be necessary.

While most of the work we have just described was carried out without realizing the close connection with DP, Werbös took DP as the starting point and suggested a number of methods that make use of approximation to bypass the curse of dimensionality (see, e.g., [Wer77]). For example, "heuristic dynamic programming," as described in [Wer77], is closely related to the approximate value iteration method of Section 6.5, and the TD(0) method. Drawing upon ideas from the field of differential dynamic programming [JaM70], he also suggested methods aimed at learning the partial derivatives of the cost-to-go function ("dual heuristic programming").

A major development in the field occurred when Sutton [Sut84], [Sut88], formalized temporal difference methods, which made them amenable to mathematical analysis. Subsequently, the work of Watkins [Wat89] clarified the relation between temporal difference methods and DP, and introduced Q-learning. At this point, the understanding of the connection between reinforcement learning and DP had reached a degree of maturity, and was laid out in an influential overview paper by Barto, Bradtke, and Singh [BBS95] (which circulated widely as a report in 1993).

To the above account of the history of the field, we should add that the idea of approximate DP, using a parametric representation of the cost-to-go function, goes back to Bellman and his coworkers [BeD59], [BKK73]. Since then, approximate DP has been analyzed and applied extensively, with much of the research focusing on discretizations of continuous state spaces (see, e.g., Daniel [Dan76], Whitt [Whi78], [Whi79], Chow and Tsitsiklis [ChT91], Kushner and Dupuis [KuD92]). However, most of this work was targeted on low-dimensional problems and did not consider the possibility of making simulation an integral part of the methods. A comprehensive survey of many recent research directions in approximate DP is provided by Rust [Rus96].

6.1. Bounds on the performance of greedy policies (Prop. 6.1) have been known for some time, even though it is hard to find early references. Recent proofs have been given by Williams and Baird [WiB93], Singh and Yee [SiY94], and by Tsitsiklis and Van Roy [TsV96a]. The use of a rollout policy has been suggested by Tesauro in the context of backgammon, and has been experimentally evaluated by Tesauro and Galperin [TeG96].

6.2. The error bounds for approximate policy iteration (Props. 6.2-6.3) are due to the authors and are also given by Bertsekas in [Ber95a]. Example 6.4 showing the tightness of the error bounds is new.

6.3. Temporal difference methods, in conjunction with function approximation, are due to Sutton [Sut84], [Sut88]. Example 6.5 showing that $\lambda = 1$ may be preferable for the purpose of cost-to-go approximation is due to Bertsekas [Ber95c]. The divergent example for TD(0) with nonlinear parametrizations (Example 6.6) is due to Van Roy (unpublished).

For the discounted case, the convergence results and the error bounds for TD(λ) (Props. 6.4-6.5) are due to Tsitsiklis and Van Roy, whose results extend to the case of an infinite state space [TsV96b]. For stochastic shortest path problems the convergence results in Prop. 6.6 have been obtained by Tsitsiklis and Van Roy (unpublished), and by Gurvits (unpublished) who has also proved extensions to the case of an infinite state space. Some of the key ideas, including the use of a weighted quadratic norm, can be traced back to earlier work by Gurvits, Lin, and Hanson that dealt with the lookup table case [GLH94].

The importance of sampling states based on simulated trajectories has been repeatedly stressed by Sutton (see e.g., [Sut95]). The divergent example for TD(0) with linear parametrizations and arbitrary sampling is due to Tsitsiklis and Van Roy [TsV96a]. Other divergent examples have been constructed by Baird [Bai95] and Gordon [Gor95]. Further insights can be obtained by the divergent computational experiments of Boyan and Moore [BoM95].

An alternative policy evaluation method introduced and analyzed by

Bradtke and Barto [BrB96], which is called the "least squares temporal difference method," has been shown to converge in the case of linear parametrizations and the limit is the same as for TD(0). In particular, the error bounds of Prop. 6.5 apply for this method.

6.4. Optimistic policy iteration is due to Sutton [Sut88] and has been used in many applications. The analysis and the divergent examples in Section 6.4.1, as well as the explanation of the chattering phenomenon in Section 6.4.2, are new. Bradtke [Bra94] proves convergence for a partially optimistic policy iteration method, based on Q-factors, applied to a linear quadratic problem, for the case of a quadratic approximation architecture.

6.5. Approximate value iteration can be traced back to Bellman. Since then, it has been used and analyzed a lot, mostly within the context of discretizations of continuous-state problems. The divergent examples are due to Tsitsiklis and Van Roy [TsV96a].

6.6. Q-learning is due to Watkins [Wat89]. Advantage updating has been proposed by Baird [Bai93]. An application to a simple linear quadratic problem similar to Example 6.11 is described in the latter paper. Another application, involving a linear quadratic differential game has been carried out by Harmon, Baird, and Klopf [HBK94].

6.7. The development and the results in this section follow Tsitsiklis and Van Roy [TsV96a], [Van95]. Similar results appear in subsequent work by Gordon [Gor95].

Related convergence results have been obtained by Singh, Jaakkola, and Jordan [SJJ94], who consider the discounted case and assume that decisions are constrained to be constant within each partition, motivated from problems of imperfect information. They also provided an argument that Markov, as opposed to independent, sampling would be enough to guarantee convergence. Other work by the same authors [SJJ95] provides an extension to an approximation architecture that corresponds to "soft" state aggregation.

6.8. The main result in this section is essentially proved by Van Roy and Tsitsiklis [VaT96] (that paper only considers the case of the unweighted Euclidean norm, but the extension is not difficult). The application to Q-learning for optimal stopping problems is due to Tsitsiklis and Van Roy (unpublished).

6.9. The discussion in this section follows Tsitsiklis and Van Roy [TsV96a]. However, the method, and the idea of the convergence proof we have outlined is classical and is closely related to the finite-element methods overviewed by Kushner and Dupuis [KuD92].

6.10. Bellman error methods have been discussed by many authors, a relatively early reference being the paper by Schweitzer and Seidman [ScS85]. The two-sample variant is briefly mentioned by Harmon, Baird, and Klopf [HBK94], and is also discussed by Bertsekas [Ber95a]. In earlier work, Werbös [Wer90] showed that a single sample Bellman error method can produce biased results, but did not discuss a two-sample method. The combination of the gradient direction and the TD(0) update direction has been advocated by Baird [Bai95].

6.11. The discussion in the beginning of this section proceeds along the lines of Dayan and Singh [DaS96] and is closely related to the subject of advantage updating [Bai93].

There is a large body of literature on continuous-time optimal control problems, some of which deals with the technical issues that we did not address. See, for example, Athans and Falb [AtF66], Hestenes [Hes66], Bryson and Ho [BrH75], Fleming and Rishel [FlR75], and Bertsekas [Ber95a].

The importance of learning the slope of the cost-to-go function has been emphasized by Werbös who has proposed a number of algorithms with that objective in mind; see [Wer92a] and [Wer92b] for an overview.

6.12. The approximate linear programming method we have described is due to Schweitzer and Seidman [ScS85]. Trick and Zin [TrZ93] have applied large scale linear programming methods to obtain approximate solutions of DP problems arising from the discretization of continuous-state problems.

It is hard to predict,
especially about the future.
(Niels Bohr)

7

Extensions

Contents

In this chapter, we discuss various extensions of the NDP methodology developed so far. In particular, we consider average cost per stage DP problems and dynamic games. We will see that the NDP methods for these problems are less well developed than for discounted and stochastic shortest path problems. Still, however, there are some methods that can be tried with a good chance of success. In the last section of this chapter, we discuss issues of parallel computation.

7.1 AVERAGE COST PER STAGE PROBLEMS

The methodology discussed so far applies to problems where the optimal total expected cost is finite either because of discounting or because of a cost-free termination state that the system eventually enters. In many situations, however, discounting is inappropriate and there is no natural cost-free termination state. It is then often meaningful to optimize the average cost per stage starting from a state i, which is defined for any policy $\pi = \{\mu_0, \mu_1, \ldots\}$ by

$$J^\pi(i) = \lim_{N\to\infty} \frac{1}{N} E\left[\sum_{k=0}^{N-1} g\big(i_k, \mu_k(i_k), i_{k+1}\big) \;\Big|\; i_0 = i\right],$$

assuming that the limit exists.

It is important to keep in mind that the average cost per stage of a policy primarily expresses cost incurred in the long term. Costs incurred in the early stages do not matter since their contribution to the average cost per stage is reduced to zero as $N \to \infty$; that is,

$$\lim_{N\to\infty} \frac{1}{N} E\left[\sum_{k=0}^{K} g\big(i_k, \mu_k(i_k), i_{k+1}\big)\right] = 0, \tag{7.1}$$

for any K that is either fixed or is random and has a finite expected value. Consider now a stationary policy μ and two states i and j such that the system will, under μ, eventually reach j with probability 1 starting from i. It is then intuitively clear that the average costs per stage starting from i and from j cannot be different, since the costs incurred in the process of reaching j from i do not contribute essentially to the average cost per stage. More precisely, let $K_{ij}(\mu)$ be the first passage time from i to j under μ, that is, the first index k for which $i_k = j$ starting from $i_0 = i$ under μ (see Appendix B). Then the average cost per stage corresponding to initial condition $i_0 = i$ can be expressed as

$$J^\mu(i) = \lim_{N\to\infty} \frac{1}{N} E\left[\sum_{k=0}^{K_{ij}(\mu)-1} g\big(i_k, \mu(i_k), i_{k+1}\big)\right]$$
$$+ \lim_{N\to\infty} \frac{1}{N} E\left[\sum_{k=K_{ij}(\mu)}^{N-1} g\big(i_k, \mu(i_k), i_{k+1}\big)\right].$$

If $E\big[K_{ij}(\mu)\big] < \infty$ (which is equivalent to assuming that the system eventually reaches j starting from i with probability 1; see Appendix B), then it can be seen that the first limit is zero [cf. Eq. (7.1)], while the second limit is equal to $J^\mu(j)$. Therefore,

$$J^\mu(i) = J^\mu(j), \qquad \text{for all } i, j \text{ with } E\big[K_{ij}(\mu)\big] < \infty.$$

We conclude that if under a given policy there is a state that can be reached from all other states with probability 1, then the average cost per stage of the policy is independent of the initial state.

The preceding argument suggests that the optimal cost $J^*(i)$ should also be independent of the initial state i under normal circumstances. Indeed, it can be shown that

$$J^*(i) = J^*(j), \qquad \forall\ i, j,$$

under a variety of fairly natural assumptions that guarantee the accessibility of some state from the other states (see [Ber95a], Vol. II, Section 4.2). Our analysis will be based on such an assumption.

In what follows in this section, we first present, mostly without proofs, some of the basic results and algorithms relating to the average cost problem. We then discuss the corresponding NDP methodology. Unfortunately, this methodology and its underlying theory are not as complete as for stochastic shortest path and discounted problems. One reason is that Bellman's equation for the average cost problem does not involve a weighted maximum norm contraction. Nonetheless, some of the analysis relating to stochastic shortest path problems can be brought to bear on the average cost problem thanks to a connection that we now proceed to discuss.

7.1.1 The Associated Stochastic Shortest Path Problem

We will make a connection between the average cost problem and an associated stochastic shortest path problem. For this we will need the following assumption:

Assumption 7.1: One of the states $1, \ldots, n$, call it s, is such that for some integer $m > 0$, and for all initial states and all stationary policies, s is visited with positive probability at least once within the first m stages.

Assumption 7.1 can be shown to be equivalent to assuming that each stationary policy corresponds to a Markov chain with a single recurrent class (as defined in Appendix B), and that state s belongs to the recurrent class of each stationary policy.

To motivate the connection with the stochastic shortest path problem, consider a sequence of generated states, and divide it into cycles marked by successive visits to the state s. The first cycle includes the transitions from the initial state to the first visit to state s, and the kth cycle, $k = 2, 3, \ldots$, includes the transitions from the $(k-1)$st to the kth visit to state s. Each of the cycles can be viewed as a state trajectory of a corresponding stochastic shortest path problem with the termination state being essentially s. More precisely, this stochastic shortest path problem is defined by leaving unchanged all transition probabilities $p_{ij}(u)$ for $j \neq s$, by eliminating all transitions into state s, and by introducing an artificial termination state $\bar{s}$ to which we move from each state i with probability $p_{is}(u)$; see Fig. 7.1. Note that Assumption 7.1 implies that all policies are proper, which in turn implies the Assumptions 2.1 and 2.2 of Section 2.2 under which the results of Section 2.2 on stochastic shortest path problems were shown.

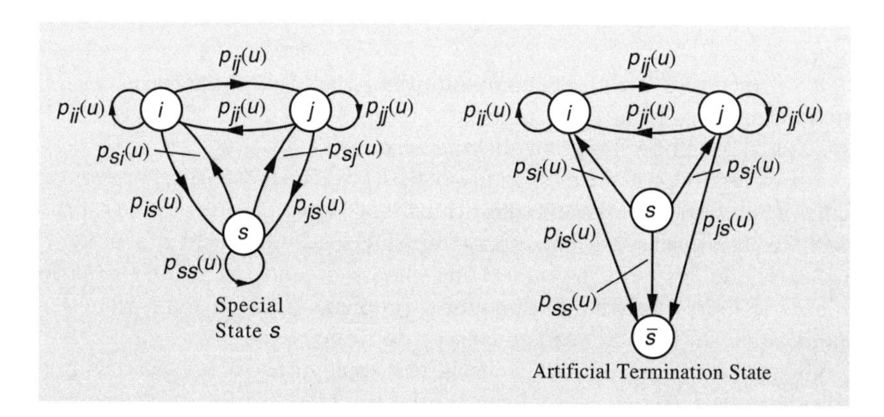

Figure 7.1: Transition probabilities for an average cost problem and its associated stochastic shortest path problem. The latter problem is obtained by introducing, in addition to $1, \ldots, n$, an artificial termination state $\bar{s}$ to which we move from each state i with probability $p_{is}(u)$, by eliminating all transitions into state s, and by leaving unchanged all other transition probabilities. In effect, a transition into s in the original problem corresponds to termination in the new problem.

We next argue that if we redefine the cost per stage to be

$$g(i, u, j) - \lambda^*,$$

where λ^* is the optimal average cost per stage, then the associated stochastic shortest path problem becomes essentially equivalent to the original average cost per stage problem. Even though λ^* is unknown, this transformation will be conceptually and analytically useful to us. In particular, Bellman's equation for the associated stochastic shortest path problem will be viewed as Bellman's equation for the original average cost per stage problem.

For a heuristic argument of why this is so, note that under all stationary policies there will be an infinite number of cycles marked by successive visits to s. From this, it can be conjectured (and it can also be shown) that the average cost problem amounts to finding a stationary policy μ that minimizes the average cycle cost

$$\lambda_\mu = \frac{C_{ss}^\mu}{N_{ss}^\mu}, \tag{7.2}$$

where

C_{ss}^μ : expected cost under μ starting from s up to the first return to s,

N_{ss}^μ : expected number of stages to return under μ to s starting from s.

The optimal average cost per stage λ^* starting from s, satisfies $\lambda^* \leq \lambda_\mu$ for all μ, so from Eq. (7.2) we obtain

$$C_{ss}^\mu - N_{ss}^\mu \lambda^* \geq 0, \tag{7.3}$$

with equality holding if μ is optimal. Thus, to attain an optimal μ, we must minimize over μ the expression $C_{ss}^\mu - N_{ss}^\mu \lambda^*$, which is the expected cost of μ starting from s in the associated stochastic shortest path problem with one-stage cost

$$g(i, u, j) - \lambda^*. \tag{7.4}$$

Let us denote by $h^*(i)$ the optimal cost of this stochastic shortest path problem when starting at the nontermination states $i = 1, \ldots, n$. Then by Prop. 2.1(a), $h^*(1), \ldots, h^*(n)$ solve uniquely the corresponding Bellman equation, which has the form

$$h^*(i) = \min_{u \in U(i)} \left[\sum_{j=1}^{n} p_{ij}(u)\big(g(i, u, j) - \lambda^*\big) + \sum_{j \neq s} p_{ij}(u) h^*(j) \right], \tag{7.5}$$

since in the stochastic shortest path problem, the transition probability from i to $j \neq s$ is $p_{ij}(u)$ and the transition probability from i to s is zero under all u. If μ^* is an optimal stationary policy for the average cost problem, then this policy must satisfy, in view of Eq. (7.2),

$$C_{ss}(\mu^*) - N_{ss}(\mu^*)\lambda^* = 0,$$

and from Eq. (7.3), this policy must also be optimal for the associated stochastic shortest path problem. It follows that we must have

$$h^*(s) = C_{ss}(\mu^*) - N_{ss}(\mu^*)\lambda^* = 0, \tag{7.6}$$

so that the missing term $p_{is}(u)h^*(s)$ in the second summation in Eq. (7.5) are zero. By including these terms, we can write this equation as

$$\lambda^* + h^*(i) = \min_{u \in U(i)} \sum_{j=1}^{n} p_{ij}(u)\big(g(i,u,j) + h^*(j)\big), \qquad i = 1, \ldots, n. \tag{7.7}$$

Note that $h^*(i)$ has the interpretation of a *relative* or *differential cost*; it is the minimum of the difference between the expected cost to reach s from i for the first time and the cost that would be incurred if the cost per stage was the average λ^*.

Equation (7.7), which is really Bellman's equation for the associated stochastic shortest path problem, will be viewed as Bellman's equation for the average cost per stage problem. The preceding argument indicates that if λ^* is given, this equation has a unique solution as long as we impose the constraint $h^*(s) = 0$. Furthermore, by minimization of its right-hand side we should obtain an optimal stationary policy. These together with some additional facts (including that the optimal average cost per stage λ^* is well-defined and is the same for all initial states) are summarized in the following proposition. A formal proof can be found in [Ber95a], Vol. I, Section 7.4. In the proposition as well as in the remainder of this section, we will use the shorthand notation involving the mappings T and T_μ, which take the form

$$(Th)(i) = \min_{u \in U(i)} \sum_{j=1}^{n} p_{ij}(u)\big(g(i,u,j) + h(j)\big), \qquad i = 1, \ldots, n,$$

$$(T_\mu h)(i) = \sum_{j=1}^{n} p_{ij}\big(\mu(i)\big)\big(g\big(i,\mu(i),j\big) + h(j)\big), \qquad i = 1, \ldots, n.$$

We will also denote by e the n-vector that has all its components equal to 1; that is, $e(i) = 1$ for all i.

Proposition 7.1: Under Assumption 7.1, the following hold for the average cost per stage problem:

(a) The optimal average cost per stage λ^* is the same for all initial states and together with some vector h^* satisfies Bellman's equation

$$\lambda^* e + h^* = Th^*. \tag{7.8}$$

Furthermore, out of all vectors h^* satisfying this equation, there is a unique vector for which $h^*(s) = 0$. In addition, if for some solution h^* of the above equation and some stationary policy μ^*, we have $T_{\mu^*} h^* = Th^*$, then μ^* is optimal.

(b) If a scalar λ and a vector h satisfy Bellman's equation, then λ is the average optimal cost per stage for each initial state.

(c) Given a stationary policy μ with corresponding average cost per stage λ_μ, there is a unique vector h_μ such that $h_\mu(s) = 0$ and

$$\lambda_\mu e + h_\mu = T_\mu h_\mu. \tag{7.9}$$

Furthermore, if a scalar λ and a vector h satisfy the equation $\lambda e + h = T_\mu h$, then $\lambda = \lambda_\mu$.

We note that Prop. 7.1 can be shown to be true under considerably weaker conditions (see Section 4.2 of [Ber95a], Vol. II). In particular, it can be proved assuming that all stationary policies have a single recurrent class, even if their corresponding recurrent classes do not have a state s in common. Proposition 7.1 can also be proved assuming that for every pair of states i and j, there exists a stationary policy under which there is positive probability of reaching j starting from i. However, we will continue to use Assumption 7.1 because it is essential to some of the subsequent algorithmic analysis.

7.1.2 Value Iteration Methods

The most natural version of the value iteration method for the average cost problem is simply to select arbitrarily a terminal cost function, say J_0, and to generate successively the corresponding optimal k-stage costs $J_k(i)$, $k = 1, 2, \ldots$ This can be done by executing the DP algorithm starting with J_0, that is, by using the recursion

$$J_{k+1} = TJ_k. \tag{7.10}$$

It is natural to expect that the ratios $J_k(i)/k$ converge to the optimal average cost per stage λ^* as $k \to \infty$, that is,

$$\lim_{k\to\infty} \frac{J_k(i)}{k} = \lambda^*, \qquad i = 1, \ldots, n.$$

Indeed, it can be shown that $J_k(i)/k$ converges to λ^* under any conditions that guarantee that Bellman's equation (7.8) holds for some vector h^* (see [Ber95a], Vol. I, Section 7.4).

The value iteration method just described is simple and straightforward, but it is numerically cumbersome, since typically some of the components of J_k diverge to ∞ or $-\infty$. Furthermore, the method is not well-suited to simulation-based approximations. In what follows in this section, we discuss several value iteration methods that, to a great extent, do not suffer from these drawbacks.

Relative Value Iteration

One way to bypass the difficulty with J_k diverging to ∞ is to subtract the same constant from all components of the vector J_k, so that the difference remains bounded. In particular, we can introduce an algorithm that iterates on the differences

$$h_k(i) = J_k(i) - J_k(s), \qquad i = 1, \ldots, n, \tag{7.11}$$

which have the meaning of cost relative to the reference state s. (Actually, any state can be used in place of s, as can be verified from the following calculations.) By using Eq. (7.10), we obtain

$$h_{k+1} = J_{k+1} - J_{k+1}(s)e = TJ_k - (TJ_k)(s)e. \tag{7.12}$$

On the other hand, from the definition $h_k = J_k - J_k(s)e$, we have the two relations

$$Th_k = TJ_k - J_k(s)e$$

and

$$(Th_k)(s)e = (TJ_k)(s)e - J_k(s)e,$$

which when subtracted, yield

$$TJ_k - (TJ_k)(s)e = Th_k - (Th_k)(s)e. \tag{7.13}$$

By combining Eqs. (7.12) and (7.13), we obtain

$$h_{k+1} = Th_k - (Th_k)(s)e. \tag{7.14}$$

The above algorithm, known as *relative value iteration*, is mathematically equivalent to the value iteration method (7.10) that generates the sequence J_k. The iterates generated by the two methods merely differ by a constant [cf. Eq. (7.11)], and the minimization problems involved in the corresponding iterations of the two methods are essentially the same. However, under Assumption 7.1, it can be shown that the iterates $h_k(i)$ generated by the relative value iteration method are bounded.

Note from Eq. (7.14), that if the sequence h_k converges to some vector h, then we have

$$(Th)(s)e + h = Th.$$

By Prop. 7.1(b), this implies that $(Th)(s)$ is the optimal average cost per stage for all initial states, and h is an associated differential cost vector.

On the other hand, Assumption 7.1 is not sufficient to guarantee that the sequence h_k converges to some vector. For this, it is necessary to introduce an additional assumption that in effect implies that the Markov chains corresponding to the stationary policies of the problem are aperiodic. We refer to [Ber95a] (Vol. II, Section 4.3) for a statement of this assumption and for a proof that it implies convergence of h_k. We mention parenthetically that we can bypass the need for this assumption by a slight modification of the relative value iteration algorithm. This modified algorithm is given by

$$h_{k+1} = (1 - \beta)h_k + T(\beta h_k) - \big(T(\beta h_k)\big)(s)e, \tag{7.15}$$

where β is a scalar such that $0 < \beta < 1$. Note that as $\beta \to 1$, we obtain the relative value iteration (7.14). The variant (7.15) of relative value iteration can be shown to converge under just Assumption 7.1 (see Section 4.3 of [Ber95a], Vol. II).

We mention also another variant of relative value iteration that has the form

$$h_{k+1} = Th_k - h_k(s)e, \tag{7.16}$$

This iteration will be useful to us later. It can be shown to be equivalent to the relative value iteration method (7.14) for the slightly modified average cost per stage problem that involves one additional state from which the system moves with probability 1 to state s at cost 0 under all policies [the transition probabilities $p_{ij}(u)$ for all other states are unaffected]. We leave the verification of this fact for the reader.

Contracting Value Iteration and the λ-SSP

Contrary to the value iteration methods for discounted and stochastic shortest path problems, relative value iteration does not involve a weighted maximum norm contraction or even a contraction of any kind. It is possible to develop a value iteration method that does involve a weighted maximum norm contraction by exploiting the connection with the stochastic shortest path problem that we developed earlier (cf. Fig. 7.1). We consider this problem for a one-stage cost equal to $g(i,u,j) - \lambda$, where λ is a scalar parameter, and we refer to it as the λ-SSP.

Let $h_{\mu,\lambda}(i)$ be the cost of stationary policy μ for the λ-SSP, starting from state i; that is, $h_{\mu,\lambda}(i)$ is the total expected cost incurred starting from state i up to the first positive time that we reach the termination state s, which is the fixed reference state of Assumption 7.1. Let $h_\lambda(i) = \min_\mu h_{\mu,\lambda}(i)$ be the corresponding optimal cost of the λ-SSP. Then the following can be shown (see Fig. 7.2):

(a) For all μ, and all scalars λ and $\overline{\lambda}$, we have

$$h_{\mu,\lambda}(i) = h_{\mu,\overline{\lambda}}(i) + (\overline{\lambda} - \lambda)N_\mu(i), \qquad i = 1,\ldots,n,$$

where $N_\mu(i)$ is the expected value of the first positive time that s is reached under μ starting from state i. This is because the only difference between the λ-SSP and the $\overline{\lambda}$-SSP is that the one-stage costs in the λ-SSP are offset from the one-stage costs in the $\overline{\lambda}$-SSP by $\overline{\lambda} - \lambda$. Thus, in particular, for all scalars λ, we have

$$h_{\mu,\lambda}(i) = h_{\mu,\lambda_\mu}(i) + (\lambda_\mu - \lambda)N_\mu(i), \qquad i = 1,\ldots,n, \tag{7.17}$$

where λ_μ is the average cost per stage of μ. Furthermore, because $h_{\mu,\lambda_\mu}(s) = 0$ [compare with Eq. (7.6)], we have

$$h_{\mu,\lambda}(s) = (\lambda_\mu - \lambda)N_\mu(s). \tag{7.18}$$

(b) The functions

$$h_\lambda(i) = \min_\mu h_{\mu,\lambda}(i), \qquad i = 1,\ldots,n, \tag{7.19}$$

are concave, monotonically decreasing, and piecewise linear as functions of λ, and we have

$$h_\lambda(s) = 0 \qquad \text{if and only if} \qquad \lambda = \lambda^*. \tag{7.20}$$

Furthermore, the vector h_{λ^*}, together with λ^*, satisfies Bellman's equation $\lambda^* e + h_{\lambda^*} = Th_{\lambda^*}$.

We can use Fig. 7.2 to provide several different algorithms for computing λ^* using the solution of a sequence of λ-SSPs, corresponding to different values of λ (see [Ber95a], Vol. II, pp. 236-239). One possibility is to update λ by an iteration of the form

$$\lambda_{k+1} = \lambda_k + \beta_k h_{\lambda_k}(s), \tag{7.21}$$

where β_k is a positive stepsize parameter. This iteration is motivated by Fig. 7.2 where it is seen that $\lambda^* < \lambda$ if and only if $h_\lambda(s) < 0$. Indeed, it can

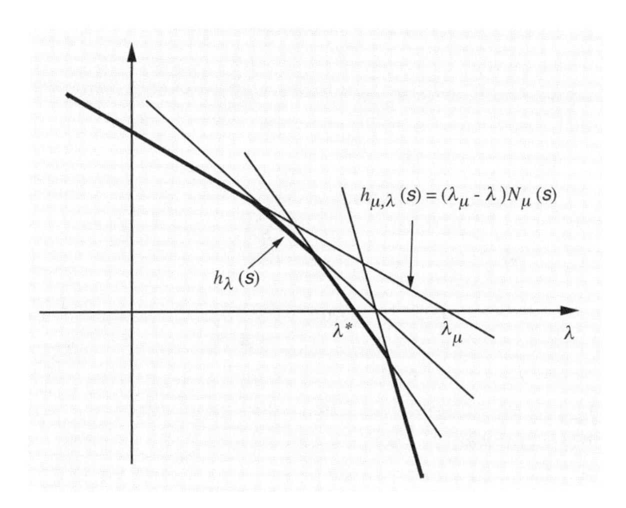

Figure 7.2: Relation of the costs of stationary policies in the average cost problem and the λ-SSP. Here, $h_{\mu,\,\lambda}$ is the cost-to-go vector of policy μ in the λ-SSP, while h_λ is the corresponding optimal cost-to-go vector; that is, $h_\lambda(i) = \min_\mu h_{\mu,\,\lambda}(i)$ for all i. Note that $h_{\mu,\,\lambda}$ is linear in λ, cf. Eq. (7.17), and that $h_\lambda(i)$ is piecewise linear and concave as a function of λ. Furthermore, for the reference state s, we have

$$h_{\mu,\,\lambda}(s) = (\lambda_\mu - \lambda)N_\mu(s),$$

since $h_{\mu,\,\lambda_\mu}(s) = 0$ [cf. Eq. (7.18)].

be seen that the sequence λ_k thus generated converges to λ^* provided the stepsize β_k is the same for all iterations and does not exceed the threshold value $1/\max_\mu N_\mu(s)$. Such a stepsize is sufficiently small to guarantee that once the difference $\lambda - \lambda^*$ becomes positive, it does not change sign during the remainder of the algorithm (7.21). Note that each λ-SSP can be solved by value iteration, which has the form

$$h_{k+1}(i) = \min_{u\in U(i)} \left[\sum_{j=1}^{n} p_{ij}(u)g(i,u,j) + \sum_{j\neq s} p_{ij}(u)h_k(j)\right] - \lambda, \; i = 1,\ldots,n, \tag{7.22}$$

with λ kept fixed throughout the value iteration method.

The preceding value iteration process can be approximated by using an iteration of the form (7.21), but with $h_{\lambda_k}(s)$ replaced by the current value iterate $h_{k+1}(s)$. Such an algorithm may be viewed as a *value iteration algorithm for a stochastic shortest path problem with slowly varying one-stage costs*. It has the form

$$h_{k+1}(i) = \min_{u\in U(i)} \left[\sum_{j=1}^{n} p_{ij}(u)g(i,u,j) + \sum_{j\neq s} p_{ij}(u)h_k(j)\right] - \lambda_k, \; i = 1,\ldots,n, \tag{7.23}$$

$$\lambda_{k+1} = \lambda_k + \beta_k h_{k+1}(s), \tag{7.24}$$

where β_k is a positive stepsize. The motivation for this method is that it involves a weighted maximum norm contraction. In particular, the mapping $F : \Re^n \to \Re^n$ with components given by

$$(Fh)(i) = \min_{u \in U(i)} \left[\sum_{j=1}^{n} p_{ij}(u) g(i,u,j) + \sum_{j \neq s} p_{ij}(u) h_k(j) \right], \quad i = 1, \ldots, n, \tag{7.25}$$

is a contraction with respect to some weighted maximum norm as shown in Prop. 2.2. For this reason, we refer to iteration (7.23)-(7.24) as the *contracting value iteration method*. Convergence of this method can be proved for a variety of stepsize rules, including the case where β_k is a sufficiently small constant (see Bertsekas [Ber95f]).

We note that the issue of stepsize selection is crucial for the fast convergence of the contracting value iteration. In particular, if β_k is chosen constant but very small, or diminishing at the rate of $1/k$ (as is common in stochastic approximation algorithms), then λ changes slowly relative to h, which leads to slow convergence. On the other hand, if β_k is too large, λ_k will oscillate and diverge. One possibility that seems to work reliably and efficiently is to start with a fairly large stepsize (a typical value is $\beta_0 = 1$), and gradually diminish it if the value $h_k(s)$ changes sign frequently; for example, we may use

$$\beta_k = \frac{1}{m_k}, \tag{7.26}$$

where m_k is equal to 1 plus the number of times that $h(s)$ has changed sign up to iteration k. The contracting value iteration can also benefit by projecting λ onto an interval defined by some upper and lower bounds on λ^*, due to Odoni [Odo69], that are obtained in the natural course of the computation with little computational overhead. We refer to [Ber95f] for a discussion of this and other related computational issues.

The maximum norm contraction structure is helpful when Q-learning variants of the contracting value iteration are considered. Furthermore, it is responsible for the validity of Gauss-Seidel and asynchronous variants of the method, as we now discuss.

Asynchronous Variants of Value Iteration

In the value iteration methods discussed so far, all the components of h are simultaneously updated. One may consider asynchronous variants where the components are updated one-at-a-time in arbitrary order as in the corresponding asynchronous value iteration algorithm of Section 2.2. Unfortunately, no convergent asynchronous version of the relative value iteration (7.14) is known. In fact, simple counterexamples show that the straightforward asynchronous variant of this method may diverge (for one such counterexample, see [Ber82a]).

On the other hand, the weighted maximum norm contraction property of the mapping F of Eq. (7.25) can be exploited to construct a valid asynchronous variant of the contracting value iteration, where the components of h are updated sequentially. In particular, one can show convergence for the Gauss-Seidel version of the method, given by

$$h_{k+1}(i) = G_i(h_k, \lambda_k), \qquad i = 1, \ldots, n,$$

$$\lambda_{k+1} = \lambda_k + \beta_k h_{k+1}(s),$$

where $G : \Re^{n+1} \to \Re^n$ has components given by

$$G_1(h, \lambda) = \min_{u \in U(1)} \left[\sum_{j=1}^{n} p_{1j}(u) g(1, u, j) + \sum_{j \neq s} p_{1j}(u) h_k(j) \right] - \lambda,$$

$$G_i(h, \lambda) = \min_{u \in U(i)} \left[\sum_{j=1}^{n} p_{ij}(u) g(i, u, j) + \sum_{j=1,\, j \neq s}^{i-1} p_{ij}(u) G_j(h, \lambda) \right.$$
$$\left. + \sum_{j=i,\, j \neq s}^{n} p_{ij}(u) h(j) \right] - \lambda, \qquad i = 2, \ldots, n.$$

For a proof of this and other related results, we refer to Bertsekas [Ber95f]. Generally, the Gauss-Seidel variant of the contracting value iteration tends to be faster than the original version of Eqs. (7.23) and (7.24).

7.1.3 Policy Iteration

It is possible to use a policy iteration algorithm for the average cost problem. This algorithm is similar to the policy iteration algorithms of Ch. 2: given a stationary policy, we obtain an improved policy by means of a minimization process, and we proceed sequentially until no further improvement is possible. In particular, at the typical step of the algorithm, we have a stationary policy μ_k. We then perform a *policy evaluation* step; that is, we obtain corresponding average and differential costs λ_k and $h_k(i)$ satisfying

$$\lambda_k e + h_k = T_{\mu_k} h_k, \tag{7.27}$$

$$h_k(s) = 0, \tag{7.28}$$

where s is the fixed reference state of Assumption 7.1. We subsequently perform a *policy improvement* step; that is, we find a stationary policy μ_{k+1}, where

$$T_{\mu_{k+1}} h_k = T h_k. \tag{7.29}$$

If $\lambda_{k+1} = \lambda_k$ and $h_{k+1}(i) = h_k(i)$ for all i, the algorithm terminates; otherwise, the process is repeated with μ_{k+1} replacing μ_k.

To prove that the policy iteration algorithm terminates, it is sufficient that each iteration make some irreversible progress towards optimality, since there are finitely many stationary policies. Indeed, by using an argument similar to the one used in the proof of Prop. 2.4, it can be shown that for all k we have $\lambda_{k+1} \leq \lambda_k$. Thus, for each k there are two possibilities: either $\lambda_{k+1} < \lambda_k$, in which case there is irreversible progress in the value of λ, or else $\lambda_{k+1} = \lambda_k$, in which case we essentially revert to policy iteration for the λ_k-SSP, making irreversible progress towards the solution of that problem. In this way we can prove the following proposition (see [Ber95a], Vol. I, Section 7.4).

Proposition 7.2: Under Assumption 7.1, in the policy iteration algorithm, for each k we either have

$$\lambda_{k+1} < \lambda_k, \tag{7.30}$$

or else we have

$$\lambda_{k+1} = \lambda_k, \qquad h_{k+1}(i) \leq h_k(i), \quad i = 1, \ldots, n. \tag{7.31}$$

Furthermore, the algorithm terminates and the policy obtained upon termination is optimal.

We note that policy iteration can be shown to terminate under less restrictive conditions than Assumption 7.1, such as for example, that every stationary policy corresponds to a Markov chain with a single recurrent class (see [Ber95a], Vol. II, Section 4.3).

7.1.4 Linear Programming

A linear programming-based solution method is also possible for the average cost problem. Consider the following linear program in the variables λ and $h(i)$,

$$\begin{aligned} &\text{maximize} \quad \lambda \\ &\text{subject to} \quad \lambda + h(i) \leq \sum_{j=1}^{n} p_{ij}(u)\big(g(i,u,j) + h(j)\big), \qquad \forall\ i,\ u \in U(i). \end{aligned}$$

It can be shown that if λ^* is the optimal average cost and h^* is a corresponding differential cost vector, then (λ^*, h^*) is an optimal solution of the above linear program. Furthermore, in any optimal solution $(\overline{\lambda}, \overline{h})$ of this linear program, we have $\overline{\lambda} = \lambda^*$ (see [Ber95a], Vol. II, pp. 221-222).

When the number of states is very large, we may consider finding an approximation to the differential cost vector, which can be used in turn to obtain a (suboptimal) policy by minimizing in Bellman's equation. One possibility is to approximate the differential costs $h(i)$ with the *linear* form

$$\tilde{h}(i,r) = \sum_{k=1}^{m} r(k)\phi_k(i),$$

where $r = \big(r(1),\ldots,r(m)\big)$ is a vector of parameters, and ϕ_k are some given basis functions. We then obtain a corresponding linear program in the parameter vector r, similar to Section 6.12.

7.1.5 Simulation-Based Value Iteration and Q-Learning

We now discuss how the value iteration methods discussed earlier can form the basis for constructing Q-learning algorithms.

Q-Learning Based on Relative Value Iteration

To derive a Q-learning algorithm for the average cost problem, we argue as in Section 5.6. We form an auxiliary average cost problem by augmenting the original system with one additional state for each possible pair (i,u) with $u \in U(i)$. The probabilistic transition mechanism from the original states is the same as for the original problem, while the probabilistic transition mechanism from an auxiliary state (i,u) is that we move only to states j of the original problem with corresponding probabilities $p_{ij}(u)$ and costs $g(i,u,j)$. It can be seen that the auxiliary problem has the same optimal average cost per stage λ as the original, and that the corresponding Bellman equation is

$$\lambda + h(i) = \min_{u\in U(i)} \sum_{j=1}^{n} p_{ij}(u)\big(g(i,u,j) + h(j)\big), \qquad i = 1,\ldots,n, \tag{7.32}$$

$$\lambda + Q(i,u) = \sum_{j=1}^{n} p_{ij}(u)\big(g(i,u,j) + h(j)\big), \qquad i = 1,\ldots,n,\ u \in U(i), \tag{7.33}$$

where $Q(i,u)$ is the differential cost corresponding to (i,u). Taking the minimum over u in Eq. (7.33) and substituting in Eq. (7.32), we obtain

$$h(i) = \min_{u\in U(i)} Q(i,u), \qquad i = 1,\ldots,n. \tag{7.34}$$

Substituting the above form of $h(i)$ in Eq. (7.33), we obtain Bellman's equation in a form that exclusively involves the Q-factors:

$$\lambda + Q(i,u) = \sum_{j=1}^{n} p_{ij}(u)\left(g(i,u,j) + \min_{v\in U(j)} Q(j,v)\right),\ i = 1,\ldots,n,\ u \in U(i). \tag{7.35}$$

Let us now apply to the auxiliary problem the following variant of relative value iteration

$$h_{k+1} = Th_k - h_k(s)e,$$

[see Eq. (7.16)]. We then obtain the iteration

$$h_{k+1}(i) = \min_{u\in U(i)} \sum_{j=1}^{n} p_{ij}(u)\big(g(i,u,j) + h_k(j)\big) - h_k(s), \qquad i = 1,\ldots,n, \tag{7.36}$$

$$Q_{k+1}(i,u) = \sum_{j=1}^{n} p_{ij}(u)\big(g(i,u,j)+h_k(j)\big)-h_k(s), \qquad i = 1,\ldots,n,\ u\in U(i). \tag{7.37}$$

From these equations, we have that

$$h_k(i) = \min_{u\in U(i)} Q_k(i,u), \qquad i = 1,\ldots,n, \tag{7.38}$$

and by substituting the above form of h_k in Eq. (7.37), we obtain the following relative value iteration for the Q-factors

$$Q_{k+1}(i,u) = \sum_{j=1}^{n} p_{ij}(u)\left(g(i,u,j) + \min_{v\in U(j)} Q_k(j,v)\right) - \min_{v\in U(s)} Q_k(s,v). \tag{7.39}$$

This iteration is analogous to the value iteration for the Q-factors in the stochastic shortest path context. The sequence $\min_{u\in U(s)} Q_k(s,u)$ is expected to converge to the optimal average cost per stage and the sequences $\min_{u\in U(i)} Q_k(i,u)$ are expected to converge to the optimal differential costs $h^*(i)$.

A small-step version of the preceding iteration that involves a positive stepsize γ is given by

$$Q(i,u) := (1-\gamma)Q(i,u) + \gamma\left(\sum_{j=1}^{n} p_{ij}(u)\Big(g(i,u,j) + \min_{v\in U(j)} Q(j,v)\Big) - \min_{v\in U(s)} Q(s,v)\right). \tag{7.40}$$

A natural form of the Q-learning method based on this iteration is obtained by replacing the expected value above by a single sample, i.e.,

$$Q(i,u) := (1-\gamma)Q(i,u) + \gamma\Big(g(i,u,j) + \min_{v\in U(j)} Q(j,v) - \min_{v\in U(s)} Q(s,v)\Big), \tag{7.41}$$

where j and $g(i,u,j)$ are generated from the pair (i,u) by simulation.

Q-Learning Based on the Contracting Value Iteration

We now consider an alternative Q-learning method, which is based on the contracting value iteration method discussed in Section 7.1.2 [cf. Eqs. (7.23) and (7.24)]. If we apply this value iteration method to the auxiliary problem used above, we obtain the following analogs of Eqs. (7.36) and (7.37):

$$h_{k+1}(i) = \min_{u \in U(i)} \left[\sum_{j=1}^{n} p_{ij}(u) g(i, u, j) + \sum_{j \neq s} p_{ij}(u) h_k(j) \right] - \lambda_k, \qquad (7.42)$$

$$Q_{k+1}(i, u) = \sum_{j=1}^{n} p_{ij}(u) g(i, u, j) + \sum_{j \neq s} p_{ij}(u) h_k(j) - \lambda_k, \qquad (7.43)$$

$$\lambda_{k+1} = \lambda_k + \beta_k h_{k+1}(s).$$

From these equations, we have that

$$h_k(i) = \min_{u \in U(i)} Q_k(i, u),$$

and by substituting the above form of h_k in Eq. (7.43), we obtain the following relative value iteration for the Q-factors and for λ:

$$Q_{k+1}(i, u) = \sum_{j=1}^{n} p_{ij}(u) g(i, u, j) + \sum_{j \neq s} p_{ij}(u) \min_{v \in U(j)} Q_k(j, v) - \lambda_k,$$

$$\lambda_{k+1} = \lambda_k + \beta_k \min_{v \in U(s)} Q_{k+1}(s, v).$$

This iteration parallels the relative value iteration (7.39) for the Q-factors.

A small-step version of the preceding iteration is given by

$$Q(i, u) := (1 - \gamma) Q(i, u) + \gamma \left(\sum_{j=1}^{n} p_{ij}(u) g(i, u, j) + \sum_{j \neq s} p_{ij}(u) \min_{v \in U(j)} Q(j, v) - \lambda \right),$$

$$\lambda := \lambda + \beta \min_{v \in U(s)} Q(s, v),$$

where γ and β are positive stepsizes. A natural form of the Q-learning method based on this iteration is obtained by replacing the expected values by a single sample, i.e.,

$$Q(i, u) := (1 - \gamma) Q(i, u) + \gamma \Big(g(i, u, j) + \min_{v \in U(j)} \hat{Q}(j, v) - \lambda \Big), \qquad (7.44)$$

$$\lambda := \lambda + \beta \min_{v \in U(s)} Q(s, v), \tag{7.45}$$

where

$$\hat{Q}(j, v) = \begin{cases} Q(j, v), & \text{if } j \neq s, \\ 0, & \text{otherwise,} \end{cases} \tag{7.46}$$

and j and $g(i, u, j)$ are generated from the pair (i, u) by simulation. Here the stepsizes γ and β should be diminishing, but β should diminish "faster" than γ; that is, the ratio of the stepsizes β/γ should converge to zero. For example, we may use $\gamma = C/k$ and $\beta = c/k \log k$, where C and c are positive constants and k is the number of iterations performed on the corresponding pair (i, u) or λ, respectively.

The algorithm has two components: the iteration (7.44), which is essentially a Q-learning method that aims to solve the λ-SSP for the current value of λ, and the iteration (7.45), which updates λ towards its correct value λ^*. However, λ is updated at a slower rate than Q, since the stepsize ratio β/γ converges to zero. The effect of this is that the Q-learning iteration (7.44) is fast enough to keep pace with the slower changing λ-SSP.

Convergence Analysis of Q-Learning Algorithms

Let us now discuss briefly the convergence of the Q-learning algorithms (7.41) and (7.44)-(7.46). Unfortunately, the general convergence results of Ch. 4 do not apply, because these algorithms involve mappings that are neither monotone nor contracting. [Note that while iteration (7.44) involves a contraction, its companion iteration (7.45) does not.] As a result, we need an entirely different line of analysis than the one given for stochastic shortest path and discounted problems in Section 5.6. In particular, we turn to the ODE approach, which was introduced in Section 4.4. Convergence proofs based on this approach have been developed by Abounadi, Bertsekas, and Borkar [ABB96]. These proofs rely on a sophisticated ODE-type analysis for asynchronous stochastic iterative algorithms due to Borkar [Bor95]. We will only sketch the assumptions and the convergence analysis, and we refer to [ABB96] for a detailed account.

Consider first a formal description of the Q-learning algorithm (7.41), based on relative value iteration. Using the type of notation established in Section 5.6, we have

$$Q_{t+1}(i, u) = \big(1 - \gamma_t(i, u)\big)Q_t(i, u) + \gamma_t(i, u)\Big(g(i, u, \bar{i}) + \min_{v \in U(\bar{i})} Q_t(\bar{i}, v) - \min_{v \in U(s)} Q_t(s, v)\Big), \tag{7.47}$$

where $\bar{i}$ is the next state generated from the pair (i, u) according to the transition probabilities $p_{i\bar{i}}(u)$. Here we assume that the stepsizes $\gamma_t(i, u)$

satisfy the following conditions [for a set of more general conditions on $\gamma_t(i,u)$, we refer to Borkar [Bor95]]:

(a) Each Q-factor $Q(i,u)$ is updated at an infinite set of times $T^{i,u}$, and $\gamma_t(i,u) = 0$ for all $t \notin T^{i,u}$.

(b) For some positive constants c and d, we have

$$\gamma_t(i,u) = \frac{c}{\nu(t,i,u) + d},$$

where $\nu(t,i,u)$ is the cardinality of the set $\{k \le t \mid k \in T^{i,u}\}$, which is the number of updates of $Q(i,u)$ until time t.

(c) There exists a (deterministic) scalar $\delta > 0$ such that with probability 1 we have

$$\liminf_{t\to\infty} \frac{\nu(t,i,u)}{t} \ge \delta, \qquad \forall\ (i,u).$$

Under the above assumptions, an analysis of Borkar [Bor95] can be used to establish a relation between the Q-learning algorithm (7.47) and the following ODE

$$\dot{Q} = HQ - Q - \min_{v\in U(s)} Q(s,v)e, \tag{7.48}$$

where Q is the vector with components the Q-factors $Q(i,u)$, and H is the mapping defined by

$$(HQ)(i,u) = \sum_{j=1}^{n} p_{ij}(u)\left(g(i,u,j) + \min_{v\in U(j)} Q(j,v)\right), \quad i = 1,\ldots,n,\ u \in U(i). \tag{7.49}$$

The convergence results in [Bor95] state that if the ODE (7.48) has a unique globally stable equilibrium Q^* and H has certain special properties, and also if the generated sequences $Q_t(i,u)$ are guaranteed to be bounded with probability 1, then $Q_t(i,u)$ converges to $Q^*(i,u)$ with probability 1 for all (i,u).

The convergence proof given in [ABB96] shows that the ODE (7.48) has the stability property just described and that the mapping H of Eq. (7.49) has the required properties. In particular, the unique globally stable equilibrium of the ODE (7.48) is the unique optimal Q-factor vector of the average cost problem that satisfies the normalization condition

$$\min_{v\in U(s)} Q^*(s,v) = 0.$$

The proof of this is based on a remarkable result of Borkar and Soumyanath [BoS93], which asserts that if a mapping H is nonexpansive with respect to the maximum norm and has a nonempty set of fixed points, then the solution of the ODE $\dot{Q} = HQ - Q$ converges to some fixed point of H (which depends on the initial condition). The proof of asymptotic stability of the ODE (7.48) uses the fact that its solution differs from the solution of the ODE $\dot{Q} = HQ - Q$ by a multiple of the vector $e = (1, \ldots, 1)$ (assuming the two ODEs start at the same initial condition). A complete convergence proof also requires that boundedness of the generated sequence $Q_t(i, u)$ be shown rather than assumed. Whether this is possible is a question that is presently being investigated.

The ODE approach can also be used to prove convergence of the Q-learning algorithm (7.44)-(7.46), that is based on the contracting value iteration method. The stepsize $\gamma_t(i, u)$ used in iteration (7.44) is required to satisfy the conditions (a)-(c) given above. The stepsize β_t used in iteration (7.45) must satisfy

$$\lim_{t\to\infty} \frac{\beta_t}{\min_{\{(i,u) \mid \gamma_t(i,u)>0\}} \gamma_t(i, u)} = 0,$$

as well as $\sum_{t=0}^{\infty} \beta_t = \infty$. Here it is possible to prove boundedness with probability 1 of the generated sequences $Q_t(i, u)$, once we make sure that the sequence λ_t is kept bounded. This can be done by projection on an interval of real numbers that is known to contain λ^*. Once this projection device is introduced, the boundedness proof is similar to the one for stochastic shortest path problems assuming that all policies are proper (cf. Prop. 5.5 in Section 5.6). We refer to [ABB96] for a detailed account.

Q-Learning with Function Approximation

Let us now consider the case where we have approximate Q-factors of the form $\tilde{Q}(i, u, r)$, where r is a parameter vector. One possibility is to fix λ, apply Q-learning with function approximation to the corresponding λ-SSP (cf. Section 6.6), and then adjust λ as indicated by Fig. 7.2. On the other hand, as discussed in Section 6.6, there is no real mathematical justification for this type of algorithm, even when λ is fixed. Alternatively, one may be tempted, following the heuristic reasoning of Section 6.6, to use a method of the form [cf. Eqs. (7.44)-(7.46)]

$$r := (1 - \gamma)r + \gamma \nabla \tilde{Q}(i, u, r)\Big(g(i, u, j) + \min_{v \in U(j)} \hat{Q}(j, v, r) - \lambda\Big),$$

$$\lambda := \lambda + \beta \min_{v \in U(s)} \tilde{Q}(s, v, r),$$

where

$$\hat{Q}(j, v, r) = \begin{cases} \tilde{Q}(j, v, r), & \text{if } j \neq s, \\ 0, & \text{otherwise.} \end{cases}$$

Then the situation becomes even less clear, because λ is now changing, and if the Q-factor approximations $\tilde{Q}(s, v, r)$ are poor, the iteration of λ is strongly affected. For example, if $\tilde{Q}(s, v, r)$ remains bounded from below by a positive number, we see that λ diverges to ∞. Thus the proper use of Q-learning in conjunction with function approximation for average cost problems is unclear at present.

7.1.6 Simulation-Based Policy Iteration

We now describe how the simulation-based policy iteration methods of Chs. 5 and 6 can be adapted to work for average cost problems. Given a stationary policy μ, we may accurately calculate λ_μ by simulating the system under policy μ over a very large number of stages and by then dividing the total accumulated cost by the number of stages. Once we have the value of λ_μ for the policy μ, the corresponding differential costs can be obtained as the costs-to-go of the associated λ_μ-SSP. In the case of a lookup table representation, these costs can be calculated exactly by using the simulation methods of Ch. 5. If function approximation is involved, one can use methods described in Ch. 6, such as Monte-Carlo simulation followed by a least squares fit, or temporal difference methods.

Note that it is not necessary to strictly separate the calculation of λ_μ and the corresponding differential costs. For example to calculate the differential costs by Monte-Carlo simulation, one may obtain sample costs starting from various states up to reaching the termination/reference state s, record the corresponding number of stages to reach s, and then subtract out the appropriate multiple of λ_μ once an accurate enough value of λ_μ has been calculated via the simulation. Note also a potential weakness of the method: the termination state s must be reached sufficiently often from the various states in order to accumulate enough cost samples in reasonable time. In problems with many states there may not exist a natural termination state with this property.

To analyze the practically more interesting case where function approximation is used for the differential costs, we consider a method that involves approximate policy evaluation and approximate policy improvement. In particular, this method generates a sequence of stationary policies μ_k, a corresponding sequence of average costs λ_{μ_k}, and a sequence of approximate differential cost functions h_k satisfying

$$h_k(s) = 0, \qquad \max_{i=1,\ldots,n} \left| h_k(i) - h_{\mu_k,\lambda_k}(i) \right| \leq \epsilon, \qquad k = 0, 1, \ldots$$

where

$$\lambda_k = \min_{m=0,1,\ldots,k} \lambda_{\mu_m},$$

$h_{\mu_k,\lambda_k}(i)$ is the cost-to-go from state i to the reference state s for the λ_k-SSP under policy μ_k, and ϵ is a positive scalar quantifying the accuracy of evaluation of the cost-to-go function of the λ_k-SSP. Note that we assume *exact* calculation of the average costs λ_{μ_k}. Note also that we may calculate approximate differential costs $\tilde{h}_k(i,r)$ that depend on a parameter vector r without regard to the reference state s (for example, by using the Bellman equation error method to be discussed in Section 7.1.7). These differential costs may then be replaced by

$$h_k(i) = \tilde{h}_k(i,r) - \tilde{h}(s,r), \qquad i = 1,\ldots,n,$$

so that the normalization condition $h_k(s) = 0$ holds.

As in the analysis of Ch. 6, we assume that policy improvement is carried out by approximate minimization in the DP mapping. In particular, we assume that there exists a tolerance $\delta > 0$ such that for all i and k, $\mu_{k+1}(i)$ attains the minimum in the expression

$$\min_{u\in U(i)} \sum_{j=1}^n p_{ij}(u)\big(g(i,u,j) + h_k(j)\big),$$

within a tolerance $\delta > 0$.

We now note that since λ_k is monotonically nonincreasing and is bounded below by λ^*, it must converge to some scalar $\overline{\lambda}$. Since λ_k can take only one of the finite number of values λ_μ corresponding to the finite number of stationary policies μ, we see that λ_k must converge finitely to $\overline{\lambda}$; that is, for some $\overline{k}$, we have

$$\lambda_k = \overline{\lambda}, \qquad \forall\, k \geq \overline{k}.$$

Let $h_{\overline{\lambda}}(s)$ denote the optimal cost-to-go from state s in the $\overline{\lambda}$-SSP. Then, by using Prop. 6.3 in Section 6.2, we have

$$\limsup_{k\to\infty}\big(h_{\mu_k,\overline{\lambda}}(s) - h_{\overline{\lambda}}(s)\big) \leq \frac{n(1-\rho+n)(\delta+2\epsilon)}{(1-\rho)^2}, \tag{7.50}$$

where

$$\rho = \max_{i=1,\ldots,n,\,\mu} \mathrm{P}\big(i_k \neq s,\ k=1,\ldots,n \mid i_0 = i, \mu\big),$$

and i_k denotes the state of the system after k stages. On the other hand, as can also be seen from Fig. 7.3, the relation $\overline{\lambda} \leq \lambda_{\mu_k}$ implies that

$$h_{\mu_k,\overline{\lambda}}(s) \geq h_{\mu_k,\lambda_{\mu_k}}(s) = 0.$$

It follows, using also Fig. 7.3, that

$$h_{\mu_k,\overline{\lambda}}(s) - h_{\overline{\lambda}}(s) \geq -h_{\overline{\lambda}}(s) \geq -h_{\mu^*,\overline{\lambda}}(s) = (\overline{\lambda} - \lambda^*)N_{\mu^*}, \tag{7.51}$$

where μ^* is an optimal policy for the λ^*-SSP (and hence also for the original average cost per stage problem) and N_{μ^*} is the expected number of stages

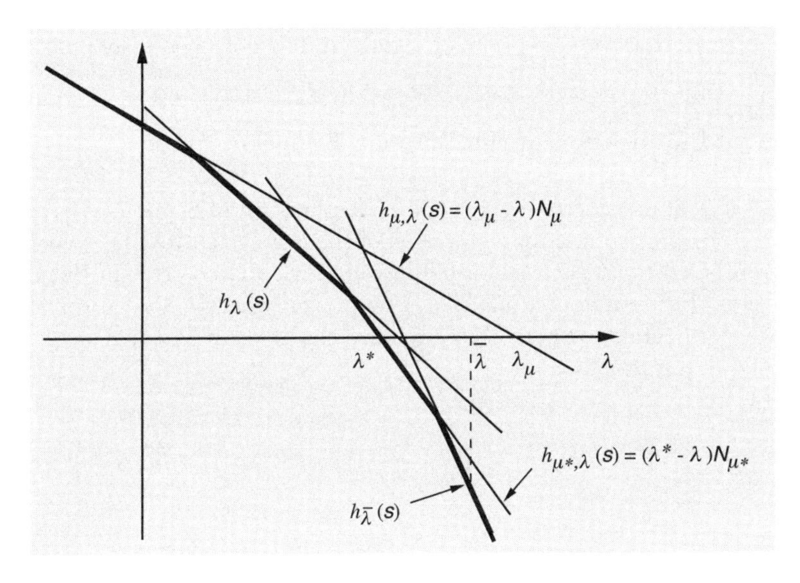

Figure 7.3: Relation of the costs of stationary policies for the λ-SSP in the approximate policy iteration method (compare with Fig. 7.2). Here, N_μ is the expected number of stages to return to state s, starting from s and using μ. Since $\lambda_{\mu_k} \geq \overline{\lambda}$, we have

$$h_{\mu_k,\overline{\lambda}}(s) \geq h_{\mu_k,\lambda_{\mu_k}}(s) = 0.$$

Furthermore, if μ^* is an optimal policy for the λ^*-SSP, we have

$$h_{\overline{\lambda}}(s) \leq h_{\mu^*,\overline{\lambda}}(s) = (\lambda^* - \overline{\lambda})N_{\mu^*}.$$

to return to state s, starting from s and using μ^*. Thus, from Eqs. (7.50) and (7.51), we have

$$\overline{\lambda} - \lambda^* \leq \frac{n(1-\rho+n)(\delta+2\epsilon)}{N_{\mu^*}(1-\rho)^2}. \tag{7.52}$$

This relation provides, an estimate on the steady state error of the approximate policy iteration method.

We finally note that the form of an appropriate optimistic version of the preceding approximate policy iteration method is unclear at present. The reason is our assumption that the average cost λ_μ of every generated policy μ is exactly calculated; in an optimistic policy iteration method the current policy μ does not remain constant for sufficiently long time to estimate accurately λ_μ. We may consider schemes where an optimistic version of policy iteration is used to solve the λ-SSP for a fixed λ. The value of λ may occasionally be adjusted downward by calculating "exactly" through simulation the average cost λ_μ of some of the (presumably most promising) generated policies μ, and by then updating λ according to $\lambda :=$

$\min\{\lambda, \lambda_\mu\}$. However, this type of algorithm has not been tested as of the present writing.

7.1.7 Minimization of the Bellman Equation Error

There is a straightforward extension of the method of Section 6.10 for obtaining an approximate representation of the average cost λ and associated differential costs $h(i)$, based on minimizing the squared error in Bellman's equation. Here we approximate $h(i)$ by $\tilde{h}(i,r)$, where r is a vector of unknown parameters/weights. We minimize the error in Bellman's equation by solving the problem

$$\min_{\lambda,r} \sum_{i\in\tilde{S}} \left(\lambda + \tilde{h}(i,r) - \min_{u\in U(i)} \sum_{j=1}^{n} p_{ij}(u)\big(g(i,u,j) + \tilde{h}(j,r)\big) \right)^2,$$

where $\tilde{S}$ is a suitably chosen subset of "representative" states. This minimization may be attempted by using some gradient method of the type discussed in Section 6.10. Much of our discussion in Section 6.10 remains applicable.

The Bellman equation error approach may also be used for approximate evaluation of a given stationary policy in the context of approximate policy iteration. Within this context it has an advantage over alternative methods in that it does not require a termination/reference state. Furthermore, the corresponding least squares problem can be solved by the two-sample gradient method discussed in Section 6.10. However, there are no error bounds such as the estimate (7.52) for the corresponding approximate policy iteration method.

7.2 DYNAMIC GAMES

In this section, we briefly consider generalizations of the discounted and stochastic shortest path DP models to the case where there is an opponent who, at each state i and simultaneously with the choice of u, chooses another control v from a finite constraint set $V(i)$ with the aim of maximizing the cost-to-go. To be specific, let us consider the discounted cost problem with the difference that the transition probabilities and the one-stage costs have the form

$$p_{ij}(u,v), \qquad g(i,u,v,j).$$

We follow the game theoretic terminology where u and v are chosen by players that are referred to as the *minimizer* and the *maximizer*, respectively.

We introduce *randomized strategies* for the choice of u and v. In particular, we assume that at a given state i, the minimizer chooses a probability distribution $y = \{y_u \mid u \in U(i)\}$ over the set $U(i)$ and the maximizer chooses a probability distribution $z = \{z_v \mid v \in V(i)\}$ over the set $V(i)$. Here, y_u and z_v are the probabilities of selecting u and v, respectively. The system moves from state i to state j with probability

$$\sum_{u\in U(i)} \sum_{v\in V(i)} y_u z_v p_{ij}(u,v),$$

and the incurred expected cost is

$$G(i,y,z) = \sum_{u\in U(i)} \sum_{v\in V(i)} y_u z_v \sum_{j=1}^{n} p_{ij}(u,v) g(i,u,v,j).$$

A policy $\pi_M = \{\mu_0, \mu_1, \ldots\}$ for the minimizer consists of functions μ_k that choose at state i the probability distribution $\mu_k(i)$ over the set $U(i)$. A policy $\pi_N = \{\nu_0, \nu_1, \ldots\}$ for the maximizer is analogously defined. Stationary policies for the minimizer and the maximizer have the form $\{\mu, \mu, \ldots\}$ and $\{\nu, \nu, \ldots\}$, and are referred to as the stationary policies μ and ν, respectively. The cost-to-go corresponding to an initial state i and a pair of policies (π_M, π_N) is

$$J^{\pi_M,\pi_N}(i) = \sum_{k=0}^{\infty} \alpha^k E\Big[G\big(i_k, \mu_k(i_k), \nu_k(i_k)\big) \mid i_0 = i\Big],$$

where i_k denotes the state of the system at the kth stage, and either α is equal to 1 or else α is a discount factor from the interval $(0,1)$. The cost-to-go corresponding to an initial state i and a pair of stationary policies (μ, ν) is denoted by $J^{\mu,\nu}(i)$.

We consider the min-max and max-min costs

$$\underline{J}(i) = \min_{\pi_M} \max_{\pi_N} J^{\pi_M,\pi_N}(i),$$

and

$$\overline{J}(i) = \max_{\pi_N} \min_{\pi_M} J^{\pi_M,\pi_N}(i),$$

which correspond to the different orders in which the maximizer and minimizer choose their policies. A question of primary interest is whether $\underline{J}(i)$ and $\overline{J}(i)$ are equal to a common value $J^*(i)$ that can be viewed as the *value* or *equilibrium value* of the game.

7.2.1 Discounted Games

Let us consider the discounted case where $0 < \alpha < 1$. For this case, it was shown in an important early paper by Shapley [Sha53] that indeed $\underline{J}(i)$ and $\overline{J}(i)$ are equal to an equilibrium value $J^*(i)$, and that there exist stationary policies μ^* and ν^* such that for all $i = 1, \ldots, n$, we have

$$J^{\mu^*,\nu^*}(i) = \max_{\nu} J^{\mu^*,\nu}(i) = \min_{\mu} J^{\mu,\nu^*}(i) = \underline{J}(i) = \overline{J}(i) = J^*(i). \quad (7.53)$$

Generally, as long as one considers randomized policies, the results and algorithms of Ch. 2 for the one-player discounted case have counterparts for the two-player discounted case. In particular, there are analogs of Bellman's equation, of value iteration, and of policy iteration, which we now proceed to discuss.

For stationary policies μ and ν, and states i, the components of the corresponding probability distributions $\mu(i)$ and $\nu(i)$ will be denoted by $\mu_u(i)$, $u \in U(i)$, and $\nu_v(i)$, $v \in V(i)$, respectively. We introduce the mappings

$$(T_{\mu,\nu}J)(i) = \sum_{u \in U(i)} \sum_{v \in V(i)} \mu_u(i)\nu_v(i) \sum_{j=1}^{n} p_{ij}(u,v)\big(g(i,u,v,j) + \alpha J(j)\big),$$

$$(\underline{T}_{\mu}J)(i) = \max_{\nu}(T_{\mu,\nu}J)(i),$$

$$(\overline{T}_{\nu}J)(i) = \min_{\mu}(T_{\mu,\nu}J)(i),$$

$$(TJ)(i) = \min_{\mu} \max_{\nu}(T_{\mu,\nu}J)(i) = \max_{\nu} \min_{\mu}(T_{\mu,\nu}J)(i).$$

The minima and maxima in the above relations are taken over the corresponding unit simplices, that is, the sets of probability distributions on $U(i)$ and $V(i)$, respectively. The interchange of "min" and "max" in the last relation is justified by a fundamental result of game theory, the minimax theorem (see e.g., [Ber95b], Section 5.4.3), which applies because $(T_{\mu,\nu}J)(i)$ is linear in $\mu(i)$ when $\nu(i)$ is held fixed, and is also linear in $\nu(i)$ when $\mu(i)$ is held fixed.

A key property of the mappings T, $\underline{T}_{\mu}$, $\overline{T}_{\nu}$, and $T_{\mu,\nu}$ is that they are all contraction mappings with respect to the maximum norm, and with contraction modulus α. This is fairly straightforward to show along the lines of the corresponding proof for the DP (one-player) contraction property of the mapping T. The contraction property shows that Bellman's equation, which is $J^* = TJ^*$ has a unique solution, and is responsible for the following proposition, which we state without proof.

Proposition 7.3: The following hold for the discounted dynamic game problem:

(a) There exist stationary policies μ^* and ν^* such that for all $i = 1, \ldots, n$, we have

$$J^{\mu^*,\nu^*}(i) = \max_{\nu} J^{\mu^*,\nu}(i) = \min_{\mu} J^{\mu,\nu^*}(i) = \underline{J}(i) = \overline{J}(i). \quad (7.54)$$

Furthermore, the common value of the vectors $\underline{J}$ and $\overline{J}$, denoted by J^*, satisfies

$$J^* = TJ^*,$$

and is the only solution of Bellman's equation $J = TJ$.

(b) Stationary policies μ^* and ν^* satisfy the equilibrium relation (7.54) if and only if

$$\underline{T}_{\mu^*} J^* = \overline{T}_{\nu^*} J^* = TJ^*.$$

The contraction property of the mappings T, $\underline{T}_\mu$, $\overline{T}_\nu$, and $T_{\mu,\nu}$ also implies the validity of various value iteration methods, such as the method that starts from some vector J and sequentially generates $TJ, T^2J, \ldots$ We have

$$\lim_{k \to \infty} T^k J = J^*. \quad (7.55)$$

There are also Gauss-Seidel and asynchronous variants of this value iteration method, which also draw their validity from the contraction property of T.

Finally, one can show the validity of the natural policy iteration algorithm. Here, we start with a stationary policy μ_0 and we generate a sequence of stationary policies μ_k. At the generic step, we have μ_k and we evaluate the corresponding cost-to-go to the minimizer using μ_k, assuming that the maximizer knows μ_k and plays optimally; that is, we find the vector J^{μ_k} defined by

$$J^{\mu_k}(i) = \max_{\nu} J^{\mu_k,\nu}(i), \qquad i = 1, \ldots, n.$$

Note that J^{μ_k} is the optimal reward-to-go vector for a maximization DP problem: the α-discounted (single player) problem that has a one-stage reward equal to $g\big(i, \mu_k(i), v, j\big)$, and involves maximization of $J^{\mu_k,\nu}$ over ν. This problem can be solved using some form of value iteration or policy iteration and J^{μ_k} satisfies the Bellman equation

$$J^{\mu_k}(i) = (\underline{T}_{\mu_k} J^{\mu_k})(i) = \max_{\nu} (T_{\mu_k,\nu} J^{\mu_k})(i), \qquad i = 1, \ldots, n. \quad (7.56)$$

Once J^{μ_k} has been evaluated, we obtain μ_{k+1} by a policy improvement step, whereby

$$\mu_{k+1}(i) = \arg\min_{\mu}(\underline{T}_{\mu} J^{\mu_k})(i), \qquad i = 1, \ldots, n. \tag{7.57}$$

This equation requires solving a (static) game problem for each state i.

It can be shown that this policy iteration algorithm is convergent, in the sense that $\lim_{k\to\infty} J^{\mu_k} = J^*$. Note that there is also a symmetrical version of the policy iteration algorithm, where we take the point of view of the maximizer. In particular, in this version we generate a sequence ν_k such that

$$\nu_{k+1}(i) = \arg\max_{\nu}(\overline{T}_{\nu} J^{\nu_k})(i), \qquad \forall\, i = 1, \ldots, n,$$

where $J^{\nu_k} = \overline{T}_{\nu_k} J^{\nu_k}$.

7.2.2 Stochastic Shortest Path Games

Consider now the case where $\alpha = 1$ and there exists a cost-free and absorbing termination state. If we assume that this state is reached with probability 1 under all policies for the minimizer and the maximizer, a small adaptation of the proof of Prop. 2.2 shows that the mappings T, $\underline{T}_{\mu}$, and $\overline{T}_{\nu}$ are all contraction mappings with respect to some weighted maximum norm (see Patek and Bertsekas [PaB96a] for a proof). Furthermore, the theory and algorithms outlined above for discounted games go though, with the obvious notational changes.

A more challenging situation arises when the maximizer can force the game to be prolonged indefinitely for some policies of the minimizer; that is, with appropriate choice of the minimizer's and maximizer's policies, there is positive probability that the termination state will never be reached. Then it is necessary to introduce an appropriately modified notion of proper policies and to provide an analysis that parallels the one given for stochastic shortest path problems in Section 2.2. This has been done by Patek and Bertsekas [PaB96a], to which we refer for a detailed development. In what follows in this section we will focus on discounted and stochastic shortest path problems for which the mappings T, $\underline{T}_{\mu}$, and $\overline{T}_{\nu}$ are all contraction mappings. This will allow us to provide a common analysis for both types of problems.

7.2.3 Sequential Games, Policy Iteration, and Q-Learning

We now focus on an important subclass of dynamic games, where the minimizer and the maximizer take turns in selecting u and v (rather that selecting u and v simultaneously). The choice of each player is selected with full knowledge of the preceding choice of the other player and the state transition resulting from that choice (as is done in board games like chess and backgammon). We refer to this type of games as a *sequential game*.

We can embed sequential games within the general framework given earlier by using the following device: we introduce a state space of the form $\underline{S} \cup \overline{S}$. At a state i of $\underline{S}$, only the minimizer is allowed to choose a $u \in U(i)$ and the system then moves to a state j with probability $p_{ij}(u, \overline{v})$, where $\overline{v}$ is a special "dummy" selection for the maximizer. Similarly, at a state i of $\overline{S}$, only the maximizer is allowed to choose a $v \in V(i)$ and the system then moves to a state j with probability $p_{ij}(\underline{u}, v)$, where $\underline{u}$ is a special "dummy" selection for the maximizer. We continue to assume that the sets $\underline{S}$, $\overline{S}$, $U(i)$, and $V(i)$ are all finite.

This special structure has some important ramifications. A policy for the minimizer (or the maximizer) need only be defined on the corresponding portion $\underline{S}$ (or $\overline{S}$, respectively) of the state space. Furthermore, the mappings $\underline{T}_\mu$ and $\overline{T}_\nu$ are only defined on the corresponding portions of the state space, and do not involve a maximization over ν and a minimization over μ, respectively. In particular, we have

$$(\underline{T}_\mu J)(i) = \sum_{u \in U(i)} \mu_u(i) \sum_{j \in \underline{S} \cup \overline{S}} \underline{p}_{ij}(u)\big(\underline{g}(i, u, j) + \alpha J(j)\big), \qquad \forall\, i \in \underline{S},$$

$$(\overline{T}_\nu J)(i) = \sum_{v \in V(i)} \nu_v(i) \sum_{j \in \underline{S} \cup \overline{S}} \overline{p}_{ij}(v)\big(\overline{g}(i, v, j) + \alpha J(j)\big), \qquad \forall\, i \in \overline{S},$$

where we use the notations

$$\underline{p}_{ij}(u) = p_{ij}(u, \overline{v}), \qquad \overline{p}_{ij}(v) = p_{ij}(\underline{u}, v),$$

$$\underline{g}(i, u, j) = g(i, u, \overline{v}, j), \qquad \overline{g}(i, v, j) = g(i, \underline{u}, v, j).$$

A major consequence of this fact is that the minimum of $(\underline{T}_\mu J)(i)$ over all probability distributions $\mu(i)$ is attained by a single-point distribution. In particular, there is a distribution that attains the minimum and that assigns probability 1 to a single u and probability 0 to all other $u \in U(i)$; this is the $u \in U(i)$ that minimizes the expression

$$\sum_{j=1}^{n} \underline{p}_{ij}(u)\big(\underline{g}(i, u, j) + \alpha J(j)\big).$$

There is also a symmetric property that holds for the maximizer. As a result, the minimization over probability distributions $\mu(i)$ can be replaced by minimization over all $u \in U(i)$, and the maximization over probability distributions $\nu(i)$ can be replaced by maximization over all $v \in V(i)$. Thus, the mapping T can be defined by

$$(TJ)(i) = \begin{cases} \min_{u \in U(i)} \sum_{j \in \underline{S} \cup \overline{S}} \underline{p}_{ij}(u)\big(\underline{g}(i,u,j) + \alpha J(j)\big), & \text{if } i \in \underline{S}, \\ \max_{v \in V(i)} \sum_{j \in \underline{S} \cup \overline{S}} \overline{p}_{ij}(v)\big(\overline{g}(i,v,j) + \alpha J(j)\big), & \text{if } i \in \overline{S}, \end{cases} \tag{7.58}$$

thereby greatly simplifying the value iteration algorithm.

Policy Iteration

There is also simplification in the policy iteration algorithm of Eqs. (7.56) and (7.57). In particular, since the set of choices available at each state is finite, the number of stationary policies is also finite, and this can be used to show that the policy iteration algorithm terminates in a finite number of iterations.

We note, however, that there is an intuitively appealing version of the policy iteration algorithm that *does not* necessarily work. In this version, we start with a pair of stationary policies (μ_0, ν_0) and we generate a sequence of stationary policies (μ_k, ν_k) as follows: at the generic step, we have (μ_k, ν_k) and we evaluate the corresponding cost-to-go J^{μ_k,ν_k} by solving the system of linear equations

$$J^{\mu_k,\nu_k}(i) = \begin{cases} \sum_{j \in \underline{S} \cup \overline{S}} \underline{p}_{ij}\big(\mu_k(i)\big)\big(\underline{g}\big(i,\mu_k(i),j\big) + \alpha J^{\mu_k,\nu_k}(j)\big), & \text{if } i \in \underline{S}, \\ \sum_{j \in \underline{S} \cup \overline{S}} \overline{p}_{ij}\big(\nu_k(i)\big)\big(\overline{g}\big(i,\nu_k(i),j\big) + \alpha J^{\mu_k,\nu_k}(j)\big), & \text{if } i \in \overline{S}. \end{cases} \tag{7.59}$$

We then obtain (μ_{k+1}, ν_{k+1}) using a policy "improvement" step

$$\mu_{k+1}(i) = \arg \min_{u \in U(i)} \sum_{j \in \underline{S} \cup \overline{S}} \underline{p}_{ij}(u)\big(\underline{g}(i,u,j) + \alpha J^{\mu_k,\nu_k}(j)\big), \qquad \forall\, i \in \underline{S}, \tag{7.60}$$

$$\nu_{k+1}(i) = \arg \max_{v \in V(i)} \sum_{j \in \underline{S} \cup \overline{S}} \overline{p}_{ij}(v)\big(\overline{g}(i,v,j) + \alpha J^{\mu_k,\nu_k}(j)\big), \qquad \forall\, i \in \overline{S}. \tag{7.61}$$

It has been shown by Pollatschek and Avi-Itzhak [PoA69] that this method can be interpreted as Newton's method for solving the Bellman equation $J = TJ$. (This interpretation, however, stretches the definition of Newton's method because TJ is not differentiable everywhere as a function of J.) If this method terminates [i.e., we have $(\mu_{k+1}, \nu_{k+1}) = (\mu_k, \nu_k)$], it can be seen that the pair (μ_k, ν_k) is an equilibrium pair. However, while the method frequently terminates, there are examples where it fails to do so. Nonetheless, this version of policy iteration may be worth trying, since it is much simpler than the provably correct version of Eqs. (7.56) and (7.57); instead of solving a maximizer's DP problem at each iteration, it involves the much simpler policy evaluation/solution of the linear Eq. (7.59).

It is also possible to introduce a positive stepsize γ_k in the above policy iteration algorithm, so that it takes the form

$$J_{k+1} = J_k + \gamma_k (J^{\mu_k,\nu_k} - J_k), \tag{7.62}$$

where J^{μ_k,ν_k} is obtained by solving the policy evaluation equation (7.59), and the policy pair (μ_k, ν_k) is obtained from the equations

$$\mu_k(i) = \arg\min_{u \in U(i)} \sum_{j \in \underline{S} \cup \overline{S}} \underline{p}_{ij}(u)\big(\underline{g}(i,u,j) + \alpha J_k(j)\big), \qquad \forall\, i \in \underline{S}, \tag{7.63}$$

$$\nu_k(i) = \arg\max_{v \in V(i)} \sum_{j \in \underline{S} \cup \overline{S}} \overline{p}_{ij}(v)\big(\overline{g}(i,v,j) + \alpha J_k(j)\big), \qquad \forall\, i \in \overline{S}. \tag{7.64}$$

When $\gamma_k = 1$ for all k, we have $J_{k+1} = J^{\mu_k,\nu_k}$, and the algorithm (7.62)-(7.64) reduces to the algorithm (7.59)-(7.61).

The convergence properties of the stepsize-based policy iteration algorithm (7.62)-(7.64) are not known at present. However, there has been encouraging experimentation with a version of the algorithm where γ_k is chosen to reduce the square of the norm of the Bellman equation error; see Filar and Tolwinski [FiT91], who view the algorithm (7.62)-(7.64) as a form of damped Newton's method.

Q-Learning

It is possible to derive a Q-learning algorithm for sequential games similar to the DP cases we have discussed so far in Sections 5.6 and 7.1.5. We form an auxiliary sequential game by augmenting the original system with one additional state for each possible pair (i,u) with $i \in \underline{S}$ and $u \in U(i)$, and one additional state for each possible pair (i,v) with $i \in \overline{S}$ and $v \in V(i)$. The probabilistic transition mechanism from the original states is the same as for the original problem. The probabilistic transition mechanism from an auxiliary state (i,u) with $i \in \underline{S}$ is that we move only to states j of the original problem with corresponding probabilities $\underline{p}_{ij}(u)$ and costs $\underline{g}(i,u,j)$, while from an auxiliary state (i,v) with $i \in \overline{S}$ we move only to states j of the original problem with corresponding probabilities $\overline{p}_{ij}(v)$ and costs $\overline{g}(i,v,j)$. Similar to our earlier developments, we can obtain a Bellman equation satisfied by the Q-factors $Q(i,u)$, $i \in \underline{S}$, and by the Q-factors $Q(i,v)$, $i \in \overline{S}$. It has the form

$$Q(i,u) = \sum_{j \in \underline{S} \cup \overline{S}} \underline{p}_{ij}(u)\left(\underline{g}(i,u,j) + \alpha J(j)\right), \qquad i \in \underline{S},\ u \in U(i), \tag{7.65}$$

$$Q(i,v) = \sum_{j \in \underline{S} \cup \overline{S}} \overline{p}_{ij}(v)\left(\overline{g}(i,v,j) + \alpha J(j)\right), \qquad i \in \overline{S},\ v \in V(i), \tag{7.66}$$

where

$$J(j) = \begin{cases} \min_{u \in U(j)} Q(j,u), & \text{if } j \in \underline{S}, \\ \max_{v \in V(j)} Q(j,v), & \text{if } j \in \overline{S}. \end{cases}$$

The natural form of the Q-learning algorithm for the sequential game problem is thus

$$Q(i,u) := Q(i,u) + \gamma\big(\underline{g}(i,u,j) + \alpha J(j) - Q(i,u)\big), \qquad i \in \underline{S},\ u \in U(i),$$

$$Q(i,v) := Q(i,v) + \gamma\big(\overline{g}(i,v,j) + \alpha J(j) - Q(i,v)\big), \qquad i \in \overline{S},\ v \in V(i),$$

where j and $\underline{g}(i,u,j)$ or $\overline{g}(i,v,j)$ are generated from the pair (i,u) or (i,v), respectively, by simulation. Convergence results for this method can be obtained, similar to the cases of discounted and stochastic shortest path problems, by appealing to the general convergence theory of Ch. 4 for stochastic iterative methods involving maximum norm contractions. The only difference from the analysis of Section 5.6 is that the minimization becomes maximization for some of the states. However, the underlying algorithmic mapping is still a maximum norm contraction. As a result, the proof of Prop. 5.5 applies nearly verbatim, and shows convergence with probability 1 of the sequence of Q-factors to their optimal values.

A further simplification occurs in *symmetric sequential games*. These are games like chess and backgammon, where there is a one-to-one correspondence between the two portions of the state space $\underline{S}$ and $\overline{S}$, and the maximizer and the minimizer have the same controls available and face transition costs of opposite signs when they are at the corresponding states of $\underline{S}$ and $\overline{S}$, respectively. Then if $\underline{i}$ and $\overline{i}$ are corresponding states of $\underline{S}$ and $\overline{S}$, respectively, we have $U(\underline{i}) = V(\overline{i})$, and it can be seen that $J^*(\underline{i}) = -J^*(\overline{i})$, and $Q(\underline{i},u) = -Q(\overline{i},v)$ when $u = v$. This reduces by a factor of two the number of cost-to-go values and Q factors that one has to calculate, and accordingly simplifies the value iteration and Q-learning algorithms.

7.2.4 Function Approximation Methods

The NDP methodology with cost-to-go function approximation is not very well developed at present for dynamic games. However, there have been reports of success with some of the natural extensions of the one-player NDP methods to sequential games. An interesting and straightforward possibility is to implement an approximate version of the (flawed) policy iteration algorithm of Eqs. (7.59)-(7.61). In particular, we may replace the policy evaluation of Eq. (7.59) with an approximate evaluation $\tilde{J}^{\mu_k,\nu_k}(\cdot, r_k)$, involving a function approximation architecture. This evaluation involves the stationary Markov chain corresponding to the policy pair (μ_k, ν_k), and can be performed using Monte-Carlo simulation or TD(λ), in exactly the same way as for discounted and stochastic shortest path problems. The "damped" version (7.62) of the policy iteration algorithm then takes the form

$$J_{k+1} = J_k + \gamma_k(\tilde{J}^{\mu_k,\nu_k} - J_k). \tag{7.67}$$

Unfortunately, there are no results that are comparable to the ones of Section 6.2 regarding the steady-state accuracy of the approximate version.

There are also optimistic versions of the approximate policy iteration scheme just discussed. For example, a TD(λ)-based method that parallels the first method given in Section 6.4 operates as follows. At the typical iteration, we are at some state i_k and we have a current parameter vector r_k. If $i_k \in \underline{S}$, we select $u_k \in U(i_k)$ such that

$$u_k = \arg \min_{u \in U(i_k)} \sum_{j \in \underline{S} \cup \overline{S}} \underline{p}_{i_k j}(u)\big(\underline{g}(i_k, u, j) + \alpha \tilde{J}(j, r_k)\big),$$

we generate the next state i_{k+1} by simulating a transition according to the transition probabilities $\underline{p}_{i_k j}(u_k)$, and finally we perform the TD(λ) update

$$r_{k+1} = r_k + \gamma_k d_k \sum_{m=0}^{k} (\alpha\lambda)^{k-m} \nabla \tilde{J}(i_m, r_m),$$

where d_k is the temporal difference

$$d_k = \underline{g}(i_k, u_k, i_{k+1}) + \alpha \tilde{J}(i_{k+1}, r_k) - \tilde{J}(i_k, r_k).$$

If $i_k \in \overline{S}$, we use the corresponding formulas where u_k is replaced by

$$v_k = \arg \max_{v \in V(i_k)} \sum_{j \in \underline{S} \cup \overline{S}} \overline{p}_{i_k j}(v)\big(\overline{g}(i_k, v, j) + \alpha \tilde{J}(j, r_k)\big),$$

$\underline{g}(i_k, u_k, j)$ is replaced by $\overline{g}(i_k, v_k, j)$, and the next state i_{k+1} is generated by simulating a transition according to the transition probabilities $\overline{p}_{i_k j}(v_k)$. Similar to the one-player case, no guarantees of success can be offered for this type of method.

Finally, it is possible to implement a cost function approximation method based on minimizing the squared residual of the Bellman equation $J = TJ$ [cf. Eq. (7.58)] using some gradient-like method of the type discussed in Section 6.10. The Bellman equation error approach may also be used for approximate evaluation of given stationary policies in the context of approximate policy iteration. The corresponding least squares problem can be solved by the two-sample gradient method discussed in Section 6.10.

7.3 PARALLEL COMPUTATION ISSUES

It is well-known that Monte-Carlo simulation is very well-suited for parallelization; one can simply carry out multiple simulation runs in parallel and occasionally merge the results. Also several DP-related methods are well-suited for parallelization; for example, each value iteration can be parallelized by executing the cost-to-go updates at different states in different parallel processors. In fact the parallel updates can be asynchronous. By this we mean that different processors may execute cost updates as fast as they can, without waiting to acquire the most recent updates from other processors; these latter updates may be late in coming because some of the other processors may be slow or because some of the communication channels connecting the processors may be slow. Asynchronous parallel value iteration can be shown to have the same convergence properties as its synchronous counterpart, and is often substantially faster. We refer to [Ber82a] and [BeT89] for an extensive discussion.

There are similar parallelization possibilities in approximate DP, in several off-line and on-line contexts. These include the generation of trajectories by simulation to use either for training architectures or for implementation of rollout policies, the minimization over all controls in the right-hand side of Bellman's equation, and the parallelization of various training algorithms. For example, approximate policy iteration with Monte Carlo simulation may be viewed as a combination of two operations:

(a) *Simulation*, which produces many pairs $\big(i, c(i)\big)$ of states i and sample costs $c(i)$ associated with a policy μ.

(b) *Training*, which obtains the state-sample cost pairs produced by the simulator and uses them in the least squares optimization of the parameter vector r of the approximate cost-to-go function $\tilde{J}^\mu(\cdot, r)$.

The simulation operation can be parallelized in the usual way by executing multiple independent simulations in multiple processors. The training operation can also be parallelized to a great extent. For example, one may parallelize the gradient iteration

$$r := r - \gamma \sum_{k=1}^{N} \nabla \tilde{J}^\mu(i_k, r)\big(\tilde{J}^\mu(i_k, r) - c(i_k)\big),$$

that is used for training. There are two possibilities here:

(1) To assign different components of r to different processors and to execute the component updates in parallel.

(2) To parallelize the computation of the sample gradient

$$\sum_{k=1}^{N} \nabla \tilde{J}^\mu(i_k, r)\big(\tilde{J}^\mu(i_k, r) - c(i_k)\big)$$

in the gradient iteration, by assigning different blocks of state-sample cost pairs to different processors.

There are several straightforward versions of these parallelization methods, and it is also valid to use asynchronous versions of them (see [BeT89], Ch. 7).

There is still another parallelization approach for the training process when a partitioned architecture is used (cf. Section 3.1.3). In this case, the state space is divided into several subsets S_m, $m = 1, \ldots, M$, and a different approximation $\tilde{J}^\mu(i, r_m)$ is calculated for each subset S_m. The parameter vectors r_m can then be obtained by a parallel training process using the applicable simulation data, that is, the state-sample cost pairs $\big(i, c(i)\big)$ with $i \in S_m$.

7.4 NOTES AND SOURCES

7.1. The theory of the average cost DP problem is discussed in all DP textbooks. The line of development used here, in terms of the associated stochastic shortest path problem, was introduced in Bertsekas [Ber95a]. The relative value iteration method is due to White [Whi63] and the contracting value iteration method is due to Bertsekas [Ber95f]. Another value iteration algorithm that may be particularly relevant to the simulation context is given by Jalali and Ferguson [JaF92]. Q-learning algorithms based on the relative value iteration have been proposed by Schwartz [Sch93], Singh [Sin94], and Mahadevan [Mah96]. The version given here differs slightly from the ones in these references. The Q-learning algorithm based on the contracting iteration was given in Bertsekas [Ber95a]. None of the above references on Q-learning contains a convergence analysis. The convergence results described in Section 7.1.5 are due to Abounadi, Bertsekas, and Borkar [ABB96]. The error bound of Section 7.1.6 for the approximate policy iteration method is new.

7.2. Dynamic games with finite state space and discounted cost were introduced by Shapley [Sha53]. The properties of policy iteration for these problems were analyzed by Pollatschek and Avi-Itzhak [PoA69], who proposed the simplified policy iteration (7.59) under some restrictive conditions. A counterexample to the convergence of this iteration was given by van der Wal [Van78], and a modification that improves the convergence properties is proposed by Filar and Tolwinski [FiT91]. For an extensive literature survey on dynamic games, we refer to Raghavan and Filar [RaF91]. Stochastic shortest path games were analyzed by Patek and Bertsekas [PaB96a]. Undiscounted dynamic games that are more general than stochastic shortest path games were studied by Kumar and Shiau [KuS81]. The convergence of Q-learning algorithms for discounted dynamic games was proved by Littman [Lit96].

Whether white or black,
it is a good cat if it catches mice.
(Deng Xiao Ping)

8

Case Studies

Contents

Our discussion so far has focused primarily on theoretical analysis. However, success with the solution of challenging problems also requires skill in applying the NDP methodology. In this chapter we describe some experiences of ourselves and of others using NDP algorithms. We also illustrate the interplay between problem structure, the choice of approximation architecture, and the choice of training methods, through the use of examples and case studies.

All of the case studies are organized roughly along the following sequence:

(a) Formulation of the DP problem, insight into its optimal solution, and description of the difficulties in solving the problem optimally.

(b) Description of the approximation architectures used with NDP methods.

(c) Description of the NDP training methods implemented.

(d) Evaluation of the accuracy of the solutions obtained, including a comparison with the optimal performance when possible.

(e) Evaluation of the effectiveness of the NDP methodology through comparisons with alternative heuristic solution methods when appropriate.

We have chosen a fairly broad cross-section of case studies and NDP methods to illustrate many of the major ideas of the preceding chapters. The computational results presented here are a small but representative sample of our experimentation. We have made an effort to illustrate some of the failures as well as some of the successes of the NDP methodology. Hopefully, this mixture of success and failure conveys to the reader our own overall impression that the NDP methodology is capable of impressive achievement, yet requires a lot of time-consuming trial and error experimentation with various methods and parameter settings. Vitally important for the success of this trial and error process is the understanding of the properties and limitations of the various NDP methods, as well as the exploitation of whatever insight can be developed about the practical problem at hand.

8.1 PARKING

Our first case study involves a well-known academic example. Our objective is to demonstrate some of the aspects of approximate policy iteration in a simple and intuitive problem setting.

A driver is looking for inexpensive parking on the way to his destination. The parking area contains N spaces. The driver starts at space N and traverses the parking spaces sequentially; that is, from space i he goes next to space $i-1$, etc. The destination corresponds to parking space 0. Each parking space is free with probability p independently of whether other parking spaces are free or not. The driver can observe whether a parking space is free only when he reaches it, and then, if it is free, he makes a decision to park in that space or not. If he parks in space i, he incurs a cost $c(i) > 0$. If he reaches the destination without having parked, he must park in the destination's garage, which is expensive and costs $C > 0$. The problem is to characterize the optimal parking policy and to approximate it computationally.

We formulate the problem as a stochastic shortest path problem. In addition to the termination state, we have state 0 that corresponds to reaching the expensive garage, and states (i, F) and $(i, \overline{F})$, $i = 1, \ldots, N$, where (i, F) [or $(i, \overline{F})$] corresponds to space i being free (or not free, respectively). The second component of the states (i, F) and $(i, \overline{F})$ is uncontrollable (cf. Section 2.4), so we can write Bellman's equation in terms of the reduced state i:

$$J^*(i) = p \min\big\{c(i),\ J^*(i-1)\big\} + (1-p)J^*(i-1), \qquad i = 1, \ldots, N, \tag{8.1}$$

$$J^*(0) = C. \tag{8.2}$$

The exact costs-to-go $J^*(i)$ can be easily calculated recursively from these equations, starting from $i = 0$ and proceeding to higher values of i. Furthermore, an optimal policy has the form

$$\mu^*(i) = \begin{cases} \text{park}, & \text{if space } i \text{ is free and } c(i) \leq J^*(i-1), \\ \text{do not park}, & \text{otherwise.} \end{cases}$$

From Eq. (8.1), we have for all i,

$$\begin{aligned} J^*(i) &= p \min\big\{c(i),\ J^*(i-1)\big\} + (1-p)J^*(i-1) \\ &\leq pJ^*(i-1) + (1-p)J^*(i-1) \\ &= J^*(i-1). \end{aligned}$$

Thus $J^*(i)$ is monotonically nonincreasing in i. From this it follows that if $c(i)$ is monotonically increasing in i, there exists an integer i^* such that it is optimal to park at space i if and only if i is free and $i \leq i^*$, as illustrated in Fig. 8.1.

To illustrate the approximate policy iteration method, we consider the case where

$$p = 0.05, \quad c(i) = i, \quad C = 100, \quad N = 200, \tag{8.3}$$

and we introduce two approximation architectures for J^*. The first is the quadratic architecture

$$\tilde{J}(i, r) = r(0) + r(1)i + r(2)i^2, \tag{8.4}$$

and the second is the piecewise linear/quadratic architecture

$$\tilde{J}(i,r) = \begin{cases} r(0) + r(1)\big(r(2) - i\big)^2, & \text{if } i \le r(2), \\ r(0), & \text{if } i > r(2). \end{cases} \tag{8.5}$$

The choice of the latter architecture is somewhat artificial, because it requires insight into the form of the optimal cost-to-go function. We have used this architecture to contrast the corresponding results with those for the quadratic architecture, which is less capable of approximating the optimal cost-to-go function J^*.

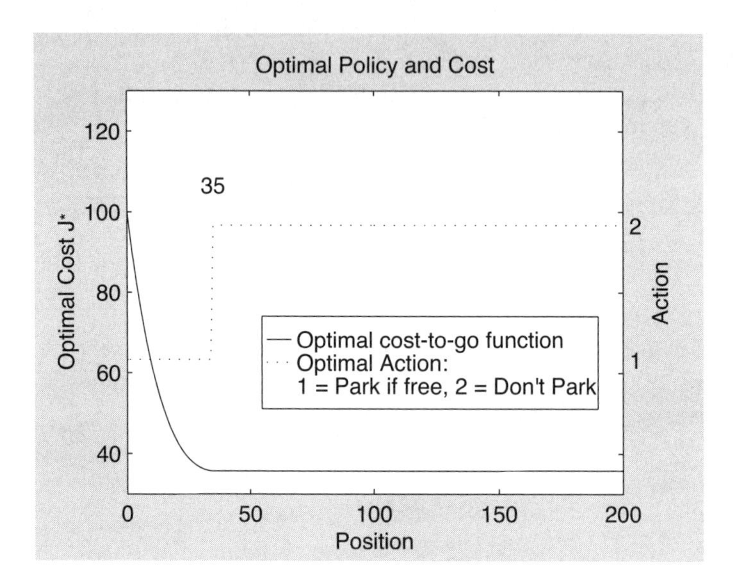

Figure 8.1: Optimal cost-to-go and optimal policy for the parking problem with the data in Eq. (8.3). The optimal policy is to park at the first available space after the threshold space 35 is reached.

We trained these architectures using approximate policy iteration and Monte-Carlo simulation. Each policy was evaluated using cost samples from 1000 trajectories. Each trajectory visited a number of states and generated one cost sample for each of these states. The initial state of each trajectory was randomly chosen from the range [1,200]. The parameter vector r corresponding to a policy was determined by solving the corresponding least squares problem by a gradient-like method. The starting policy was the (poor) policy of parking at the first free space encountered. All the policies generated by the approximate policy iteration method turned out to be threshold policies, that is, for some threshold $\bar{i}$, they park at the first free space after reaching space $\bar{i}$.

Figure 8.2 illustrates the sequence of approximate cost-to-go functions $\tilde{J}(i, r_k)$ and corresponding policies generated using the quadratic architec-

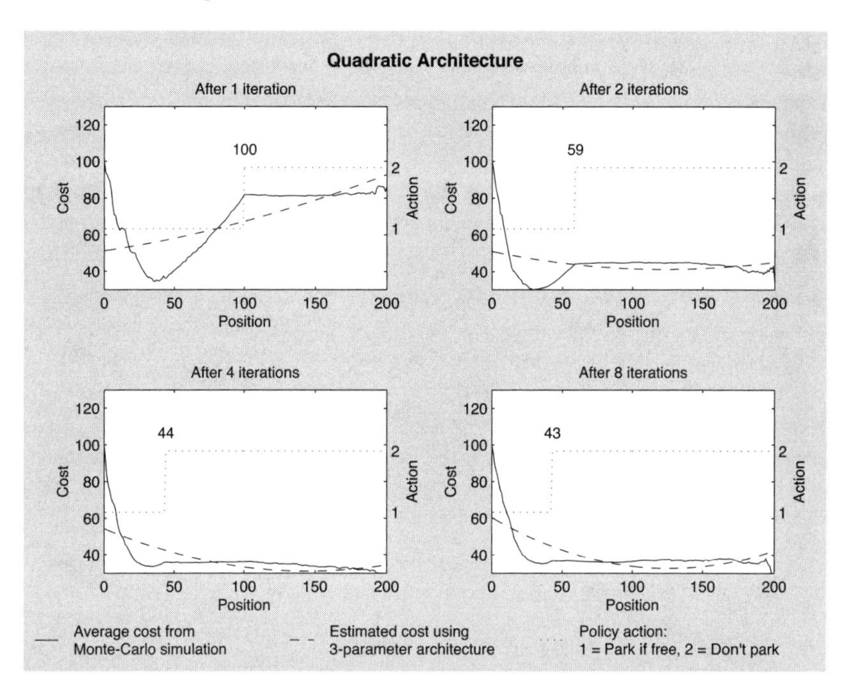

Figure 8.2: Sequence of approximate cost-to-go functions $\tilde{J}(i, r_k)$ and corresponding policies generated using approximate policy iteration, in conjunction with Monte-Carlo simulation, and the quadratic architecture. The actual cost-to-go of the policy (as estimated by Monte-Carlo simulation) and the approximate cost-to-go (as estimated by the quadratic architecture) are shown as functions of position. Each generated policy is a threshold policy, that is, a policy that parks at the first available space after a certain threshold space is reached. The sequence of generated thresholds is 200, 100, 59, 45, 44, 42, 43, 44, 43, 44. The threshold 200 corresponds to the initial policy that parks at the first available space. The irregularities of the average cost (solid) curve for i near 200 are due to the way that trajectories are initialized, which results in fewer sample costs for i near 200.

ture, while Fig. 8.3 shows the corresponding sequences for the piecewise linear/quadratic architecture. The quadratic architecture cannot provide a good approximation of J^* and the thresholds of the generated policies oscillate between 43 and 44. On the other hand the piecewise linear/quadratic architecture approximates J^* much better and the thresholds of the generated policies oscillate between 34 and 37. (Recall that the optimal threshold is 35.) We observe here a generic feature of approximate policy iteration: the generated policies oscillate. However, the performance of the policies involved in the oscillation depends on the power of the architecture. When the policy evaluation steps involve substantial error, the oscillation may occur far from the optimum, as illustrated in Fig. 8.2 for the quadratic architecture.

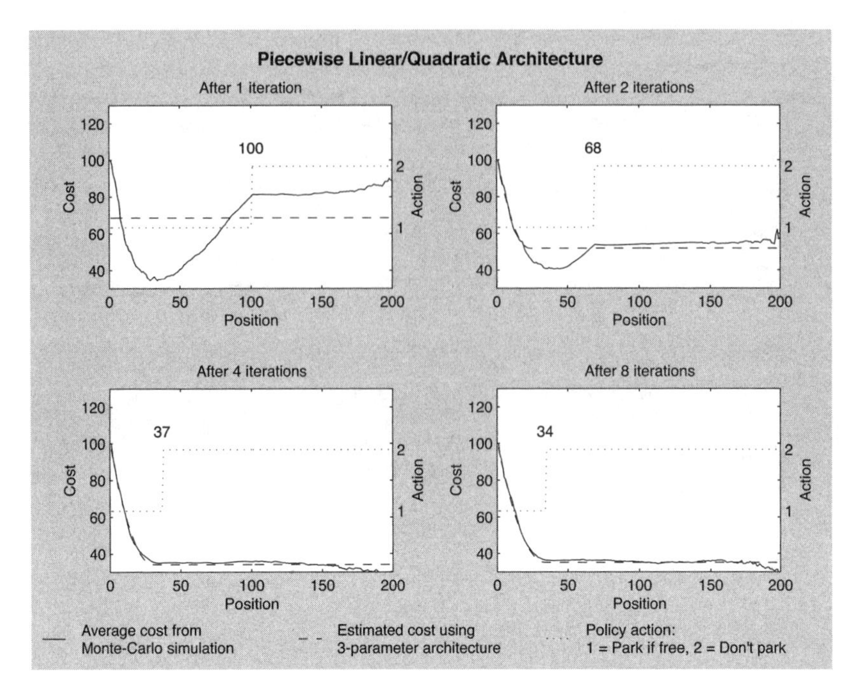

Figure 8.3: Sequence of approximate cost-to-go functions $\tilde{J}(i, r_k)$ and corresponding policies generated using approximate policy iteration, in conjunction with Monte-Carlo simulation, and using the piecewise linear/quadratic architecture. The actual cost-to-go of the policy (as estimated by Monte-Carlo simulation) and the approximate cost-to-go (as estimated by the linear/quadratic architecture) are shown as functions of position. Each generated policy is a threshold policy, that is, a policy that parks at the first available space after a certain threshold space is reached. The sequence of generated thresholds is 200, 100, 68, 49, 37, 35, 37, 36, 34, 36. The threshold 200 corresponds to the initial policy that parks at the first available space.

8.2 FOOTBALL

In this section, we discuss a simplified version of the game of American football, which we cast as a stochastic shortest path problem. We focus on a single offensive drive by one of the teams. The objective is to obtain a play selection policy that maximizes the expected score during the drive minus the estimated score of the opposing team at the next drive. We used two different types of architectures and several versions of policy iteration to obtain reward-to-go approximations and corresponding play selection policies. Here are some of the salient features of this case study:

(a) Despite the substantial number of states (of the order of 15,000), it is possible to compute the optimal reward-to-go function and an optimal policy exactly, and to use them to assess the performance of the NDP methods.

(b) We partitioned the state space in four subsets (one for each "down"), and we used a separate approximation architecture within each subset. We experimented with two different types of architectures: (1) a multilayer perceptron, and (2) a quadratic polynomial architecture with feature iteration, whereby we progressively added features to be used, together with the state, in the quadratic function approximation.

(c) We used various forms of approximate and optimistic policy iteration. The performance of the best policies obtained with these methods was very close to being optimal and was also substantially better than the performance of well-designed heuristic policies. We also implemented rollout policies based on some of the policies obtained by the policy iteration methods, and we verified that rollout can lead to substantial performance improvements.

Problem Formulation

We will assume that the reader is somewhat familiar with the game of American football. The state here is the triplet (x, y, d), where x is the number of yards to the goal (field position ranging from 1 to 100), y is the number of yards to go until the next 1st down (ranging from 1 to x), and d is the down number (which can be 1, 2, 3, or 4). We restrict x and y to take integer values only. The team starts at some field position with a 1st down and 10 yards to go to the next 1st down, unless $x < 10$ in which case we set $y = x$. At each state the quarterback must choose one out of four play options: run, pass attempt, punt, and field-goal attempt. If a play yields a smaller number of yards than y, the down number is incremented; otherwise the team starts again with a 1st down and 10 yards to go to the next 1st down, unless $x < 10$ in which case we set $y = x$. There is a termination state, which is reached when a touchdown or a field goal is scored. The termination state is also reached when the ball is turned over to the opposing team through a punt, or an interception, or when the team fails to reach a 1st down on a 4th down play. Upon termination, there is a corresponding reward if a touchdown is scored (6.8 points, that is, 6 points for the touchdown itself and 1 point for an extra point with probability 0.8) or a field goal is scored (3 points) or a safety occurs (-2 points). There is also a terminal cost that depends on the field position at which the opposing team obtains the ball. This latter cost represents our estimate of the expected score of the opposing team at the next drive and is assumed to be known. The transitions from state to state are governed by known probabilities, which depend on the play selected (the precise details of the model can be found in [PaB96b]; generally we guessed numbers that seemed reasonable). With our model, termination is inevitable under all policies, and in fact, depending on the initial state, the maximum number of transitions to termination is of the order of 40.

The problem is small enough to be solved exactly (the total number of states is 15,100). On a 120 MHz Pentium PC-type machine, it took about 2.5 minutes to compute the optimal reward-to-go function and the optimal policy shown in Fig. 8.4 using policy iteration. To evaluate the reward-to-go function for each iterate policy, we used value iteration until the maximum norm difference between successive iterates was less than 10^{-6} football points. Starting from an initial policy that may be described as "always run," six policy iterations were required to determine an optimal policy.

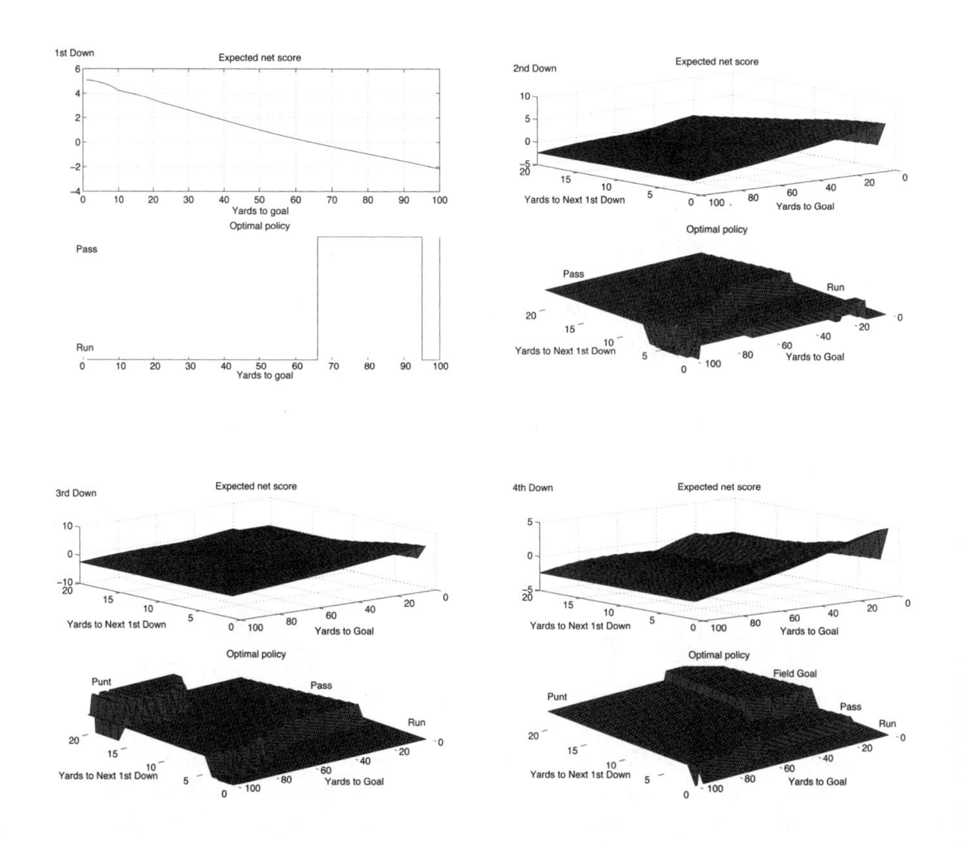

Figure 8.4: Optimal reward-to-go and optimal policy for each of the four downs in the football problem.

Approximation Architectures

For a given policy μ, the approximate reward-to-go function $\tilde{J}^{\mu}(i, r)$ was realized as the output of one of the four independently trained parametric approximators $\tilde{J}_d^{\mu}(i, r_d)$, where i is the state and $d \in \{1, 2, 3, 4\}$ is the down number. Thus, we can think of r as consisting of the four components r_1, r_2, r_3, r_4.

One of the two architectures consisted of a multilayer perceptron (MLP). The MLP corresponding to down d accepts as input the field position x and the yards-to-next-first down y, produces as output $\tilde{J}_d^{\mu}(x, y, r_d)$, and involves a single hidden layer consisting of 20 hyperbolic tangent sigmoidal units.

The second architecture we used was based on a *quadratic polynomial architecture* and *feature iteration*. In the simplest version of this architecture, for each down $d = 2, 3, 4$, the corresponding approximate reward-to-go function $\tilde{J}_d^{\mu}(x, y, r_d)$ was given by a quadratic polynomial in x and y. Thus for each $d = 2, 3, 4$, the vector of parameters r_d had 6 components, corresponding to the 6 coefficients of the quadratic polynomial. For $d = 1$ we can only have $y = 10$, so the architecture was quadratic in x only, and involved 3 parameters. This is an extremely simple, linear architecture, which did not produce a performance comparable to the one obtained with a multilayer perceptron. In an effort to improve the performance, we used feature iteration; that is, we augmented the state with extra features, which were the approximate reward-to-go functions of several past policies. In particular, we fixed an integer $m > 0$, and we evaluated the approximate reward-to-go of the current policy μ_k as a quadratic polynomial of x, y, *and* the approximate reward-to-go functions of the past $\hat{m}$ policies $\tilde{J}_d^{\mu_{k-1}}(x, y, r_d^{k-1}), \ldots, \tilde{J}_d^{\mu_{k-\hat{m}}}(x, y, r_d^{k-\hat{m}})$, where $\hat{m} = \min\{k, m\}$. We thus obtained a richer architecture, which in the end resulted in better performance. This architecture is still linear, and leads to linear least squares problems that can be solved with exact algorithms such as the SVD (cf. Section 3.2.2).

Training Methods

The methods used for training the architectures were the approximate policy iteration methods of Sections 6.2-6.4. There is a broad range of possibilities here. In particular, one can use several different methods for policy evaluation, and one can also use different amounts of training data (numbers of simulated system trajectories) between policy updates.

We experimented with the following methods for policy evaluation.

(a) For the multilayer perceptron architecture, we used TD(λ) for $\lambda = 0$, $\lambda = 0.5$, and $\lambda = 1$. We also used the minimization of the Bellman equation error method with the two-sample gradient method described in Section 6.10.

(b) For the quadratic polynomial architecture, we formulated the policy evaluation problem as a linear least squares problem (cf. Section 3.2.2), which we solved with the SVD method or with variants of the Kalman filtering algorithm.

Each of these methods, was implemented in two basic versions:

(1) Approximate policy iteration, which involved using a very large number of simulated trajectories to train for each generated policy. Here, to economize in computation time in the case of the MLP architecture, we used the device of *recycling trajectories*, that is, generating a fairly large number of trajectories and reusing them in the training process several times.

(2) Optimistic policy iteration, which involved using a relatively small number of simulated trajectories to train for each generated policy.

The method for choosing the initial state for the simulated trajectories proved to be significant. We tried several possibilities:

(i) Start from a limited set of initial states [1st down, 10 yards to go to the next 1st down, and a range of yards to go to the goal (around 80 yards)].

(ii) Sequentially step through each element of the state space, ensuring that all states are used as initial states of trajectories at least a fixed number of times.

(iii) Randomly select initial states according to some probability distribution, which gives preference to regions of the state space that "need" special attention.

The performance obtained with the first method for choosing the initial state was not as good as with the other methods. Regarding the second method, it seems that the ordering itself may cause the approximation results to become biased. This problem can be avoided with the third method (the precise probability distribution of the initial state selection is described in [PaB96b]). All of the subsequent experimental results were obtained using method (iii) for initial state sampling.

Experimental Results

Figure 8.5 gives results obtained with the MLP architecture and approximate policy iteration. The figure shows the expected net score achieved by successive policies starting with a 1st down at the 80-yard line (as computed by Monte-Carlo simulation with 8000 sample trajectories). (The optimal score starting with a 1st down at the 80-yard line was calculated to be -0.945.) An oscillation fairly close to the optimum can be observed. The figure also shows an essentially exact evaluation of the error between the expected score of the "best" of the approximate policies generated and

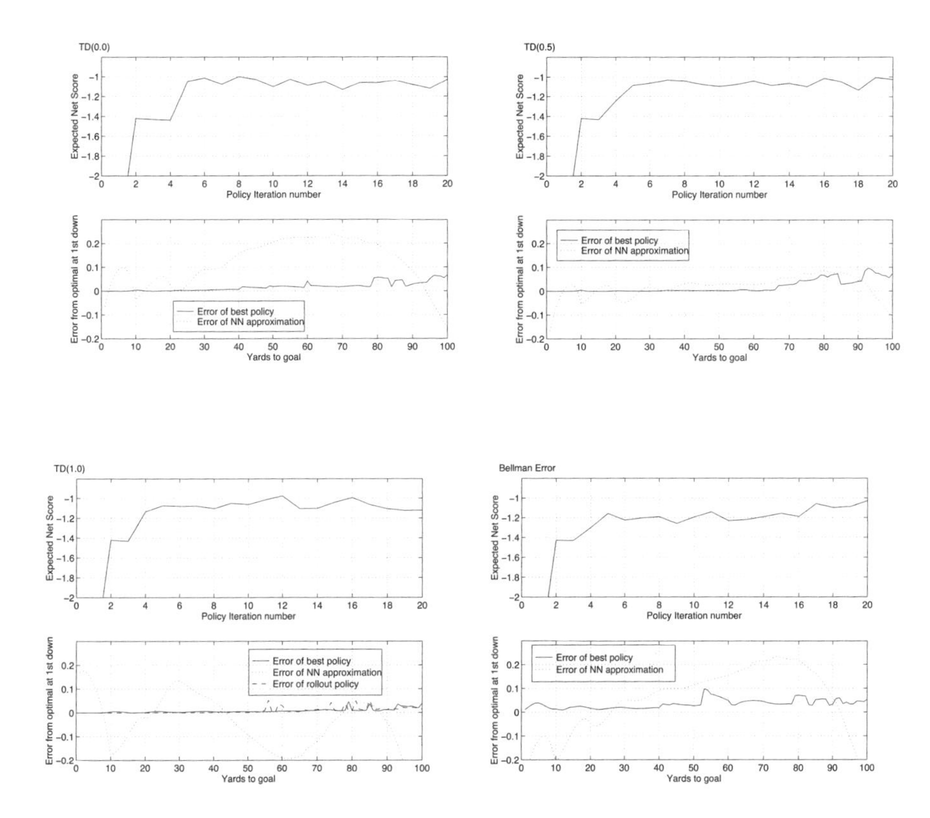

Figure 8.5: Performance of approximate policy iteration with the MLP architecture using TD(λ) for $\lambda = 0, 0.5, 1$, and the Bellman equation error method. Each approximate policy evaluation was based on 10000 trajectories, which were recycled 100 times each. Each figure shows the expected net score from a 1st down, and an initial field position of $x = 80$, as a function of policy number. The figures also show the difference between the average score of one of the policies (judged to be the "best" generated by the corresponding training method) and the optimal average score, for various starting field positions and a 1st down. The dotted plots give the average score of the "best" policy, as predicted by the neural network.

the optimal policy, as a function of field position. It can be seen that all methods produce policies that come within 0.08 of the optimal score. The best policy, produced by TD(1), is very close to optimal. We also generated a rollout policy based on that policy, which as shown in the figure, is also very close to optimal.

Figure 8.6 gives results that are analogous to the ones of Fig. 8.5, except that optimistic policy iteration was used in place of approximate

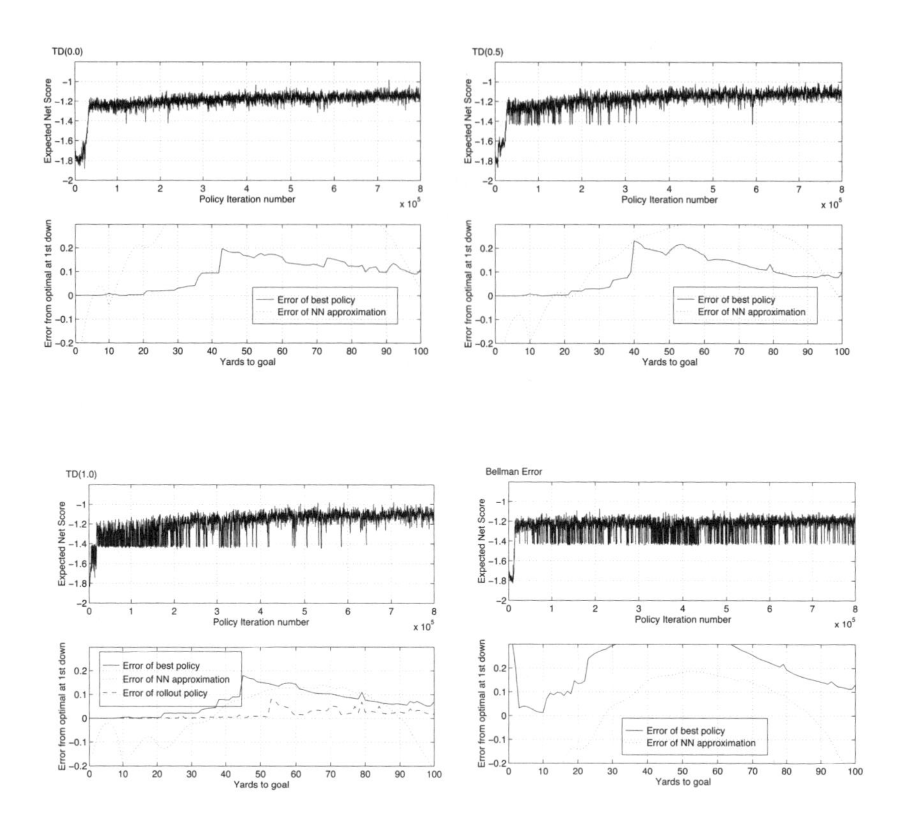

Figure 8.6: Performance of optimistic policy iteration with the MLP architecture using TD(λ) for $\lambda = 0, 0.5, 1$, and the Bellman equation error method. Each policy evaluation was based on a single trajectory. Note the improvement obtained using a rollout policy (bottom left figure).

policy iteration. The plots shown in the top figures for each method show fairly accurate performance estimates of the policies (using Monte-Carlo simulation with 8000 samples). Chattering is clearly occurring here. The performance achieved was quite good (except perhaps for the Bellman error method), but not as good as the performance achieved using approximate policy iteration. Note that using a rollout policy substantially improves the performance, as shown in the bottom left figure.

Figures 8.7 and 8.8 parallel Figs. 8.5 and 8.6, respectively, for the case of the quadratic architecture. The performance, however, is not as good as for the MLP architecture. On the other hand, substantial improvement is obtained when feature iteration is introduced, as indicated in Figs. 8.9 and 8.10.

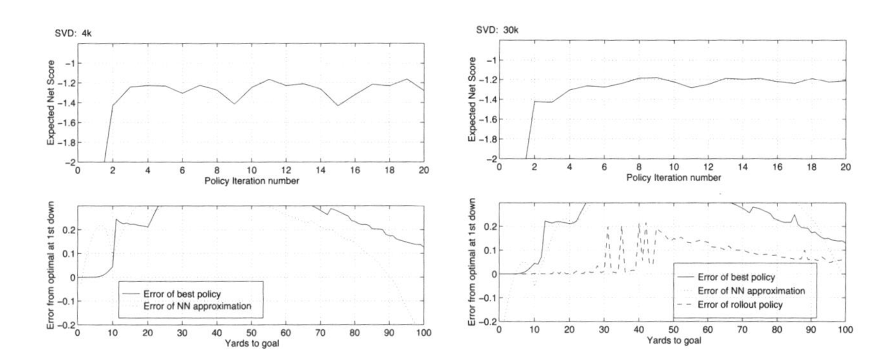

Figure 8.7: Performance of approximate policy iteration with the quadratic polynomial architecture using the SVD method of training. Each policy evaluation was based on 4000 trajectories and also on 30000 trajectories. Note the improvement obtained using a rollout policy (bottom right figure).

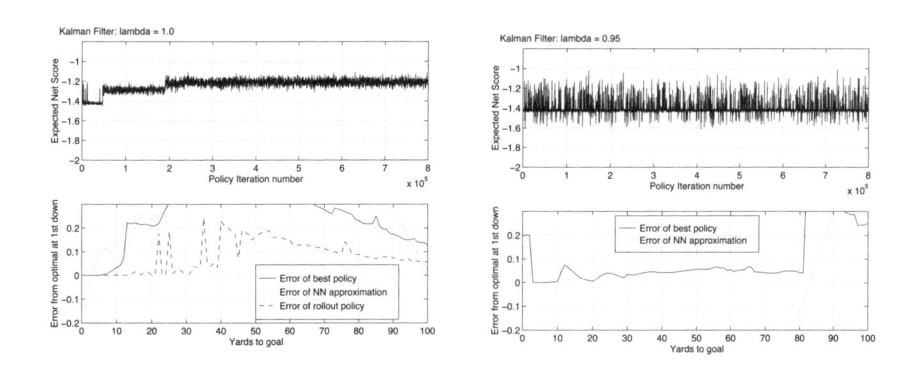

Figure 8.8: Performance of optimistic policy iteration with the quadratic polynomial architecture using a Kalman filtering method for training, with the exponential fading factor set at $\lambda = 0.95$ and $\lambda = 1$. Each policy evaluation was based on a single trajectory. Note the improvement obtained using a rollout policy (bottom left figure).

Comparison with a Heuristic Policy

We also compared the policies obtained with NDP methods with some reasonable heuristic policies. In particular, we introduced a parametric class of policies and we optimized over the free parameters. Basically,

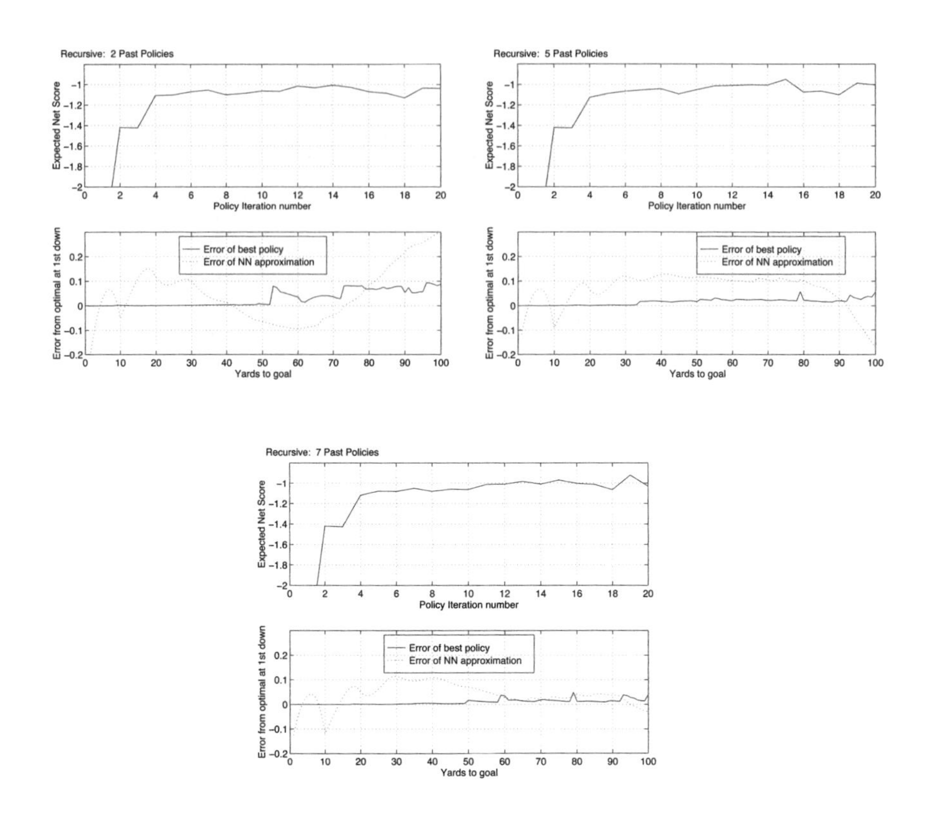

Figure 8.9: Performance of approximate policy iteration with the quadratic polynomial architecture using the SVD method of training, and feature iteration using the last 2, 5, and 7 policies. Each policy evaluation was based on 30000 trajectories. Comparing with the results of Fig. 8.7, we see that feature iteration significantly improves the performance.

the parameters consisted of thresholds for yards-to-go to the 1st down separating the regions within which a particular type of play (run, pass, etc.) was used. Separate thresholds were chosen for each down and a total of 1286 heuristic policies were generated. We evaluated these policies using Monte-Carlo simulation, and we selected the one that achieves the best expected score, starting with a 1st down at the 80-yard line. This policy is shown in Fig. 8.11 together with its expected score. Starting with a 1st down at the 80-yard line, this policy achieves a score of about -1.261, which is substantially smaller than the ones obtained with the NDP generated policies.

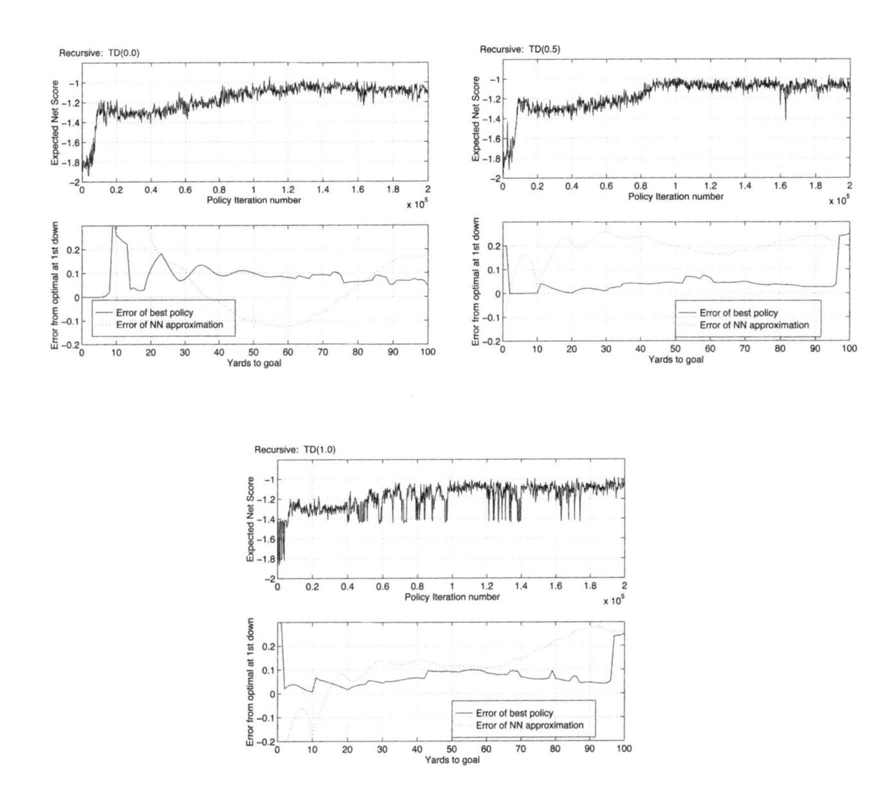

Figure 8.10: Performance of optimistic policy iteration with the quadratic polynomial architecture using TD(λ) for $\lambda = 0, 0.5, 1$, and feature iteration. Each policy evaluation was based on a single trajectory. A new feature was added every 40000 iterations. This feature was the approximate reward function of one of the policies generated within the preceding 40000 iterations (this policy was selected by a special screening procedure).

8.3 TETRIS

This case study relates to the game of tetris, which was described in Example 2.3 of Section 2.4. As discussed in that example, the problem can be formulated as a stochastic shortest path problem with a very large number of states. Since some effective features of the state are easily identifiable, it is natural to consider a linear feature-based approximation architecture. After some experimentation, the following features were used:

(a) The height h_k of the kth column of the wall. There are w such features, where w is the wall's width.

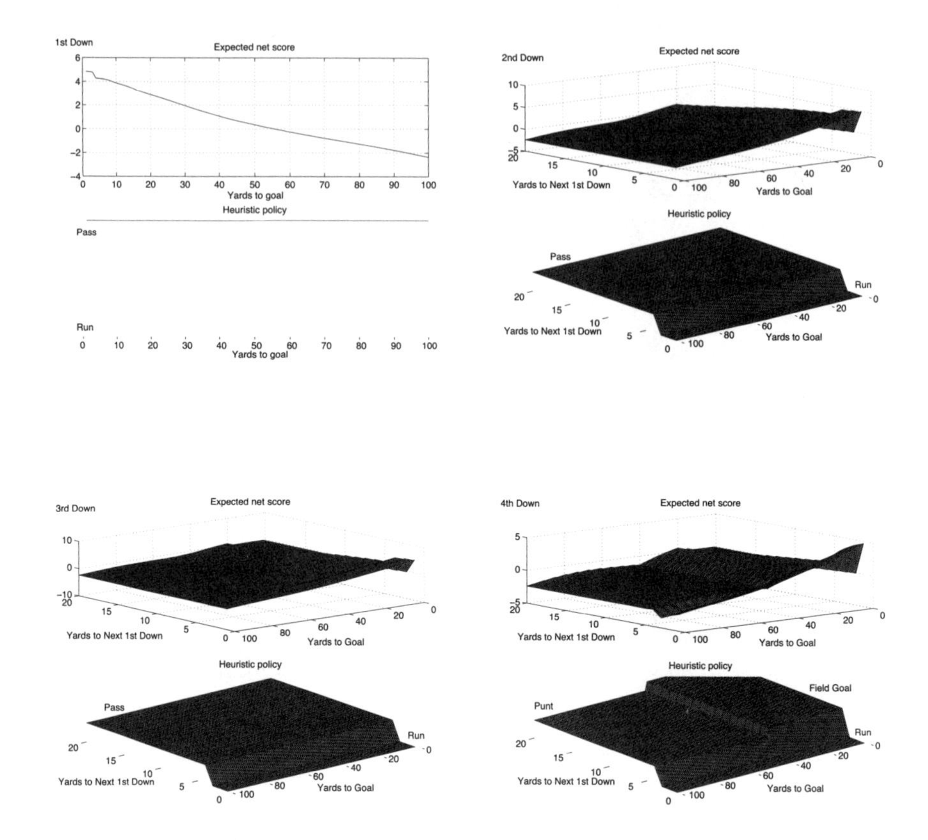

Figure 8.11: Reward-to-go of the best heuristic policy that we have been able to construct.

(b) The absolute difference $|h_k - h_{k+1}|$ between the heights of the kth and the $(k+1)$st column, $k = 1, \ldots, w-1$.

(c) The maximum wall height $\max_k h_k$.

(d) The number of holes L in the wall, that is, the number of empty positions of the wall that are surrounded by full positions.

Thus, there are $2w + 1$ features, which together with a constant offset, require $2w + 2$ weights in a linear architecture of the form

$$\tilde{J}(i, r) = r(0) + \sum_{k=1}^{w} r(k) h_k + \sum_{k=1}^{w-1} r(k+w)|h_k - h_{k+1}| \\ + r(2w) \max_k h_k + r(2w+1)L.$$

We tried an approximate version of the λ-policy iteration method, discussed in Section 2.3. To describe the typical iteration of this method, let r_t be the weight vector after t iterations, and let μ_t be the greedy policy with respect to r_t. The policy μ_t is evaluated using a batch of M (of the order of 100) games, and the corresponding weight vector r_{t+1} is obtained as

$$r_{t+1} = \arg\min_r \sum_{m=1}^{M} \sum_{k=0}^{N_m} \Big(\tilde{J}(i_{m,k}, r) - \tilde{J}(i_{m,k}, r_t) - \sum_{s=k}^{N_m - 1} \lambda^{s-k} d(i_{m,s}, i_{m,s+1}) \Big)^2, \tag{8.6}$$

where $(i_{m,0}, i_{m,1}, \ldots, i_{m,N_m-1}, i_{m,N_m})$ is the sequence of states comprising the mth game in the batch, with i_{m,N_m} being equal to the termination state, and

$$d(i_{m,s}, i_{m,s+1}) = g\big(i_{m,s}, \mu_t(i_{m,s}), i_{m,s+1}\big) + \tilde{J}(i_{m,s+1}, r_t) - \tilde{J}(i_{m,s}, r_t)$$

are the corresponding temporal differences with $\tilde{J}(i_{m,N_m}, r_t)$ being equal to the terminal value 0. All games were started from the empty board position (things did not change much when the initial board position was chosen more randomly).

We also tried an optimistic version of the λ-policy iteration method, whereby a weight vector $\hat{r}_t$ was calculated by solving the minimization problem in Eq. (8.6), with the number M of games in the batch being relatively small (of the order of 5). The new weight vector r_{t+1} was then computed by interpolation between r_t and $\hat{r}_t$, i.e.,

$$r_{t+1} = r_t + \gamma_t(\hat{r}_t - r_t), \tag{8.7}$$

where γ_t is a stepsize that satisfies $0 < \gamma_t \leq 1$ and is diminishing with t.

We now describe some of the results of the computational experimentation. The wall width w was taken to be 10, the wall height was taken to be 20, and the types of falling objects were the 7 possible shapes that consist of 4 pieces. Each falling object was chosen with equal probability from the 7 possible shapes, independently of the shapes of the preceding objects. The starting set of weights in our experiments was $r(2w) = 10$, $r(2w+1) = 1$, and $r(k) = 0$ for $k < 2w$ (this set of weights was derived from those used by Van Roy [Van95]). With the weights fixed at these initial values, the corresponding greedy policy scores in the low tens.

The approximate λ-policy iteration method quickly gave playing policies that score in the thousands, except when $\lambda = 1$, in which case the method failed to make satisfactory progress. Generally, the maximum score achieved depended on the value of λ. Table 8.1 gives some illustrative results with different values of λ. Figure 8.12 shows the sequence of tetris scores obtained during training for the case where $\lambda = 0$, $\lambda = 0.3$, $\lambda = 0.5$, and $\lambda = 0.7$. These results were obtained using training with 100 games per policy update. Similar results were obtained using 300 games per update. An interesting and somewhat paradoxical observation is that a high performance is achieved after relatively few policy iterations, but the performance gradually drops significantly. We have no explanation for this intriguing phenomenon, which occurred with all of the successful methods that we tried.

λ	0.0	0.1	0.2	0.3	0.4	0.5	0.6	0.7	0.8	0.9
Score	2909	2818	2730	2968	3014	2786	3183	1941	2103	1054

Table 8.1: Average scores of the highest scoring policies obtained using different values of λ after 15 policy/weight updates, with 100 games between policy updates. Each data point is the average of the scores of 100 games.

The optimistic version of the method produced similar results to the nonoptimistic version. Figure 8.13 shows the sequence of tetris scores obtained during training for the case where 5 games were used per policy iteration and $\lambda = 0.6$. The stepsize γ_t used was of the form $c/(t + d)$, where t is the policy index and c and d are positive scalars. The proper value of c strongly depends on λ. In particular, smaller values of c are required for λ close to 1, since the weight update increments tend to be larger as λ increases. On the whole, however, it was not difficult to obtain reasonable stepsize parameter values by trial and error.

Note that the variance of the scores obtained with various policies (and particularly the high scoring ones) is very high. This may explain the relatively poor performance of the methods for values of λ that are near 1.

We finally mention two additional approaches that were used for training tetris players. The first was a policy iteration approach, where the weights of the architecture were updated using the TD(λ) method described in Section 6.4. This approach ran into serious difficulties, typically failing to make substantial progress, and was abandoned in favor of the much better performing λ-policy iteration method.

In the second approach, the problem was altered by imposing a cost structure that provides an incentive for not losing quickly rather than for achieving a high score. In particular, the objective was reformulated so that maximization of the average game score, was replaced by minimization of the total number of wall height increases in the course of placing the next N falling objects, where N is a fixed integer (values between 10 and 100 were used). Thus, for a given stage where the maximum wall height changes

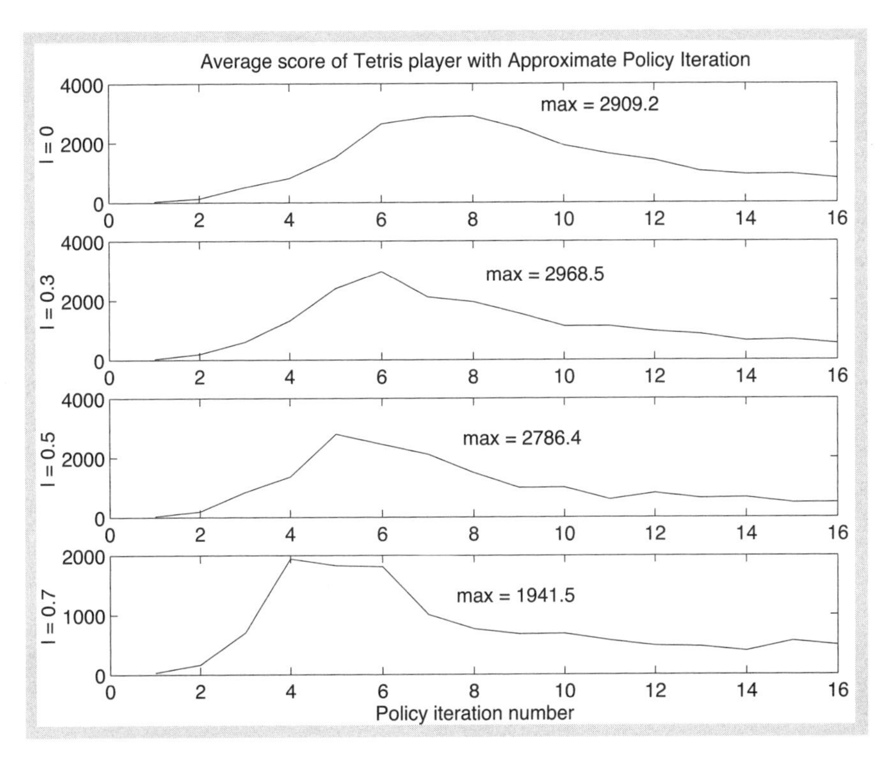

Figure 8.12: Sequence of tetris scores of the policies generated using approximate λ-policy iteration, for four different values of λ. Each data point is the average of the scores of 100 games.

from a value of h to a value of $\overline{h}$, the cost is $\max\{0, \overline{h} - h\}$. Also, to discourage termination of the game, a fixed terminal cost was introduced (values between 5 and 20 were used). Thus the problem was transformed to a stochastic shortest path problem, where the termination occurs in at most N stages. The two problem formulations are substantially different, but it was reasoned that a good policy under one formulation should also be good for the other. Note that it is essential to fix the number of stages within which to count height increases, because otherwise a good tetris playing policy, which achieves a high score, would perform very poorly in terms of number of height increases over a long horizon. Note also that because the horizon is limited, the cost samples obtained through simulation have much lower variance than the game scores, and this appears to be significant for the effectiveness of the training process.

Based on the alternative problem formulation just described, an approximate and an optimistic version of the λ-policy iteration method of Section 2.3 were tried. The results obtained were similar but somewhat more favorable than the ones presented in Table 8.1. This occurred uni-

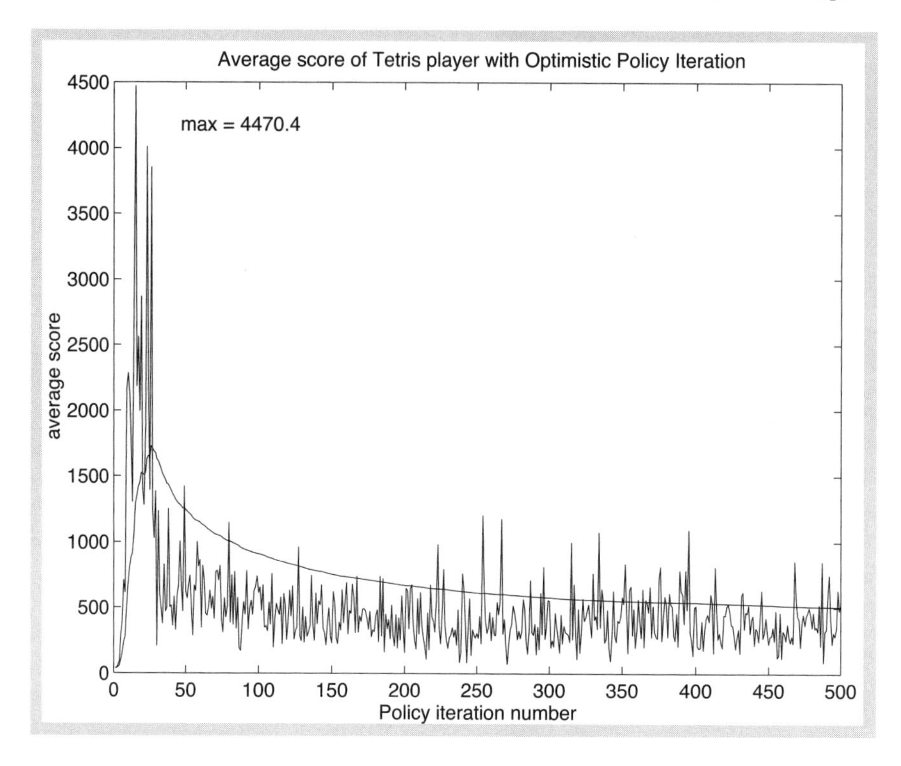

Figure 8.13: Sequence of tetris scores of the policies generated using optimistic λ-policy iteration, where $\lambda = 0.6$. Each data point is the average of the scores of 5 games. The cumulative average score is also shown.

formly for all values of λ. In particular, a maximum score of 3554 was obtained for $\lambda = 0.3$, which should be compared with the best score of 3183 given in Table 8.1 ($\lambda = 0.6$).

8.4 COMBINATORIAL OPTIMIZATION – MAINTENANCE AND REPAIR

We discussed in Example 2.6 of Section 2.4 and also in Section 3.1.4 how NDP techniques can be used to address deterministic discrete optimization problems, by first converting them to DP problems. In this section, we provide an example of this solution process. In particular, we consider a two-stage version of the maintenance problem of Example 2.4 in Section 2.4. There is a simple optimal policy for the second stage, so this problem is in reality a one-stage problem. Still, the first stage problem can be a difficult combinatorial problem, so we apply the approach in Section 3.1.4, which uses heuristics as feature extraction mappings in an architecture that is trained using various NDP methods.

Referring to Example 2.4 in Section 2.4, we have T types of machines. A machine of type t that is operational at the beginning of a stage breaks down during that stage with probability p_t, independently of other breakdowns, and may be repaired at the end of the stage, so that it is operational at the beginning of the next stage. Each repair requires the use of one spare part. The problem is to find the repair policy that maximizes the total value of surviving machines at the end of the two stages, using a fixed number of spare parts. Equivalently, the objective is to minimize the total cost of machines that break down and do not get repaired. The essence of the problem is to trade off repairing the first stage breakdowns with leaving enough spare parts to repair the most expensive of the second stage breakdowns. We note that the maintenance problem of this section is typical of a broad and important class of two-stage problems known as *stochastic programming problems*.

The state has the form (m, y, s); here s is the number of available spare parts at the beginning of the first stage, and m and y are the vectors

$$m = (m_1, \ldots, m_T), \qquad y = (y_1, \ldots, y_T),$$

where m_t, $t = 1, \ldots, T$, is the number of working machines of type t at the beginning of the first stage, and y_t, $t = 1, \ldots, T$, is the number of breakdowns of machines of type t during the first stage. The control is

$$u = (u_1, \ldots, u_T),$$

where u_t is the number of spare parts used to repair breakdowns of machines of type t at the end of the first stage. We note that during the second stage, it is optimal to use the remaining spare parts to repair the machines that break in the order of their cost (that is, repair the most expensive broken machines, then if spare parts are left over, consider the next most expensive broken machines, etc). Thus, if we start the second stage with $\overline{s}$ spare parts, and $\overline{m}_t$ machines of type $t = 1, \ldots, T$, and during the second stage, $\overline{y}_t$ machines of type t break, $t = 1, \ldots, T$, the optimal cost-to-go of the second stage, which is denoted by $G(\overline{m}, \overline{y}, \overline{s})$, where

$$\overline{m} = (\overline{m}_1, \ldots, \overline{m}_T), \qquad \overline{y} = (\overline{y}_1, \ldots, \overline{y}_T),$$

can be calculated analytically. We will not give the formula for G, because it is quite complicated, although it can be easily programmed for computation.

Let us denote by R the expected value, over the second stage breakdowns, of the second stage cost

$$R(\overline{m}, \overline{s}) = E_{\overline{y}}\big[G(\overline{m}, \overline{y}, \overline{s})\big].$$

Then in the first stage, the problem is to find $u = (u_1, \ldots, u_T)$ that solves the problem

$$\begin{aligned} &\text{minimize} \quad \sum_{t=1}^{T} C_t(y_t - u_t) + V_{(m,y,s)}(u) \\ &\text{subject to} \quad \sum_{t=1}^{T} u_t \leq s, \qquad 0 \leq u_t \leq y_t, \quad t = 1, \ldots, T, \end{aligned}$$

where

$$V_{(m,y,s)}(u) = R\left(m_1 - y_1 + u_1, \ldots, m_T - y_T + u_T, s - \sum_{t=1}^{T} u_t\right),$$

and C_t is the cost of a machine of type t. For any given values of the vectors (m, y, s) and u, the second stage expected cost $V_{(m,y,s)}(u)$ is computed using a relatively simple, if time-consuming, routine.

We reformulate this first stage problem as a DP problem, as discussed in Section 6.1.4. In the reformulated problem, at any state where $\sum_{t=1}^{T} y_t > 0$, the control choices are to select a particular breakdown type, say t, with $y_t > 0$, and then select between two options:

(1) Leave the breakdown unrepaired, in which case the state evolves to

$$(m_1, \ldots, m_{t-1}, m_t - 1, m_{t+1}, \ldots, m_T, y_1, \ldots, y_{t-1}, y_t - 1, y_{t+1}, \ldots, y_T, s)$$

and the cost C_t of permanently losing the corresponding machine is incurred.

(2) Repair the breakdown, in which case the state evolves to

$$(m_1, \ldots, m_T, y_1, \ldots, y_{t-1}, y_t - 1, y_{t+1}, \ldots, y_T, s - 1),$$

and no cost is incurred.

Once we reach a state of the form $(\overline{m}_1, \ldots, \overline{m}_T, 0, \ldots, 0, \overline{s})$, where $y_1 = \cdots = y_T = 0$, there is no decision to make, and we simply pay the optimal cost-to-go of the second stage, $R(\overline{m}_1, \ldots, \overline{m}_T, \overline{s})$, and terminate.

We consider an approximation architecture based on heuristic algorithms. We used the following two heuristics, which given the state (m, y, s), produce a first stage solution u:

(1) *Proportional Heuristic*: In this heuristic, given the state, we compute an estimate $\overline{N}$ of the total number of second stage breakdowns based on the probabilities p_t of breakdown of individual machines of type t, and on a preliminary estimate of the number of remaining machines

at the start of the second stage. We form the estimated ratio of first stage to second stage breakdowns,

$$f = \frac{\sum_{t=1}^{T} y_t}{\overline{N}}.$$

We then fix the number of spare parts to be used in the first stage to

$$s_1 = afs,$$

where a is a positive parameter. The first stage problem is then solved by allocating the s_1 spare parts to machines of type t in the order of the costs $C_t(1-p_t)$. (The factor of $1-p_t$ is used to account for the undesirability of repairing machines that are likely to break again.) The constant a provides a parametrization of this heuristic. In particular, when $a < 1$, the heuristic is conservative, allocating more spare parts to the second stage than the projected ratio of breakdowns suggests, while if $a > 1$, the heuristic is more myopic, giving higher priority to the breakdowns that have already occurred in the first stage. The cost of the solution produced by the proportional heuristic for a given value of a is denoted by $H_a(m, y, s)$.

(2) *Value-Based Heuristic*: In this heuristic, given the state, we assign values of C_t and $C_t(1-p_t)$ to each spare part used to repair a machine of type t in the second stage and the first stage respectively. Note that a repair of a machine in the first stage is valued less than a repair of the same machine in the second stage, since a machine that is repaired in the first stage may break down in the second stage and require the use of an extra spare part. We rank-order the values C_t and $C_t(1-p_t)$, $t = 1, \ldots, T$, and we repair broken down machines in decreasing order of value, using the estimate $p_t(m_t - y_t)$ for the number of machines to break down in the second stage. The cost of the solution produced by the value-based heuristic is denoted by $H_v(m, y, s)$.

The architectures we used involve affine combinations of the costs $H_a(m, y, s)$ of the solutions of various proportional heuristics, involving different values of the proportionality parameter a, and the cost $H_v(m, y, s)$. of the solution of the value-based heuristic (cf. Section 3.1.4). In particular, we considered the *single proportional heuristic architecture*

$$\tilde{J}(m, y, s, r) = r(0) + r(a)H_a(m, y, s),$$

in which a single fixed value of a was used; the *multiple proportional heuristic architecture*

$$\tilde{J}(m, y, s, r) = r(0) + \sum_{k=1}^{K} r(a_k)H_{a_k}(m, y, s),$$

where $a_1, \ldots, a_K$ are K different values of a; the *single value-based heuristic architecture*

$$\tilde{J}(m, y, s, r) = r(0) + r(v)H_v(m, y, s);$$

and the *combined proportional and value-based heuristic architecture*

$$\tilde{J}(m, y, s, r) = r(0) + r(a)H_a(m, y, s) + r(v)H_v(m, y, s),$$

in which a single fixed value of a was used. In the above representations, $r(0)$, $r(a_i)$, $r(a)$, and $r(v)$ are tunable weights.

An architecture, together with a set of weights, defines a policy that generates a suboptimal solution for any given problem by looking at the given first stage breakdowns, and deciding which of them to repair. We refer to this as the *regular NDP solution.* An improvement over the regular NDP solution is the *extended NDP solution.* This solution is obtained by augmenting the decision space of the DP problem so that at any one state, in addition to selecting a part type and deciding whether to repair it or not, we have available some additional decisions, namely whether to adopt the solution obtained by one of the heuristic algorithms from that state and then terminate (see the discussion at the end of Section 3.1.4). The expanded decision space is used in the process of training the architecture. When calculating a solution to a given problem, after the architecture is trained, we generated the NDP solution in $\sum_{t=1}^{T} y_t$ steps, and we stored the best solution obtained by completing all the intermediate partial solutions using the heuristics. The extended NDP solution is obtained at the end, by selecting either the NDP solution or the best of the intermediately obtained solutions using the heuristics, whichever results in smaller cost. It can be seen that given any cost-to-go approximation, when a problem is submitted for solution by the corresponding greedy policy, the extended NDP solution will never be worse than the solution obtained by any one of the heuristic algorithms for the same problem. Thus the extended NDP solution offers a minimum performance guarantee that is not afforded by the regular NDP solution. At the same time, the extended NDP solution method requires very little additional computation both for training and for solving individual problems after training. In our experiments to be discussed shortly, we have tested both the regular and the extended NDP solution methods.

We implemented an NDP solution approach for a class of problems involving 5 machine types. The costs of these machine types are 2, 4, 6, 8, and 10, respectively. A problem from the class considered involves 0 to 10 machines of each type, a number of spare parts from 0 to the total number of machines, and a breakdown probability for each machine type randomly drawn from the range [0,1]. We used approximate policy iteration, with policy evaluation performed by Monte Carlo simulation. The initial policy was the greedy policy that repairs as many breakdowns as possible in the first stage, in the order of their cost. Each generated policy was trained

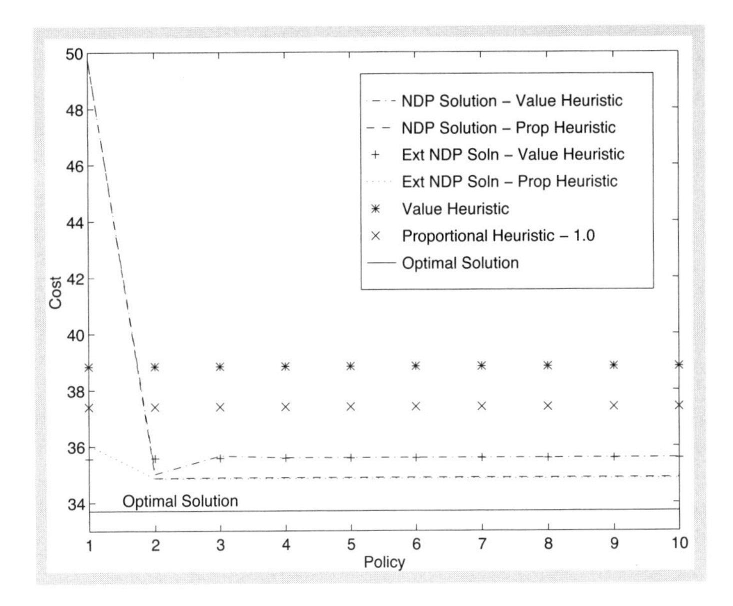

Figure 8.14: Approximate policy iteration results using single heuristic architectures. Each data point is an average over a 5000-problem test set of the costs of the solutions generated by the corresponding policy. The horizontal lines correspond to the performances of each of the heuristics applied to the original problem without training. The proportional heuristic uses a value $a = 1$. The optimal cost (averaged over the 5000 test problems) is 33.71.

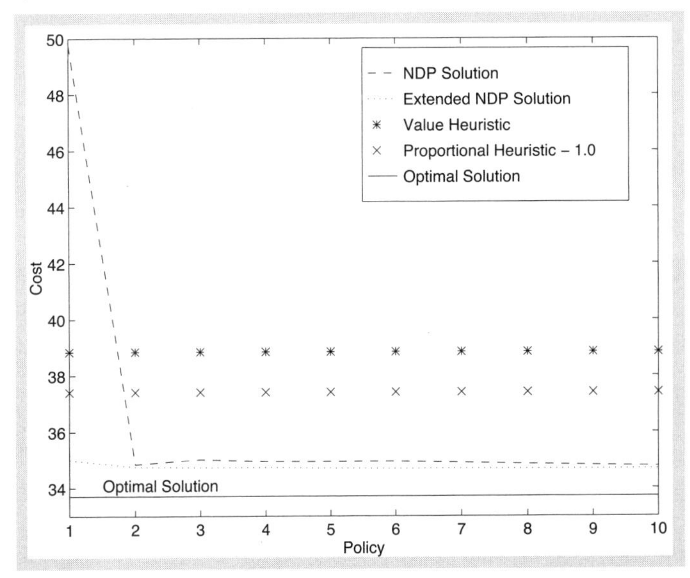

Figure 8.15: Approximate policy iteration results using the two-heuristic architecture involving the value-based heuristic and the proportional heuristic for $a = 1$.

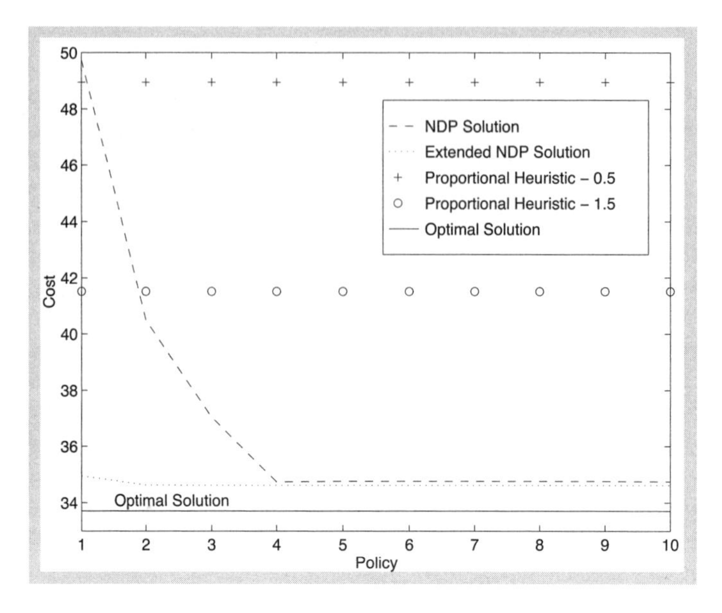

Figure 8.16: Approximate policy iteration results using the two-heuristic architecture involving the proportional heuristic for $a = 0.5$ and $a = 1.5$.

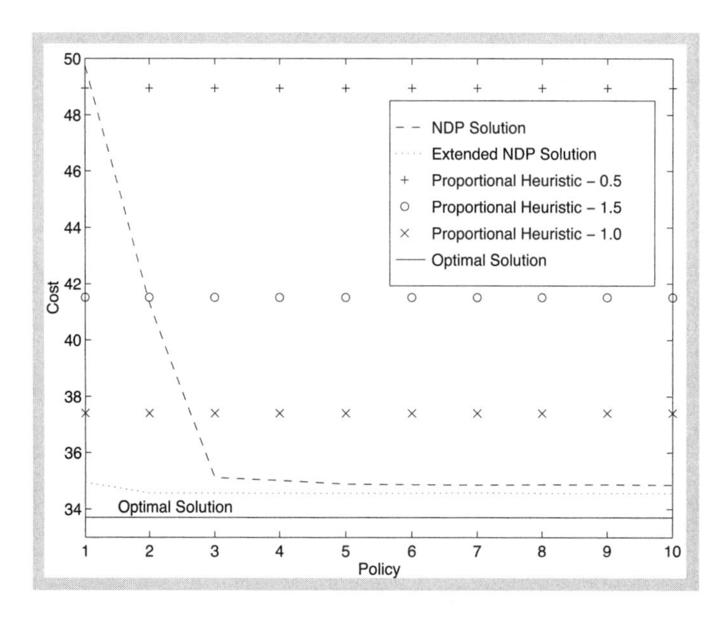

Figure 8.17: Approximate policy iteration results using the three-heuristic architecture involving the proportional heuristic for $a = 0.5$, $a = 1$, and $a = 1.5$.

using a fixed set of 5000 problems that were randomly chosen, and was evaluated using a different but fixed set of 5000 problems, also randomly chosen. The SVD algorithm was used to solve the corresponding linear least squares problem.

The average costs (over the 5000-problem test set) corresponding to the policies generated by the approximate policy iteration method are shown in Figs. 8.14-8.17 for various architectures, and are compared with the average costs corresponding to the various heuristics, and also with the average optimal cost. This latter cost, which is equal to 33.71, was obtained by optimally solving each of the 5000 test problems using a form of exhaustive enumeration of the feasible solutions. In these figures, the horizontal lines correspond to the performances of each of the heuristics applied to the original problem without training. Generally, the NDP methodology uniformly produced improvements to the performance of the various heuristics. A significant part of these improvements, however, can be attributed to the multistage application of the heuristics (one breakdown at a time). This is suggested by the excellent performance of the extended NDP solution, even without any training. In particular, our experimental findings are as follows:

(a) The extended NDP solutions using a single-heuristic architecture are substantially better than the solutions obtained when that heuristic is applied to the original problem without training (see Fig. 8.14).

(b) The NDP solutions using multiple-heuristic architectures are better than the solutions obtained when each of the heuristics is applied to the original problem without training (see Figs. 8.15-8.17).

(c) The NDP solutions using the multiple-heuristic architectures are somewhat better on the average that the ones obtained using the one-heuristic architectures.

(d) For the multiple-heuristic architectures, the extended NDP solution produced a small improvement over the regular NDP solution.

We have also tried more powerful architectures such as the ones discussed in Section 3.1.4, which use feature-dependent weights $r(k)$. However, these architectures produced NDP solutions of roughly the same quality as the corresponding architectures that use state independent weights $r(k)$. This is probably due to the fact that the heuristics we employed are apparently very effective for the class of problems considered.

As this example illustrates, the NDP methodology can be a powerful supplement to heuristics in combinatorial optimization. However, much more research and computational experimentation is needed to provide an accurate assessment of the potential of NDP methods within the broad and application-rich context of discrete optimization.

8.5 DYNAMIC CHANNEL ALLOCATION

In this section we discuss the use of NDP methods to solve the channel allocation problem discussed in Example 2.5 of Section 2.4. We have trained a feature-based linear architecture using optimistic policy iteration in conjunction with TD(0). Our experimentation shows that the policies obtained outperform the best heuristic that we could find in the literature.

The dynamic channel allocation problem can be formulated as a DP problem, as discussed in Section 2.4. Here, we will restrict ourselves to the case where there are no handoffs. Thus, state transitions occur when channels become free due to user departures, or when a user arrives at a given cell and wishes to be assigned a channel. The state at each time consists of two components:

(1) The list of occupied and unoccupied channels at each cell. We call this the *configuration* of the cellular system.

(2) The event that causes the state transition (arrival or departure). This component of the state is uncontrollable.

The decision/control applied at the time of a call departure is the rearrangement of the channels in use with the aim of creating a more favorable channel packing pattern among the cells (one that will leave more channels free for future assignments). We restrict this rearrangement to just the cell where the current call departure occurs. The decision exercised at the time of a call arrival is the assignment of a free channel, or the blocking of the call if no free channel is currently available. (A channel is defined to be free in a given cell if it can be used in that cell without violating the reuse constraint.) Note that we never block a call if it can be accepted, that is, if there is a free channel in the cell where the call originates. The objective is to obtain a policy that assigns channels so as to maximize

$$E\left[\int_{t=0}^{\infty} e^{-\beta t} n(t) dt\right],$$

where $n(t)$ is the number of ongoing calls at time t, and β is a discount factor that makes immediate profit more valuable than future profit. Maximizing this cost is equivalent to minimizing the (discounted) number of blocked calls.

Note that this is a continuous-time problem. It turns out, however, that such problems can be converted to discrete-time problems with a state-dependent discount factor (see e.g., [Ber95a], Vol. II, Ch. 5). In particular, the methodology presented in previous chapters is valid with only notational changes to account for a state-dependent discount factor.

We have introduced a linear feature-based approximation architecture using two sets of features that depend only on the configuration of the system, which is the controllable part of the state: one availability feature for each cell and one packing feature for each cell-channel pair. The availability feature for a cell is the number of free channels in that cell, while the packing feature for a cell-channel pair is the number of times that channel is used within a radius equal to the channel reuse distance. This architecture has the advantage that even though training is centralized, the policy obtained is decentralized because the features are local: to choose decisions in a given cell, only the weights of the features from the neighboring cells (those within the reuse distance) are needed.

The architecture was trained using optimistic policy iteration and TD(0). We used a single, infinitely long trajectory for training. As the weights of the architecture were trained, the reward-to-go approximation was used to generate decisions as follows:

> *Call Arrival*: When a call arrives in a given cell, evaluate the next configuration for each free channel in that cell, and assign the channel that leads to the configuration with the largest estimated value. If there is no free channel at all, the call is blocked and no decision has to be made.
>
> *Call Termination*: When a call terminates in a given cell, each ongoing call in that cell is considered for reassignment to the just freed channel; the resulting configurations are evaluated and compared to the value of not doing any reassignment at all. The decision that leads to the highest value configuration is then chosen.

In our experimentation, all arrivals were simulated as Poisson processes with a separate mean for each cell, and call durations were simulated with an exponential distribution. We compared the NDP solution with a highly regarded heuristic, which is the method of Borrowing with Directional Channel Locking (BDCL) of Zhang and Yum [ZhY89]. This heuristic partitions and assigns the channels to cells taking into account the reuse constraint. The channels assigned to a cell are its nominal channels. The channels are ordered by their identification number. The idea of the BDCL heuristic is to try to use low-numbered channels within their assigned cells, and to allow borrowing of the nominal channels of a given cell by neighboring cells, starting with the high-numbered channels. In particular, if a nominal channel is available when a call arrives in a cell, the smallest numbered such channel is assigned to the call. If no nominal channel is available, then the set of free channels for the cell (those that are not in use within the reuse distance) is determined, and the neighbor with the most nominal channels that are free is selected. The largest numbered nominal channel that is free is then borrowed from that neighbor and is assigned to the call. The call is blocked if there are no free channels at all. When a call terminates in a cell and the channel so freed is a nominal channel, say numbered i, of that cell, then if there is a call in that cell on a borrowed channel, the call on the smallest numbered borrowed channel is reassigned to i and the borrowed channel is returned to the appropriate cell. On the other hand, if there is no call on a borrowed channel, then if there is a call on a nominal channel numbered larger than i, the call on the highest numbered nominal channel is reassigned to i. If the call just terminated was itself on a borrowed channel, the call on the smallest numbered borrowed channel is reassigned to it and that channel is returned to the cell from which it was borrowed. Notice that when a borrowed channel is returned to its original cell, a nominal channel becomes free in that cell and triggers a reassignment. Thus, in the worst case a call termination in one cell can sequentially cause reassignments in arbitrarily far away cells. BDCL is quite sophisticated and is widely viewed as one of the best if not the best practical dynamic channel assignment algorithm. In particular, the simulations of [ZhY89] show that BDCL is superior to all of its main competitors in terms of blocking probability performance.

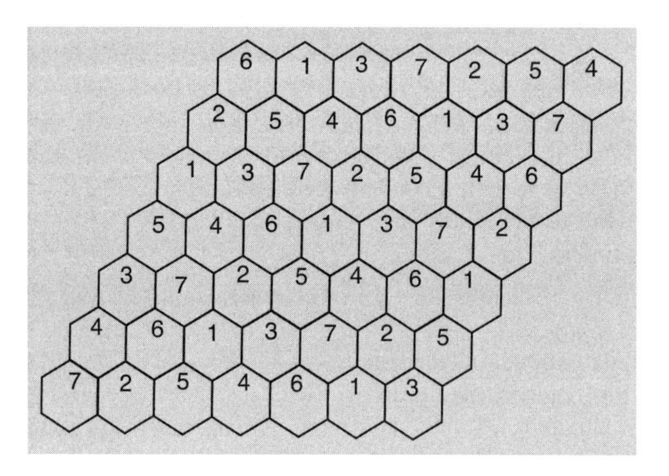

Figure 8.18: Cellular system used in our experiments. Cells with the same number are at the minimum distance at which the simultaneous use of the same channel is possible.

The first set of results relates to the 7 by 7 cellular array of Fig. 8.18 with 70 channels and a channel reuse constraint of 3 (this problem is borrowed from the paper by Zhang and Yum [ZhY89]). In Fig. 8.19, we give results for the case of uniform call arrival rates of 150, 200, and 350 calls/hr respectively in each cell. The mean call duration for all the experiments reported here is 3 minutes. Each plot gives the empirical cumulative blocking probability during training as a function of simulated time. In particular, each data point is the percentage of system-wide calls that were blocked up until that point in time. Thus, for the NDP case the data points do not correspond to performance of any one policy, but rather provide a time average of the blocking probability corresponding to the channel assignment

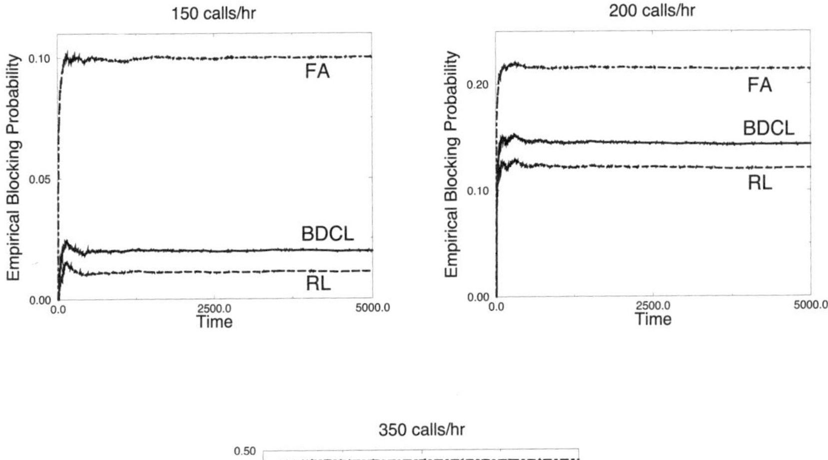

Figure 8.19: These figures compare performance of the NDP solution (denoted RL), the fixed channel allocation (FA), and BDCL on the 7 by 7 cellular array of Fig. 8.18. The means of the call arrival (Poisson) processes are shown in the graph titles. Each curve presents the cumulative empirical blocking probability as a function of simulated time measured in minutes.

policies generated during the training interval. We compare the performance of the NDP policies (denoted by RL for "reinforcement learning") with BDCL and with a nearly optimal *fixed* allocation policy denoted FA. The RL curves in Fig. 8.19 show the empirical blocking probability while training. Note that the performance improvement is quite rapid. As the mean call arrival rate is increased, the relative difference between the three algorithms decreases. In fact, it can be shown that FA is asymptotically optimal as call arrival rates increase because with many calls in every cell, there are no short-term fluctuations to exploit. However, as demonstrated in Fig. 8.19, for practical traffic rates, the NDP solution outperforms FA and to a smaller but significant extent BDCL.

Figure 8.20 presents multiple sets of bar graphs of asymptotic blocking probabilities for the three algorithms for a system that consists of a 20 by 1 cellular array with 24 channels and a channel reuse constraint of 3. For each set, the average per-cell call arrival rate is the same (120 calls/hr; mean duration of 3 minutes), but the pattern of call arrival rates across the 20 cells is varied. The patterns are shown on the left of the bar-graphs and are explained in the caption. The NDP algorithm was trained using the uniform traffic pattern, but was used without any modification of the feature weights for various nonuniform traffic patterns. It can be seen that the NDP algorithm is much less sensitive to varied patterns of non-uniformity than both BDCL and FA.

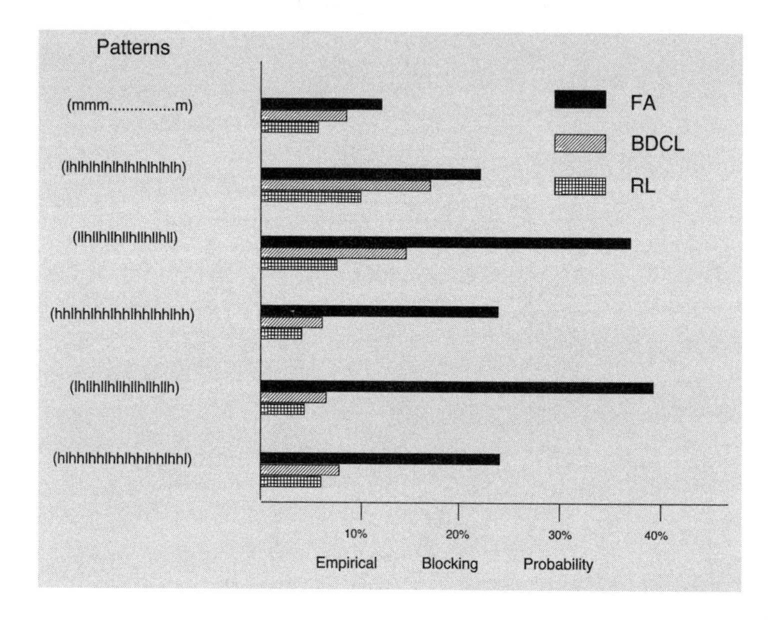

Figure 8.20: Sensitivity of channel assignment methods to non-uniform traffic patterns. This figure plots asymptotic empirical blocking probability for the three algorithms for a 20 by 1 linear array of cells with a channel reuse constraint of 3. For each set of bars, the pattern of mean call arrival rates varied, but the per-cell mean call arrival rate was the same (120 calls/hr). The traffic patterns are indicated on the left of each set of bars. For example, the topmost set of bars are for a uniform pattern with medium traffic intensity, and the second from top set of bars are for the pattern low-high repeated.

8.6 BACKGAMMON

The development of a world-class computer backgammon player ("TD-gammon") by Tesauro has been the most spectacular use of NDP methods to date. Besides providing insights into the behavior of NDP in practice, it has also been a source of optimism about the potential for success in dealing with truly complex problems.

Backgammon is an old two-player board game. Each one of two players (call them white and black) has control of 15 checkers that move in opposite directions along an essentially linear track. Each player tries to move the checkers into the last quadrant and then "collect" them, according to some rules. (For the purposes of this exposition, we will disregard certain complicating aspects of the game, such as "gammons" and "doubling.") The two players alternate in rolling a pair of dice and then selecting a legal move. The set of available legal moves depends on the outcome of the dice roll and the board configuration. The game is complex because the position of the checkers of one player can restrict the moves available to the opponent. In addition, single ("unprotected") checkers of one player can be "hit" by the other player and sent back to the first quadrant. Despite the randomness in the outcome of the dice rolls, good play requires sophisticated tactics and strategy, so that backgammon is mostly a game of skill rather than a game of chance.

The game can be modeled as a sequential Markov game of the type discussed in Section 7.2. A typical state consists of a description of the board configuration (the position of the 30 checkers), the outcome of the current dice roll, and the identity of the player who is to make a move. The state space is huge, its cardinality estimated to be of the order of 10^{20}. An additional feature of this game, that distinguishes it from chess, is a large branching factor: there are 21 possible dice roll outcomes and for each one of them there are typically about 20 legal moves. Thus, the branching factor is of the order of 400, and unlike chess, only a shallow lookahead is possible.

Earlier approaches to computer backgammon had involved one of two methods:

(a) *Supervised learning*
Here, an approximation architecture such as a multilayer perceptron is fed with a large database of example states together with the move chosen by an expert human player. Then, the approximation architecture is "trained" and learns how to emulate the moves of the human trainer. A backgammon player of this type ("Neurogammon"), also developed by Tesauro [Tes89], won the 1989 Computer Olympiad, but did not seem able to challenge the best human players.

(b) *Handcrafted position evaluators*
Many commercial backgammon programs employ a "position evaluator" and each move is selected so as to lead to a state with the best evaluation. Such an evaluation function is often a linear combination of handcrafted features, judged to be important by expert players, and its parameters are usually hand-tuned.

The centerpiece of TD-gammon is also a position evaluator (or cost-to-go function), except that its parameters are selected automatically, based on self-play and a temporal difference training method. In the course of self-play, there is no teacher available to advise the program as to the desirability of each particular move and learning is only based on the outcomes of the different games. Figuring out which moves in the course of a game were responsible for a win or a loss, and how to take this information into account in order to update the cost-to-go function, seems like a formidable task that has been referred to as the "credit assignment problem" in the artificial intelligence literature. While exact DP provides a perfect – but theoretical – solution to the credit assignment problem, it was far from clear whether NDP could provide an effective solution.

We now provide an overview of the method employed by Tesauro. Let i stand for a generic board configuration. Let $J^*(i,0)$ be the probability that white wins, given that the current board position is i, and that it is white's turn to roll the dice, assuming that both players follow optimal policies. Similarly, we define $J^*(i,1)$ as the probability that white wins, given that it is black's turn to roll the dice. The function $J^*(i,l)$ is approximated by the output of a multilayer perceptron $\tilde{J}(i,l,r)$, with inputs i and $l \in \{0,1\}$, and a vector of weights r.

Tesauro has experimented extensively with a variety of MLP architectures, by varying the number of hidden layers and the number of units at each layer, with 40 being a representative choice for the number of hidden units. In early work, the input to the MLP was a raw encoding of the board configuration i, and this was enough to obtain a backgammon player outperforming all other computer programs. In later versions, a number of handcrafted functions of the state (features) were also fed as separate inputs, which led to improved performance.

The training method employed was optimistic TD(λ). In more detail, the algorithm is as follows. Consider a typical stage in the game where the parameter vector is r_t, the board configuration is i_t, and it is the turn of player $l_t \in \{0,1\}$. The player rolls the dice, and then chooses a legal move for which the next position i_{t+1} has the best possible value of $\tilde{J}(i_{t+1}, l_{t+1}, r_t)$. Upon the completion of a move, we compute the temporal difference

$$d_t = \tilde{J}(i_{t+1}, l_{t+1}, r_t) - \tilde{J}(i_t, l_t, r_t),$$

and we are ready to proceed to the next stage. The only exception to the above formula is if a move ends the game in which $\tilde{J}(i_{t+1}, l_{t+1}, r_t)$ is replaced by 1 or 0, depending on whether white or black was the winning player. After each move, and once the temporal difference d_t is generated, the parameter vector r_t is updated according to the on-line TD(λ) rule, which is

$$r_{t+1} = r_t + \gamma d_t \sum_{k=0}^{t} \lambda^{t-k} \nabla \tilde{J}(i_k, l_k, r_k).$$

At the end of a game (i.e., at the end of a trajectory, in our terminology), a new game is started, with r_0 set at the value obtained at the end of the preceding game.

The parameter vector r was initialized randomly. The stepsize γ was usually kept to a constant value, chosen through experimentation. Also, several values for λ were tried, although the choice of λ was not reported as being particularly important. Training was carried out for a large number of games ranging from 200,000 to more than a million, and the amount of CPU time spent ranged from a few hours to several weeks, depending on the architecture and the number of games.

Once training is completed, the available cost-to-go function $\tilde{J}(i, l, r)$ leads to a greedy policy that can be implemented for playing against an opponent. While the greedy policy corresponds to one-stage lookahead, even better performance was obtained by performing multistage lookahead. Due to the high branching factor, such lookahead has to be limited to a couple of dice rolls, but it is nevertheless beneficial. Another improvement involved the use of a rollout policy, as discussed in Section 6.1. This requires the ability to simulate up to completion a fair number of games before each move is chosen, and a parallel computer was deployed in order to keep the time between moves at an acceptable level [TeG96].

An interesting aspect of the final cost-to-go function "learned" by TD-gammon is that it provides a fairly inaccurate estimate of the probability of winning. Nevertheless, the comparison of different candidate states (which is all that is needed for move selection) turned out to be much more reliable. The reasons why training with TD(λ) results in this desirable property are not understood.

Backgammon has some special features that seem to have helped to the success of TD-gammon. First, there is a fair amount of randomness, which eliminates the need for explicit exploration. Thus, despite the strict adherence to a greedy policy in the course of self-play, most of the important parts of the state space were apparently explored. A second aspect, again due to randomness, is that the cost-to-go function is a relatively "smooth" function of position, which often makes approximation easier. This is in contrast to deterministic problems like chess, where the function J^* can only take values 1 (for a win), 0 (for a draw), and -1 (for a loss), and which can be strongly affected by seemingly minor changes in the board configuration. Finally, the game has a certain directionality that drives the state towards termination, no matter what policy is used. (All meaningful policies – if not all policies – are proper.) In a sense, we are close to a situation of a layered state space, which is in general favorable to the effectiveness of DP.

As of this writing, TD-gammon has been assessed to play "at a strong grandmaster level and is extremely close to equaling the world's best human players" [Tes95]. It is not at all unlikely that it will outperform all human players in the near future.

8.7 NOTES AND SOURCES

8.1. The experimental results of this section were obtained by Keith Rogers.

8.2. The material on football comes from the paper by Patek and Bertsekas [PaB96b], which contains further analytical and experimental information.

8.3. The NDP approach and the implementation used for tetris was based on unpublished work by Sergey Ioffe.

8.4. The material on the maintenance problem is based on experiments conducted by Cynara Wu.

8.5. The material on channel allocation is due to Singh and Bertsekas [SiB96].

8.6. Tesauro's backgammon work is described in [Tes89], [Tes92], [Tes94], [Tes95], and [TeG96]. The exposition given here is mostly based on [Tes95].

We briefly mention two more case studies that involve truly large scale problems that arise in real-world applications. Zhang and Dietterich [ZhD95] consider a job-shop scheduling problem that arises in the planning of NASA space shuttle missions. Any particular instance of the problem is solved by using a sequence of modifications ("repairs") of a candidate (possibly infeasible) schedule. The question is which particular type of modification to apply, given a current schedule. This leads to a DP formulation similar to the one described in Example 2.7 of Section 2.4. Zhang and Dietterich used a multilayer perceptron architecture and applied optimistic TD(λ). They report obtaining schedules that outperform the best existing conventional methods. See also [ZhD96] for further results using a more sophisticated neural network architecture.

Crites and Barto [CrB96] considered the problem of dispatching a number of elevators that serve a multi-story building. They applied Q-learning, with a separate set of Q-factors associated with each elevator, and they reported improvements over the best available heuristics for some cases. In their approach, the parametric representation of the approximate Q-factors consisted of a multilayer perceptron whose input was the entire state of the system. Pepyne, Looze, Cassandras and Djaferis [PLC96] also used Q-learning on the elevator dispatching problem and obtained a dispatcher that outperformed a commercial one. In contrast to the approach of Crites and Barto, they used state aggregation to obtain a problem with a manageable number of states. They then applied Q-learning with a lookup table representation, which is closely related to the method that was described in Section 6.7.7.

APPENDIX A: Mathematical Review

Contents

The purpose of this appendix is to provide notation, mathematical definitions, and results that are used frequently in the text. For detailed discussions of linear algebra and analysis, see references [HoK61], [Roy68], [Rud64], and [Str76].

A.1 SETS

If x is a member of a set S, we write $x \in S$. We write $x \notin S$ if x is not a member of S. A set S may be specified by listing its elements within braces. For example, by writing $S = \{x_1, x_2, \ldots, x_n\}$ we mean that the set S consists of the elements $x_1, x_2, \ldots, x_n$. A set S may also be specified in the generic form

$$S = \{x \mid x \text{ satisfies } P\},$$

as the set of elements satisfying property P. For example,

$$S = \{x \mid x \text{ real}, \; 0 \leq x \leq 1\},$$

denotes the set of all real numbers x satisfying $0 \leq x \leq 1$.

The *union* of two sets S and T is denoted by $S \cup T$, and the *intersection* of S and T is denoted by $S \cap T$. The union and intersection of a sequence of sets $S_1, S_2, \ldots, S_k, \ldots$ is denoted by $\cup_{k=1}^{\infty} S_k$ and $\cap_{k=1}^{\infty} S_k$, respectively. If S is a subset of T (i.e., if every element of S is also an element of T), we write $S \subset T$ or $T \supset S$.

Finite and Countable Sets

A set S is said to be *finite* if it consists of a finite number of elements. It is said to be *countable* if the elements of S can be put into one-to-one correspondence with a subset of the nonnegative integers. Thus, according to our definition, a finite set is also countable but not conversely. A countable set S that is not finite may be represented by listing its elements $x_0, x_1, x_2, \ldots$ (i.e., $S = \{x_0, x_1, x_2, \ldots\}$).

Sets of Real Numbers

If a and b are real numbers or $+\infty$, $-\infty$, we denote by $[a, b]$ the set of numbers x satisfying $a \leq x \leq b$ (including the possibility $x = +\infty$ or $x = -\infty$). A rounded, instead of square, bracket denotes strict inequality in the definition. Thus $(a, b]$, $[a, b)$, and (a, b) denote the set of all x satisfying $a < x \leq b$, $a \leq x < b$, and $a < x < b$, respectively.

If S is a set of real numbers that is bounded above, then there is a smallest real number y such that $x \leq y$ for all $x \in S$. This number is called the *least upper bound or supremum* of S and is denoted by $\sup\{x \mid x \in S\}$.

Similarly, the greatest real number z such that $z \le x$ for all $x \in S$ is called the *greatest lower bound or infimum* of S and is denoted by $\inf\{x \mid x \in S\}$. If S is unbounded above, we write $\sup\{x \mid x \in S\} = +\infty$, and if it is unbounded below, we write $\inf\{x \mid x \in S\} = -\infty$. If S is the empty set, then by convention we write $\inf\{x \mid x \in S\} = +\infty$ and $\sup\{x \mid x \in S\} = -\infty$.

A.2 EUCLIDEAN SPACE

The set of all n-tuples of real numbers $x = \big(x(1), \ldots, x(n)\big)$ constitutes the *n-dimensional Euclidean space*, denoted by $\Re^n$. The elements of $\Re^n$ are referred to as n-dimensional vectors or simply vectors when confusion cannot arise. The one-dimensional Euclidean space $\Re^1$ consists of all the real numbers and is denoted by $\Re$. Vectors in $\Re^n$ can be added by adding their corresponding components. They can be multiplied by some scalar by multiplication of each component by that scalar.

A set of vectors $a_1, a_2, \ldots, a_k$ is said to be *linearly dependent* if there exist scalars $\lambda_1, \lambda_2, \ldots, \lambda_k$, not all zero, such that $\sum_{i=1}^k \lambda_i a_i = 0$. If no such set of scalars exists, the vectors are said to be *linearly independent*.

The *inner product* of two vectors $x = \big(x(1), \ldots, x(n)\big)$ and $y = \big(y(1), \ldots, y(n)\big)$ is denoted by $x'y$ and is equal to $\sum_{i=1}^n x(i)y(i)$. Two vectors in $\Re^n$ are said to be *orthogonal* if their inner product is equal to zero.

A *norm* $\|\cdot\|$ on $\Re^n$ is a mapping that assigns a scalar $\|x\|$ to every $x \in \Re^n$ and that has the following properties:

(a) $\|x\| \ge 0$ for all $x \in \Re^n$.

(b) $\|cx\| = |c| \cdot \|x\|$ for every $c \in \Re$ and every $x \in \Re^n$.

(c) $\|x\| = 0$ if and only if $x = 0$.

(d) $\|x + y\| \le \|x\| + \|y\|$ for all $x, y \in \Re^n$.

The *Euclidean norm* of a vector $x = \big(x(1), \ldots, x(n)\big) \in \Re^n$ is denoted by $\|x\|$ and is equal to $(x'x)^{1/2} = \big(\sum_{i=1}^n x(i)^2\big)^{1/2}$. The Schwartz inequality applies to the Euclidean norm and states that for any two vectors x and y, we have

$$|x'y| \le \|x\| \cdot \|y\|,$$

with equality holding if and only if $x = \alpha y$ for some scalar α. The *maximum norm* $\|\cdot\|_\infty$ (also called *sup-norm* or *ℓ_∞-norm*) is defined by

$$\|x\|_\infty = \max\big\{|x(1)|, |x(2)|, \ldots, |x(n)|\big\}.$$

A generalization of the maximum norm, called *weighted maximum norm*, is defined by

$$\|x\|_\xi = \max\left\{\frac{|x(1)|}{\xi(1)}, \frac{|x(2)|}{\xi(2)}, \ldots, \frac{|x(n)|}{\xi(n)}\right\},$$

where $\xi = \big(\xi(1), \xi(2), \ldots, \xi(n)\big)$ is a vector with positive components. Given any two norms $\|\cdot\|_1$ and $\|\cdot\|_2$ in $\Re^n$, there exists a constant c (that depends on these norms) such that $\|x\|_1 \leq c\|x\|_2$ for all $x \in \Re^n$. This is also called the *norm equivalence* property of $\Re^n$.

A.3 MATRICES

An $m \times n$ *matrix* is a rectangular array of numbers, referred to as elements, which are arranged in m rows and n columns. If $m = n$ the matrix is said to be *square*. The element in the ith row and jth column of a matrix A is denoted by a subscript ij, such as a_{ij}, in which case we write $A = [a_{ij}]$. The $n \times n$ *identity matrix*, denoted by I, is the matrix with elements $a_{ij} = 0$ for $i \neq j$ and $a_{ii} = 1$, for $i = 1, \ldots, n$. The *sum* of two $m \times n$ matrices A and B is written as $A + B$ and is the matrix whose elements are the sum of the corresponding elements in A and B. The *product of a matrix A and a scalar λ*, written as λA or $A\lambda$, is obtained by multiplying each element of A by λ. The *product* AB of an $m \times n$ matrix A and an $n \times p$ matrix B is the $m \times p$ matrix C with elements $c_{ij} = \sum_{k=1}^{n} a_{ik} b_{kj}$. If b is an $n \times 1$ matrix (i.e., an n-dimensional column vector) and A is an $m \times n$ matrix, then Ab is an m-dimensional (column) vector. We follow the convention, that unless otherwise explicitly stated, a vector is treated as a column vector.

The *transpose* of an $m \times n$ matrix A is the $n \times m$ matrix A' with elements $a'_{ij} = a_{ji}$. A square matrix A is *symmetric* if $A' = A$. An $n \times n$ matrix A is called *nonsingular* or *invertible* if there is an $n \times n$ matrix called the *inverse* of A and denoted by A^{-1}, such that $A^{-1}A = I = AA^{-1}$, where I is the $n \times n$ identity matrix. An $n \times n$ matrix is nonsingular if and only if the n vectors that constitute its rows are linearly independent or, equivalently, if the n vectors that constitute its columns are linearly independent. Thus, an $n \times n$ matrix A is nonsingular if and only if the relation $Av = 0$, where $v \in \Re^n$, implies that $v = 0$.

Rank of a Matrix

The *rank* of a matrix A is equal to the maximum number of linearly independent rows of A. It is also equal to the maximum number of linearly independent columns. Thus, the rank of an $m \times n$ matrix is at most equal to the minimum of the dimensions m and n. An $m \times n$ matrix is said to be of *full rank* if its rank is maximal, that is, if it is equal to the minimum of m and n. A square matrix is of full rank if and only if it is nonsingular.

Eigenvalues

Given a square $n \times n$ matrix A, the determinant of the matrix $\gamma I - A$, where I is the $n \times n$ identity matrix and γ is a scalar, is an nth degree polynomial. The n roots of this polynomial are called the *eigenvalues* of A. Thus, γ is an eigenvalue of A if and only if the matrix $\gamma I - A$ is singular, or equivalently, if and only if there exists a nonzero vector v such that $Av = \gamma v$. Such a vector v is called an *eigenvector* corresponding to γ. The eigenvalues and eigenvectors of A can be complex even if A is real. The matrix A is singular if and only if it has an eigenvalue that is equal to zero. If A is nonsingular, then the eigenvalues of A^{-1} are the reciprocals of the eigenvalues of A. The eigenvalues of A and A' coincide.

If $\gamma_1, \ldots, \gamma_n$ are the eigenvalues of A, then the eigenvalues of $cI + A$, where c is a scalar and I is the identity matrix, are $c + \gamma_1, \ldots, c + \gamma_n$. The eigenvalues of A^k, where k is any integer, are equal to $\gamma_1^k, \ldots, \gamma_n^k$. From this it follows that $\lim_{k\to\infty} A^k = 0$ if and only if all the eigenvalues of A lie strictly within the unit circle of the complex plane. Furthermore, if the latter condition holds, the iteration $x_{k+1} = Ax_k + b$, where b is a given vector, converges to $\overline{x} = (I - A)^{-1}b$, which is the unique solution of the equation $x = Ax + b$.

If the n eigenvalues of A are distinct, there exists a set of corresponding linearly independent eigenvectors. In this case, if $\gamma_1, \ldots, \gamma_n$ are the eigenvalues and $v_1, \ldots, v_n$ are these eigenvectors, every vector $x \in \Re^n$ can be decomposed as

$$x = \sum_{i=1}^{n} \xi_i v_i,$$

where the ξ_i are some unique (possibly complex) numbers. Furthermore, we have for all positive integers k,

$$A^k x = \sum_{i=1}^{n} \gamma_i^k \xi_i v_i.$$

A symmetric $n \times n$ matrix A has real eigenvalues and a set of n real linearly independent eigenvectors, which are orthogonal (the inner product of any pair is 0). If $\gamma_1, \ldots, \gamma_n$ are the eigenvalues of a symmetric matrix A, we have

$$\max\{\gamma_1, \ldots, \gamma_n\} = \max_{x \neq 0} \frac{x'Ax}{\|x\|^2},$$

$$\max\{|\gamma_1|, \ldots, |\gamma_n|\} = \max_{x \neq 0} \frac{\|Ax\|}{\|x\|},$$

where $\|\cdot\|$ denotes the Euclidean norm. If in addition A has nonnegative eigenvalues, we have

$$\max\{\gamma_1, \ldots, \gamma_n\} = \max_{x \neq 0} \frac{x'Ax}{\|x\|^2} = \max_{x \neq 0} \frac{\|Ax\|}{\|x\|}.$$

Positive Definite and Semidefinite Matrices

A square $n \times n$ matrix A is said to be *positive semidefinite* if $x'Ax \geq 0$ for all $x \in \Re^n$. It is said to be *positive definite* if $x'Ax > 0$ for all nonzero $x \in \Re^n$. The matrix A is said to be *negative semidefinite (definite)* if $-A$ is *positive semidefinite (definite)*. It can be shown that if A is positive definite, then there exists a positive scalar β such that $x'Ax \geq \beta x'x$ for all $x \in \Re^n$.

A positive (negative) definite matrix is invertible and its inverse is also positive (negative) definite. Also, an invertible positive (negative) semidefinite matrix is positive (negative) definite. If A and B are $n \times n$ positive semidefinite (definite) matrices, then the matrix $\lambda A + \mu B$ is also positive semidefinite (definite) for all $\lambda > 0$ and $\mu > 0$. If A is an $n \times n$ positive semidefinite matrix and C is an $m \times n$ matrix, then the matrix CAC' is positive semidefinite. If A is positive definite, and C has rank m (equivalently, if $m \leq n$ and C has full rank), then CAC' is positive definite.

An $n \times n$ positive definite and symmetric matrix A can be written as CC' where C is a square invertible matrix. If A is positive semidefinite and symmetric, and its rank is m, then it can be written as CC', where C is an $n \times m$ matrix of full rank.

If A is positive semidefinite (definite) and symmetric, its eigenvalues are nonnegative (respectively, positive). A positive semidefinite and symmetric matrix A can be factored as

$$A = Q\Sigma Q',$$

where Σ is a diagonal matrix with the eigenvalues of A along the diagonal, and Q is a matrix that has as columns corresponding orthonormal eigenvectors (orthogonal eigenvectors, which are scaled so that their Euclidean norm is equal to 1).

A.4 ANALYSIS

Convergence of Sequences

A sequence of vectors $x_0, x_1, \ldots, x_k, \ldots$ in $\Re^n$, is denoted by $\{x_k\}$, or sometimes, with a slight abuse of notation, just by x_k. Let us fix a norm $\|\cdot\|$ in $\Re^n$. A sequence $\{x_k\}$ is said to converge to a *limit* x if $\|x_k - x\| \to 0$ as $k \to \infty$ (i.e., if, given any $\epsilon > 0$, there is an N such that for all $k \geq N$ we have $\|x_k - x\| < \epsilon$). If $\{x_k\}$ converges to x, we write $x_k \to x$ or $\lim_{k\to\infty} x_k = x$. We have $Ax_k + By_k \to Ax + By$ if $x_k \to x$, $y_k \to y$, and A, B are matrices of appropriate dimensions. The convergence of a sequence does not depend on the choice of norm; that is, if a sequence converges with respect to one norm, it converges with respect to all norms; this is due to the norm equivalence property of $\Re^n$.

A vector x is said to be a *limit point* of a sequence $\{x_k\}$ if there is a subsequence of $\{x_k\}$ that converges to x, that is, if there is an infinite subset $\mathcal{K}$ of the nonnegative integers such that for any $\epsilon > 0$, there is an N such that for all $k \in \mathcal{K}$ with $k \geq N$ we have $\|x_k - x\| < \epsilon$.

A sequence of real numbers $\{r_k\}$, which is monotonically nondecreasing (nonincreasing), that is, it satisfies $r_k \leq r_{k+1}$ ($r_k \geq r_{k+1}$) for all k, must either converge to a real number or be unbounded above (below). In the latter case we write $\lim_{k\to\infty} r_k = \infty$ $(-\infty)$. Given any bounded sequence of real numbers $\{r_k\}$, we may consider the sequence $\{s_k\}$, where $s_k = \sup\{r_i \mid i \geq k\}$. Since this sequence is monotonically nonincreasing and bounded, it must have a limit. This limit is called the *limit superior* of $\{r_k\}$ and is denoted by $\limsup_{k\to\infty} r_k$. The *limit inferior* of $\{r_k\}$ is similarly defined and is denoted by $\liminf_{k\to\infty} r_k$. If $\{r_k\}$ is unbounded above, we write $\limsup_{k\to\infty} r_k = \infty$, and if it is unbounded below, we write $\liminf_{k\to\infty} r_k = -\infty$.

Open, Closed, and Compact Sets

A subset S of $\Re^n$ is said to be *open* if for every vector $x \in S$ one can find an $\epsilon > 0$ such that $\{z \mid \|z - x\| < \epsilon\} \subset S$. A set S is *closed* if and only if every convergent sequence $\{x_k\}$ with elements in S converges to a point that also belongs to S. A set S is said to be *compact* if and only if it is both closed and bounded (i.e., it is closed and for some $M > 0$ we have $\|x\| \leq M$ for all $x \in S$). A set S is compact if and only if every sequence $\{x_k\}$ with elements in S has at least one limit point that belongs to S. Any bounded sequence is contained in a compact set and therefore has at least one limit point.

Continuous Functions

A function f mapping a set S_1 into a set S_2 is denoted by $f : S_1 \mapsto S_2$. A function $f : \Re^n \mapsto \Re^m$ is said to be *continuous* if for all x, $f(x_k) \to f(x)$ whenever $x_k \to x$. Equivalently, f is continuous if, given $x \in \Re^n$ and $\epsilon > 0$, there is a $\delta > 0$ such that whenever $\|y - x\| < \delta$, we have $\|f(y) - f(x)\| < \epsilon$. The function

$$(a_1 f_1 + a_2 f_2)(\cdot) = a_1 f_1(\cdot) + a_2 f_2(\cdot)$$

is continuous for any two scalars a_1, a_2 and any two continuous functions $f_1, f_2 : \Re^n \mapsto \Re^m$. If S_1, S_2, S_3 are any sets and $f_1 : S_1 \mapsto S_2$, $f_2 : S_2 \mapsto S_3$ are functions, the function $f_2 \circ f_1 : S_1 \mapsto S_3$ defined by $(f_2 \circ f_1)(x) = f_2\big(f_1(x)\big)$ is called the *composition* of f_1 and f_2. If $f_1 : \Re^n \mapsto \Re^m$ and $f_2 : \Re^m \mapsto \Re^p$ are continuous, then $f_2 \circ f_1$ is also continuous.

Weierstrass' Theorem asserts that a continuous function $f : \Re^n \mapsto \Re$ attains a minimum over any nonempty compact set A, i.e., there exists $x^* \in A$ such that $f(x^*) = \inf_{x \in A} f(x)$.

Derivatives

Let $f : \Re^n \mapsto \Re$ be some function. For a fixed $x \in \Re^n$, the first partial derivative of f at the point x in the ith coordinate is defined by

$$\frac{\partial f(x)}{\partial x(i)} = \lim_{\alpha \to 0} \frac{f(x + \alpha e_i) - f(x)}{\alpha},$$

where e_i is the ith unit vector. If the partial derivatives with respect to all coordinates exist, f is called *differentiable at* x and its *gradient* at x is defined to be the column vector

$$\nabla f(x) = \begin{pmatrix} \frac{\partial f(x)}{\partial x(1)} \\ \vdots \\ \frac{\partial f(x)}{\partial x(n)} \end{pmatrix}.$$

The function f is called *differentiable* if it is differentiable at every $x \in \Re^n$. If $\nabla f(x)$ exists for every x and is a continuous function of x, f is said to be *continuously differentiable*. Such functions admit the first order Taylor expansion

$$f(x + y) = f(x) + y'\nabla f(x) + o(\|y\|),$$

where $o(\|y\|)$ is a function of y with the property $\lim_{\|y\| \to 0} o(\|y\|)/\|y\| \to 0$. A related result is the *mean value theorem*, which states that if $f : \Re^n \mapsto \Re$ is continuously differentiable, then for every $x, y \in \Re^n$, there exists some $\alpha \in [0, 1]$ such that

$$f(y) - f(x) = \nabla f\big(x + \alpha(y - x)\big)'(y - x).$$

If the gradient $\nabla f(x)$ is itself a differentiable function, then f is said to be twice differentiable. We denote by $\nabla^2 f(x)$ the Hessian matrix of f at x, that is, the matrix

$$\nabla^2 f(x) = \left[\frac{\partial^2 f(x)}{\partial x(i) \partial x(j)}\right]$$

the elements of which are the second partial derivatives of f at x.

If f is twice continuously differentiable, that is, if $\nabla^2 f(x)$ exists and is continuous, then we have second order versions of the Taylor expansion and the mean value theorem. In particular, for all x and y, we have

$$f(x + y) = f(x) + y'\nabla f(x) + \tfrac{1}{2} y'\nabla^2 f(x) y + o(\|y\|^2),$$

and there exists some $\alpha \in [0, 1]$ such that

$$f(x + y) = f(x) + y'\nabla f(x) + \tfrac{1}{2}y'\nabla^2 f(x + \alpha y)y.$$

A vector-valued function $f : \Re^n \mapsto \Re^m$ is called differentiable (respectively, continuously differentiable) if each component f_i of f is differentiable (respectively, continuously differentiable). The *gradient matrix* of f, denoted by $\nabla f(x)$, is the $n \times m$ matrix whose ith column is the gradient $\nabla f_i(x)$ of f_i. Thus,

$$\nabla f(x) = \Big[\nabla f_1(x) \cdots \nabla f_m(x)\Big].$$

The transpose of ∇f is the *Jacobian* of f, that is, the matrix whose ijth entry is equal to the partial derivative $\partial f_i / \partial x(j)$.

Let $f : \Re^k \mapsto \Re^m$ and $g : \Re^m \mapsto \Re^n$ be continuously differentiable functions, and let $h(x) = g\big(f(x)\big)$. The *chain rule* for differentiation states that

$$\nabla h(x) = \nabla f(x) \nabla g\big(f(x)\big), \qquad \text{for all } x \in \Re^k.$$

If $f : \Re^n \mapsto \Re^m$ is of the form $f(x) = Ax$, where A is an $m \times n$ matrix, we have $\nabla f(x) = A'$. Also, if $f : \Re^n \mapsto \Re$ is of the form $f(x) = x'Ax/2$, where A is a symmetric $n \times n$ matrix, we have $\nabla f(x) = Ax$ and $\nabla^2 f(x) = A$.

A.5 CONVEX SETS AND FUNCTIONS

A subset C of $\Re^n$ is said to be *convex* if for every $x, y \in C$ and every scalar α with $0 \leq \alpha \leq 1$, we have $\alpha x + (1 - \alpha)y \in C$. In words, C is convex if the line segment connecting any two points in C belongs to C. A function $f : C \mapsto \Re$ defined over a convex subset C of $\Re^n$ is said to be *convex* if for every $x, y \in C$ and every scalar α with $0 \leq \alpha \leq 1$ we have

$$f\big(\alpha x + (1 - \alpha)y\big) \leq \alpha f(x) + (1 - \alpha) f(y).$$

The function f is said to be *concave* if $(-f)$ is convex, or, equivalently, if for every $x, y \in C$ and every scalar α with $0 \leq \alpha \leq 1$, we have

$$f\big(\alpha x + (1 - \alpha)y\big) \geq \alpha f(x) + (1 - \alpha) f(y).$$

If $f : C \mapsto \Re$ is convex, then the sets $\Gamma_\lambda = \big\{x \mid x \in C,\ f(x) \leq \lambda\big\}$ are also convex for every scalar λ. An important property is that a real-valued convex function on $\Re^n$ is continuous.

If $f_1, f_2, \ldots, f_m$ are convex functions over a convex subset C of $\Re^n$ and $\alpha_1, \alpha_2, \ldots, \alpha_m$ are nonnegative scalars, then the function $\alpha_1 f_1 + \cdots + \alpha_m f_m$ is also convex over C. If $f : \Re^m \mapsto \Re$ is convex, A is an $m \times n$ matrix, and b is a vector in $\Re^m$, the function $g : \Re^n \mapsto \Re$ defined by $g(x) = f(Ax + b)$ is also convex. If $f : \Re^n \mapsto \Re$ is convex, then the function $g(x) = E\big[f(x+w)\big]$, where w is a random vector in $\Re^n$, is a convex function provided the expected value is finite for every $x \in \Re^n$.

For functions $f : \Re^n \mapsto \Re$ that are differentiable, there are alternative characterizations of convexity. Thus, the function f is convex if and only if

$$f(y) \geq f(x) + \nabla f(x)'(y - x), \qquad \text{for all } x, y \in \Re^n.$$

If f is twice differentiable, then f is convex if and only if $\nabla^2 f(x)$ is a positive semidefinite symmetric matrix for every $x \in \Re^n$. A quadratic function of the form $f(x) = x'Ax + b'x$ is convex if and only if the matrix A is positive semidefinite.

For more material on convex functions and their role in optimization, we refer to the classic treatise [Roc70] and also to the simpler presentation in Appendix B of [Ber95b].

APPENDIX B: On Probability Theory and Markov Chains

Contents

This appendix lists selectively some of the basic probabilistic notions we will be using. Its main purpose is to familiarize the reader with some of the terminology we will adopt. It is not meant to be exhaustive, and the reader should consult references such as [Ash70], [Fel68], [Pap65], and [Ros85] for detailed treatments, particularly regarding operations with random variables, conditional probability, Bayes' rule, and so on. For fairly accessible treatments of measure theoretic probability, see [AdG86] and [Ash72]. For detailed presentations on finite-state Markov chains, see [Ash70], [Chu60], [KeS60], [Ros83], and [Ros85].

B.1 PROBABILITY SPACES

A *probability space* consists of:

(a) A set Ω.

(b) A collection $\mathcal{F}$ of subsets of Ω, called *events*, which includes Ω and has the following properties:

(1) If A is an event, then the complement $\overline{A} = \{\omega \in \Omega \mid \omega \notin A\}$ is also an event. (The complement of Ω is the empty set and is considered to be an event.)

(2) If $A_1, A_2, \ldots, A_k, \ldots$ are events, then $\cup_{k=1}^{\infty} A_k$ and $\cap_{k=1}^{\infty} A_k$ are also events.

(c) A function $\mathrm{P}(\cdot)$ assigning to each event A a nonnegative number $P(A)$, called the *probability of the event* A, and satisfying

(1) $\mathrm{P}(\Omega) = 1$.

(2) $\mathrm{P}(\cup_{k=1}^{\infty} A_k) = \sum_{k=1}^{\infty} \mathrm{P}(A_k)$ for every sequence of mutually disjoint events $A_1, A_2, \ldots, A_k, \ldots$

The function P is referred to as a *probability measure*.

Convention for Finite and Countable Probability Spaces

When Ω is finite or countable, we implicitly assume that the associated collection of events is the collection of *all* subsets of Ω (including Ω and the empty set). Then, if Ω is a finite set $\Omega = \{\omega_1, \omega_2, \ldots, \omega_n\}$, the probability space is specified by the probabilities $p_1, p_2, \ldots, p_n$, where p_i denotes the probability of the event consisting of ω_i. Similarly, if $\Omega = \{\omega_1, \omega_2, \ldots, \omega_k, \ldots\}$, the probability space is specified by the corresponding probabilities $p_1, p_2, \ldots, p_k, \ldots$ In either case we refer to $(p_1, p_2, \ldots, p_n)$ or $(p_1, p_2, \ldots, p_k, \ldots)$ as a *probability distribution over* Ω.

B.2 RANDOM VARIABLES

Given a probability space $(\Omega, \mathcal{F}, P)$, a *random variable* on the probability space is a function $x : \Omega \mapsto \Re$ such that for every scalar λ the set

$$\{\omega \in \Omega \mid x(\omega) \leq \lambda\}$$

is an event (i.e., belongs to the collection $\mathcal{F}$). An n-dimensional *random vector* $x = \big(x(1), \ldots, x(n)\big)$ is an n-tuple of random variables $x(1), \ldots, x(n)$, each defined on the same probability space.

We define the *distribution function* $F : \Re \mapsto \Re$ of a random variable x by

$$F(z) = \mathrm{P}\big(\{\omega \in \Omega \mid x(\omega) \leq z\}\big),$$

that is, $F(z)$ is the probability that the random variable takes a value less than or equal to z. We define the distribution function $F : \Re^n \mapsto \Re$ of a random vector $x = \big(x(1), \ldots, x(n)\big)$ by

$$F\big(z(1), \ldots, z(n)\big) = \mathrm{P}\Big(\big\{\omega \in \Omega \mid x(1, \omega) \leq z(1), \ldots, x(n, \omega) \leq z(n)\big\}\Big).$$

Given the distribution function of a random vector $x = \big(x(1), \ldots, x(n)\big)$, the (marginal) distribution function F_i of each random variable $x(i)$ is obtained from

$$F_i\big(z(i)\big) = \lim_{z(j) \to \infty,\, j \neq i} F\big(z(1), \ldots, z(n)\big).$$

The random variables $x(1), \ldots, x(n)$ are said to be *independent* if

$$F\big(z(1), \ldots, z(n)\big) = F_1\big(z(1)\big) \cdots F_n\big(z(n)\big),$$

for all scalars $z(1), \ldots, z(n)$.

The *expected value of a random variable* x with distribution function F is defined as

$$E[x] = \int_{-\infty}^{\infty} z \, dF(z)$$

provided the integral is well defined. The *expected value of a random vector* $x = \big(x(1), \ldots, x(n)\big)$ is the vector

$$E[x] = \big(E[x(1)], \ldots, E[x(n)]\big).$$

B.3 CONDITIONAL PROBABILITY

We shall restrict ourselves to the case where the underlying probability space Ω is a countable (possibly finite) set and the set of events is the set of all subsets of Ω.

Given two events A and B with $\mathrm{P}(A) > 0$, we define the *conditional probability of B given A* by

$$\mathrm{P}(B \mid A) = \frac{\mathrm{P}(A \cap B)}{\mathrm{P}(A)}.$$

If $B_1, B_2, \ldots$ are a countable (possibly finite) collection of mutually exclusive and exhaustive events (i.e., the sets B_i are disjoint and their union is Ω) and A is an event, then we have

$$\mathrm{P}(A) = \sum_i \mathrm{P}(A \cap B_i).$$

From the two preceding relations it is seen that

$$\mathrm{P}(A) = \sum_i \mathrm{P}(B_i)\mathrm{P}(A \mid B_i).$$

We thus obtain for every k,

$$\mathrm{P}(B_k \mid A) = \frac{\mathrm{P}(A \cap B_k)}{\mathrm{P}(A)} = \frac{\mathrm{P}(B_k)\mathrm{P}(A \mid B_k)}{\sum_i \mathrm{P}(B_i)\mathrm{P}(A \mid B_i)},$$

provided $\mathrm{P}(A) > 0$. This relation is referred to as *Bayes' rule*.

Consider now two random vectors x and y on the (countable) probability space taking values in $\Re^n$ and $\Re^m$, respectively [i.e., $x(\omega) \in \Re^n$, $y(\omega) \in \Re^m$ for all $\omega \in \Omega$]. Given two subsets X and Y of $\Re^n$ and $\Re^m$, respectively, we denote

$$\mathrm{P}(X \mid Y) = \mathrm{P}\Big(\{\omega \mid x(\omega) \in X\} \;\Big|\; \{\omega \mid y(\omega) \in Y\}\Big).$$

For a fixed vector $v \in \Re^n$, we define the *conditional distribution function* of x given v by

$$F(z \mid v) = \mathrm{P}\Big(\{\omega \mid x(\omega) \le z\} \;\Big|\; \{\omega \mid y(\omega) = v\}\Big),$$

and the *conditional expectation* of x given v by

$$E[x \mid v] = \int_{\Re^n} z \, dF(z \mid v),$$

provided the integral is well defined. Note that $E[x \mid v]$ is a function mapping v into $\Re^n$, and is itself a random variable. Its expectation is $E[x]$, that is,

$$E\big[E[x \mid v]\big] = E[x].$$

Furthermore, if w is another random variable, then

$$E\big[E[x \mid v, w] \mid w\big] = E[x \mid w].$$

B.4 STATIONARY MARKOV CHAINS

A square $n \times n$ matrix $[p_{ij}]$ is said to be a *stochastic* matrix if all of its elements are nonnegative, that is, if $p_{ij} \geq 0$, $i, j = 1, \ldots, n$, and the sum of the elements of each of its rows is equal to 1, that is, $\sum_{j=1}^{n} p_{ij} = 1$ for all $i = 1, \ldots, n$.

Suppose we are given a stochastic $n \times n$ matrix P together with a finite set of states $S = \{1, \ldots, n\}$. The pair (S, P) will be referred to as a *stationary finite-state Markov chain*. We associate with (S, P) a process whereby an initial state $x_0 \in S$ is chosen in accordance with some initial probability distribution

$$\pi_0 = \big(\pi_0(1), \pi_0(2), \ldots, \pi_0(n)\big).$$

Subsequently, transitions are made sequentially from a current state x_k to a successor state x_{k+1} in accordance with a probability distribution specified by P as follows. The probability that the successor state will be j is equal to p_{ij} whenever the current state is i, independently of the preceding states; that is

$$\mathrm{P}(x_{k+1} = j \mid x_k = i, x_{k-1}, \ldots, x_0) = \mathrm{P}(x_{k+1} = j \mid x_k = i) = p_{ij}, \qquad \text{(B.1)}$$

for all i and j. The probability that after the kth transition the state x_k will be j, given that the initial state x_0 is i, is denoted by

$$p_{ij}^k = \mathrm{P}(x_k = j \mid x_0 = i), \qquad i, j = 1, \ldots, n. \qquad \text{(B.2)}$$

A straightforward calculation shows that these probabilities are equal to the elements of the matrix P^k (P raised to the kth power), in the sense that p_{ij}^k is the element in the ith row and jth column of P^k:

$$P^k = [p_{ij}^k]. \qquad \text{(B.3)}$$

Given the initial probability distribution π_0 of the state x_0 (viewed as a row vector in $\Re^n$), the probability distribution of the state x_k after k transitions

$$\pi_k = \big(\pi_k(1), \pi_k(2), \ldots, \pi_k(n)\big),$$

(viewed again as a row vector) is given by

$$\pi_k = \pi_0 P^k, \qquad k = 1, 2, \ldots \qquad \text{(B.4)}$$

This relation follows from Eqs. (B.2) and (B.3) once we write

$$\pi_k(j) = \sum_{i=1}^{n} \mathrm{P}(x_k = j \mid x_0 = i)\pi_0(i) = \sum_{i=1}^{n} p_{ij}^k \pi_0(i).$$

Finally, if $g(x_k)$ is a function of the state x_k (such as a cost function in the context of dynamic programming), the expected value $E\big[g(x_k) \mid x_0 = i\big]$ is the ith component of the vector $P^k g$, where g is the vector with components $g(1), \ldots, g(n)$.

B.5 CLASSIFICATION OF STATES

Given a stationary finite-state Markov chain (S, P), we say that two states i and j *communicate* if there exist two positive integers k_1 and k_2 such that $p_{ij}^{k_1} > 0$ and $p_{ji}^{k_2} > 0$. In words, states i and j communicate if one can be reached from the other with positive probability. A subset of states $\tilde{S}$ is said to be a *recurrent class*, if all states in $\tilde{S}$ communicate, and for all $i \in \tilde{S}$ and $j \notin \tilde{S}$, we have $p_{ij}^k = 0$ for all k.

If S forms by itself a recurrent class (i.e., if all states communicate with each other), then we say that the Markov chain is *irreducible*. It is possible that there exist several recurrent classes. It is also possible to prove that at least one recurrent class must exist. States that do not belong to any recurrent class are called *transient*. We have

$$\lim_{k \to \infty} p_{ii}^k = 0 \text{ if and only if } i \text{ is transient.}$$

In other words, if the process starts at a transient state, the probability of returning to the same state after k transitions diminishes to zero as k tends to infinity.

The definitions imply that if the process starts within a recurrent class, it stays within that class. If it starts at a transient state, it eventually (with probability one) enters a recurrent class after a number of transitions and subsequently remains there.

B.6 LIMITING PROBABILITIES

An important property of any stochastic matrix P is that the matrix P^* defined by

$$P^* = \lim_{N \to \infty} \frac{1}{N} \sum_{k=0}^{N-1} P^k \tag{B.5}$$

exists [in the sense that the sequences of the elements of $(1/N) \sum_{k=0}^{N-1} P^k$ converge to the corresponding elements of P^*]. The elements p_{ij}^* of P^* satisfy

$$p_{ij}^* \geq 0, \qquad \sum_{j=1}^{n} p_{ij}^* = 1, \qquad i, j = 1, \ldots, n.$$

That is, P^* is a stochastic matrix. (For a proof see Prop. 1.1, Ch. 4 of [Ber95a], Vol. II.)

Note that the (i, j)th element of the matrix P^k is the probability that the state will be j after k transitions starting from state i. With this in mind, it can be seen from the definition (B.5) that p_{ij}^* can be interpreted as the long term expected fraction of time that the state is j given that the initial state is i. This suggests that for any two states i and i' in the same recurrent class we have $p^*_{ij} = p^*_{i'j}$, and this can indeed be proved. In particular, if a Markov chain is irreducible, the matrix P^* has identical rows. Also, if j is a transient state, we have

$$p^*_{ij} = 0, \qquad \text{for all } i = 1, \ldots, n,$$

so the columns of the matrix P^* corresponding to transient states are identically zero.

A vector $\pi^* = \big(\pi^*(1), \pi^*(2), \ldots, \pi^*(n)\big)$ such that $\pi^* P = \pi^*$, that is,

$$\sum_{i=1}^{n} \pi^*(i) P_{ij} = \pi^*(j), \qquad \forall\, j = 1, \ldots, n,$$

is called an *invariant* or *steady-state* distribution. If the probability distribution of the initial state x_0 is π^*, then the probability distribution of the state x_k after any number k of transitions will also be π^* [cf. Eq. (B.4)].

B.7 FIRST PASSAGE TIMES

Let us denote by q^k_{ij} the probability that the state will be j for the first time after exactly $k \geq 1$ transitions given that the initial state is i, that is,

$$q^k_{ij} = \mathrm{P}\big(x_k = j,\, x_m \neq j,\, 1 \leq m < k \mid x_0 = i\big).$$

Denote also, for fixed i and j,

$$K_{ij} = \min\{k \geq 1 \mid x_k = j,\, x_0 = i\}.$$

Then K_{ij} is a random variable, called the *first passage time from i to j*. We have, for every $k = 1, 2, \ldots,$

$$\mathrm{P}(K_{ij} = k) = q^k_{ij},$$

and we write

$$\mathrm{P}(K_{ij} = \infty) = \mathrm{P}\big(x_k \neq j,\, k = 1, 2, \ldots \mid x_0 = i\big) = 1 - \sum_{k=1}^{\infty} q^k_{ij}.$$

Note that it is possible that $\sum_{k=1}^{\infty} q^k_{ij} < 1$. This will occur, for example, if j cannot be reached from i in which case $q^k_{ij} = 0$ for all $k = 1, 2, \ldots$ The *mean first passage time* from i to j is the expected value of K_{ij}:

$$E[K_{ij}] = \begin{cases} \sum_{k=1}^{\infty} k q^k_{ij}, & \text{if } \sum_{k=1}^{\infty} q^k_{ij} = 1, \\ \infty, & \text{if } \sum_{k=1}^{\infty} q^k_{ij} < 1. \end{cases}$$

It can be proved that if i and j belong to the same recurrent class then

$$E[K_{ij}] < \infty.$$

In fact if there is only one recurrent class and t is a state of that class, the mean first passage times $E[K_{it}]$ are the unique solution of the following linear system of equations

$$E[K_{it}] = 1 + \sum_{j \neq t} p_{ij} E[K_{jt}], \qquad i = 1, \ldots, n,\ i \neq t;$$

(see Section 2.4, Example 2.1). If i and j belong to two different recurrent classes, then $E[K_{ij}] = E[K_{ji}] = \infty$. If i belongs to a recurrent class and j is transient, we have $E[K_{ij}] = \infty$.

References

[ABB96] Abounadi, J., Bertsekas, D. P., and Borkar, V. S., 1996. "ODE Analysis for Q-Learning Algorithms," Lab. for Info. and Decision Systems Draft Report, Massachusetts Institute of Technology, Cambridge, MA.

[AdG86] Adams, M., and Guillemin, V., 1986. Measure Theory and Probability, Wadsworth and Brooks, Monterey, CA.

[Ash70] Ash, R. B., 1970. Basic Probability Theory, Wiley, N. Y.

[Ash72] Ash, R. B., 1972. Real Analysis and Probability, Wiley, N. Y.

[AtF66] Athans, M., and Falb, P., 1966. Optimal Control, Mc-Graw Hill, N. Y.

[BBS95] Barto, A. G., Bradtke, S. J., and Singh, S. P., 1995. "Learning to Act Using Real-Time Dynamic Programming," Artificial Intelligence, Vol. 72, pp. 81-138.

[BKK73] Bellman, R., Kalaba, R., and Kotkin, B., 1973. "Polynomial Approximation – A New Computational Technique in Dynamic Programming: Allocation Processes," Mathematical Computation, Vol. 17, pp. 155-161.

[BMP90] Benveniste, A., Metivier, M., and Priouret, P., 1990. Adaptive Algorithms and Stochastic Approximations, Springer-Verlag, N. Y.

[BSA83] Barto, A. G., Sutton, R. S., and Anderson, C. W., 1983. "Neuronlike Elements that Can Solve Difficult Learning Control Problems," IEEE Trans. on Systems, Man, and Cybernetics, Vol. 13, pp. 835-846.

[BSS93] Bazaraa, M. S., Sherali, H. D., and Shetty, C. M., 1993. Nonlinear Programming Theory and Algorithms (2nd Ed.), Wiley, N. Y.

[Bai93] Baird, L. C., 1993. "Advantage Updating," Report WL-TR-93-1146, Wright Patterson AFB, OH.

[Bai95] Baird, L. C., 1995. "Residual Algorithms: Reinforcement Learning with Function Approximation," in Machine Learning: Proceedings of the Twelfth International Conference, Morgan Kaufmann, San Francisco, CA.

[Bar93] Barnard, E., 1993. "Temporal Difference Methods and Markov Models," IEEE Trans. on Systems, Man, and Cybernetics, Vol. 23, pp. 357-365.

[BeD59] Bellman, R. E., and Dreyfus, S. E., 1959. "Functional Approximations and Dynamic Programming," Mathematical Tables and Other Aids to Computation, Vol. 13, pp. 247-251.

[BeI96] Bertsekas, D. P., and Ioffe, S., 1996. "Temporal Differences-Based Policy Iteration and Applications in Neuro-Dynamic Programming," Lab. for Info. and Decision Systems Report LIDS-P-2349, Massachusetts Institute of Technology, Cambridge, MA.

[BeT89] Bertsekas, D. P., and Tsitsiklis, J. N., 1989. Parallel and Distributed Computation: Numerical Methods, Prentice-Hall, Englewood Cliffs, N. J.

[BeT91a] Bertsekas, D. P., and Tsitsiklis, J. N., 1991. "An Analysis of Stochastic Shortest Path Problems," Mathematics of Operations Research, Vol. 16, pp. 580-595.

[BeT91b] Bertsekas, D. P., and Tsitsiklis, J. N., 1991. "Some Aspects of Parallel and Distributed Iterative Algorithms – A Survey," Automatica, Vol. 27, pp. 3-21.

[Ber82a] Bertsekas, D. P., 1982. "Distributed Dynamic Programming," IEEE Trans. on Automatic Control, Vol. AC-27, pp. 610-616.

[Ber82b] Bertsekas, D. P., 1982. Constrained Optimization and Lagrange Multiplier Methods, Academic Press, N. Y.

[Ber95a] Bertsekas, D. P., 1995. Dynamic Programming and Optimal Control, Vols. I and II, Athena Scientific, Belmont, MA.

[Ber95b] Bertsekas, D. P., 1995. Nonlinear Programming, Athena Scientific, Belmont, MA.

[Ber95c] Bertsekas, D. P., 1995. "A Counterexample to Temporal Differences Learning," Neural Computation, Vol. 7, pp. 270-279.

[Ber95d] Bertsekas, D. P., 1995. "Incremental Least Squares Methods and the Extended Kalman Filter," Lab. for Info. and Decision Systems Report LIDS-P-2237, Massachusetts Institute of Technology, Cambridge, MA; to appear in SIAM J. on Optimization.

[Ber95e] Bertsekas, D. P., 1995. "A Hybrid Incremental Gradient Method for Least Squares Problems," Lab. for Info. and Decision Systems Report LIDS-P-2257, Massachusetts Institute of Technology, Cambridge, MA; to appear in SIAM J. on Optimization.

[Ber95f] Bertsekas, D. P., 1995. "A New Value Iteration Method for the Average Cost Dynamic Programming Problem," Lab. for Info. and Decision Systems Report LIDS-P-2307, Massachusetts Institute of Technology, Cambridge, MA; to appear in SIAM J. on Control and Optimization.

[Bis95] Bishop, C. M, 1995. Neural Networks for Pattern Recognition, Oxford University Press, N. Y.

[BoM95] Boyan, J. A., and Moore, A. W., 1995. "Generalization in Reinforcement Learning: Safely Approximating the Value Function," in Advances in Neural Information Processing Systems 7, MIT Press, Cambridge, MA.

[BoS93] Borkar, V. S., and Soumyanath, K., 1993. "A New Parallel Scheme for Fixed Point Computation Part I: Theory," unpublished report.

[Bor95] Borkar, V. S., 1995. "Asynchronous Stochastic Approximations," unpublished report.

[BrB96] Bradtke, S. J., and Barto, A. G., 1996. "Linear Least-Squares Algorithms for Temporal Difference Learning," Machine Learning, Vol. 22, pp. 33-57.

[BrH75] Bryson, A. E., and Ho, Y.-C., 1975. Applied Optimal Control, Hemisphere Publishing Corp., Washington, D.C.

[Bra94] Bradtke, S. J., 1994. Incremental Dynamic Programming for On-Line Adaptive Optimal Control, Ph. D. thesis, University of Massachusetts, Amherst, MA.

[CGM95] Cybenko, G., Gray, R., and Moizumi, K., 1995. "Q-Learning: A Tutorial and Extensions," unpublished report, presented at Mathematics of Artificial Neural Networks, Oxford University, England, July 1995.

[ChK86] Christensen, J., and Korf, R. E., 1986. "A Unified Theory of Heuristic Evaluation Functions and its Application to Learning," in Proceedings AAAI-86, pp. 148-152.

[ChR92] Chong, E. K. P., and Ramadge, P. J., 1992. "Convergence of Recursive Optimization Algorithms Using Infinitesimal Perturbation Analysis Estimates," Discrete Event Dynamic Systems: Theory and Applications, Vol. 1, pp. 339-372.

[ChT91] Chow, C.-S., and Tsitsiklis, J. N, 1991. "An Optimal One-Way Multigrid Algorithm for Discrete-Time Stochastic Control," IEEE Trans. on Automatic Control, Vol. 36, pp. 898-914.

[ChT94] Cheng, B., and Titterington D. M., 1994. "Neural Networks: A Review from a Statistical Perspective," Statistical Science, Vol. 9, pp. 2-54.

[Chu60] Chung, K. L., 1960. Markov Chains with Stationary Transition Probabilities, Springer-Verlag, Berlin and N. Y.

[CrB96] Crites, R. H., and Barto, A. G., 1996. "Improving Elevator Performance using Reinforcement Learning," in Advances in Neural Information

Processing Systems 8, MIT Press, Cambridge, MA, pp. 1017-1023.

[Cyb89] Cybenko, 1989. "Approximation by Superpositions of a Sigmoidal Function," Math. of Control, Signals, and Systems, Vol. 2, pp. 303-314.

[DaS94] Dayan, P., and Sejnowski, T. J., 1994. "TD(λ) Converges with Probability 1," Machine Learning, Vol. 14, pp. 295-301.

[DaS96] Dayan, P., and Singh, S., 1996. "Improving Policies without Measuring Merits," in Advances in Neural Information Processing Systems 8, MIT Press, Cambridge, MA, pp. 1059-1065.

[Dan76] Daniel, J. W., 1976. "Splines and Efficiency in Dynamic Programming," Journal of Mathematical Analysis and Applications, Vol. 54, pp. 402-407.

[Dav76] Davidon, W. C., 1976. "New Least Squares Algorithms," J. Optimization Theory and Applications, Vol. 18, pp. 187-197.

[Day92] Dayan, P., 1992. "The Convergence of TD(λ) for General λ," Machine Learning, Vol. 8, pp. 341-362.

[Fel68] Feller, W., 1968. An Introduction to Probability Theory and Its Applications, Wiley, N. Y.

[FiT91] Filar, J. A., and Tolwinski, B., 1991. "On the Algorithm of Pollatschek and Avi-Itzhak," in Raghavan, T. et al. (eds.), Stochastic Games and Related Topics, In Honor of Professor L. S. Shapley, Kluwer Academic Publishers, Dordrecht, The Netherlands.

[FlR75] Fleming, W. H., and Rishel, R. W., 1975. Deterministic and Stochastic Optimal Control, Springer-Verlag, N. Y.

[Fun89] Funahashi, K., 1989. " On the Approximate Realization of Continuous Mappings by Neural Networks," Neural Networks, Vol. 2, pp. 183-192.

[GLH94] Gurvits, L., Lin, L.-J., and Hanson, S. J., 1994. "Incremental Learning of Evaluation Functions for Absorbing Markov Chains: New Methods and Theorems," unpublished report.

[Gai94] Gaivoronski, A. A., 1994. "Convergence Analysis of Parallel Backpropagation Algorithm for Neural Networks," Optimization Methods and Software, Vol. 4, pp. 117-134.

[Gla91] Glasserman, P., 1991. Gradient Estimation via Perturbation Analysis, Kluwer Academic Publishers, Norwell, MA.

[Gor95] Gordon, G. J., 1995. "Stable Function Approximation in Dynamic Programming," in Machine Learning: Proceedings of the Twelfth International Conference, Morgan Kaufmann, San Francisco, CA.

[Gri94] Grippo, L., 1994. "A Class of Unconstrained Minimization Methods for Neural Network Training," Optimization Methods and Software, Vol.

4, pp. 135-150.

[HBK94] Harmon, M. E., Baird, L. C., and Klopf, A. H., 1994. "Advantage Updating Applied to a Differential Game," unpublished report, presented at the 1994 Neural Information Processing Systems Conference, Denver, CO.

[HSW89] Hornick, K., Stinchcombe, M., and White, H., 1989. "Multilayer Feedforward Networks are Universal Approximators," Neural Networks, Vol. 2, pp. 359-159.

[Hag88] Hager, W. W., 1988. Applied Numerical Linear Algebra, Prentice-Hall, Englewood Cliffs, N. J.

[Hay94] Haykin, S., 1994. Neural Networks: A Comprehensive Foundation, McMillan, N. Y.

[Hes66] Hestenes, M. R., 1966. Calculus of Variations and Optimal Control Theory, Wiley, N. Y.

[HoC91] Ho, Y.-C., and Cao, X.-R., 1991. Perturbation Analysis of Discrete Event Dynamic Systems, Kluwer Academic Publishers, Norwell, MA.

[HoK61] Hoffman, K., and Kunze, R., 1961. Linear Algebra, Prentice-Hall, Englewood Cliffs, N. J.

[Hol86] Holland, J. H., 1986. "Escaping Brittleness: the Possibility of General-Purpose Learning Algorithms Applied to Rule-Based Systems," in Machine Learning: An Artificial Intelligence Approach, Michalski, R. S., Carbonell, J. G., and Mitchell, T. M., (eds.), Morgan Kaufmann, San Mateo, CA, pp. 593-623.

[JJS94] Jaakkola, T., Jordan, M. I., and Singh, S. P., 1994. "On the Convergence of Stochastic Iterative Dynamic Programming Algorithms," Neural Computation, Vol. 6, pp. 1185-1201.

[JaF92] Jalali, A., and Ferguson, M. J., 1992. "On Distributed Dynamic Programming," IEEE Trans. on Automatic Control, Vol. AC-37, pp. 685-689.

[JaM70] Jacobson, D. H., and Mayne, D. Q., 1970. Differential Dynamic Programming, Elsevier, N. Y.

[Jon90] Jones, L. K., 1990. "Constructive Approximations for Neural Networks by Sigmoidal Functions," Proceedings of the IEEE, Vol. 78, pp. 1586-1589.

[KeS60] Kemeny, J. G., and Snell, J. L., 1960. Finite Markov Chains, Van Nostrand-Reinhold, N. Y.

[Koh74] Kohonen, T., 1974. "An Adaptive Associative Memory Principle," IEEE Trans. on Computers, Vol. C-23, pp. 444-445.

[Kor90] Korf, R. E., 1990. "Real-Time Heuristic Search," Artificial Intelligence, Vol. 42, pp. 189-211.

[KuC78] Kushner, H. J., and Clark, D. S., 1978. Stochastic Approximation Methods for Constrained and Unconstrained Systems, Springer-Verlag, Berlin.

[KuD92] Kushner, H. J., and Dupuis, P., 1992. Numerical Methods for Stochastic Control in Continuous Time, Springer-Verlag, N. Y.

[KuS81] Kumar, P. R, and Shiau, T. H., 1981. "Zero-Sum Dynamic Games," in C. T. Leondes, (ed.), Academic Press, N. Y., pp. 1345-1378.

[KuY93] Kushner, H. J., and Yang, J., "Stochastic Approximation with Averaging: Optimal Asymptotic Rates of Convergence for General Processes," SIAM J. Control and Optimization, Vol. 31, pp. 1045-1062.

[LBH96] Logan, D. A., Bertsekas, D. P., Homer, M. L., Looze, D. P., Patek, S. D., Pepyne, D., Sandell, N. R., 1996. "Application of Neural Networks to Command Centers," Report TR-757, Alphatech, Inc., Burlington, MA.

[LeC85] Le Cun, Y., 1985. "Une Procédure d' Apprentissage pour Réseau à Seuil Assymétrique," in Cognitiva 85, à la Frontière de l'Intelligence Artificielle des Sciences de la Connaissance des Neurosciences, CESTA, Paris, pp. 599-604.

[Lit96] Littman, M. L., 1996. "Algorithms for Sequential Decision Making," Ph. D. thesis, Brown University, Providence, R. I.

[Lju77] Ljung, L., 1977. "Analysis of Recursive Stochastic Algorithms," IEEE Trans. on Automatic Control, Vol. 22, pp. 551-575.

[Lju79] Ljung, L., 1979. "Asymptotic Behavior of the Extended Kalman Filter as a Parameter Estimator for Linear Systems," IEEE Trans. on Automatic Control, Vol. 24, pp. 36-50.

[Lju94] Ljung, L., 1994. "Aspects on Accelerated Convergence in Stochastic Approximation Schemes," Proceedings of the 33d IEEE Conference on Decision and Control, Lake Buena Vista, FL.

[LuT94] Luo, Z. Q., and Tseng, P., 1994. "Analysis of an Approximate Gradient Projection Method with Applications to the Backpropagation Algorithm," Optimization Methods and Software, Vol. 4, pp. 85-101.

[Lue69] Luenberger, D. G., 1969. Optimization by Vector Space Methods, Wiley, N. Y.

[Lue84] Luenberger, D. G., 1984. Introduction to Linear and Nonlinear Programming, (2nd Ed.), Addison-Wesley, Reading, MA.

[Luo91] Luo, Z. Q., 1991. "On the Convergence of the LMS Algorithm with Adaptive Learning Rate for Linear Feedforward Networks," Neural Computation, Vol. 3, pp. 226-245.

[MaS94] Mangasarian, O. L., and Solodov, M. V., 1994. "Serial and Parallel Backpropagation Convergence Via Nonmonotone Perturbed Minimization," Optimization Methods and Software, Vol. 4, pp. 103-116.

[Mah96] Mahadevan, S., 1996. "Average Reward Reinforcement Learning: Foundations, Algorithms, and Empirical Results," Machine Learning, Vol. 22, pp. 1-38.

[Nev75] Neveu, J., 1975. Discrete Parameter Martingales, North-Holland, Amsterdam.

[Odo69] Odoni, A. R., 1969. "On Finding the Maximal Gain for Markov Decision Processes," Operations Research, Vol. 17, pp. 857-860.

[PLC96] Pepyne, D. L., Looze, D. P., Cassandras, C. G., and Djaferis, T. E., 1996. "Application of Q-Learning to Elevator Dispatching," unpublished report.

[PaB96a] Patek, S., and Bertsekas, D. P., 1996. "Stochastic Shortest Path Games," Lab. for Info. and Decision Systems Report LIDS-P-2319, Massachusetts Institute of Technology, Cambridge, MA.

[PaB96b] Patek, S., and Bertsekas, D. P., 1996. "Play Selection in Football: a Case Study in Neuro-Dynamic Programming," Lab. for Info. and Decision Systems Report LIDS-P-2350, Massachusetts Institute of Technology, Cambridge, MA.

[Pap65] Papoulis, A., 1965. Probability, Random Variables and Stochastic Processes, McGraw-Hill, N. Y.

[PeW94] Peng, J., and Williams, R. J., 1994. "Incremental Multi-Step Q-Learning," Proceedings of the Eleventh International Conference on Machine Learning, Morgan Kaufmann, San Franscisco, CA, pp. 226-232.

[PoA69] Pollatschek, M., and Avi-Itzhak, B., 1969. "Algorithms for Stochastic Games with Geometrical Interpretation," Management Science, Vol. 15, pp. 399-415.

[PoJ92] Polyak, B. T., and Juditsky, A. B., 1992. "Acceleration of Stochastic Approximation by Averaging," SIAM J. Control and Optimization, Vol. 30, pp. 838-855.

[PoT73] Poljak, B. T., and Tsypkin, Y. Z., 1973. "Pseudogradient Adaptation and Training Algorithms," Automation and Remote Control, Vol. 12, pp. 83-94.

[Pol64] Poljak, B. T., 1964. "Some Methods of Speeding up the Convergence of Iteration Methods," Z. VyČisl. Mat. i Mat. Fiz., Vol. 4, pp. 1-17.

[Pol87] Poljak, B. T., 1987. Introduction to Optimization, Optimization Software Inc., N. Y.

[PuS78] Puterman, M. L., and Shin, M. C., 1978. "Modified Policy Iteration Algorithms for Discounted Markov Decision Problems," Management Science, Vol. 24, pp. 1127-1137.

[Put94] Puterman, M. L., 1994. Markovian Decision Problems, Wiley, N. Y.

[RHW86] Rumelhart, D. E., Hinton, G. E., and Williams, R. J., 1986. "Learning Representations by Back-Propagating Errors," Nature, Vol. 323, pp. 533-536.

[RaF91] Raghavan, T. E. S., and Filar, J. A., 1991. "Algorithms for Stochastic Games – A Survey," ZOR – Methods and Models of Operations Research, Vol. 35, pp. 437-472.

[Roc70] Rockafellar, R. T., 1970. Convex Analysis, Princeton University Press, Princeton, N. J.

[Ros83] Ross, S. M., 1983. Introduction to Stochastic Dynamic Programming, Academic Press, N. Y.

[Ros85] Ross, S. M., 1985. Probability Models, Academic Press, Orlando, FL.

[Roy68] Royden, H. L., 1968. Principles of Mathematical Analysis, McGraw-Hill, N. Y.

[Rud64] Rudin, W., 1964. Real Analysis, McGraw-Hill, N. Y.

[Rus96] Rust, J., 1996. "Numerical Dynamic Programming in Economics," in Amman, H., Kendrick, D., and Rust, J., (eds.), Handbook of Computational Economics, Elsevier, Amsterdam, Chapter 14, pp. 614-722.

[SBC93] Saarinen, S., Bramley, R., and Cybenko, G., 1993. "Ill-Conditioning in Neural Network Training Problems," SIAM J. Scientific Computation, Vol. 14, pp. 693-714.

[SJJ94] Singh, S. P., Jaakkola, T., and Jordan, M. I., 1994. "Learning without State-Estimation in Partially Observable Markovian Decision Processes," Proceedings of the Eleventh Machine Learning Conference, pp. 284-292.

[SJJ95] Singh, S. P., Jaakkola, T., and Jordan, M. I., 1995. "Reinforcement Learning with Soft State Aggregation," in Advances in Neural Information Processing Systems 7, MIT Press, Cambridge, MA.

[Sam59] Samuel, A. L., 1959. "Some Studies in Machine Learning Using the Game of Checkers," IBM Journal of Research and Development, pp. 210-229.

[Sam67] Samuel, A. L., 1967. "Some Studies in Machine Learning Using the Game of Checkers. II – Recent Progress," IBM Journal of Research and Development, pp. 601-617.

[ScS85] Schweitzer, P. J., and Seidman, A., 1985. "Generalized Polynomial Approximations in Markovian Decision Processes," Journal of Mathematical Analysis and Applications, Vol. 110, pp. 568-582.

[ScW96] Schapire, R. E., and Warmuth, M. K., 1996. "On the Worst-Case Analysis of Temporal-Difference Learning Algorithms," Machine Learning, Vol. 22.

[Sch93] Schwartz, A., 1993. "A Reinforcement Learning Method for Maximizing Undiscounted Rewards," Proceedings of the Tenth Machine Learning Conference, pp. 298-305.

[Sha50] Shannon, C., 1950. "Programming a Digital Computer for Playing Chess," Phil. Mag., Vol. 41, pp. 356-375.

[Sha53] Shapley, L. S., 1953. "Stochastic Games," Proceedings of the National Academy of Sciences, Mathematics, Vol. 39, pp. 1095-1100.

[SiB96] Singh, S. P., and Bertsekas, D. P., 1996. "Reinforcement Learning for Dynamic Channel Allocation in Cellular Telephone Systems," submitted to the 1996 Neural Information Processing Systems Conference.

[SiD96] Singh, S. P., and Dayan, P., 1996. "Analytical Mean Squared Error Curves in Temporal Difference Learning," unpublished report.

[SiS96] Singh, S. P., and Sutton, R. S., 1996. "Reinforcement Learning with Replacing Eligibility Traces," Machine Learning, Vol. 22, pp. 123-158.

[SiY94] Singh, S. P., and Yee, R. C., 1994. "An Upper Bound on the Loss from Approximate Optimal Value Functions," Machine Learning, Vol. 16, pp. 227-233.

[Sin94] Singh, S. P., 1994. "Reinforcement Learning Algorithms for Average-Payoff Markovian Decision Processes," Proceedings of the 12th National Conference on Artificial Intelligence, pp. 202-207.

[Str76] Strang, G., 1976. Linear Algebra and its Applications, Academic Press, N. Y.

[Sut84] Sutton, R. S., 1984. Temporal Credit Assignment in Reinforcement Learning, Ph. D. thesis, University of Massachusetts, Amherst, MA.

[Sut88] Sutton, R. S., 1988. "Learning to Predict by the Methods of Temporal Differences," Machine Learning, Vol. 3, pp. 9-44.

[Sut95] Sutton, R. S., 1995. "On the Virtues of Linear Learning and Trajectory Distributions," Proceedings of the Workshop on Value Function Approximation, Boyan, J. A., Moore, A. W., and Sutton, R. S., (eds.), Report CMU-CS-95-206, Carnegie Mellon University, Pittsburgh, PA.

[TeG96] Tesauro, G., and Galperin, G. R., 1996. "On-Line Policy Improvement Using Monte Carlo Search," unpublished report.

[Tes89] Tesauro, G. J., 1989. "Neurogammon Wins Computer Olympiad," Neural Computation, Vol. 1, pp. 321-323.

[Tes92] Tesauro, G. J., 1992. "Practical Issues in Temporal Difference Learning," Machine Learning, Vol. 8, pp. 257-277.

[Tes94] Tesauro, G. J., 1994. "TD-Gammon, a Self-Teaching Backgammon Program, Achieves Master-Level Play," Neural Computation, Vol. 6, pp. 215-219.

[Tes95] Tesauro, G. J., 1995. "Temporal Difference Learning and TD-Gammon," Communications of the ACM, Vol. 38, pp. 58-68.

[TrZ93] Trick, M. A., and Zin, S. E., 1993. "A Linear Programming Approach to Solving Stochastic Dynamic Programs," preprint.

[TsV96a] Tsitsiklis, J. N., and Van Roy, B., 1996. "Feature-Based Methods for Large-Scale Dynamic Programming," Machine Learning, Vol. 22, pp. 59-94.

[TsV96b] Tsitsiklis, J. N., and Van Roy, B., 1996. "An Analysis of Temporal-Difference Learning with Function Approximation," Lab. for Info. and Decision Systems Report LIDS-P-2322, Massachusetts Institute of Technology, Cambridge, MA.

[Tsi89] Tsitsiklis, J. N., 1989. "A Comparison of Jacobi and Gauss-Seidel Parallel Iterations," Applied Mathematics Letters, Vol. 2, pp. 167-170.

[Tsi94] Tsitsiklis, J. N., 1994. "Asynchronous Stochastic Approximation and Q-Learning," Machine Learning, Vol. 16, pp. 185-202.

[Tsi96] Tsitsiklis, J. N., 1996. "On the Convergence of Optimistic Policy Iteration," unpublished report.

[VaT96] Van Roy, B., and Tsitsiklis, J. N., 1996. "Stable Linear Approximations to Dynamic Programming for Stochastic Control Problems with Local Transitions," in Advances in Neural Information Processing Systems 8, MIT Press, Cambridge, MA, pp. 1045-1051.

[Van76] van Nunen, J. A. E. E., 1976. "A Set of Successive Approximation Methods for Discounted Markovian Decision Problems," Z. Oper. Res., Vol. 20, pp. 203-208.

[Van78] van der Wal, J., 1978. "Discounted Markov Games: Successive Approximation and Stopping Times," International J. of Game Theory, Vol. 6, pp. 11-22.

[Van95] Van Roy, B., 1995. "Feature-Based Methods for Large Scale Dynamic Programming," Lab. for Info. and Decision Systems Report LIDS-TH-2289, Massachusetts Institute of Technology, Cambridge, MA.

[WGM73] Widrow, B., Gupta, N. K., and Maitra, S., 1973. "Punish/Reward: Learning with a Critic in Adaptive Threshold Systems," IEEE Trans. on

Systems, Man, and Cybernetics, Vol. 3, pp. 455-465.

[WaD92] Watkins, C. J. C. H., and Dayan, P., 1992. "Q-Learning," Machine Learning, Vol. 8, pp. 279-292.

[Wat89] Watkins, C. J. C. H., 1989. "Learning from Delayed Rewards," Ph.D. Thesis, Cambridge University, Cambridge, England.

[Wer74] Werbös, P. J, 1974. "Beyond Regression: New Tools for Prediction and Analysis in the Behavioral Sciences," Ph.D. Thesis, Harvard University, Cambridge, MA.

[Wer77] Werbös, P. J., 1977. "Advanced Forecasting Methods for Global Crisis Warning and Models of Intelligence," General Systems Yearbook, Vol. 22, pp. 25-38.

[Wer90] Werbös, P. J., 1990. "Consistency of HDP Applied to a Simple Reinforcement Learning Problem," Neural Networks, Vol. 3, pp. 179-189.

[Wer92a] Werbös, P. J, 1992. "Approximate Dynamic Programming for Real-Time Control and Neural Modeling," in D. A. White and D. A. Sofge, (eds.), Handbook of Intelligent Control, Van Nostrand, N. Y.

[Wer92b] Werbös, P. J, 1992. "Neurocontrol and Supervised Learning: an Overview and Valuation," in D. A. White and D. A. Sofge, (eds.), Handbook of Intelligent Control, Van Nostrand, N. Y.

[WhS92] White, D. A., and Sofge, D. A., (eds.), 1992. Handbook of Intelligent Control, Van Nostrand, N. Y.

[Whi63] White, D. J., 1963. "Dynamic Programming, Markov Chains, and the Method of Successive Approximations," Journal of Mathematical Analysis and Applications, Vol. 6, pp. 373-376.

[Whi78] Whitt, W., 1978. "Approximations of Dynamic Programs I," Mathematics of Operations Research, Vol. 3, pp. 231-243.

[Whi79] Whitt, W., 1979. "Approximations of Dynamic Programs II," Mathematics of Operations Research, Vol. 4, pp. 179-185.

[Whi89] White, H., 1989. "Some Asymptotic Results for Learning in Single Hidden-Layer Feedforward Network Models," Journal of the American Statistical Association, Vol. 84, pp. 1003-1013.

[WiB93] Williams, R. J., and Baird, L. C., 1993. "Analysis of Some Incremental Variants of Policy Iteration: First Steps Toward Understanding Actor-Critic Learning Systems," Report NU-CCS-93-11, College of Computer Science, Northeastern University, Boston, MA.

[WiH60] Widrow, B., and Hoff, M. E., 1960. "Adaptive Switching Circuits," Institute of Radio Engineers, Western Electronic Show and Convention, Convention Record, part 4, pp. 96-104.

[Yin92] Yin, G., 1992. "On Extensions of Polyak's Averaging Approach to Stochastic Approximation," Stochastics, Vol. 36, pp. 245-264.

[ZhD95] Zhang, W., and Dietterich, T. G., 1995. "A Reinforcement Learning Approach to Job-Shop Scheduling," Proceedings of the Fourteenth International Joint Conference on Artificial Intelligence, pp. 1114-1120.

[ZhD96] Zhang, W., and Dietterich, T. G., 1996. "High-Performance Job-Shop Scheduling with a Time-Delay TD(λ) Network," in Advances in Neural Information Processing Systems 8, MIT Press, Cambridge, MA, pp. 1024-1030.

[ZhY89] Zhang, M., and Yum, T.-S. P., 1989. "Comparisons of Channel-Assignment Strategies in Cellular Mobile Telephone Systems," IEEE Trans. Vehic. Technol., Vol. 38, pp. 211-215.

INDEX

U

V

W